PRIMORDIAL IMMUNITY: FOUNDATIONS FOR THE VERTEBRATE IMMUNE SYSTEM

ANNALS OF THE NEW YORK ACADEMY OF SCIENCES

Volume 712

PRIMORDIAL IMMUNITY: FOUNDATIONS FOR THE VERTEBRATE IMMUNE SYSTEM

Edited by Gregory Beck, Edwin L. Cooper, Gail S. Habicht, and John J. Marchalonis

The New York Academy of Sciences
New York, New York
1994

⊗ *The paper used in this publication meets the minimum requirements of American National Standard for Information Sciences—Permanence of Paper for Printed Library Materials, ANSI Z39.48-1984.*

Cover: Henri Matisse, *Beasts of the Sea,* Ailsa Mellon Bruce Fund, ©1994 National Gallery of Art, Washington, 1950, paper collage on canvas, 2.955 × 1.540 (116 3/8 × 60 5/8).

Library of Congress Cataloging-in-Publication Data

Primordial immunity : foundations for the vertebrate immune system / edited by Gregory Beck . . . [et al.].
p. cm. -- (Annals of the New York Academy of Sciences, ISSN 0077-8923 : v. 712)
Contains the proceedings of the first meeting on primordial immunity sponsored by the New York Academy of Sciences. Cf. Pref.
Includes bibliographical references and index.
ISBN 0-89766-839-1 (cloth : acid-free paper). -- ISBN 0-89766-840-5 (paper : acid-free paper)
1. Immune system--Evolution--Congresses. 2. Vertebrates--Evolution--Congresses. I. Beck, Gregory. II. New York Academy of Sciences. III. Series.
Q11.N5 vol. 712
[QR184.6]
500 s--dc20
[591.2'9] 94-4703
CIP

&/PCP
Printed in the United States of America
ISBN 0-89766-839-1 (cloth)
ISBN 0-89766-840-5 (paper)
ISSN 0077-8923

ANNALS OF THE NEW YORK ACADEMY OF SCIENCES

Volume 712
April 15, 1994

PRIMORDIAL IMMUNITY: FOUNDATIONS FOR THE VERTEBRATE IMMUNE SYSTEM[a]

Editors and Conference Organizers
GREGORY BECK, EDWIN L. COOPER, GAIL S. HABICHT,
and JOHN L. MARCHALONIS

CONTENTS

[a]This volume is the result of a conference entitled **Primordial Immunity: Foundations for the Vertebrate Immune System,** which was held by the New York Academy of Sciences on May 2 through 5, 1993 in Woods Hole, Massachusetts.

Part III. Establishment of a Humoral Defense

Part IV. Was a Cellular Defense the First Defense?

Part V. Recognition in Invertebrates

Part VI. Effector Mechanisms in Relation to the Internal and External Milieux

Part VII. Poster Papers

Financial assistance was received from:

Supporter

- NATIONAL SCIENCE FOUNDATION

Contributor

- ICI PHARMACEUTICALS GROUP

Preface

GAIL S. HABICHT

Department of Pathology
State University of New York
Stony Brook, New York 11794-8691

This volume of the *Annals* contains the proceedings of the first meeting on primordial immunity ever sponsored by the New York Academy of Sciences. The conference was conceived by Gregory Beck, who wanted to create a forum for discussion of the cellular and molecular strategies employed by invertebrates for host defenses: He recruited Gail Habicht, John Marchalonis, and Edwin Cooper to make up the organizing committee. The program brought together scientists from four continents, united by a common desire to understand the origins of the mechanisms of host defense. What are the cellular and molecular ancestors of both anticipatory and nonanticipatory defenses used by modern vertebrates? What are the defense strategies unique to the invertebrate world? Can these be exploited for human benefit? What can be learned about mammalian host defenses through a thorough understanding of the defense strategies employed by animals lower on the evolutionary tree? The organizing committee sought speakers whose work addresses these basic questions.

The life cycle of each animal imposes unique requirements on its host defense system. Single-celled amoebae must rely on phagocytosis: Short-lived animals have no need for an inducible defense system for which induction takes longer than the life span, and homeotherms need mechanisms that can respond rapidly to match the increased capacity of pathogens to multiply at elevated temperatures. In spite of these diverse requirements, there has been remarkable conservation throughout evolution of families of host defense cells and molecules—in some cases as far back as the protozoa. In seeking to understand these defenses, it should be remembered that there is no need to force invertebrate host defense systems into the strategic paradigms used by vertebrates.

There is no argument that vertebrates and invertebrates share diverse strategies for dealing with foreigners—in some cases there is remarkable conservation of the molecular structures involved. These nonanticipatory defenses involve molecules whose specificity is encoded in the germline. Anticipatory immune responses employ molecules whose specificity is achieved by gene rearrangement, and general agreement exists that these responses are found only in the vertebrates, although there may not be general agreement on the terminology. No convincing data are extant to show that invertebrates possess the power to generate diverse recognition molecules by gene rearrangements, yet they are able to distinguish self from many nonself components using a variety of their own recognition molecules. Many invertebrates possess diverse lectins that act as recognition units. These may be engaged singly or in combinations to send different signals to the host defense system. Thus, invertebrates possess combinatorial but not rearranging capabilities for generating diversity of their recognition units. The Big Bang of gene amplification occurred around the time of separation of the echinoderms and tunicates from the rest of the chordates to produce the families of recognition molecules in use today. The primordial gene for the immunoglobulin domain preceded this big bang—it is present in multiple copies in hemolin from *Manducca* and neuroglian from *Drosophila*.

Invertebrates also possess some of the molecular signaling systems that are major features of higher eukaryote immune systems. NF-κB (nuclear factor-κB) is a mammalian transcription factor regulating a number of genes involved in immune, inflammatory, and acute-phase responses. An insect κB-like factor that is regulated much the same way NF-κB is regulated and that binds to similar DNA sequences has been described. Another κB-like factor has been found to play a role in differentiation of *Drosophila* larvae. It is of interest that the latter factor is regulated by a membrane receptor with sequence homology to the mammalian interleukin 1 (IL-1) receptor.

Invertebrate coelomic fluids demonstrate the presence of bioactivities characteristic of mammalian proinflammatory cytokines. These molecules act on both mammalian and invertebrate cells, suggesting that there is great conservation of the three-dimensional structure needed for interaction with cytokine receptors. In agreement with this conclusion is the observation that echinoderm IL-1 is immunoreactive with neutralizing antibodies specific for mammalian IL-1. Proinflammatory biological activities characterize the invertebrate cytokines. For example, echinoderm IL-1 promotes phagocytosis by coelomocytes, an activity that would be invaluable in echinoderm host defense. This biological activity may have been lost through evolution in that mammalian IL-1 has little or no effect on macrophage phagocytosis.

New parallels between vertebrate and invertebrate cytokines have been established not only in terms of their functions and their ability to bind to membrane receptors but also in terms of their regulation. That sea urchin mRNAs possess several copies of an AU-rich sequence that is responsible for cytokine mRNA instability suggests that several proinflammatory cytokines will be identified in these animals. The wisdom of choosing relevant stimuli for the induction of these putative cytokine mRNAs cannot be overstated. *E. coli* or trinitrophenol may never come in contact with invertebrate host defenses!

It is gratifying to find that defenses that are considered to be primitive in vertebrate species are indeed phylogenetically ancient. α_2-Macroglobulin and C-reactive protein (CRP) have been characterized in several invertebrate species and are structurally conserved. Invertebrates possess a variety of host defense–related molecules that have been conserved in the vertebrates. These molecules share sequence homology with their vertebrate correlates, but it must be remembered that structural and functional homology may derive from tertiary and quaternary protein structures as well. These need to be investigated separately.

Invertebrates make use of both unique and conserved effector mechanisms for host defense. Unique invertebrate effector systems include elaboration of a variety of peptide and nonpeptide toxins. Some of these are being exploited for human benefit.

Nevertheless, it is the cellular defense systems that are most highly conserved among the members of the animal kingdom. Elie Metchnikoff made this observation over a century ago when he described the parallels between starfish coelomocyte and vertebrate macrophage phagocytosis. Phagocytosis is the effector mechanism for most invertebrate host defense systems where it is regulated by cytokines and enhanced by opsonins. Other cellular effector strategies such as encapsulation are also found in both higher and lower animals.

Exceptional advances have been made in studies of crustaceans. The clotting cascade in *Limulus* has been defined, and its members cloned and sequenced to reveal multiple homologies to vertebrate lectins, epidermal growth factor (EGF), and serine proteases. Defense-related binding proteins for microbial components have been purified and characterized.

Invertebrate microbial killing strategies may be fundamentally different. Peptide/receptor interactions are not the only means of attack. Peptide amphipathic helicies and nonprotein antimicrobials are among the diverse defense systems employed by lower vertebrates.

This conference marks one small step in our efforts to understand the diverse systems by which animals recognize themselves, each other, and foreigners, and once the recognition is achieved, mount an appropriate response. It is gratifying to see the progress—to the way in which so many unique systems have developed over the recent past. Phenomena that have been meticulously characterized over years, even decades, are now being understood at the molecular level.

We know—but we need to be convincing to others—that studies of primordial immunity have tremendous importance:

for the clues they give us to higher systems;

for their unique strategies—some of which may be exploitable for human benefit—not just in medicine but in agriculture where biological pest control is a desirable alternative to toxic pesticides;

but, most importantly, for their own sake.

The value of maintaining the biodiversity of our planet should be obvious: The gene pool possessed by even the lowliest creatures has the potential for enormous benefits to mankind.

We thank the New York Academy of Sciences, especially the conference subcommittee members Julia Phillips-Quagliata and Susanna Cunningham-Rundles who provided scientific guidance, and Academy staff Geraldine Busacco, Lynn Serra, and Sherryl Greenberg of the Conference Department and Sheila Kane and Linda Mehta of the Editorial Department. LouAnn King, conference director for the Marine Biological Laboratory at Woods Hole where the meeting was held, provided a delightful setting. A more fitting venue could not have been found: The marine species discussed by several speakers were on display in nearby aquaria, and many participants for whom Woods Hole has been an important component of their training experienced a nostalgic "Alumni Week."

OVERVIEW

Development of an Immune System[a]

J. J. MARCHALONIS AND S. F. SCHLUTER

Department of Microbiology and Immunology
University of Arizona
College of Medicine
Tucson, Arizona 85724

INTRODUCTION

There has always been considerable interest in understanding the evolutionary origins of the immune system,[1–6] and this quest has recently been given impetus by the application of advanced methods in recombinant DNA technology[7–14] and immunochemistry.[15–18] An understanding of the genetic mechanisms underlying the capacity of vertebrates to respond to a potentially enormous (greater than 10^7) set of antigenic markers associated with pathogens or cancers that might never have been presented to the animal during its evolutionary development is of general theoretical importance for the understanding of anticipatory mechanisms[19] and of practical value in modulating the immune response in autoimmunity or amplifying it in cases of immunodeficiency. The overall conclusion is that antibodies of all gnathostomes ranging from elasmobranchs (sharks, rays, and their kin) to mammals are constructed in the same general fashion of light and heavy polypeptide chains that show V-region diversity imparting specificity for particular antigens.[3] The extant jawless vertebrates, the cyclostomes, are more primitive forms that are derived from ostracoderms, rather than from the placoderms, which were ancestral to jawed vertebrates.[20–22] Although numerous studies indicate that lampreys[23–26] and hagfish[27] produce antibodies when immunized with foreign antigens, only a limited degree of characterization data for these antibodies is currently available.[24,27–30] Proteins homologous to complement components of higher vertebrates have been isolated, and gene sequence documents their homologies.[31,32] Despite the gaps in detailed knowledge regarding the structures of antibodies of cyclostomes, it is usually considered that all vertebrates have the capacity to generate diverse sets of Igs in response to challenge with pathogens and foreign antigens. There is an extensive literature on immune-like phenomena in invertebrates,[33,34] with a strong recent focus on insects.[35,36] Many invertebrates show inducible cellular or humoral defense reactions following stimulation with bacterial products including endotoxins, but these phenomena do not appear to be directly related to the vertebrate-type immune response.[19,37]

[a]This research was supported in part by National Science Foundation Grant DCB 9106934 and National Institutes of Health Grant GM-42437.
[b]Address for correspondence: John J. Marchalonis, Ph.D., Department of Microbiology and Immunology, College of Medicine, University of Arizona, Tucson, AZ 85724.

CELLULAR RECOGNITION REACTIONS

Cellular recognition and aggressive reactions against foreign cells occur in diverse invertebrates, including annelids,[34] echinoderms,[38] and protochordate tunicates.[39] These animals also have innate constitutive mechanisms that include phagocytosis and recognition by lectins, which are either associated with cells or freely circulating.[40] Key issues that must be addressed in analyses of these cellular phenomena is whether the molecules involved are directly homologous to molecules of vertebrates that recognize either self (e.g., integrins, selectins) or "non-self" (antibodies, Tcr), and whether the cells, likewise, can be identified as homologues of lymphocytes and contain genetic programs capable of generating combining site diversities. Until extremely recently, it was generally accepted that the capacity to reject "allografts" was widely spread throughout the animal kingdom[41,42] and that a sophisticated "integrated" immune system involving the full range of T-cell, B-cell, and accessory cell function including the range of regulatory cytokines was the result of a process of Darwinian gradualism. The concept that allograft rejection in lower chordates (e.g. echinoderms) is a primordial T-cell/MHC recognition system has come under serious attack.[37] Moreover, the great discrepancies in gene arrangement between light[43,44] and heavy[45] chains of two distinct species of sharks and those of mammals raises questions regarding the regulation of antibody production in these species[46] even though the structures of the serum antibodies are remarkably similar. It has further been questioned whether sharks have MHC-restricted T-cells bearing α/β heterodimeric receptors[37,47]; and whether, in point of fact, they even have a need for such cells.[37,46]

PHYLOGENETIC EMERGENCE OF RECOGNITION–DEFENSE MOLECULES

Observing cellular phenomena is clearly not an adequate approach to solving these fundamental questions. Definitive solutions must come from characterization of the molecules that perform the recognition functions and from analyzing their gene arrangements. In cyclostomes and tunicates, the task must be carried out using a combination of protein immunochemistry and recombinant DNA technology, because the levels of antibody-like proteins in these species are extremely low compared with the situation in mammals and sharks. Sharks, for example, contain about 40 mg per milliliter of serum protein and approximately one-half of this is Ig.[46] Lampreys, by contrast, are agammaglobulinemic if one applies standards traditionally used for mammals and sharks, because the level of antibody-like material in lampreys is less than 2% of the amount found in gnathostomes. Furthermore, lamprey antibody shows a heterogeneity in charge comparable to that shown for other polyclonal antibody populations.[29] FIGURE 1 presents a simplified phylogenetic tree summarizing animal evolution arranged on the basis of developmental mechanisms and body plans[21,48] that serves to focus the problems associated with the origins of vertebrate-type immunity. The divergence of ancestral protostomes and deuterostomes occurred 500 to 600 million years ago,[21] a time at which all of the extant variants of animal body forms had appeared.[49,50] The diagram depicts the distribution

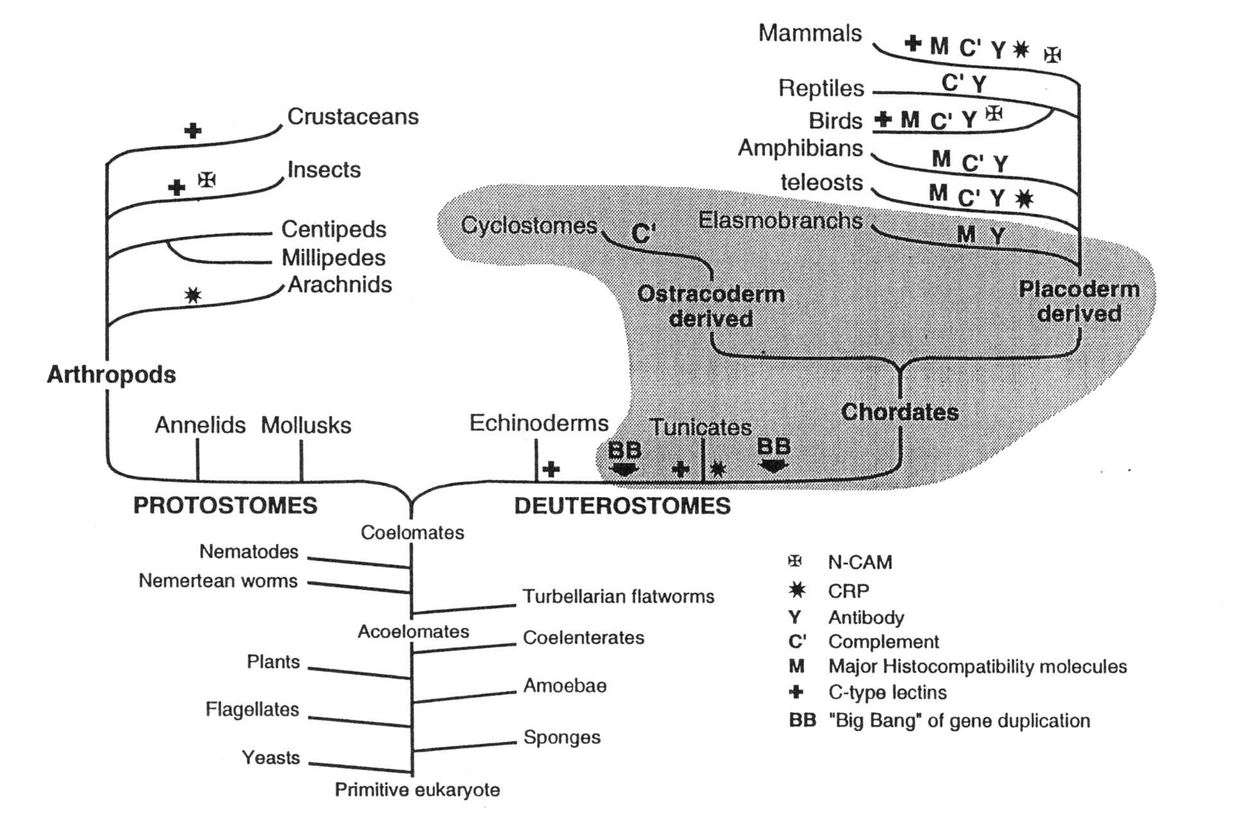

FIGURE 1. Phylogenetic tree summarizing evolutionary relationships organized on the basis of developmental mechanisms and body plans. The divergence between ancestral protostomes and deuterstomes occurred 500 to 600 million years ago. Documented presence of recognition molecules, including antibodies and effectors such as complement, are mapped onto the diagram. **N-CAM,** neural cell adhesion molecules; **CRP,** C-reactive proteins (pentraxin family); other molecules are defined in the figure.

of characterized recognition molecules including the C-reactive protein/pentraxin family,[51,52] lectins showing similarities to vertebrate hepatic lectins,[53] cytokines such as interleukin-1,[54] products of the major histocompatibility complex,[14,55] classical Igs, and members of the Ig superfamily (related to N-CAMs[36,56–58]). The vast majority of living forms are protostomes that differ from the deuterostomes in basic modes of embryonic differentiation including mechanisms of blood cell formation.[21,37,48,59] It is noteworthy that molecules of the C-reactive protein family occur in arthropods such as the ancient arachnoid *Limulus polyphemus,* the horseshoe crab.[51] An interesting observation, which most probably reflects convergent evolution, is the sharing of idiotypic markers characteristics of the binding site for phosphorylcholine among murine antibodies, mammalian C-reactive protein and the horseshoe crab sialic acid–binding lectin.[60] Members of the Ig superfamily related to neural cell adhesion molecules occur in higher vertebrates and in the neural development of insects.[57,58] Moreover, injection of cecropia larvae with a killed bacterial vaccine induces the presence in hemolymph of a molecule homologous to the N-CAMs.[36] In addition to lectins related to the CRP/pentraxin family, a number of calcium-dependent galactose-specific lectins have been found to be widely distributed.[53] Drickamer[62] organized animal lectins into two categories; namely, C- and S-type lectins with the C-type animal lectins having a characteristic carbohydrate-recognition domain comprising 120 to 130 residues. Examples of this type of lectin occur in mammals,[63] chickens,[64] arthropods (barnacles[65]), echinoderms (sea urchins[66]), and the tunicate *Polyandrocarpa misakiensis.*[53] Lymphocyte surface-homing molecules (selectins) contain C-type lectin domains as well as other structures. Individual tunicates can express diverse families of lectins, and there is considerable variation from species to species.[67] A galactose-binding lectin we isolated from the tunicate *Didemnun*[68,69] is sufficiently different in sequence and general properties to represent a family distinct from the C-type lectin of *Polyandrocarpa.* The most primitive species in which genes specifying MHC products have yet been identified are the MHC class II α genes of elasmobranchs.[14] It was once argued that the driving force for the generation of the immune response was allorecognition involving MHC products and primordial T-cells.[41] However, contemporary interpretations of the basis of MHC restriction suggest that MHC molecules serve more as vehicles for the presentation of peptide antigens[70] rather than themselves acting as critical nonself antigens. In any case, the interpretation of T-cell immunity arising in lower deuterostomes such as echinoderms and tunicates on the basis of cellular rejection phenomena is currently the target of severe criticism.[19,37] This lectin, which is probably related to cellular homing molecules, is found in a primitive true vertebrate, the lamprey, as well as in a protochordate.

GENE ARRANGEMENT AND IMMUNOGLOBULIN DIVERSITY

The arrangement of shark immunoglobulin gene segments is completely distinct from that of mammals and challenges accepted mechanisms for clonal restriction. We have extensive characterization of shark IgM/λ Igs and are in the process of completing our characterization of germline genes specifying the λ light chains and the μ heavy chains.[12,43,46] Genes from two species of sharks, the sandbar shark

(Carcharhinus plumbeus) studied by us and the horned shark *(Heterodontus francheschii)* studied by Litman and his colleagues,[7,44,45,71] have now been characterized in some detail. The ancestors of these two sharks diverged from one another more than 180 million years ago,[72] and the degree of sequence divergence between λ light chains of the two species is consistent with this early tine of divergence (approximately 50% identity). Comparison of the μ chain sequences indicates that some C-region domains diverged less than did others; for example, Cμ4 domains of the two sharks are greater than 80% identical, whereas the Cμ3 domains are only about 60% identical.[46] Interestingly, the best match in computer database searches of one of our sandbar shark V_L clones is a human VλVI sequence, which shows greater than 50% identity.[43,46] The predicted 3-D structure of shark Igs is extremely similar, if not identical, to those of human Igs.[12] Although the protein and gene sequences of the coding regions of shark and mammalian Ig light chains shows unquestionable homology, the gene arrangement differs markedly between these two groups of vertebrates. There is a distinction between birds and other vertebrates, because birds depend heavily on gene conversion to generate diversity,[73] but the other vertebrate classes (teleosts[10] and amphibians[74]) have the mammalian-type gene arrangement in which there are a number of V region segments, a small number of joining segments (in the case of heavy chains and Tcr β chains a small set of diversity segments), and a few C regions. Immunological commitment follows from rearrangement of particular V and J segments, and the distance between the set of V region genes (usually hundreds) and the J segments is large—greater than 100 kilobases. By contrast, the V, J, and C segments specifying both sandbar shark and *Heterodontus* shark light chains occur as small clusters with the total distance approximately 3 to 6 kb.[43,44] Each gene cluster contains V, J, and C elements that differ in sequence from those in other clusters, and the distance between clusters is large, at a minimum >20 kilobases, but the exact distance remains to be determined. In the two germline gene clusters sequenced for the horned shark, the V and J segments are separated from one another by approximately 0.4 kilobases, and the intron contains heptamers and nonomers needed for recombination. Thus far, we have sequence data on four germline clones from *Carcharhinus.* In all four, the V and J segments are fused in register in the germline. In the shark, the primary basis for V-region diversity could not be the type of recombination mechanism found in mammals and other vertebrates. This conclusion is buttressed by the finding that sandbar shark light chain V-J segments are fused in the germline, so there may be no need for recombination to generate diversity and ensure clonal commitment. The diversity would most probably be the result of the existence of large numbers of V_n, J_n, and C_n in cassettes. A parallel situation exists for the horned shark heavy chain genes.[71] The fundamental question with respect to the origins of the immune system is whether the multiple gene arrangement found in sharks is the primordial one for Igs. It could be speculated that the particular Ig clusters or translocons of vertebrates that arose subsequent to sharks followed from the selection of individual clusters and the amplification of V regions by tandem duplication. The situation in chickens, in which both light chains and heavy chains[73,75] have only one functional V segment but a number of pseudogenes and generate variability by gene conversion, might be a degenerate form of the mammalian–amphibian–teleost arrangement.

The degree of Ig diversity can be estimated to some degree by isoelectric focusing analysis. Light chains isolated from polyclonal serum Igs of three species of *Carcharhine* sharks are as heterogeneous in isoelectric focusing as are polyclonal Igs of mice and humans.[76] Lamprey Ig-like light chains detected by antibody against the Jβ peptide also show substantial charge heterogeneity in isoelectric focusing.[29] The shark result is consistent with the variation in V-region gene sequence that we have observed. The lamprey results indicate that, whatever the gene organization, the molecules show charge heterogeneity comparable to that of classical Igs. By contrast, the μCRM of tunicates is extremely restricted in isoelectric focusing, even though it is of a sufficient size to contain V, J, and C domains.[77]

Because of the multiple gene arrangement of shark Igs into "mega isotypes,"[46] counting genes alone is not sufficient to establish the mechanism of diversification. For this reason, we also propose to determine the presence of recombination activating genes[78,79] in elasmobranchs and in lower deuterostomes. Our approach is to develop PCR probes based on the sequence of RAG1 genes in man, mouse, and chickens and to analyze germline DNA of carcharhine sharks and the horned shark. The procedure, if we are successful at this stage, will parallel our approach with Ig molecules; namely, we will use sequences obtained from elasmobranchs to search for homologous molecules in cyclostomes and tunicates. The mechanism underlying the vertebrate type of immune system arose in deuterostomes and most probably in the lower chordates.[80] Although lamprey antibodies and their genes remain to be characterized, the circumstantial evidence to date suggests that these primitive true vertebrates have the capacity to generate classical vertebrate antibodies. The gene arrangement may prove to be either sharklike or typical of most vertebrate classes. It is unlikely that the vertebrate-type T-cell recognition system was present in echinoderms,[37] but it is still possible that the system emerged either before the origins of contemporary tunicates or after that ancestral divergence but before the emergence of ancestral cyclostomes.

WHENCE THE IMMUNE RESPONSE?

Study of the mammalian immune system arose largely in the context of microbial infections and its greatest triumph has been the near eradication of major infectious diseases including smallpox and polio. Because of this association with disease, it has been widely accepted that the function of the immune response was to protect people against infectious diseases.[81] On reflection, however, it is apparent that virtually all living organisms are subject to microbial infections, and these organisms often show induced protective responses. For example, plants respond to fungal infections by producing toxic organic molecules,[82] and insects respond to bacterial infections by producing a series of bacteriostatic molecules.[33,35] Thus, many organisms that do not have the capacity to mount a vertebrate-type immune response can protect themselves by inducible defense mechanisms using molecules distinct from immunoglobulins. Even within man, there are ancient defense mechanisms shared with protostome invertebrates. One of the best examples of these is the presence of "acute phase" molecules of the C-reactive protein family in horseshoe crabs[51] and in man. The level of the molecule in the horseshoe crab hemolymph is constitutive at about 1 to 2 mg per milliliter. In man, this protein is usually undetectable in the serum of healthy persons, but is induced by infection with bacteria such as pneumococci and reaches the level of a few milligrams per milliliter. The C-reactive protein has a specificity for certain bacteria and for small organic molecules such as phosphorylcholine. Moreover, it can interact with a number of molecules of the immune system such as complement and Fc receptors on lymphocytes,[83] although it is clearly not related to immunoglobulin on the basis of protein or gene sequence. Therefore, a marked specificity and inducibility do not themselves indicate a vertebrate type of immune response. Parallel arguments have been made for a correlation between the emergence of cancer and vertebrate-type immunity. However, neoplasms are widely distributed among protostomes and lower vertebrates.[84]

The real issue underlying the generation of the vertebrate-type immune response is the capacity for the generation of multiple V-region combining sites allowing the complementary recognition of non-self molecules. The mechanism for the generation of diversity is a critical evolutionary event in the emergence of the vertebrate immune system. Because in mammals, birds, and amphibians, large arrays of V regions exist and a recombination event is required for clonal restriction, it is appropriate to focus attention on the emergence of recombinase activator genes,[78,79] with the suggestion that the insertion of these into a primitive chordate genome might have led to the generation of diversity. However, the gene cluster arrangement in elasmobranchs where light chain V and C genes can exist fused in register in the germ line[43] shows that considerable diversity can be generated without the apparent need for a recombination mechanism. Therefore, this suggests that other mechanisms may be necessary to bring about clonal restriction in B lymphocytes of the shark.

We believe that B lymphocytes in the shark show clonal restriction because immunization with various antigens[85–88] can lead to increases in titer of greater than 10,000-fold without increasing the total serum concentration of serum immunoglobulin. This contrasts with the situation for acute-phase proteins in which the level of the C-reactive protein might increase from undetectable up to a few milligrams per milliliter. Another feature that distinguishes sharks from higher vertebrates is that secondary immunization does not produce either an increase in titer (memory) over the primary response or affinity maturation in which the later antibodies bind antigen better than those of the initial immunization.[46,47,87] This difference most probably correlates with the fact that sharks only have IgM immunoglobulins and cannot show a switch to another immunoglobulin class defined by another heavy chain. The shark most probably would express only germline genes, although Litman presented evidence for mutation in shark V regions.[89] Its extent, however, is apparently insufficient for secondary maturation of affinity to occur. The shark immune system is a promising model for approaching an alternative adaptive argument regarding the origins of the immune system.

The concept that the immune response arose as an antimicrobial defense mechanism has attractive features, particularly because microbes can provide second signals stimulating the immune system.[81] Moreover, viruses have evolved defense mechanisms using elements of the immune system.[90] This may reflect a long-term coevolution of parasite and host in which changes in microbial antigenicity have provided stimuli for enhancing mutational and selective rates among the lymphocyte population.[3] However, a concept is developing that a large part of the immune system is directed against self, where antibodies can carry out a physiological homeostatic role such as removal of aged red cells[91] or DNA or other cell products that might be floating around in serum,[92] as well as in a process of immunoregulation directed against combining sites of Igs and Tcrs. It was recognized first by Ehrlich[94] and subsequently expounded by Jerne that natural antibodies provide the basis for recognition in the immune system.[95] Burnet argued that individual cells in the lymphocyte population are clonally restricted with respect to capacity to recognize particular antigens because they express surface Ig receptors with unique specificity. Contact with the right antigen would select only for those cells capable of responding.[96] Healthy animals, ranging from man[97] to a variety of lower species,[46,93,98] naturally contain antibodies directed against many self components. Sharks have high levels of natural antibodies against DNA, thyroglobulin, and immunoglobulins,[46] as well as against proteins such as ovalbumin,[46] with which they have never been purposefully immunized. The shark, thus, has untapped potential for studying IgM-mediated autoimmunity where germline genes are expressed, and there is little need for T-cell help.[46] It has recently been hypothesized that the B cells of sharks most probably resemble the $CD5^+$ cells of man, which produce polyspecific IgM often showing autoantibody activity,[46,47] and that their T cells resemble $\gamma\delta$ T cells rather than the α/β Tcr–bearing helper cells.[47] Studies with man and animals indicate that a large part of the natural antibody repertoire is directed against self-recognition molecules including idiotypes of antibodies[93,97] and variable region markers of T-cell receptors.[99] A large part of the immune response may be used for regulation of self recognition and focused on V regions.[100] Internal pressure for diversification may arise from the necessity for antibodies to recognize combining site and regulatory epitopes of other antibodies and corresponding regions of T-cell receptors. In the situation represented by sharks and fetal mammals, in which IgM antibodies expressing germline V_H genes constitute the dominant (or only) antibodies, a high degree of polyreactivity of individual antibodies and autoreactivity would occur. Regulatory autoreactivity[99–102] may prove to be the definitive feature of the vertebrate immune response, which gives it the diversity to carry out autologous physiological roles as well as function in defense in a unique fashion not yet shown in protostomes or primitive deuterostomes.

SUMMARY

Minimally, an immune response is an induced cellular and/or humoral defense mechanism specific for the challenging agent. The system is a cognitive one inasmuch as a second stimulus with the same antigen can specifically induce either an enhanced response (memory) or diminished response (tolerance). The cells responsible for the initial antigen-specific recognition in higher vertebrates are clonally restricted T and B lymphocytes. Accessory cells are necessary for the processing and presentation of antigen, and physiologic mediators (cytokines) are essential for proliferation, interaction, and regulation of the system. Although it now appears that the recombination mechanisms essential for the anticipatory immune response occurred late in the deuterostome stream leading to vertebrates, molecules required for cell adhesion and regulation are widely spread in phylogeny. Their emergence must have preceded the divergence between ancestral protostomes and deuterostomes.

Genetic mechanisms underlying the generation of diversity in the light and heavy chains of antibodies of mammals may be quite distinct in primitive vertebrates, particularly elasmobranchs, the ancestors of which diverged from those of mammals more than 400 million years ago. Despite this, clonal selection of antigen receptors of lymphocytes is most probably universal within the vertebrates. There is no need to force induced recognition in protostomes (e.g. insects) or lower deuterostomes (e.g. echinoderms) into mammalian models of immunity.

ACKNOWLEDGMENTS

We would like to thank Ms. Diana Humphreys for assistance in the preparation of this manuscript.

REFERENCES

1. Metchnikoff, E. 1905. Immunity in Infective Diseases. Cambridge University Press. Cambridge, England.
2. Burnet, F. M. 1971. Nature (London) **232:** 230–235.
3. Marchalonis, J. J. 1977. Immunity in Evolution. Harvard Press. Cambridge, MA. 316 pp.
4. Hildemann, W. H. 1974. Nature (London) **250:** 116–119.
5. Noguchi, H. 1903. Centralbl. Balst. Abt. Orig. **33:** 353–362.
6. Marchalonis, J. J. & C. L. Reinisch, Eds. 1990. Defense Molecules. Wiley-Liss. New York.
7. Hinds, K. R. & G. W. Litman. 1986. Nature (London) **320:** 546–549.
8. Litman, G. W., K. Murphy, L. Berger, R. Litman, K. Hinds & B. W. Erickson. 1985. Proc. Natl. Acad. Sci. USA **82:** 844–848.
9. Schwager, J., C. A. Mikoryak & L. A. Steiner. 1988. Proc. Natl. Acad. Sci. USA **85:** 2245–2249.
10. Wilson, M. R. & G. W. Warr. 1992. Ann. Rev. Fish Dis. **2:** 201–221.
11. Zezza, D. J., S. E. Stewart & L. A. Steiner. 1992. J. Immunol. **149:** 3968–3977.
12. Schluter, S. F., V. S. Hohman, A. B. Edmundson & J. J. Marchalonis. 1990. Proc. Natl. Acad. Sci. USA **86:** 9961–9965.
13. Marchalonis, J. J. & S. F. Schluter. 1989. FASEB J. **3:** 2469–2479.
14. Kasahara, M., M. Vazquez, K. Sata, E. C. McKinney & M. F. Flajnik. 1992. Proc. Natl. Acad. Sci. USA **89:** 6688–6692.
15. Schluter, S. F., I. L. Rosenshein, R. A. Hubbard & J. J. Marchalonis. 1987. Biochem. Biophys. Res. Commun. **145:** 699–705.
16. Eshhar, Z., O. Gigi, D. Givol & Y. Ben-Neriah. 1983. Eur. J. Immunol. **13:** 533–540.
17. Marchalonis, J. J., S. F. Schluter, H.-Y. Yang, V. S. Hohman, K. McGee & L. Yeaton. 1992. Comp. Biochem. Physiol **101A**(No. 4): 675–687.
18. Marchalonis, J. J., V. S. Hohman, H. Kaymaz & S. F. Schluter. 1993. Comp. Biochem. Physiol. **105B(3/4):** 423–441.
19. Klein, J. 1989. Scand. J. Immunol. **29:** 499–505.
20. Romer, A. S. 1966. Vertebrate Paleontology, 3rd edit. University of Chicago Press. Chicago.
21. Loomis, W. F. 1988. Four billion years: An essay on the evolution of genes and organisms. Sinuaer Associates. Sunderland, MA.
22. Forey, P. & P. Janvier. 1993. Nature **361:** 129–134.
23. Finstad, J. & R. A. Good. 1964. J. Exp. Med. **120:** 1151–1168.
24. Marchalonis, J. J. & G. M. Edelman. 1968. J. Exp. Med. **127:** 891–914.

25. FUJII, T. 1982. J. Morphol. **173:** 87–100.
26. HAGEN, M., M. F. FILOSAN & J. H. YOUSON. 1985. Comp. Biochem Physiol **82A:** 207–210.
27. RAISON, R. L., C. J. HULL & W. H. HILDEMANN. 1978. Proc. Natl. Acad. Sci. USA **75:** 5679–5783.
28. KOBAYASHI, K., S. TOMONAGA & K. HAGIWARA. 1985. Mol. Immunol. **22:** 1091–1097.
29. MARCHALONIS, J. J. & S. F. SCHLUTER. 1989. Phylogenetic studies with rearranging immunoglobulins. *In* UCLA Symposium on Defense Molecules. J. J. Marchalonis & C. L. Reinisch, Eds.: 265–280. Alan R. Liss. New York.
30. VARNER, J., P. NEAME & G. W. LITMAN. 1991. Proc. Natl. Acad. Sci. USA **88:** 1746–1750.
31. NONAKA, M. & M. TAKAHASHI. 1992. J. Immunol. **148:** 3290–3295.
32. ISHIGURO, H., K. KOBAYASHI, M. SUZUKI, K. TITANI, S. TOMONAGA & Y. KUROSAWA. 1992. EMBO J. **11:** 829–837.
33. BREHELIN, M. 1986. Immunity in Invertebrates. Springer-Verlag. New York.
34. COOPER, E. L. 1982. An overview. *In* The Reticuloendothelial System, A Comprehensive Treatise. Vol. III. N. Cohen & M. M. Seigel, Eds.: 1–35. Plenum Press. New York.
35. DUNN, P. E. 1990. BioScience **40:** 738–744.
36. SUN, S. C., I. LINDSTROM, H. G. BOMAN, I. FAYE & O. SCHMIDT. 1990. Science **250:** 1729–1732.
37. SMITH, L. C. & E. H. DAVIDSON. 1992. Immunol. Today **13:** 356–361.
38. KARP, R. D. & W. H. HILDEMANN. 1976. Transplant **22:** 434–438.
39. RAFTOS, D. A., N. N. TAIT & D. A. BRISCOE. 1987. Dev. Comp. Immunol. **11:** 343–351.
40. CASSELS, F. J., J. J. MARCHALONIS & G. R. VASTA. 1986. Comp. Biochem. Physiol. **85B:** 23–30.
41. HILDEMANN, W. H., E. A. CLARK & R. L. RAISON. 1981. Comprehensive Immunogenetics. Elsevier Press. New York.
42. COOPER, E. L., B. RINKEOVICH, G. UHLENBRUCK & P. VALEMBOIS. 1992. Scand. J. Immunol. **35:** 247–251.
43. HOHMAN, V. S., S. F. SCHLUTER & J. J. MARCHALONIS. 1992. Proc. Natl. Acad. Sci. USA **89:** 276–280.
44. SHAMBLOTT, M. J. & G. W. LITMAN. 1989. EMBO J. **8:** 3733–3739.
45. LITMAN, G. W. & K. R. HINDS. 1987. *In* Evolution and Vertebrate Immunity. G. Kelsoe & D. H. Schulze, Eds.: 35–51. University of Texas Press. Austin, TX.
46. MARCHALONIS, J. J., V. S. HOHMAN, C. THOMAS & S. F. SCHLUTER. 1993. Dev. Comp. **17:** 41–53.
47. MCKINNEY, E. C. 1992. Ann. Rev. Fish Dis. **2:** 43–51.
48. DAVIDSON, E. H. 1986. Gene Activity in Early Development. Academic Press. Orlando, FL.
49. LEVINTON, J. 1992. Sci. Am.: 84–91.
50. GOULD, S. J. 1989. Wonderful Life: The Burgess Shale and the Nature of History. W. W. Norton Press. New York.
51. NGUYEN, N. Y., A. SUZUKI, R. A. BOYKINS & T. Y. LIU. Biol. Chem. **261:** 10456–10460.
52. YING, S.-C., J. J. MARCHALONIS, A. T. GEWURZ, J. N. SIEGEL, H. JIANG, B. E. GEWURZ & H. GEWURZ. 1992. Immunology **76:** 324–330.
53. SUZUKI, T., T. TAKAGI, T. FURUKOHRI, K. KAWAMURA & M. NAKAUCHI. 1990. J. Biol. Chem. **265:** 1274–1281.
54. BECK, G., G. R. VASTA, J. J. MARCHALONIS & G. S. HABICHT. 1989. Comp. Biochem. Physiol. **92B:** 93–98.
55. KAUFMAN, J., K. SKJOEDT & J. SALOMONSEN. 1990. Immunol. Rev. **113:** 83–117.
56. EDELMAN, GERALD M. 1987. Immunol. Rev. **100:** 12–45.
57. HARRELSON, A. L. & C. S. GOODMAN. 1988. Science **242:** 700–708.
58. SEEGER, M. A., L. HAFFLEY & T. C. KAUFMAN. 1988. Cell **55:** 489–600.
59. RATCLIFF, N. A. & D. A. MILLAR. 1988. *In* Vertebrate Blood Cells. A. F. Rowley & N. A. Ratcliff, Eds.: 1–17. Cambridge University Press. Cambridge, England.

60. VASTA, G. R., J. J. MARCHALONIS & H. KOHLER. 1984. J. Exp. Med. **159:** 1270–1276.
62. DRICKAMER, K. 1988. J. Biol. Chem. **263:** 9557–9560.
63. SPIESS, M. & H. F. LODISH. 1985. Proc. Natl. Acad. Sci. USA **82:** 6465–6469.
64. DRICKAMER, K. 1981. J. Biol. Chem. **256:** 5827–5839.
65. KAMIYA, H. J., K. MURAMOTO & R. GOTO. 1987. Dev. Comp. Immunol. **11:** 297–307.
66. GIGA, Y., A. IKAI & K. TAKAHASHI. 1987. J. Biol. Chem. **262:** 6197–6203.
67. VASTA, G. R., G. W. WARR & J. J. MARCHALONIS. 1982. Comp. Biochem. Physiol. **73B:** 887–900.
68. VASTA, G. R., J. HUNT, J. J. MARCHALONIS & W. W. FISH. 1986. J. Biol. Chem. **261:** 9174–9181.
69. VASTA, G. R. & J. J. MARCHALONIS. 1986. J. Biol. Chem. **261:** 9182–9186.
70. ROTHBARD, J. & M. L. GEFTER. 1991. Ann. Rev. Immunol. **9:** 527–565.
71. KOKUBU, F., K. HINDS, R. LITMAN, M. J. SHAMBLOTT & G. W. LITMAN. 1988. EMBO J. **7:** 1979–1988.
72. COMPAGNO, L. J. V. 1988. Sharks of the Order Carcharhiniformes. Princeton University Press. Princeton, NJ.
73. MCCORMACK, W. T., L. W. TJOELKER & C. B. THOMPSON. 1991. Ann. Rev. Immunol. **9:** 219–241.
74. DU PASQUIER, L., J. SCHWAGER & M. F. FLAJNIK. 1989. Ann. Rev. Immunol. **7:** 251–275.
75. REYNAUD, A.-C., A. DAHAN, V. ANQUEZ & J.-C. WEILL. 1989. Cell **59:** 171–183.
76. MARCHALONIS, J. J., S. F. SCHLUTER, I. L. ROSENSHEIN & A. C. WANG. 1988. Dev. Comp. Immunol. **12:** 65–74.
77. SCHLUTER, S. F., E. WANG, J. SCHROEDER & J. J. MARCHALONIS. 1993. Ann. N.Y. Acad. Sci. This volume.
78. SCHATZ, D. G. & D. BALTIMORE. 1988. Cell **53:** 107–115.
79. SCHATZ, D. G., M. A. OETTINGER & D. BALTIMORE. 1989. Cell **59:** 1035–1048.
80. MARCHALONIS, J. J. & S. F. SCHLUTER. 1990. BioScience, **40**(No. 10): 758–768.
81. JANEWAY, C. A. JR. 1989. Cold Spring Harbor Symp. Quant. Biol. **LIV:** 1–13.
82. ANDERSON, A. J. 1990. *In* Defense Molecules. J. J. Marchalonis & C. L. Reinisch, Eds.: 17–32. Wiley-Liss. New York..
83. VOLANAKIS, E., X. YUANYUAN & K. J. MACON. 1990. *In* Defense Molecules. J. J. Marchalonis & C. L. Reinisch, Eds.: 161–175. Wiley-Liss. New York.
84. HARSHBARGER, J. C., A. M. CHARLES & P. M. SPERO. 1981. *In* Phyletic Approaches to Cancer. C. J. Dawe, Ed.: 357–384. Japan Sci. Soc. Press. Tokyo.
85. MÄKELÄ, O. & G. W. LITMAN. 1980. Nature **287:** 639–642.
86. SHANKEY, T. V. & L. W. CLEM. 1980. Mol. Immunol. **17:** 365–375.
87. CLEM, L. W. & M. M. SIGEL. 1966. *In* Phylogeny of Immunity. R. T. Smith, P. A. Miescher & R. A. Good, Eds.: 190–198. University Presses of Florida. Gainesville, FL.
88. MARCHALONIS, J. J. & G. M. EDELMAN. 1965. J. Exp. Med. **122:** 601–618.
89. LITMAN, G. W., R. N. HAIRE, K. R. HINDS, C. T. AMEMIYA, J. P. RAST & M. E. HULST. 1992. Ann. N.Y. Acad. Sci. **651:** 360–368.
90. KOTWAL, G. J., R. MCKENZIE, S. N. ISAACS, A. W. HUGIN, N. M. FRANK & B. MOSS. 1990. *In* Defense Molecules. J. J. Marchalonis & C. L. Reinisch, Eds.: 149–159. Wiley-Liss. New York.
91. KAY, M. M. B., J. J. MARCHALONIS, J. HUGHES, K. WATANABE & S. F. SCHLUTER. 1990. Proc. Natl. Acad. Sci. USA **87:** 5734–5738.
92. ZOUALI, M., B. D. STOLLAR & R. S. SCHWARTZ. 1985. Immunol. Rev. **105:** 137–141.
93. ADIB, M., J. RAGIMBEAU, S. AVRAMEAS & T. TERNYNCK. 1990. J. Immunol. **145:** 3807–3812.
94. EHRLICH, P. 1900. Proc. R. Soc. London Ser. B **65:** 424–430.
95. JERNE, N. K. 1955. Proc. Natl. Acad. Sci. USA **11:** 849–857.
96. BURNET, F. M. 1959. The Clonal Selection Theory of Acquired Immunity. Vanderbilt University Press. Nashville, TN.
97. HUREZ, V., S.-V. KAVERI & M. D. KAZATCHKINE 1993. Eur. J. Immunol. **23:** 783–789.
98. GONZALEZ, R., J. CHARLEMAGNE, W. MAHANA & S. AVRAMEAS. 1988. Immunology **63:** 31–36.

99. Marchalonis, J. J., H. Kaymaz, F. Dedeoglu, S. F. Schluter, D. E. Yocum & A. B. Edmundson. 1992. Proc. Natl. Acad. Sci. USA **89:** 3325–3329.
100. Stewart, J. 1992. Immunol. Today **13:** 396–399.
101. Jerne, N. 1974. Ann. Inst. Pasteur Immunol. **125C:** 435–441.
102. Cohen, I. R. & D. B. Young. 1991. Immunol. Today **12:** 105–110.

MHC Evolution and Development of a Recognition System

SUSUMU OHNO

Beckman Research Institute of the City of Hope
1450 East Duarte Road
Duarte, California 91010-0269

Extremely polymorphic loci are used by a number of diverse organisms to distinguish self from non-self. For example, discrimination against the self is practiced by mustard plants via pollen–stigma incompatibility glycoprotein to avoid self fertilization. Discrimination against non-selves, on the other hand, is a way of life for corals and colony-forming tunicates that have to maintain individual integrity in the midst of chaotic communal living.

Because MHC antigens were originally discovered as the cause of graft rejections, the temptation has always been to regard MHC antigens as members of the ancient stock that antedated establishment of the adaptive immune system in vertebrates. Here, I would like to play the role of devil's advocate in trying to advance a contrary argument that regards MHC antigens as late appendages to the immune system.

ABSENCE OF NONIMMUNOLOGICAL SELF–NON-SELF DISCRIMINATION IN BIRDS AND MAMMALS

A segment of the quail neural tube transplanted to replace a portion of the host neural tube in chick embryos distributes its descendants to the chicken body as does the host neural tube. A series of extremely elegant embryological findings by Nicole Douraine is the result.

Two H-2 different mouse blastocysts can be fused to form one embryo. The resulting chimeras are allophenic mice that are healthy in every aspect. The above seem to show that even with regard to ubiquitously expressed class I MHC antigens, allelic differences do not hinder cell–cell interaction that culminate in organogenesis. In essence, no role, aside from the immunological one, can be assigned to MHC antigens.

VERY EFFECTIVE IMMUNE RESPONSE NOT REQUIRING SELF–NON-SELF DISCRIMINATION

Phosphocholines are exposed on the surface of the bacterial plasma membrane, whereas the same is sequestered inside in the case of vertebrate plasma membrane. Thus, anti-phosphocholine antibodies are the very effective bacteriocidal agent that can be generated without fear of self destruction. Certain sugar transferases are unique to specific bacteria. Antibodies directed against glycoproteins whose sugar residues have been placed by such bacteria-specific transferases would do no harm to the host.

Although nature is more often than not irrational, one rational approach to starting the adaptive immune system might have been to generate the above type of specific antibodies that bring only benefits without cost.

ANTIVIRAL RESPONSES HAD TO BE ANTI-PROTEIN RESPONSES, HENCE THE NEED FOR SELF–NON-SELF DISCRIMINATION

With the possible exception of those with unusually large genomes, viruses use host sugar transferases for glycosylation of their proteins. Accordingly, an antiviral response has to be an antiprotein response. It would appear that the need for self–nonself discrimination arose when the decision was made to generate antigen-specific responses against viruses.

Inasmuch as class II MHC antigens are expressed only by antigen-presenting macrophages, antibody-producing B cells, and certain activated T cells, their role, being strictly immunological, is self evident. Thus, I shall deal only with class I antigens that are ubiquitously expressed by all cell types of the body. Needless to say, class I MHC antigens are targets of cytotoxic T cells. Being such targets, class I MHC antigens are also involved in control of B cells. Although the failure to provide effective T-cell help to relevant B-cell clones is the cause of each recessive non-responder status that is class II MHC linked, the dominant nonresponder status is linked to class I as well as class II MHC antigen. Active suppression of relevant B-cell clones is due to their destruction by cytotoxic T cells.

DIFFERENT ALLELES OF CLASS I MHC ANTIGEN PRESENT DIFFERENT SETS OF NONAPEPTIDES

During their parasitic existence inside the host cell, a typical virus in propagation produces only a few proteins but in large amounts. Relevant T epitopes that provoke a vigorous cytotoxic T-cell response should be found among these few viral proteins. As though to insure the presence of at least one T epitope in every viral protein, different allelic products of class I MHC antigen apparently developed binding preference to different sets of nonapeptides. The above appears to be a sufficient *raison d'entre* for extreme polymorphism as well as multiple loci for class I MHC antigens.

For example, human HLA-A2 shows the preference for hydrophobic nonapeptides, having Leu or Met at the second position and Val or Leu in their sixth and ninth positions,[1] whereas human HLA-B27 prefers basic nonapeptides with Arg or Lys in first and ninth and Arg in their second positions.[2,3] In the case of mice, H-2K^b is unique in preferring aromatic residue-containing octapeptides. Preferred octapeptides have either Tyr or Phe at the fifth position and another Tyr is often found at the third position.[1]

TOO RIGID A PREFERENCE FOR AN UNUSUAL NONAPEPTIDE MAKES THAT MHC IRRELEVANT

A number of minor histocompatibility antigens has come to be known in the mouse. The detection of minors requires the availability of diverse inbred strains. Hence, minors are well established only in the mouse. In view of the nonapeptide-presenting function of class I MHC antigens, one realizes that each minor is a result of two very unlikely events combined. First of all, one particular nonapeptide should enjoy a virtual monopoly over one particular MHC, allowing practically no access to other peptide fragments. Furthermore, that monopolizing nonapeptide should also be polymorphic, enabling cytotoxic T cells to differentiate substituted residues.

The seminal study by Fischer Lindahl and her colleagues unraveled the following on the maternally transmitted histocompatibility antigen (Mta) of the mouse.[4] Mta is the amino-terminal nonapeptide derived from ND1 protein (a mitochondrion-encoded 30-kDa hydrophobic subunit of NADH dehydrogenase). This nonapeptide is presented by one of the class I MHC dehydrogenases). This nonapeptide is presented by one of the class I MHC antigens, Hmt. Allelic polymorphism of MD1 protein involves its sixth position where Ile has been substituted by either Ala, Val, or Thr in other alleles.[4]

As shown in FIGURE 1, Hmt class I MHC antigen prefers nonapeptides that have formyl-methione at the first position. Inasmuch as the use of formyl-methionine for the translation initiation is a practice maintained only by the translation machinery of prokaryotes and mitochondria, this preference explains the virtual monopoly enjoyed by the amino-terminal nonapeptide derived from ND1 over the Hmt class I MHC antigen. By this particular preference, however, Hmt appears to have forfeited its *raison d'etre,* for the simple reason that the amino-terminal Met of viruses are never formylated. To be sure, there are certain bacteria such as *Lysteria* that can lead an intracellular existence. Yet even bacterial proteins incorporate plain Met in all but the NH_2-terminal position.

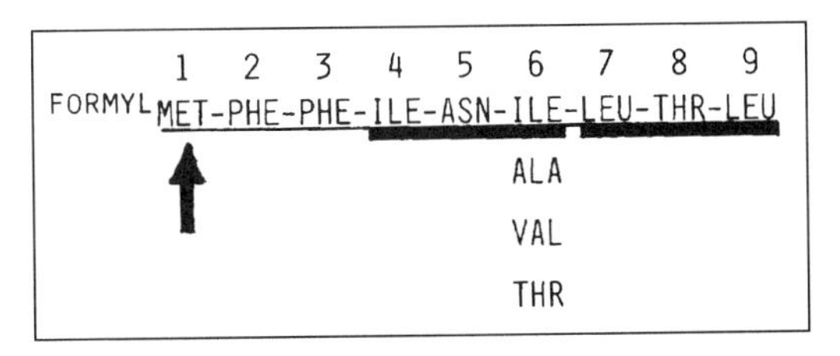

FIGURE 1. The amino-terminal nonapeptide derived from the mitochondrial ND1 subunit of NADH dehydrogenase monopolizing the mouse Hmt class I MHC antigen. Because of allelic polymorphism at its sixth position, this nonapeptide has been known as the maternally inherited minor histocompatibility antigen.[4] This class I MHC antigen cannot be a useful one, because of its preference for nonapeptides having formyl-methionine at the first position. Two successive tripeptidic palindromes within this nonapeptide are underlined by solid bars.

THE PREFERENCE BY PROMINENT CLASS I MHC ANTIGENS IS FOR PALINDROMIC NONAPEPTIDES

HLA-A2 is the most common allelic form of class I MHC antigens in all human populations. Such a prominence likely reflects its functional versatility. Indeed, each of the more common class I MHC antigens appears capable of finding at least one effective T epitope in one out of every three viral proteins, so as to provoke vigorous cytotoxic T-cell response against infected cells. This shows their preference for common nonapeptides, and common nonapeptides that are found in unrelated proteins are those containing palindromes.[5]

FIGURE 1 shows that even a mitochondrial nonapeptide preferred by the Hmt class I MHC antigen of the mouse contains two tripeptidic palindromes in succession: Ile-Asn-Ile followed by Leu-Thr-Leu. Were it not for its preference for formylmethionine at the first position, it would have been a versatile class I MHC antigen. Out of the nine peptide fragments of influenza A virus matrix protein tested, only one provoked HLA-A2-restricted cytotoxic T-cell response.[6] As shown in FIGURE 2, this T epitope again contained two tripeptidic palindromes in succession: Phe-Val-Phe followed by Thr-Leu-Thr. FIGURE 2 also shows that a nonapeptide derived from vesicular stomatitis virus N protein presented by mouse class I MHC H-2K^{b} contains the tripetidic palindrome Tyr-Val-Tyr.[7] All the H-2K^{b}-restricted cytotoxic T-cell responses were directed against this T epitope. In view of the latter finding,[1] however, this T epitope was probably an octapeptide instead of a nonapeptide, Lys of the ninth position becoming irrelevant.

```
INFLUENZA A VIRUS MATRIX PROTEIN: 252 RESIDUES

        OF 8 PEPTIDE FRAGMENTS TESTED:

                ONLY EFFECTIVE HLA-A2-PRESENTED NONAPEPTIDE TARGET

                        OF CYTOTOXIC T CELLS WAS:

                  1   2   3   4   5   6   7   8   9
THR-LYS-GLY-ILE-LEU-GLY-PHE-VAL-PHE-THR-LEU-THR-PRO-SER-GLU-ARG-GLY

VESICULAR STOMATITIS VIRUS N PROTEIN: 422 RESIDUES

                ONLY EFFECTIVE H-2Kb-PRESENTED NONAPEPTIDE TARGET

                        OF CYTOTOXIC T CELLS WAS:

                  1   2   3   4   5   6   7   8   9
                  ARG-GLY-TYR-VAL-TYR-GLN-GLY-LEU-LYS
```

FIGURE 2. The presence of tripeptidic palindromes in both proven T epitopes. The first is from influenza A virus matrix protein presented by human HLA-A2,[6] whereas the second is from vesicular stomatitis virus N protein presented by mouse H2-K^{b}.[7] The position of each peptide in a protein is shown by numbers placed above the first and the last residues.

FIGURE 3 shows that by showing preference for palindromic nonapeptides, class I MHC antigens are in constant danger of choosing viral T epitopes that are homologous with self. Palindrome-containing oligopeptides have a way of appearing in totally unrelated proteins.[5,8,9] Three pairs of such examples are shown in FIGURE 3.

SUCCESSFUL SELF–NON-SELF DISCRIMINATION BASED ON DIFFERENTIAL AFFINITY

In uninfected cells, peptide fragments derived from 5000 or so diverse intracellular proteins of the host must compete for the fixed number of sites provided by each class I MHC antigen type. Under this circumstance, only those with the highest binding affinity have a chance of appearing on the plasma membrane accompanied by a particular class I MHC antigen.

Among those that managed to be represented on the plasma membrane, many will fail to achieve the minimal target density required to be recognized by T cell receptors, which is 0.1%.[10] The net result is that with regard to each MHC antigen type, the thymic education of self to cytotoxic T cells would result in acquiring tolerance only for the absolute maximum of 1,000 different highest affinity nonapeptides.[8] The actual number should be far less, because a few nonapeptides appear to acquire target densities of 15% or more.[3]

Sequence	Protein
(170–177) LEU-GLY-PHE-LEU-ALA-LEU-VAL-THR-THR-GLY	BACTERIAL (*Caulobacter Crescentus*) HISTIDINE KINASE
(174–181) CYS-GLY-TYR-LEU-ALA-LEU-VAL-THR-SER-LEU	HUMAN PHOSPHOFRUCTOKINASE
(988–985) VAL-PRO-ARG-ARG-LYS-ALA-LYS-ILE-ILE-ARG	HIV REVERSE TRANSCRIPTASE
(131–138) ARG-PRO-ARG-ARG-LYS-ALA-LYS-MET-LEU-PRO	HUMAN Y-LINKED TESTIS-DETERMINING NUCLEAR PROTEIN
(387–394) TYR-VAL-ARG-LEU-ALA-VAL-ALA-ASP-LEU-VAL	MUSTARD PLANT (*Brassica oleraceo*) POLLEN-STIGMA INCOMPATIBILITY PROTEIN
(93–100) LEU-VAL-ASN-LEU-ALA-VAL-ALA-ASP-LEU-ALA	HUMAN RED OPSIN

FIGURE 3. Three examples of the ubiquity of palindrome-containing oligopeptides. The first pair is contributed by bacterial histidine kinase,[11] on the one hand, and human muscle phosphofructokinase, on the other.[12] Two octapeptides differ only by two conservative substitutions: Phe for Tyr, Thr for Ser. The second pair of octapeptides are contributed by HIV reverse transcriptase[13] and human Y-linked putative testis-determining protein.[14] The two share the identical hexapeptide. The third pair of octapeptides are from plant and human, one from pollen–stigma incompatibility protein of the mustard plant[15] and the other from human red opsin in the retina.[16] These two again share the identical hexapeptide.

CYTOTOXIC T CELLS SHOULD ATTACK INFECTED CELLS ONLY WHEN VIRUSES ARE ACTIVELY PROPAGATING

In hepatocytes of persons infected with hepatitis B virus, for example, there appear to be long intervals during which viruses remain quiescent, thus enjoying a sort of symbiosis with the host. In this state, a few viral proteins produced in small amounts would be swamped by 5,000 or more host proteins. Thus, a viral nonapeptide has a small chance of gaining sufficient target density on the plasma membrane to provoke cytotoxic T-cell attack. Indeed, cytotoxic T-cell attack launched in this state would be utter folly, serving only to further reduce the number of already hard-pressed hepatocytes that are performing required functions.

Only when viruses begin to propagate would a small variety of viral proteins produced in large quantity be able to swamp host proteins; thus, even those viral nonapeptides with relatively lower affinity can serve as effective T epitopes to provoke cytotoxic T-cell response. The beauty here is that for those relatively low-affinity viral nonapeptides, self–non-self becomes totally irrelevant, because homologous nonapeptides in host proteins shall never gain sufficient access to the plasma membrane of uninfected cells.[9] Such appears to be the design and working of class I MHC antigens as the appendage to the adaptive immune system.

SUMMARY

The unrestricted viability of allophenic mice shows that MHC-different cell lines have no problem engaging in organogenesis together. Thus, outside of the immune system, mammals appear to have no self–non-self discrimination mechanism based on polymorphic and ubiquitously expressed class I MHC antigens. Here, it should be pointed out that even within the immune system, certain responses require no self–non-self discrimination, for example, antiphosphocholine response and certain antipolysaccharide responses that exploit differences between bacterial and host sugar transferases.

Thus, the self–non-self discrimination via peptide fragments presented by ubiquitously expressed class I MHC antigens can be viewed as the late addition that enabled the adaptive immune system to cope with intracellular parasites that are primarily viruses. The preference for different types of peptide fragments suffices to explain extensive polymorphism as well as multiple gene loci for class I and possibly also class II MHC antigens. Yet, a too specialized class I MHC allele that presents a very unusual peptide fragment is of no use, for such a peptide fragment is not likely to be found among viral proteins. Effective MHC antigens are those that prefer common enough peptide fragments, so that at least one T epitope can be found in one out of every three viral proteins. Yet, such common peptide fragments are also likely to be present among multitudes of intracellular proteins that are the self. The immune system appears to have solved the above dilemma by mounting a vigorous cytotoxic T-cell response only when viruses are actively propagating by synthesizing a few of their own proteins in large amounts, thereby suppressing the host protein synthesis. To attack infected cells in which viruses are in the quiescent state of symbiosis with the host is the ultimate folly.

REFERENCES

1. Falk, K., O. Rotzsche, S. Stevanovic, G. Jung & H-G. Rammensee. 1991. Nature **351:** 290–296.
2. Jardetsky, T. S., W. S. Lane, R. A. Robinson, D. R. Madden & D. C. Wiley. 1991. Nature (London) **353:** 326–329.
3. Madden, D. R., J. C. Gorga, J. L. Strominger & D. C. Wiley. 1991. Nature (London) **353:** 321–325.
4. Fischer-Lindahl, K., E. Hermel, B. E. Loveland, S. Richards, C-R. Wang & H. Yonekawa. 1989. Cold Spring Harbor Symp. Quant. Biol. Vol. **LIV:** 563–569.
5. Ohno, S. 1992. Hum. Genet. **90:** 342–345.
6. Gotch, F., J. Rothbard, K. Howland, A. Townsend & A. McMichael. 1987. Nature **326:** 881–882.
7. Van Bleek, G. M. & S. G. Natheson. 1990. Nature **348:** 213–216.
8. Ohno, S. 1992. Immunogenetics **36:** 22–27.
9. Ohno, S. 1992. Proc. Natl. Acad. Sci. USA **89:** 4643–4647.
10. Christinick, R. E., M. A. Luscher, B. H. Barber & D. V. Williams. 1991. Nature **352:** 67–70.
11. Ohta, N., T. Lane, E. G. Ninfa, J. M. Sommer & E. Newton. 1992. Proc. Natl. Acad. Sci. USA **89:** 10297–10301.
12. Nakajima, H., T. Noguchi & T. Yamasaki. 1987. FEBS Lett. **223:** 113–116.
13. Ratner, L., W. P. Haseltine, R. Patarca, K. J. Livak, B. Starcich, S. F. Josephs, E. R. Doran, J. A. Rafalski, E. A. Whitehorn, K. Baumeister, L. Ivanhoff, S. R. Petteway, Jr., M. L. Pearson, J. A. Lautenberger, T. S. Papas, J. Ghraybeb, N. T. Chang, R. C. Gallo & F. Wong-Staal. 1985. Nature (London) **313:** 277–284.
14. Sinclair, A. H., P. Berta, M. S. Palmer, J. R. Hawkins, B. L. Griffiths, M. J. Smith, J. W. Foster, A. M. Frischauf, R. Lovell-Badge & P. N. Goodfellow. 1990. Nature **346:** 240–244.
15. Nasrallah, J. B., T. H. Kao, C-H. Chen, M. L. Goldberg & M. E. Nasrallah. 1987. Nature **326:** 617–619.
16. Nathans, J., D. Thomas & D. S. Hogness. 1986. Science **232:** 193–202.

Cell Surface Recognition and the Immunoglobulin Superfamily[a]

J. J. MARCHALONIS,[b,c] V. S. HOHMAN,[c] H. KAYMAZ,[c]
S. F. SCHLUTER,[c] AND A. B. EDMUNDSON[d]

[c]*Department of Microbiology and Immunology*
University of Arizona, College of Medicine
Tucson, Arizona 85724
[d]*Harrington Cancer Center*
Amarillo, Texas 79106-1794

INTRODUCTION

It was speculated for some time that molecules that were not antibodies might be related to Igs and constitute an "extended family"[1,2] or "superfamily."[3–5] Once protein and gene sequences became available for β2 microglobulin, MHC products, and a variety of cell membrane–associated proteins, Williams[3,4] recognized a large number of molecules showing various degrees of homology to immunoglobulin variable or constant domains. Usually, families are defined on the basis of having the members share at least 40% identity, but Doolittle has stated that molecules showing at least 25% identity are most probably related.[6] The members of the immunoglobulin superfamily range from classical immunoglobulins to neural cell adhesion molecules (N-CAMs) and to induced N-CAM-related hemolymph molecules of insects[7–9] as well as to viral hemagglutinins.[10] It is important to distinguish among three sets of members within the large superfamily; the first is members of the family that include the classical serum immunoglobulin chains and the antigen-specific T-cell receptors,[11] a set closely related to the family based on statistical analysis of amino acid sequences containing the immunoglobulin-like domains of MHC antigens and β2 microglobulin, both of which resemble C domains.[4,5] The large cluster, including N-CAMs and other cell-surface molecules such as Thy-1,[3,4] show marginal degrees of identity (usually less than 20%) with any individual member of the immunoglobulin set.[2,12] Despite relatively low degrees of identity, such as 22% overall for human CD8α and V_{κ}I light chain, the crystal structure of CD8α was solved by molecular replacement using a superposition of 10 light-chain V domains as comparisons, with the conclusion that this superfamily member possessed a typical V domain immunoglobulin fold.[13]

Immunoglobulin domains consist of two types: those homologous to V regions and those homologous to C domains. The V and C domains, themselves, had a

[a]This research was supported in part by grants from the National Science Foundation (DCB 9106934), the National Institutes of Health (GM 42437), and the National Cancer Institute (CA 42049) to JJM and CA 19616 to ABE.

[b]Address for correspondence: John J. Marchalonis, Ph.D., Microbiology and Immunology, College of Medicine, University of Arizona, Tucson, AZ 85724.

common ancestor early in evolution.[1] The V domains associate pairwise where structures of form V_H/V_L or V_α/V_β generate a combining site that is complementary to an antigenic determinant expressed on another macromolecule. The C domains form homodimers via noncovalent association of conserved inner surfaces. Here, we will analyze one highly conserved segment of C-region structure that plays a major role in dimer formation among C regions of immunoglobulins and most probably carries out the same role for MHC products and N-CAMs. We apply the progressive alignment algorithm of Feng and Doolittle[14] to establish possible relationships among the most highly conserved C domains of Igs and C-like domains of members of the superfamily.

Data obtained recently on gene sequences of Igs of sharks, the ancestors of which diverged from those of mammals more than 400 million years ago,[15] allow us to make detailed comparisons regarding the homologies among Ig V and C domains in evolution. Here, we will stress comparisons between the λ light chains of the sandbar shark *(Carcharhinus plembeus)*[16,17] and human light chains.[18,19] In addition, the use of synthetic peptide technology to simulate the covalent structure of human λ light chains[19] and T-cell receptor β chains[20,21] has allowed us to correlate antigenicity with sequence in three homologous molecules of the Ig family that show maximum evolutionary distance within the set.

METHODS AND MATERIALS

Gene Sequences and Synthetic Peptides

A sequence of cDNA specifying intact sandbar shark light-chain genes[16,17] and a partial heavy-chain sequence[22] was obtained as previously described. The human Tcr β-chain sequence (Vβ8.1, Jβ2.1, Cβ1) was taken from the study of Yanagi *et al.*[23] Synthetic peptides modeling the covalent structure of light-chain Mcg[19] and the Tcr β chain were synthesized and characterized as previously described.[20,21] The comprehensive synthetic approach used was based on that of Kazim and Atassi,[24] with 16-mer peptides overlapping by five residues used to duplicate the complete covalent sequence of the λ chain and the Ig V/J/C domain structure of the Tcr β chain.

Immunochemistry

The details of our enzyme-linked immunosorbent assay (ELISA) using peptides have been published.[19–21]

Comparative Analysis of Sequence Data

Comparisons of sequence data corresponding to Ig domains with the construction of dendograms was carried out using the progressive alignment algorithm of Feng and Doolittle[14] as we have reported previously.[11] All of the sequences used here are available within the Genbank Database.

RESULTS

Comparison of Sandbar Shark Light Chain and Human Tcr α Chain with λ Mcg

T-cell receptor β chains contain segments corresponding to a V domain, a J segment, and an Ig-C domain.[11] When Tcrβ chains were initially discovered, computer searches disclosed significant homology with λ light chains.[23] More recent studies on the complete sequence of sandbar shark light-chain cDNA, likewise, showed that these molecules resembled λ light chains of man.[16,17] We felt that detailed comparisons of the shark and the Tcr molecules with a completely characterized λ light chain would provide information regarding conservation of key residues, on the predicted three-dimensional structure of the new molecules, and on antigenic similarities between the members of this λ-like set. FIGURE 1 is an alignment of the derived amino acid sequence of sandbar shark cDNA clone 5.1 with the Mcg molecule. There is an overall identity of 42% in this alignment. The overall V region match is likewise 42% with the comparison of frameworks alone giving 53% identity. Mcg is not the best match for this shark λ light chain and some others such as Sut, a human Vλ6, show >50% matches in comparison over the entire V region.

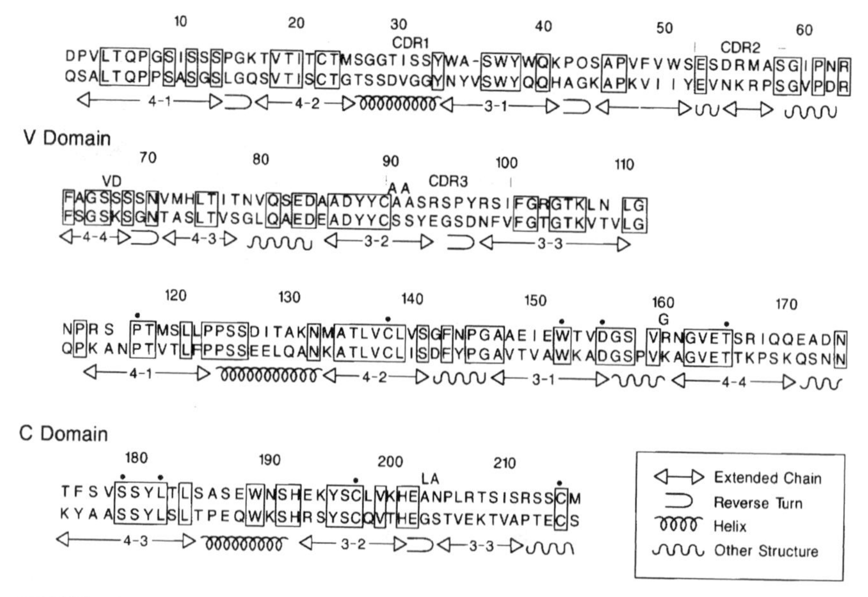

FIGURE 1. Alignment of sandbar shark light-chain sequence (cDNA clone 5.1) and human λ light chain, Mcg. The upper sequence is that of the sandbar shark; the lower is the human. The structural features including extended chain (β-band), reverse turn, helix, and other structures are based on X-ray crystallographic modeling of the human λ light chain. Identities between the shark and human sequences are indicated by boxes. The dots above particular residues indicate those that are universally conserved in all light-chain sequences.

It is apparent that the structural features such as β bands, turns and the limited amount of helix obtained from crystallographic structures of Mcg can be mapped readily on the shark light-chain sequence. This particular light chain is slightly longer in its CDR3 than is the human molecule, but this region is one in which both human and shark λ chains show variation in length.

FIGURE 2 shows a parallel alignment comparing the human Tcr β chain YT35 with the Mcg molecule. The β-chain sequence is given only for the Ig domain segments with the stem and membrane-spanning hydrophobic regions excluded. The two molecules can be aligned with the β-band structures showing comparable arrangements. In this particular comparison, the region of the molecule corresponding to CDR1 of the Vβ is considerably shorter of that of the λ light chain. The overall comparison between these two molecules shows 26% identity with the V-region frameworks showing 36% identity and the overall V regions sharing 29%. In both figures, there are two particular regions of sequence where the identities are striking. The first of these corresponds to Kabat's Fr4 region of light chains,[25] which is encoded in part by the joining segment gene. A second segment of the V-region corresponds to the Fr3 where the shark and human λ chains show 56% identity and the corresponding Tcr β/λ regions are 34% identical. A region of appreciable identity in the C region is that involving β band 4-2 and continuing into β 3-1 with residues conserved in the loop connecting the two β bands.

FIGURE 3 illustrates the three-dimensional structure of the Mcg λ chain obtained from crystallography[18] and compares it with the predicted three-dimensional struc-

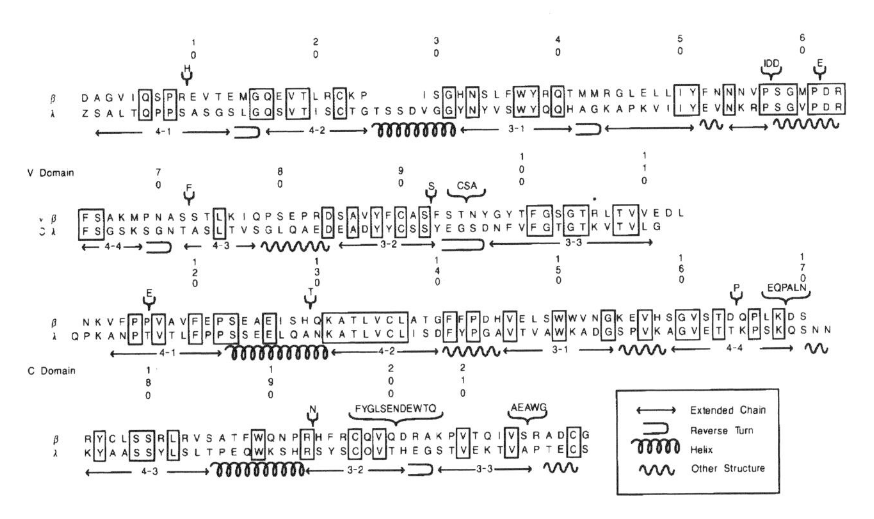

FIGURE 2. Alignment of the sequence of the human Tcr β chain YT35 with that of the human λ light chain Mcg. The structural features are based on X-ray crystallographic analysis of the λ light chain. Sequence identities are indicated by boxes. The dot above the R (arginine) and K (lysine) at position 106 indicates that the two residues are antigenically equivalent in the analysis of peptides using antibodies directed against peptides corresponding to the Jβ segment.[29]

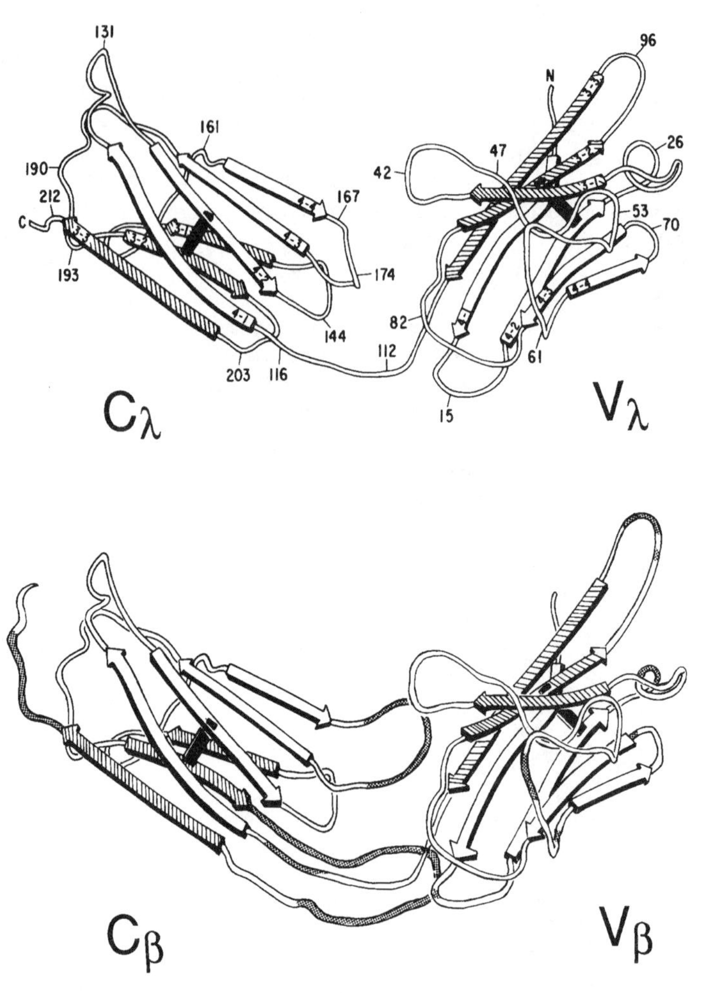

FIGURE 3. Comparison of the structure of the Mcg light chain *(upper panel)* with a predicted model of the Tcr β chain *(lower panel)*. Although this figure gives only the structure of the human λ light chain, the predicted three-dimensional structure of the shark λ light chain would be essentially equivalent so is not depicted here. Deletions and insertions between the light chain and Tcr are indicated by dotted sections in the β-chain model. Note that only a few alterations have to be made in the structure of the V domain of the λ chain to accommodate the sequence of the β chain. These changes include a shortening of the first hypervariable loop (CDR1) of the β chain, a lengthening of the third hypervariable loop (CDR3), and an extension of the framework region (Fr3) connecting the second hypervariable loop (CDR2) with the pleated-sheet strand designated 4-4. In the model of the Cβ domain, we have retained the definitive features of the immunoglobulin fold, that is, the three-stranded *(striated)* and four-stranded *(white)* β-pleated sheets depicted in the upper panel. On the basis of this assumption, two sets of additional residues in the β chain loop out from the globular domain structure into the space between the V and C domains. The "switch" region connecting the V and C domains and the COOH-terminal segment *(far left)* are elongated relative to those in the Cλ. (Adapted from Kaymaz *et al.*[21])

ture of the Tcr β chain based on the observed homologies between the two sequences. The predicted structure of the shark λ light chain would be essentially identical to that of the human molecule.[16] The Tcr β chain shows the same β-band structure and the formation of the three- and four-strand β sheets. The predicted structure differs from that of Mcg in having a smaller CDR1 region (indicated by residue #26 in the Mcg), a slightly longer CDR3 region (#96), and a slightly longer switch peptide (#112 in Mcg). The V region comparison, thus, indicates that the three-dimensional structures of the Tcr and light chain would be very similar. This conclusion is consistent with those of other workers.[26–28] Likewise, the β-sheet structure of the Cβ is extremely similar to that of Cλ. However, two major loops that project forward towards the V domains are considerably larger in the Cβ than they are in the Cλ. These are the loop connecting bands 4-3 and 4-4 and the loop connecting bands 3-2 and 3-3. The implication of this for the three-dimensional structure of the intact Vβ/Jβ/Cβ structure is that the structure of the Cβ domain can place constraints on the arrangement of the Vβ. This would impact particularly on FR3 loops and on the extra loop containing CDR2. This type of interference would not occur for the λ chain or for human Tcr α chain (Marchalonis and Edmundson, unpublished observations). Much of the peptide region specified by the joining segment would be exposed in the Vβ band 3-3. Furthermore, its continuation into the switch peptide is readily exposed to external molecules.

Sequence and Antigenic Homologies among CDR3/Fr4 Segments of Immunoglobulins

FIGURE 4 shows an alignment of the region of Ig chains and superfamily members corresponding to the J segment. Immunoglobulin light chains and Tcr chains show strong identities in the Fr4 region. Studies using synthetic peptides involving residues interchanged at various positions have established that the arginine (R) or lysine (K) at position 6 in Fr4 are antigenically equivalent,[29] but that the other substituents shown here diminish the capacity of rabbit anti-HuJβ shown here to react. We were unsuccessful in obtaining anti-Jβ antibodies by immunizing either mice or rats, but rabbits, most probably because of the glutamic acid (E) at this position, made strong antibodies to the Jβ prototype sequence. The antibodies reacted poorly, if at all, with chicken Jλ or murine J_H1. However, antibodies directed against the J_H1 sequences reacted with heavy chains of various species including sharks.[30] The marginally related superfamily members show little identity in this region, and we did not detect any reactivity of anti-Jβ antibodies in these proteins. In addition, we have not found cross-reactivity molecules in the serum or hemolymph of insects and horseshoe crabs. Schluter *et al.* (this volume) describe the use of anti-Jβ serum to identify and facilitate the isolation of Igs of lower deuterostome species.

We have recently synthesized nested sets of overlapping synthetic peptides that duplicate the covalent structure of the λ chain Mcg[19] and the T-cell receptor β chain YT35.[20,21] We used these synthetic peptides to determine whether cross-reactive peptides could be identified between human and shark λ chains and between human

	CDR3——— ———Fr4———
HuJβ	A N Y G Y T F G S G T R L T V V
MuJβ1.1	N T E V F F G K G T R L T V V
Chi Jβ	N T P L N F G Q G T R L T V L
Mu Jα	G N Y K Y V F G A G T R L K V I
Dog Jκ	S F Y P Y T F G Q G T R L E V - R
Hu Jλ	D S M S V V F G G G T R L T V L G
SbS Jλ	S P Y R S I F G R G T K L - N L G
Rab Jκ	Y A A T Y T F G G G T E V V V - K
Chi Jλ	E S S S A I F G A G T T L T V L G
Mu J_H1	G N Y F D Y W G Q G T T L T V S
- - - - -	- - - - -
Thy 1.1	V S G A N P M S S N K S I S V Y
Rat Ox-2	M C L F N M F G S G K V S G T A
Hu CD4	E V Q L L V F G L - T A N S D T
Mu Lyt-2	S N S V M Y F S S V V P V L Q K
Hu CD8	S N S I M Y F S H F V P V F L P

FIGURE 4. Alignment of "CDR3/Fr4" segments of immunoglobulins and superfamily members.

λ chains and Tcr β chains. FIGURE 5 is a graph showing ELISA binding of rabbit antiserum directed against purified shark light chains and against the synthetic Jβ peptide reacted with the overlapping synthetic peptides modeling the Mcg λ chain. The antiserum to the joining segment peptide reacts only with one peptide (#10), which is that containing the Fr4 sequence. A rabbit antiserum produced against an intact λμWaldenstrom macroglobulin reacted strongly with peptides #7 and #10, and to a lesser degree with C-region peptides #15 through 17. The reaction with the most carboxyl-terminal peptides is also probably real because some commercial antisera to human λ chains react strongly with this region.[19] It is noteworthy that the rabbit anti-shark serum also reacts strongly with peptides #7 and #10. Peptide #7 is from the third framework of the V region where shark and human show better than 50% sequence identity. The C-region reactivity is slight, but real. The key point to be

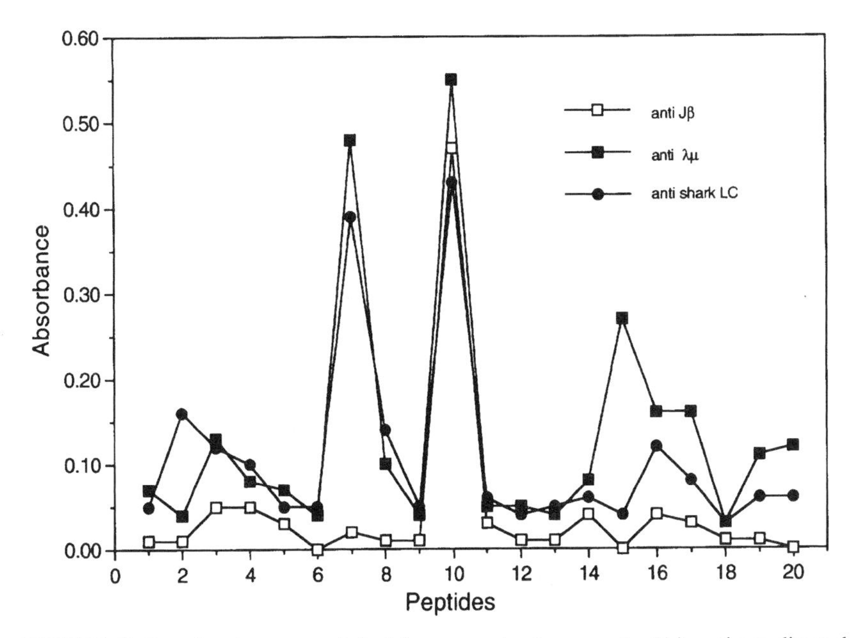

FIGURE 5. Reaction in enzyme-linked immunosorbent assay of rabbit antisera directed against a synthetic T-cell receptor Jβ peptide [-❑-], antiserum to a human λμ myeloma protein [-■-] and antiserum to purified shark light chain [-●-], with the set of synthetic, overlapping peptides duplicating the complete covalent structure of the human λ light chain Mcg. Except for the last peptide, the individual peptides were 16 residues in length and overlapped one another by five residues. The assay was carried out using serum at a dilution of 1:500.

emphasized here, for molecules that had an ancestral gene divergence more than 400 million years ago, is that the strongest serological reactions are those confined to the V region and to the J segment.

FIGURE 6 shows the peptide segments of the Tcr β chain that react in ELISA with normal rabbit serum, the antiserum against the Jβ peptide, and a commercial rabbit antiserum specific for human λ chains. In this diagram, the x axis gives the midpoints of the peptides in numerical sequence. Although binding activity is shown in the CDR1 region (around residue 30) and at the carboxyl-terminal end of the molecule (around position 230), this also occurs in normal rabbit serum. This is a real phenomenon, and we report elsewhere that normal humans have natural antibodies against particular regions of their own T-cell receptors.[20] Here, the antiserum to the synthetic Jβ peptide reacts strongly with the proper region encompassing CDR3, Fr4, and the beginning of the C region. The antiserum directed against the λ chain reacts with an Fr3 determinant and with a peptide overlapping the Jβ segment. An additional reactivity is observed with a C-region peptide around residue 190. The results of the comparison of antigenic regions of the Tcrβ segments with antiserum against human λ chain thus discloses the same portions of the molecule that were recognized in the comparison of human and shark sequences.

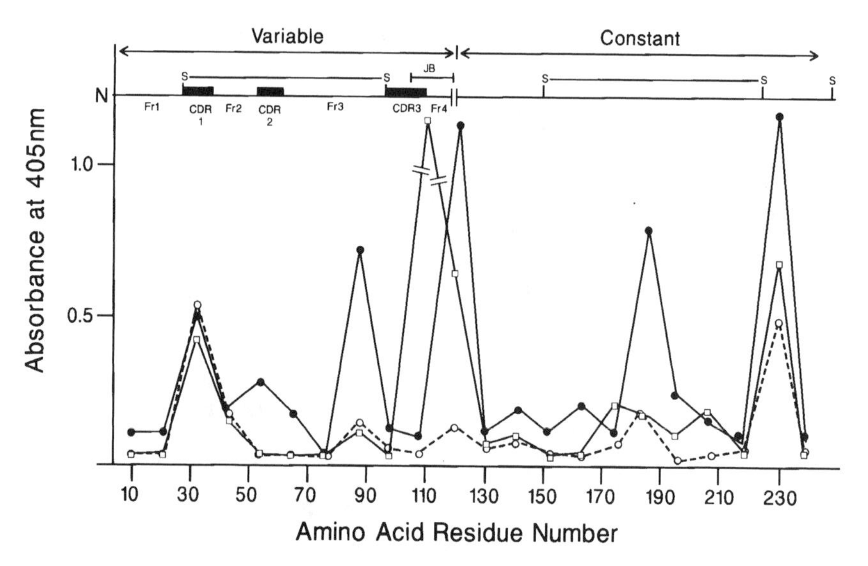

FIGURE 6. Reaction in enzyme-linked immunosorbent assay of rabbit antiserum directed against human λ light chain [-●-], antiserum against the synthetic Tcr Jβ peptide [-❑-], and normal rabbit serum [--●--] against synthetic overlapping peptides duplicating the covalent structure of the V/D/J/C/ region of Tcr β chain. The linear regions of the immunoglobulin domain structures are mapped onto the sequence.

Comparison of Constant Region Domains of Immunoglobulins and Immunoglobulin Superfamily Members

FIGURE 7 compares the sequences of the conserved segment ranging from β band 4.2 through β band 3-1 disclosed in the comparisons among shark λ chain, human λ chain, and human Tcr β chain. It is interesting that Tcr β chains express what has been characterized as a λ signature sequence[16] and share the "sentinel" tryptophan (W) at position 19 with the Ig domains. Interestingly, the human β2 microglobulin, which resembles a heavy-chain C domain in sequence and three-dimensional structure, lacks a tryptophan at that position. Tcrα chains lack a tryptophan at this position and also show the characteristic Cβ folding (Marchalonis and Edmundson, unpublished observations). The β2 microglobulin and the Ig-like domains of MHC molecules of man and the nurse shark show considerable identity to the Ig light-chain and μ-chain domains. The degree of identity between the N-CAM-related molecules of the superfamily is considerably less convincing than the degree of matches obtained in the other comparisons. Nonetheless, alignments can be made conserving the cysteine at position 5 and the tryptophan at position 19, but a gap is required.

We compared the entire sequence of the C domains of Igs and Tcrs most distal from the V region (e.g., Cλ or Cβ or Cδ3 or μ4) with the C-like domains of MHC

```
                        4-2                     3-1
SB SHARK Cλ       A T L V C L V S G F N P G A A E I E W T V D G S
Human Cλ          A T L V C L I S D F Y P G A V T V A W K A D G S
Human Cκ          A S V V C L L N N F Y P P E A K V Q W K V D N A
Bullfrog Cκ       A S T V C L V D K F Y P G G A Q V T W K G D N K
Human Cμ4         A T I T C L V T G F S P A D V F V E W M G R G E
SB Shark Cμ4      F Y L S C L V R G F S P R E I F V K W T V N D K
Human Tcrβ        A T L V C L A T G F F P D H V E L S W W V N G K
Chicken Tcrβ      A T L V C L A S G F F P D H L N L V W K V N G V
Human β2M         N F L M C Y V S G F H P S D I E V D L L K N G E
HLA-DQ-A          N T L I C L V D N I F P P V V N I T W S N G H S
HLA-DR 2B         N L L V C S V S G F Y P A S I E V R W F R N G Q
Nurse Shark IIα   N T L I C F A D G F Y P P H I T M K W R R N N E
Moth hemolin      T V L E C I I E G N D Q G V K Y S * W K K D G K
Chick-NCAM        K F F L C Q V A G E A K Y K D I S * W F S P N G
```

FIGURE 7. Conserved immunoglobulin C-domain region.

antigens and moth hemolin as shown in FIGURE 8. The Ig C domains segregate into light-chain and heavy-chain clusters. The T-cell receptor Cβs cluster with the light chains, near the beginning of the branch. The λ and κ chains form two separate clusters. Within the λ chain cluster, the expected phylogenetic relationships obtain. Proteins within the κ cluster likewise distribute in a manner consistent with expected phylogenetic relationships. Within the most distal C_H domains, the Cμ4s show a progressive evolution with the sharks most similar to one another and the human and mouse also being closest to one another. As would be expected, the *Xenopus* and chicken are intermediate. The γ chains form a separate cluster from the μ chains with the chicken and duck Y chains lying near the beginning of this branch. The *Xenopus* Y chain falls off the scheme and is the closest to the root of this particular branch. MHC Class I and II C domains form individual clusters that are consistent with phylogeny. The β2 microglobulin is in this cluster, and it is the nearest to the base. The moth hemolin, which has an N-CAM-like sequence, is quite distinct from both the Ig branches and the Ig-like domains of the MHC, consistent with its sequence.

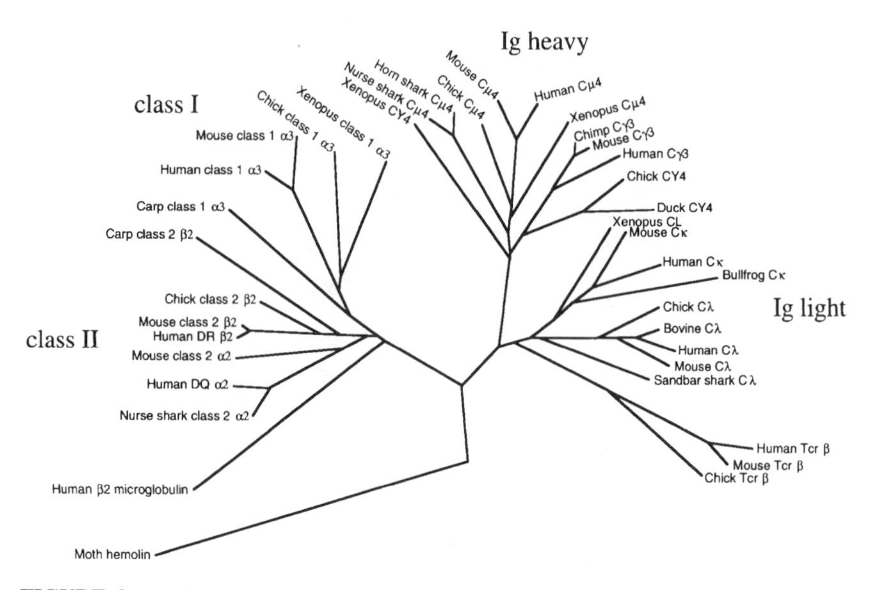

FIGURE 8. Dendogram depicting the relationships among immunoglobulin constant domains of light chains and Tcrs compared with the most V-region-distal domains of heavy chains and with IgC-like domains of MHC products and *Cecropia* moth hemolin (related to N-CAM).

DISCUSSION

The concept of an Ig superfamily is worthwhile in predicting three-dimensional structures and also in supporting possible functional interrelationships. For example, if a series of multiple C-region-like domains is expressed, it is possible that the molecules are involved in homophilic adhesions among cells. By contrast, a pairwise association of V-region-like structures, particularly if sequence heterogeneity is found, most probably would indicate a complementary recognition of non-self antigen. The superfamily concept, however, is a difficult one in attempting to determine phylogenetic relationships[2,31] because the N-CAMs, Thy-1, and many of the other molecules are so distinctly related to classical Igs that definitive assessment of homologies is not possible. The overall concept is almost too inclusive to be workable. On the other hand, the homologies between T-cell receptors and classical Igs are so clear that these can be considered to fall within an Ig family.[11] By extension, the Ig-like domains of the MHC products are of a sufficiently high statistical match with classical Ig C domains[2] to engender confidence that these are structurally very similar. X-ray crystallographic studies have shown, for example, that the α3 domain of MHC Class I has a three-dimensional structure very similar to that of β2 microglobulin.[13]

Despite the enormous sequence diversity in Ig V regions, these domains have highly conserved segments as shown in comparison among light chains[16,17] and heavy chains of sharks and man[22,32] and in comparison between Tcr β chains and light chains.[11,23] This conclusion holds particularly for determinants lying in the Fr3 of the

V-domain that are conserved in sequence and are predicted to be exposed in the three-dimensional folding of the molecules.[33] Consistent with this, formation of the V_Ha allotypes of the rabbit,[34] which are expressed as isotypes in many species including man and shark,[35] involves residues from the V_H-Fr3, in association with exposed V_HFr1 residues, to form the antigenic markers. In addition, the Vβ-Fr3 has been implicated in the recognition of superantigens[36] and is a target for natural autoantibodies.[20]

A second major antigenic region shared among light chains in evolution and Tcr is the "switch peptide,"[37] which essentially begins with Fr4 and continues into the NH_z-terminal sequence of the C region. Database searches using peptides predicted from J-segment sequence are characteristic of Igs; for example, approximately 95% of sequences in general databases showing greater than 50% identity were from Igs.[29] Building on this concept of conservation of sequence and exposure in three-dimensional structure, we produced antibodies against various synthetic peptides that act as "universal" reagents for detecting Ig light chains and T-cell receptors.[29,33,38] Similar sequence database analyses with V- and C-region peptides have not shown this restriction to Igs. Nonetheless, there are regions of the C domain, particularly that encompassing β bands 4-2 and 3-1, that show substantial conservation in sequence that is most probably reflected in the three-dimensional structures. We have not yet attempted large-scale identification of cross-reactive molecules using antibodies against this peptide sequence. However, rabbit antibodies produced against the human Cβ sequence react strongly with murine Tcrβ chains in Western blots.[39] The region defined by this sequence may have been conserved in evolution because residues present here are involved in forming the contact surface that holds C domains together as noncovalently associated dimers. In particular, residues V, at position 4, and L, at position 6 (FIG. 7), are essential for this interdomain interaction. Therefore, antibodies directed against these segments would be most useful in denaturing systems such as Western blots, but would be expected to be inefficient in reacting with the native molecules because the surfaces they define would be buried. We have found this conjecture to be verified in antigenic studies of the human λ chain Mcg.[19]

At this point in time, bonafide Ig and MHC sequences have been obtained only for gnathanstomes.[40] The only nonmammalian groups for which Tcrs have been characterized is chickens.[41] It is likely, however, that continuing application of recombinant DNA technology will shortly disclose the presence of α/β T-cell receptors in amphibians and teleosts and γ/δ molecules in elasmobranchs. We are of the opinion that there is a conjoint evolution between MHC products as presenting molecules and T-cell receptors, but that this is probably a late event in vertebrate evolution rather than the primordial event that was once hypothesized.[42] The cladistic analysis presented here is consistent with other reports that suggest that the evolutionary distinction between heavy and light chains occurred before T-cell receptors branched off from the light-chain stem.[43]

SUMMARY

Immunoglobulins serve as humoral recognition and effector molecules and as antigen-specific cell surface receptors on B and T cells. These molecules are constructed according to a characteristic domain pattern. Variable and constant domains diverged from one another early in vertebrate evolution, and they are joined by a "switch peptide" specified by the joining gene segments. Peptides specified by J-gene segments are strongly conserved in evolution in comparison among Ig light chains and T-cell receptors. Molecules less strongly related to Ig domains have been assembled into an Ig "superfamily" where the identities to classical IgC or V domains are ≤ 20%. Among these are cell surface adhesion molecules, receptors for cytokines, and Fc receptors. Moreover, MHC antigens have an Ig-like membrane-proximal domain significantly related to IgC regions. We will analyze putative evolutionary relationships among canonical Igs and members of the Ig superfamily using highly conserved sequences from light and heavy chains of primitive vertebrates (e.g., the sandbar shark) as prototypes to ascertain similarities between Ig-related molecules of vertebrates and invertebrates.

ACKNOWLEDGMENTS

We would like to thank Ms. Diana Humphreys for her assistance in the preparation of this manuscript.

REFERENCES

1. MARCHALONIS, J. J. 1977. Immunity in Evolution.: 316. Harvard University Press. Cambridge, MA.
2. MARCHALONIS, J. J., G. R. VASTA, G. W. WARR & W. C. BARKER. 1984. Immunol. Today **5:** 133–142.
3. WILLIAMS, A. F., A. G.-D. TSE & J. GAGNON. 1988. Immunogenetics **27:** 265–272.
4. WILLIAMS, A. F. & A. N. BARCLAY. 1988. Ann. Rev. Immunol. **6:** 381–391.
5. HOOD, L. M., M. KRONENBERG & T. HUNKAPILLAR. 1985. Cell **40:** 225–229.
6. DOOLITTLE, R. F. 1989. Trends Biochem. Sci. **14:** 244–249.
7. HARRELSON, A. L. & C. S. GOODMAN. 1988. Science **242:** 700–708.
8. SEEGER, M. A., L. HAFFLEY & T. C. KAUFMAN. 1988. Cell **55:** 489–600.
9. SUN, S. C., I. LINDSTROM, H. G. BOMAN, I. FAYE & O. SCHMIDT. 1990. Science **250:** 1729–1732.
10. JIN, D., Z. LI, J. QI, H. YUWEN & Y. HOW. 1989. J. Exp. Med. **170:** 571–576.
11. MARCHALONIS, J. J. & S. F. SCHLUTER. 1989. FASEB J. **3:** 2469–2479.
12. MATSUNAGA, T. & N. MORI. 1987. Scand. J. Immunol. **25:** 485–495.
13. LEAHY, D. J., R. AXEL & W. A. HENDRICKSON. 1992. Cell **68:** 1145–1162.
14. FENG, D. F. & R. F. DOOLITTLE. 1987. J. Mol. Evol. **25:** 351–360.
15. YOUNG, J. Z. 1962. The Life of Vertebrates. Oxford University Press. New York..
16. SCHLUTER, S. F., V. S. HOHMAN, A. B. EDMUNDSON & J. J. MARCHALONIS. 1990. Proc. Natl. Acad. Sci. USA **86:** 9961–9965.
17. HOHMAN, V. S., S. F. SCHLUTER & J. J. MARCHALONIS. 1992. Proc. Natl. Acad. Sci. USA **89:** 276–280.
18. EDMUNDSON, A. B., K. R. ELY, E. E. ABOLA, M. SCHIFFER & N. PANGIOTOPOULOS. 1975. Biochemistry **14:** 3953–3961.
19. MARCHALONIS, J. J., F. DEDEOGLU, H. KAYMAZ, S. F. SCHLUTER & A. B. EDMUNDSON. 1992. J. Prot. Chem. **11:** 129–137.
20. MARCHALONIS, J. J., H. KAYMAZ, F. DEDEOGLU, S. F. SCHLUTER, D. E. YOCUM & A. B. EDMUNDSON. 1992. Proc. Natl. Acad. Sci. USA **89:** 3325–3329.

21. KAYMAZ, H., F. DEDEOGLU, S. F. SCHLUTER, A. B. EDMUNDSON & J. J. MARCHALONIS. 1993. Int. Immunol. **5:** 491–502.
22. MARCHALONIS, J. J., V. S. HOHMAN, C. THOMAS & S. F. SCHLUTER. 1993. Dev. Comp. **17:** 41–53.
23. YANAGI, Y., Y. YOSHIKAI, K. LEGGETT, S. P. CLARK, I. ALEKSANDER & T. W. MAK. 1984. Nature (London) **308:** 145–149.
24. KAZIM, A. L. & M. Z. ATASSI. 1980. Biochem. J. **191:** 261–264.
25. KABAT, E. A., T. T. WU, H. M. PERRY, K. S. GOTTESMAN & C. FOELLER. 1991. Sequences of Proteins of Immunological Interest. 5th Edit. U.S. Department of Health and Human Services, Public Health Service, National Institutes of Health. Bethesda, MD..
26. BEALE, D. & J. COADWELL. 1986. Comp. Biochem. Physiol. **85B:** 205–215.
27. NOVOTNY, J., S. TONEGAWA, H. SAITO, D. M. KRANZ & H. N. EISEN. 1986. Proc. Natl. Acad. Sci. USA **83:** 742–746.
28. CHOTHIA, C., D. R. BOSELL & A. M. LESK. 1988. EMBO J. **7:** 3745–3755.
29. MARCHALONIS, J. J., S. F. SCHLUTER, R. A. HUBBARD, C. MCCABE & R. C. ALLEN. 1988. Mol. Immunol. **25:** 771–784.
30. ROSENSHEIN, I. L., S. F. SCHLUTER, G. R. VASTA & J. J. MARCHALONIS. 1985. Dev. Comp. Immunol. **9:** 783–795.
31. EDELMAN, G. M. 1987. Immunol. Rev. **100:** 12–45.
32. LITMAN, G. W. & K. R. HINDS. 1987. *In* Evolution and Vertebrate Immunity: The Antigen-Receptor and MHC Gene Families. G. Kelsoe & D. H. Schulze, Eds.: 35. University of Texas Press. Austin, TX.
33. MARCHALONIS, J. J., S. F. SCHLUTER, R. A. HUBBARD. A. DIAMANDUROS, W. C. BARKER & R. S. H. PUMPHREY. 1988. Intl. Rev. Immunol. **3:** 241–273.
34. MAGE, R. G., K. E. BERNSTEIN, N. MCCARTNEY-FRANCES, C. B. ALEXANDER, G. O. YOUNG-COOPER, E. A. PADLAN & G. H. COHEN. 1984. Mol. Immunol. **21:** 1067–1081.
35. ROSENSHEIN, I. L. & J. J. MARCHALONIS. 1985. Mol. Immunol. **22:** 1177–1183.
36. PULLEN, A. M., J. BILL, R. T. KUBO, P. MARRACK & J. W. KAPPLER. 1991. J. Exp. Med. **173:** 1183.
37. MARCHALONIS, J. J., S. F. SCHLUTER, H.-Y. YANG, V. S. HOHMAN, K. MCGEE & L. YEATON. 1992. Comp. Biochem. Physiol. **101A**(No. 4): 675–687.
38. SCHLUTER, S. F. & J. J. MARCHALONIS. 1986. Proc. Natl. Acad. Sci. USA **83:** 1872–1876.
39. DEDEOGLU, F., R. A. HUBBARD, S. F. SCHLUTER & J. J. MARCHALONIS. 1992. Exp. Clin. Immunogenet. **9:** 95–108.
40. KASAHARA, M., M. VAZQUEZ, K. SATA, E. C. MCKINNEY & M. F. FLAJNIK. 1992. Proc. Natl. Acad. Sci. USA **89:** 6688–6692.
41. TJOELKER, L. W., L. M. CARLSON, K. LEE, J. LAHTI, W. T. MCCORMACK, J. M. LEIDEN, C. H. CHEN, M. D. COOPER & C. B. THOMPSON. 1990. Proc. Natl. Acad. Sci. USA **87:** 7856–7860.
42. HILDEMANN, W. H. 1974. Nature (London) **250:** 116–119.
43. MARCHALONIS, J. J., A.-C. WANG, R. M. GALBRAITH & W. C. BARKER. 1988. *In* The Lymphocyte: Structure and Function. 2nd Edit. J. J. Marchalonis, Ed.: 307–390. Marcel Dekker. New York.

Identification of AU-Rich 3′ Untranslated Regions in mRNA from Sea Urchin Coelomocytes

MARY ANN ASSON-BATRES,[a] SANDRA L. SPURGEON,[b] AND GROVER C. BAGBY, JR.[a,c]

[a]*Department of Hematology and Medical Oncology*
Department of Medicine
Oregon Health Science University and
The Medical Research Service and
Molecular Hematopoiesis Laboratory
U.S. Veterans Administration Medical Center
Portland, Oregon 97201

[b]*Applied Biosystems Inc.*
Foster City, California 94404

AUUUA reiterations in 3′ untranslated regions of many mammalian mRNAs encode highly inducible proteins, including hematopoietic growth factors, interleukins, adhesion molecules, and protooncogenes.[1–6] 3′ AUREs destabilize mRNA molecules. Insertion of the 3′ AURE of the granulocyte-macrophage colony-stimulating factor (GM-CSF) into the 3′ untranslated region of the rabbit β-globin gene destabilizes the normally long-lived globin message,[1] and deletion of 3′ AURE from c-*fos* and c-*myc* RNAs substantially increases the half-lives of these transcripts in transfected cells.[7–10]

3′ AURE, thought to mediate mRNA instability by serving as a target for specific AU-binding factors and/or endonucleases,[11–15] are more highly conserved in mammalian mRNAs than are sequences in the protein-coding regions of the transcripts. For example, human and murine IL-3 cDNA share only 45% homology in amino acid coding regions, whereas the 3′ untranslated regions are 73% homologous and the 3′ AURE of murine and human IL-3 share 93% homology.[16]

The conservation of 3′ AURE in functionally related mammalian proteins and their potential importance in regulating the expression of proteins that govern the inflammatory response, wound healing, cell adhesion, and cell growth suggested to us that these elements may have been highly conserved throughout evolution. We tested this hypothesis by screening a sea urchin coelomocyte cDNA library for mRNAs containing 3′ AURE. After successfully identifying 124 positive clones, we isolated and sequenced six, four of which overlapped, and used one clone to examine the effects of endotoxin, heat shock, and *in vitro* culture on the expression of the native transcript. We chose the purple sea urchin, *Strongylocentrotus purpuratus,* as our invertebrate model because (a) cytokine-like proteins have been reported in the

[c]Address for correspondence: Grover C. Bagby, Jr., M.D., Professor of Medicine and Molecular and Medical Genetics, Oregon Health Sciences University, L580, 3181 SW Sam Jackson Park Road, Portland, OR 97201.

coelomocytes of a related group of echinoderms[17,18] and (b) the animals occur in abundance in the intertidal zone off the Oregon coast. The 1608 base-pair sea urchin cDNA fragment we isolated hybridizes with mRNAs (poly[A]$^+$) prepared from "stressed" sea urchin coelomocytes, but not with RNA prepared from unstressed coelomocytes. These results support our prediction that AU-rich sequences in the 3′ untranslated region of mRNA molecules are highly conserved and serve as markers of a family of stress-inducible transcripts.

METHODS

Strongylocentrotus purpuratus

S. purpuratus were collected at low tide from the intertidal zone of Boiler Bay, Depoe Bay, Oregon, and coelomocytes were obtained from them in the field. Coelomic fluid from 12 animals (480 ml) was decanted through four layers of sterile cheesecloth, diluted in ice-cold anticoagulant (filter-sterilized 0.25 M EGTA/0.25 M Tris-HCl, pH 7.6), and centrifuged (1000 g for 10 min). Lysis buffer (5 M guanidinium isothiocyanate/0.05 M Tris-HCl, pH 7.5/0.05 M EDTA, pH 8.0/0.5% sarcosyl/0.1 M β-mercaptoethanol) was added to the cell pellets. Lysates were quick frozen in a dry ice/ethanol bath, and stored at −80°C. All cells in this group (control group, "0 hr") were processed and frozen in the field within 21 min of collection of the sea urchin. Supernatants from these samples were clear and colorless to slightly pink, indicating little or no cell lysis occurred as a result of the collection process.

Human Vascular Endothelial Cells

Human umbilical vein endothelial cells were prepared to provide control samples of mRNA for RT-PCR.[19] Endothelial cells in the fourth to eighth passages were exposed to 1 ng/ml recombinant human IL-1α (kindly provided by Dr. Peter Lomedico, Hoffmann-LaRoche, Nutley, NJ) for 18 hours or exposed to medium alone.

Isolation of Total and Poly A$^+$ RNA

Total RNA was isolated from the frozen sea urchin coelomocyte lysates by ultracentrifugation through cesium chloride. mRNA was isolated using oligo-dT cellulose (Invitrogen). Total RNA was isolated from human vascular endothelial cells with guanidinium/lithium chloride.[20]

Creation of 3′ Probes Using RT-PCR

cDNA was synthesized from total RNA isolated from IL-1 induced and uninduced human vascular endothelial cells using M-MuLV RNase H$^-$ reverse transcriptase (BRL) and random hexamer primers. The RT-PCR probe/cDNA library screening strategy is outlined in FIGURE 1. cDNA amplification was carried out with an oligo dT primer containing an *Xba* 1 restriction site at the 5′ end (5′-GCTCTAGA-T_{16}-3′) and a specific ATTTA-rich upstream primer containing a *Sal* 1 restriction site at the 5′ end (5′-CTCGTCGACTATTTATTTATTTAT-3′) for 35 cycles (94°C × 1 min; 28°C × 2 min; 72°C × 2 min). Amplification products (600 bp and 350 bp) were separated on 2% NuSieve (FMC) gels and purified from gel slices using GeneClean (Bio 101). The cDNA was cut with *Sal* 1 and *Xba* 1 and cloned into pGEM11Zf(+) for sequencing (see below).

Complementary DNA was synthesized from LPS-exposed, sea urchin coelomocyte total RNA, and PCR was carried out with the primers described above. The PCR reagents and conditions were those of Showalter and Sommer.[21] Because PCR produced a heterogeneous mixture of amplification products, we reasoned that multiple 3′ cDNA fragments likely existed at the end of the reaction. Therefore, we included ([α^{32}P] dCTP) in subsequent reactions and used the entire pool of radiolabeled amplification products to screen the sea urchin coelomocyte cDNA library.

Preparation and Screening of the Sea Urchin Coelomocyte cDNA Library

Freshly-collected *S. purpuratus* were acclimated overnight in a 20-gallon aquarium filled with artificial sea water at 12°C. Coelomocytes were collected from the animals (as described above) and incubated at 12°C for 4 hr in the presence of 0.5 µg/ml LPS. Poly A^+ RNA, prepared as described above, was reverse-transcribed, and the cDNA was used to construct a library in the pcDNAII (Invitrogen). The library was plated at high colony density.[22] Colonies on replicate nylon filters were lysed and probed with radiolabeled sea urchin cDNA amplification products (6×10^6 CPM in 80 ml hybridization buffer). Sixteen of 124 clones showing a positive signal on replicate filters were picked, plated at low density (1000–2000 colonies/132-mm filter), and screened a second time. Six strongly positive clones were randomly selected for sequence analysis.

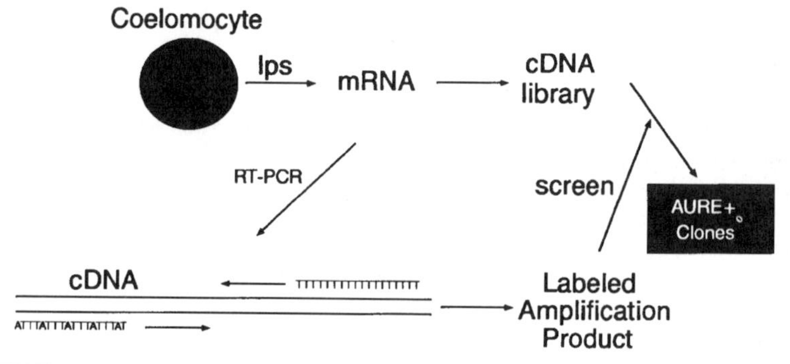

FIGURE 1. Schematic of screening strategy employed to identify mRNAs with 3′ AURE in sea urchins.

TABLE 1. Induction of Coelomocyte mRNA Accumulation: Detected with Labeled Sequence Described in FIGURE 2

Transcript Size (kb)	Inductive Stimulus	mRNA Detectable
3.3	None (rapid preparation at tide pool)	None
4.4	None (rapid preparation at tide pool)	None
3.3	*E. coli* endotoxin	Present
4.4	*E. coli* endotoxin	Present
3.3	Temperature = 18°C	Present
4.4	Temperature = 18°C	Present

DNA Sequencing

Bidirectional, manual sequencing of human endothelial cell cDNA clones was carried out with Sequenase (U.S. Biochemicals) and vector-specific primers. Automated sequencing of the sea urchin–cloned inserts was carried out with an Applied Biosystems (ABI) model 370A or, in the case of clone 1, an ABI 373A DNA sequencer. Fluorescence-labeled M13 universal forward and reverse primers (ABI) and T7 polymerase (Pharmacia) were used to partially sequence the inserts from five of the six clones. The largest of the six clones overlapped with three others. The complete sequence of this clone (FIG. 2) was determined by sequencing both strands using Taq DyeDeoxy terminators (ABI) with cycle sequencing. Primers, in this case, were synthetic oligonucleotides that were either vector-specific or complementary to the cloned insert cDNA. To sequence through difficult poly A regions, we made a degenerate primer with 25 Ts followed by a non-T base. Readable sequence beyond the poly A site can be obtained with this primer because the 3′ non-T nucleotide forces the primer to anneal to the terminus of an extended poly A tract.

Using the Clone as a Probe in Northern Blot Analysis

Coelomocytes exposed to a variety of environmental conditions were used to obtain mRNA for Northern blot analysis. Cells were processed at the tide pool, others were processed in the laboratory after exposure to 18°C, and still others were exposed to endotoxin. Total or poly A^+-selected RNA from control and experimental sea urchin coelomocytes was separated on formaldehyde agarose gels[23] and transferred to nylon membranes.[24] Cloned cDNA inserts from the four overlapping clones were excised, gel isolated, and purified (Geneclean, Bio 101 Inc.), radiolabeled with [α^{32}P] dCTP (DECAprime DNA labeling kit, Ambion), and used to probe immobilized RNA. As shown in TABLE 1, two transcripts were identified in all coelomocyte mRNA samples except those obtained from tide pool preparations.

DISCUSSION

The minimal functional sequence within 3′ AURE (that which shortens mRNA half-life when in the 3′ position) has not been defined, but available evidence

TABLE 2. Occurrence of the 13-mer "AUUUAUUUAUUUA" in Genomic or cDNA Sequences Identified as the 3′ Untranslated Regions (or Putative 3′ Untranslated Region) of Invertebrate mRNA

Invertebrates
Dictyostelium discoideum mRNA for Cap34 protein
D. discoideum mRNA for cyclic nucleotide phosphodiesterase
Plasmodium falciparum gene for S-antigen
P. falciparum gene for circumsporozoite antigen
P. falciparum mRNA for stage-specific antigen (Pfs16 gene)
P. falciparum gene for major merozoite surface antigen (P195)
P. falciparum gene for aldolase
Sarcophaga peregrina (meat fly) mRNA for sarcotoxin IIA
Drosophila melanogaster Ect gene

suggests that reiterated copies of the pentamer, AUUUA, act as recognition sequences for site-specific AU-binding factors or endonucleases. We searched all vertebrate and invertebrate entries (plants were excluded) in Genbank (Release 67) and the European Molecular Biology Library (EMBL, Release 27-67) with the query sequence, ATTTATTTATTTA, and retrieved 339 sequences from the 36,807 entries present in the data base. Of these, 60 unique mRNA sequences were identified that contain the specific AT-rich motif in the 3′ untranslated region of the molecule. The 13-mer was not found in any protein-coding domains. Invertebrate cDNA containing 3′ AURE using this particular query sequence is shown in FIGURE 1.

The results of the nucleic acid data base search supported our prediction that 3′ AURE have been conserved throughout evolution and that they are associated with differentially expressed transcripts. The query sequence is present in the 3′ untranslated region of many inducible, mammalian mRNAs encoding cytokines, growth factors, and oncogenes. It is also present in the 3′ untranslated region of 14 nonmammalian transcripts, 11 of which are inducible by bacteria, viruses, or injury, or are expressed in a developmental- or tissue-specific manner. For example, chick embryo fibroblast 9E3 mRNA is induced by Rous sarcoma virus. It encodes a secretory protein that is related to a number of inducible human proteins that are thought to be inflammatory mediators.[25,26] We felt that direct identification of such mRNAs might result in the identification of unique sequences that could be used to identify shared post-transcriptional mechanisms of gene expression.

Validation of the PCR Screening Strategy in IL-1-Induced Human Endothelial Cells

RT-PCR was performed using primers specific for the 3′ untranslated region of mRNAs containing three consecutive reiterations of ATTTA. Two specific amplification products were produced when cDNA from cells induced with IL-1α was used as a template but were undetectable when cDNA was derived from uninduced cells. DNA sequence analysis revealed that both amplification products share 100% sequence homology with the 3′ untranslated regions of transcripts known to be induced by IL-1 in endothelial cells: namely, *gro*-α (600-bp fragment[27]) and IL-1β (350-bp fragment[28]). Thus, when combined with an oligo dT sequence, 5′-TATTTATTTATTTAT-3′ can prime the specific amplification of cytokine cDNAs with 3′ AURE.[29]

DNA Sequence Analysis of AT-Rich Sea Urchin cDNAs

The labeled sea urchin RT-PCR product mix (generated as described under METHODS, above) hybridized with 124 clones in the primary screen of the sea urchin cDNA library. The AT-rich upstream primer sequence and a poly A⁺ region (situated next to vector-specific DNA) was present in the 1608-bp cDNA insert of clone 1 (FIG. 2). The clone 1 cDNA is generally rich in As and Ts (32% of the nucleotides are As and 35% are Ts) and contains 10 copies of the pentamer, ATTTA. Although none of these cDNA sequences, including that of the longest (clone 1), contains a protein coding region, they all possess a poly A⁺ tail. Moreover, when labeled as probes, these overlapping cDNAs hybridize with discrete poly A⁺ RNA molecules of 3.3 and 4.4 kb. The cDNA fragment does not share strong (>50%) sequence homology with any sequence entry in Genbank or EMBL.

We doubt that all the positive clones identified by screening the sea urchin cDNA

```
5'- GTTTATACGAATATT   ATGTTAACAGCTCTCAAGCA GTGCGTTCAAATTTCAAAAT
GTAAAAGAAAACGAAAAATA TCAAGATTTGTTTCTATGAA TTTATTTCAGACTAGAACAA     115
TACAGTTAGTAGAAAAATGC ACAGTAAACGTCATGGAACA ATGTAGAACAGTTTACATTC
TCTGCAATTTTTTCAAATAT ATGTAATGCTAAAATGCTTT CGACATACATGCCGACATTA     235
TTGCCGACTGTCATGTGATC CATATTTCCTTAAATCGTTT GAAAAGGAAACGAGTAGGAA
AAACAAAGCTATTCACTTTC TAAAATGTATATCTTTTGAA GTGGTTTAAAACTTTCATTC     355
AATGCTTCCATTTTTTACTA TTTTGTCTCTTTTACAATAA AACTATTGGCTTAATTGTAT
TCATCTTGAAAAGCGAGAAA AAAAAATATTAAGAGAGTTT TCATAATAAGTTATGAATCA     475
GGTTTTCTTTTGTATAACAT TTTTGTATTCTTAGTTTGTC ATTATTTCGAATGAATGTTT
GATGGAAAAAGAGTATGGAC AAATATCGTTTTTTAGAAAA TGGAATTGTAGATATTTAAT     595
GCTGCCTCTTTTTAAAAATG ATGATATATGTGTTTTCCTC CACGTTACATTCTTGTCAAC
GATAACAAGTATTTATATTC GCGTGCAATATAGTAGTATT ATCTTATTTACATGTTAGTT     715
CATAACCAAACAAAACTTTG GCAATACAATCAACAAAGTG AATGGTTGAGGTAAACAAAA
ACAATAACTACACCAAACTC TGTGATGGAAATTAGAATGT GTTTTGTACAGTTCATTTAG     835
CCTCACTCTGTAGTTCATAT CTGCGAAGTGAAGAAGCTGA TTTTATTTTCTTATCATATT
ACCGAGCAAAGGGATGTCAC TGTCTATATTTAGGCGGGAA AACATGAGTCACAAATTACT     955
CATCTTCAATGCACTGCCTA CGCAACCTTATGGTTCTTTC GTTCCGCATAGGCCTACAAT
CAAATTTTCAATGTGATAAT GACACACTGTTATTCTAATT ATTCAAATGTTGCGTAGTTG    1075
GCACATCTTTCAAAATGCAA CATTAACTGGTGCAAGAGCT TTCTTTGCGTCCAGTAAATT
TCAATATACACGTAGTTCCA CATTGTAGTATTCACTAGAT ACACCATGTATCAATGCATT    1195
ATCTAATATGATAACGATCA TCATTTTGTACTCGGGACTT GCCAAGCTAAAGCTTATTCG
CTTAAACAGTTTACTCAATA ACTTTGTGATTCTGTTAATT ATTTTGTACAGATTATTTTC    1315
TTTGCTGTCTAACTTATTTT CTGGTTATGTCTTCTAACAT TAGAAATTAAGATCGCTTGT
AAGTATTTATTTTCAGATAT ATTTATTTATTTATGAGTTT AAAAGAAGAATACGGATTAT    1435
GAAAGGCTGTTATTCTATTA TGTTGTTGTAAGTATTCTAT AGAATTGTGCAGAAAGTTGT
TTGGAGTATAAACATATTGC AATAATTTAATTCGTTTAAG ATTTTATGTATAGTCGTTAT    1555
GATTATTCTTAAAGCTCAAC ATAGTATTCAATAAATCTTT TATTTTATTGTG - 3'       1607
```

FIGURE 2. Nucleotide sequence of the clone 1 sea urchin coelomocyte cDNA fragment. Ten ATTTA pentamers are highlighted in bold letters; two potential polyadenylation signals are underlined. Not shown are 25 adenine residues that comprise a portion of the poly A tail. GenBank #L25749.

library represent the same gene, particularly since the partial sequence of clone 3 contains a poly A^+ tail but does not share sequence homology with the 1608-base-pair fragment described above. A poly A^+ tail is present in clone 5, but additional sequence information is not yet available.

Induction of mRNA Accumulation

Four different inducible RNAs hybridize with labeled clone 1.[29] None are detectable in coelomocytes lysed at the tide pool. However, the transcripts were present when clone 1 was used to probe total RNA isolated from coelomocytes exposed to heat shock or *E. coli* endotoxin (TABLE 2). Even the stress of removing these cells from the animal and maintaining them in artificial medium for the duration of the experiments was sufficient to induce the accumulation of both transcripts. This pattern of expression supports the notion that the expression of genes is inducible by environmental stressors.

In summary, using RT-PCR and primers designed to amplify cDNAs containing 3′ AURE, we have identified novel, invertebrate mRNA species that accumulate soon after removing coelomocytes from the animal. These results demonstrate the usefulness of this particular AT-rich sequence in identifying members of this inducible family of gene products with 3′ AURE. We fully expect that minor alterations of the specific AU-rich primer sequence used in this study may reveal additional members of this family.

REFERENCES

1. SHAW, G. & R. KAMEN. 1986. Cell **46:** 659–667.
2. AKASHI, M., G. SHAW, M. GROSS, M. SAITO & H. P. KOEFFLER. 1991. Blood **78:** 2005–2012.
3. WRESCHNER, D. H. & G. RECHAVI. 1988. Eur. J. Biochem. **172:** 333–340.
4. CAPUT, D., B. BEUTLER, K. HARTOG, R. THAYER, S. BROWN-SHIMER & A. CERAMI. 1986. Proc. Natl. Acad. Sci. USA **83:** 1670–1674.
5. BAGBY, G. C. & G. M. SEGAL. 1991. *In* Hematology: Basic Principles and Practice. R. Hoffman, E. J. Benz, S. J. Shattil, B. Furie & H. J. Cohen, Eds.: 97–121. Churchill Livingstone. New York.
6. BAGBY, G. C., G. SHAW, M. C. HEINRICH, S. HEFENEIDER, M. A. BROWN, T. G. DELOUGHERY, G. M. SEGAL & L. BAND. 1990. Prog. Clin. Biol. Res. **352:** 233–239.
7. JONES, T. R. & M. D. COLE. 1987. Mol. Cell. Biol. **7:** 4513–4521.
8. KABNICK, K. S. & D. E. HOUSMAN. 1988. Mol. Cell. Biol. **8:** 3244–3250.
9. WILSON, T. & R. TREISMAN. 1988. Nature **336:** 396–399.
10. TONOUCHI, N., K. MIWA, H. KARASUYAMA & H. MATSUI. 1989. Biochem. Biophys. Res. Commun. **163:** 1056–1062.
11. BEUTLER, B., P. THOMPSON, J. KEYES, K. HAGERTY & D. CRAWFORD. 1988. Biochem. Biophys. Res. Commun. **152:** 973–980.
12. BREWER, G. 1991. Mol. Cell. Biol. **11:** 2460–2466.
13. BOHJANEN, P. R., B. PETRYNIAK, C. H. JUNE, C. B. THOMPSON & T. LINDSTEN. 1991. Mol. Cell. Biol. **11:** 3288–3295.
14. GILLIS, P. & J. S. MALTER. 1991. J. Biol. Chem. **266:** 3172–3177.
15. MALTER, J. S. 1989. Science **246:** 664–666.
16. DORSSERS, L., H. BURGER, F. BOT, R. DELWEL, A. H. GEURTS VANKESSEL, B. LOWENBERG & G. WAGEMAKER. 1987. Gene **55:** 115–124.

17. Beck, G. & G. S. Habicht. 1986. Proc. Natl. Acad. Sci. USA **83:** 7429–7433.
18. Beck, G., R. F. O'Brien & G. S. Habicht. 1989. Bioessays **11:** 62–67.
19. Segal, G. M., E. McCall & G. C. Bagby. 1988. Blood **72:** 1364–1367.
20. Cathala, G., J. F. Savouret, B. Mendez, B. L. West, M. Karin, J. A. Martial & J. D. Baxter. 1983. DNA **2:** 329–335.
21. Schowalter, D. B. & S. S. Sommer. 1989. Anal. Biochem. **177:** 90–94.
22. Vogeli, G., E. Horn, M. Laurent & P. Nath. 1985. Anal. Biochem. **151:** 442–444.
23. Rosen, K. M., E. D. Lamperti & L. Villa-Komaroff. 1990. Biotechniques **8:** 398–403.
24. Virca, G. D., W. Northemann, B. R. Shiels, G. Widera & S. Broome. 1990. Biotechniques **8:** 370–371.
25. Kawahara, R. S. & T. F. Deuel. 1989. J. Biol. Chem. **264:** 679–682.
26. Stoeckle, M. Y. & H. Hanafusa. 1989. Mol. Cell. Biol. **9:** 4738–4745.
27. Anisowicz, A., L. Bardwell & R. Sager. 1987. Proc. Natl. Acad. Sci. USA **84:** 7188–7192.
28. Warner, S. J. C., K. R. Auger & P. Libby. 1987. J. Immunol. **139:** 1911–1917.
29. Asson-Batres, M. A., S. L. Spurgeon, J. Diaz, T. G. DeLoughery & G. C. Bagby, Jr. 1994. Proc. Natl. Acad. Sci. USA, in press.

Evolution of Teleost Antibody Genes

TAKESHI MATSUNAGA, VUOKKO TÖRMÄNEN, KAJSA KARLSSON, AND ELISABET ANDERSSON

Department for Cell and Molecular Biology
University of Umeåa
Umeåa, S 901 87, Sweden

ORIGIN OF IMMUNOGLOBULIN AND CYCLOSTOME

It is generally believed that antibody or immunoglobulin (Ig) molecules evolved in vertebrates. However, it is a mystery how and why many hundreds of diverse antibody V genes evolved all of a sudden in the primate vertebrate from an invertebrate ancestor in which there was no antibody gene. At the inception of acquisition of antibody genes by primitive vertebrates, one or a small number of antibody genes are insignificant in terms of selective advantage. First, defense mechanisms in all the invertebrates including the immediate ancestor to the vertebrate had probably been working for hundreds of millions of years. Second, facing a vast number of antigens of microorganisms and their ability for agile antigenic adaptation, evolutionary acquisition of one or a few antibody specificities by host animals would have been useless.

It was therefore proposed that the repertoire of antibody V genes evolved abruptly from T-cell receptor (TcR) genes in primitive vertebrates.[1] The gist of this idea is that, while the diversity in TcR variable-region molecules are selected in evolution by the polymorphic major histocompatibility complex (MHC) molecules (shown in FIG. 1a), an antibody does not have molecules equivalent to MHC to guide its evolution. Hence, the early vertebrates must have acquired the antibody V gene repertoire of a substantial size at a single stoke. Such an event can be accomplished readily by gene duplication. An example of genetic mechanisms that can duplicate a large number of TcR V genes is shown in FIGURE 1b.

Although it is universally believed that all vertebrates have antibodies, attempts to isolate cyclostome Ig genes have not yet produced evidence supporting the presence of antibody molecules in this most primitive vertebrate. The antibody response in cyclostomes is poor in terms of magnitude and specificities, and presumed Ig proteins of cyclostome show unusual properties not found in Ig molecules of other vertebrates. This putative Ig protein in hagfish turned out to be one of the serum complements.[2] Why is it so difficult to isolate a cyclostome Ig gene? An alternate possibility exists. If the antibody genes evolved from the TcR gene family, there may exist species that have only the TcR and MHC molecules, but not antibody molecules. Could the cyclostome possibly represent an intermediate species linking

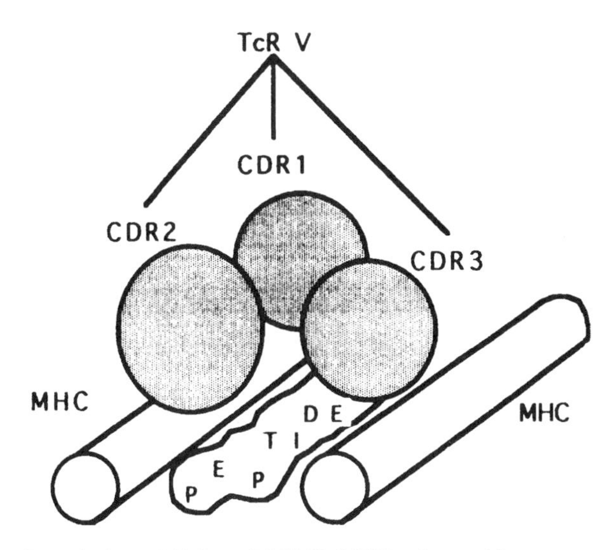

FIGURE 1a. Coevolution of TcR and MHC. MHC polymorphism, expressed as local conformational variants in the α-helical structure for peptide binding groove, can select TcR V region variants (CDR1 and 2) through complementarity. Variant of D and J segments are selected by peptides from protein antigens.

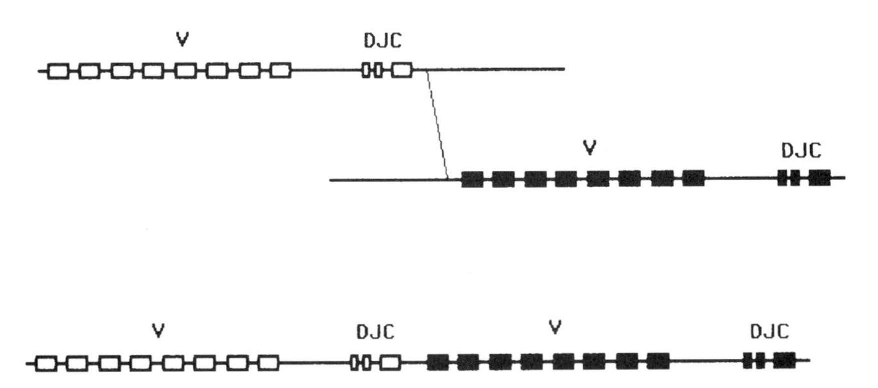

FIGURE 1b. One mechanism to duplicate the entire TcR gene repertoire. Unequal genetic exchange can duplicate a large number of V genes and D/J segments of TcR on the same chromosome. One of the two gene families can later evolve into antibody molecules with modification of the constant region gene. Thus, the V gene repertoire can be generated abruptly.

the absence of Ig in invertebrates and the evolution of Ig in shark and other higher vertebrates? This idea is presented in FIGURE 2.

MULTIPLE V GENES IN VERTEBRATE GENOME

Despite the overall structural conservation of Ig molecules among all vertebrate phyla, the antibody repertoire is generated in diverse ways. In mammals, "the primary repertoire," which is generated by many hundreds of germ line–encoded V genes and several D and/or J segments,[3] can be further modified by B-cell somatic hypermutation during the immune response.[4] Birds generate the repertoire by diversifying one or very few V genes through the process of gene conversion, using sequences of many pseudo V genes located in the upstream region.[5] Amphibian V genes also have been shown to have the capacity to hypermutate,[6] yet the mutation rate does not seem to be sufficient to immunologically meaningful, and the antibody repertoire of *Xenopus* remains rather limited.[7] The antibody repertoire in fish is also restricted (e.g. Witzel and Charlemagne[8]) and does not manifest affinity maturation during the antibody response. After all, the ontogenic generation of the antibody repertoire in cold-blooded vertebrates may depend mostly on germ line–encoded V genes[9,10] and D and J segments. It can be said that the presence of multiple V genes in the genome is one of the conserved features of vertebrate antibody genes. (The multiple pseudo V genes in bird are also counted here, because they are essential for generation of diversity.)

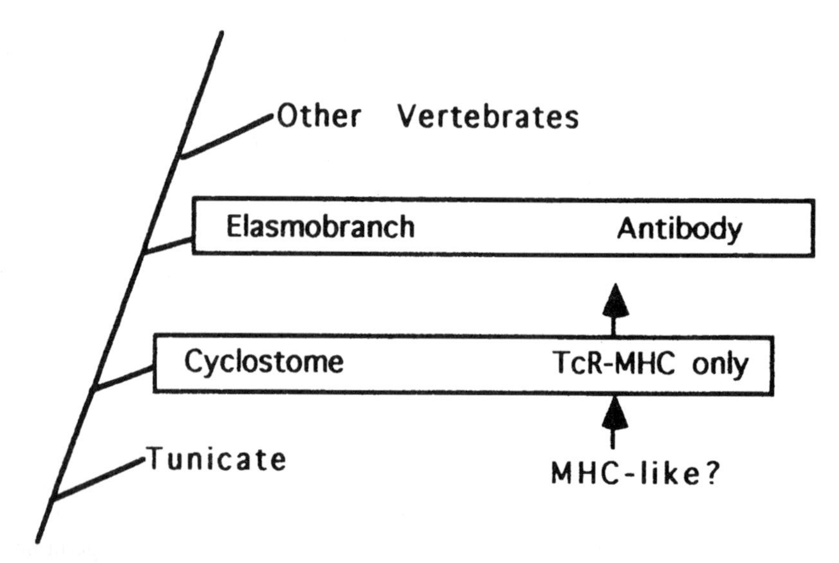

FIGURE 2. A hypothetical scheme showing that Ig molecules are not universal in all vertebrates. The TcR-MHC system is the evolutionary precursor for Ig antibody molecules. Cyclostomes, and perhaps tunicates as well, may have evolved TcR-MHC but not Ig. The Ig molecules evolved with the elasmobranch.

SPECIES-SPECIFIC RESIDUES AND HOMOGENIZATION OF V GENES

The Ig V genes in the genome belong to the multigene family. However, a typical multigene family such as rRNA or tRNA genes is characterized by the sequence homogeneity among the individual members. Antibody V genes are diverse in sequence. Two important questions may be asked: (A) What is the genetic mechanism that diversifies individual members of V genes? More specifically, what are the mechanisms that diversify complementarity-determining regions (CDR) and conserve the framework regions (FR) of V genes? and (B) What is the driving force for V gene diversity in evolution?

Nonreciprocal genetic exchanges (e.g., gene duplication and gene conversion) have been invoked as the mechanisms to homogenize sequences among multigenes (homogenization[11]). Homogenization is advantageous for the multigene family in which each member has the same or nearly the same sequences (tRNA, rRNA), since mutational deterioration occurring randomly in individual genes may be counterbalanced. Species-specific residues or sequences often found among multigene families are considered markers for the homogenizing process.[12] Such "species-specific residues," found earlier in mammalian Ig-variable region proteins[13] have been recently extended to lower vertebrates (*Xenopus, Caiman,* rainbow trout, *Heterodontus*[14]). Most of the species-specific residues are found in FRs except 34 Asn in CDR1 of rainbow trout VH (RTVH 431 and 253[15]). However, position 34 of CDR1 is one of the conserved residues, and it is buried in the β-pleated structure. Thus, the homogenization mechanisms operate throughout vertebrate Ig evolution and help to conserve FR sequences of many V genes. Species-specific residues are most likely the result of fixation of neutral amino acids in proteins, and therefore their presence is purely fortuitous. For this reason, some species-specific residues may be associated with phylogeny, that is, they can be found in closely related species.

VH GENE EVOLUTION OF SALMONID FISH

Salmonid fish diverged earlier from neoteleosts and have many related species that enable us to do comparative studies on antibody genes. If nonreciprocal genetic exchanges occur frequently, we might expect drastic change in the V gene family size of a species during evolution. However, our previous results[15] from Southern hybridization experiments involving several salmonid species and other related fish indicate that the VH family size (hybridization probe: rainbow trout, RTVH 431) is similar in several salmonid species (Atlantic salmon, arctic charr, brown trout, whitefish). The distantly related fish (pike, herring) also have some hybridization signals with the rainbow trout VH probe.[15] It may be argued that the Ig V gene family is rather stable and evolved gradually. Nevertheless, direct comparison of VH sequences in related species would be more informative. For this purpose, we have to identify homologous V genes in closely related species. We have isolated a germ-line VH gene from brown trout (*Salmo trutta,* FIG. 3) by polymerase chain reaction (PCR, see the legend of FIG. 3 for details). This brown trout VH belongs to a VH family different from the RTVH 431 family of rainbow trout, but it has a species-specific

FR1

G Q T L T E S G P V V K K P Q E S H K L T C T A S D L D [M] N

CDR1 FR2 CDR2

S Y W M A W I R Q A P G K G L E W V A T I T H D S R S T Y Y P #

FR3

S V Q G Q F T I S R D N S K R Q V Y L Q M N S L K T E D S A V Y Y C A R

FIGURE 3. Deduced amino acid sequence of a brown trout VH. This is a genomic VH clone isolated from brown trout *(Salmo trutta)* by PCR. The conserved octamer-TATA region in the 5′ VH upstream and the hexamer–nonamer 3′ region of rainbow trout RTVH 431 sequence were used for PCR primers (5′ primer, ATGCAAAGTATCCCTCCCCT; 3′ primer, CAGCCTACTGTGTCACTGTG). The PCR was run starting with two cycles (95°C, 1 min; 60°C, 2 min; 72°C, 2 min) followed by 29 cycles (95°C, 2 min; 60°C, 45 sec; 72°C, 45 sec). The PCR band (about 0.7 kb) was cut from gel and subcloned to plasmid, pBluescript SK+ vector, and the DNA sequence was obtained by the dideoxy method. In this clone, the 18 amino acid leader peptide-coding region is interrupted by an intron with the correct splicing sites. However, position 63 in CDR2 (indicated by #) has a stop codon (TAG). One possibility is that it is a VH pseudogene, although artifactual alteration of base by Taq polymerase during PCR is not excluded. The 29 methionine (boxed) in FR1 of this VH is species-specific to brown trout.

residue (Met 29 in FR1) different from those in rainbow trout (Tyr 24 and Gly 28 of FR1, Asn 34 of CDR1). The VH genes of the two species may have entirely different sets of species-specific residues, but more detailed VH information is required to warrant such a conclusion.

If the process of homogenization is too fast, V genes would not have sufficient time to diversify. If the process is too slow, the animal would accumulate too many nonfunctional V genes. Thus, there must be some kind of balancing force to evolve an optimal V gene repertoire. To investigate this issue further, it would seem fruitful to measure the speed of homogenization of V genes. If we study the preservation and change of species-specific residues as homogenization markers within and between species, we may get some idea of the speed of the homogenization process. In any event, if the divergence for rainbow trout–brown trout is approximately 10 to 15 million years ago (see later), the life span of the species-specific residues may be shorter than this time period.

ORIGIN OF SOME V GENE FAMILIES

The CDRs of V genes diversify their sequences to generate the antibody repertoire. Nevertheless, the overall conformation of CDRs is relatively well conserved. There are several conserved residues within CDRs, and residues directly involved in contact with antigenic molecules are restricted to specific regions.[16,17] Among CDRs, the CDR2 of the heavy chain is longer than other CDRs in mammals (but exceptions exist in low vertebrate Ig, see below), and often explains the fact that antigen recognition by antibody is specific to VH families of mice.[16,17] The structural conservation in CDRs is further supported by the more recent structural study of mammalian Ig proteins[18] that revealed limited patterns in CDRs conformation ("canonical structure"). How have such canonical structures evolved? Sequences similar to the mammalian canonical sequences in CDR1 can be found in rainbow trout,[14] catfish (NG 54), *Xenopus* (I′ CL, V′ CL, X′ CL), and *Heterodontus* (1315, 1207, 3083), suggesting that some CDR structures are very old and conserved. This is shown in TABLE 1.

V genes are usually grouped into families based on the sequence homology (> 80% DNA homology[19]). In the shark (*Heterodontus*[20]), there seems to be only one VH family, whereas teleosts have evolved several VH families (catfish, 5 VH families[21]; goldfish, 3 VH families[22]). We have isolated more VH genes, and their sequences show that rainbow trout has at least seven distinct VH families (to be published).

The species-specific residues are usually found with many V gene members within a VH family.[14] This suggests the possibility that the evolution of a V gene family is shaped by homogenization mechanisms; that is, close sequence similarity might reflect recent homogenization events.

Most of the CDR1 of VH is made up with five amino acid residues (position 31–35). However, CFR1s of nine amino acid residues long have been found in catfish,[21] Atlantic cod,[23] and *Xenopus*.[24] The sequence is shown in TABLE 2.

Close examination of the long CDR1 of nine *Xenopus* VHs seems to indicate that the first four residues (position 31–34) are rather conserved, whereas the other five

TABLE 1. Conservation of Canonical Structure

H2 region (CDR 2) Canonical 2	52a	b	c	53	54	55	71
KOL	D	—	—	D	G	S	R
J539	P	—	—	D	S	G	R
NQ10	S	—	—	G	S	S	R
Rainbow trout (RTVH 253)	*S*	—	—	D	S	S	R
Catfish (NG 54)	*S*	D	—	T	*S*	G	R
Xenopus (I′ CL)	*P*	—	—	D	G	S	R
Xenopus (V′ CL)	*P*	—	—	D	D	*G*	R
Xenopus (X′ CL)	*Y*	—	—	*D*	G	S	R
Heterodontus (1315)	*S*			S	*S*	G	R
Heterodontus (1207)	*S*			S	*S*	S	R
Heterodontus (3083)	*S*			P	*S*	S	R

[a]Evolutionary conservation of canonical structure in VH CDR2. There are four canonical structures in mammalian CH CDR2 and the canonical 2 sequence is represented by those from KOL, J539, and NQ10 (for details, see Cothia *et al.*[18]). The related sequences can be found in VHs of rainbow trout, catfish *Xenopus,* and *Heterodontus.* The residues identical to the mammalian sequences are underlined. The 71 Arg in FR3 contribute to the structure of CDR1.

residues (position 35–39) manifest variability. For this reason it is possible that the first four residues constitute a part of FR1 structure rather than CDR1. In any event, these extra four amino acid residues can be considered a lineage marker for VH gene family. If the divergence for teleost–amphibian occurred about 400 million years ago, the extra four residues of VH genes are as old as 400 million years.

The homogenization helps to conserve VH structures discussed above. Excessive homogenization, however, would lead to disappearance of the antibody repertoire. The classical mutations, drift, and selection provide mechanisms to diversify CDR,

TABLE 2. Long CDR1 of VH (9 Amino Acid Residues)[a]

	31				35				39
Atlantic cod	S	G	S	V	D	Y	A	T	S
Catfish									
(NG66)	D	S	S	S	H	Y	G	T	A
Xenopus									
(PLL3.4)	D	S	S	K	V	Y	A	V	D
(PLL3.1)	—	G	—	—	I	A	S	—	Q
(PLL3.2)	—	—	—	—	—	—	—	—	—
(PLL3.3)	—	—	—	—	I	L	S	—	—
(PLL3.5)	—	—	—	—	I	W	—	—	H
(PCVJ4)	—	—	—	—	—	—	S	Q	F

[a]Long CDR1s of VH that are made of nine amino acid residues. All this VH sequence information is derived from cDNA clones isolated from Atlantic cod, catfish, and *Xenopus.* Horizontal bars indicate residues identical to those of PLL3.4 (*Xenopus*).

particularly residues directly involved in antigen binding. There is evidence that CDR variability of antibody V genes are positively selected.[25]

SPEED OF IgM CONSTANT REGION GENE EVOLUTION

An Ig heavy-chain constant region gene of rainbow trout was isolated from a cDNA library from the fish spleen.[26] Southern hybridization study confirmed the finding in other teleosts that the IgM constant region gene is present as a single copy in the genome (e.g. Amemiya and Litman[27]). This result is radically different from the Ig heavy-chain gene in elasmobranch[20] in which multiple copies of IgM constant genes in association with each VH-DH-JH are present in the chromosome. The evolution from the shark-type Ig gene organization to that of the mammalian type seem to have occurred after elasmobranch–teleost divergence.

We found that the amino acid substitution rate of the IgM constant region is rather constant in the CH4 domain and to a lesser extent in the CH1 domain.[26] The amino acid sequences of 11 vertebrates from different phyla are aligned for each domain. The average number of amino acid substitutions per site (Kaa) is obtained by Kaa $= -\ln(1-p^d)$, where p^d is the percent of amino acid differences between the two peptides.[28] Each species was compared with the human IgM constant regions from which the average Kaa was calculated for the phyla or the class such as elasmobranch (shark and skate), teleosts (rainbow trout, Atlantic salmon, ladyfish, cod), amphibian, bird, and mammals (mouse, rabbit). The Kaa values were then plotted against approximate time of divergence between human and other species, which are estimated from fossil evidence.[26]

When we applied this method, we got linear relationships with various degrees of fitness. The best fitness for linearity is obtained with the CH4 domain with linear regression coefficient value of 0.92 (and CH1 with less precision), but the CH2 and CH3 domains did not give clear results. It is known that CH1 and CH4 domains of IgM molecules are more conserved than the other two domains. The substitution rate of the CH2 and CH3 domains is so high that a large sequence difference is generated rapidly after a certain period of time was allowed to pass after divergence, and thus linearity is lost.

The average amino acid substitution rate per site per year (kaa) is then obtained by kaa = Kaa/(2 × T), where T is the divergence time separating two species. We get kaa = 2.0×10^{-9} per amino acid per year for the whole IgM constant region, and kaa = 1.4×10^{-9} from the CH4 domain. These substitution rates fall in between those of cytochrome *c* (0.22×10^{-9}) and fibrinopeptide (9.0×10^{-9}), the fastest evolving proteins known.

EVOLUTION OF SALMONID FISH

When we applied the kaa for CH4 to the estimation of the divergence time for rainbow trout and Atlantic salmon using the formula T = kaa/(2 × Kaa), we obtain approximately 10 and 16 million years for the two isotypes from Atlantic salmon.[29]

The molecular studies using mitochondrial genes or t-RNA-derived retroposons[30,31] indicate that rainbow trout diverged earlier from most of the Pacific salmons, but it is closely related to species such as cutthroat salmon *(Oncorhynchus clarki)*. Nevertheless, these molecular studies did not provide us with clues on the time of divergence. Fossil evidence in North America suggests that three *Oncorhynchus*-like species (pink salmon, *O. gorbuscha;* sockeye salmon, *O. nerke;* chum salmon, *O. keta*) diverged before 5.5 to 7.6 millions years ago.[32] The genera, *Oncorhynchus, Salmo, Salvelinus, Coregonus, Hucho,* and *Thymallinus* diverged much earlier.[33] Our molecular estimate indicates that the divergence for the *Salmo-Oncorhynchus* is approximately 10 to 15 million years ago. These conclusions (shown in FIG. 4) are tentative and susceptible to revisions when more sequence information becomes available.

UNUSUAL SPLICING OF MEMBRANE IgM OF TELEOST FISH

It has been reported that catfish[34] and other teleosts (Atlantic cod[24] and Atlantic salmon[29]) have unusual mRNA splicing of the IgM gene, resulting in the deletion of the whole CH4 domain in the membrane IgM protein. An earlier report indicates that the rainbow trout also has IgM proteins of two different sizes.[35] The IgM cDNA

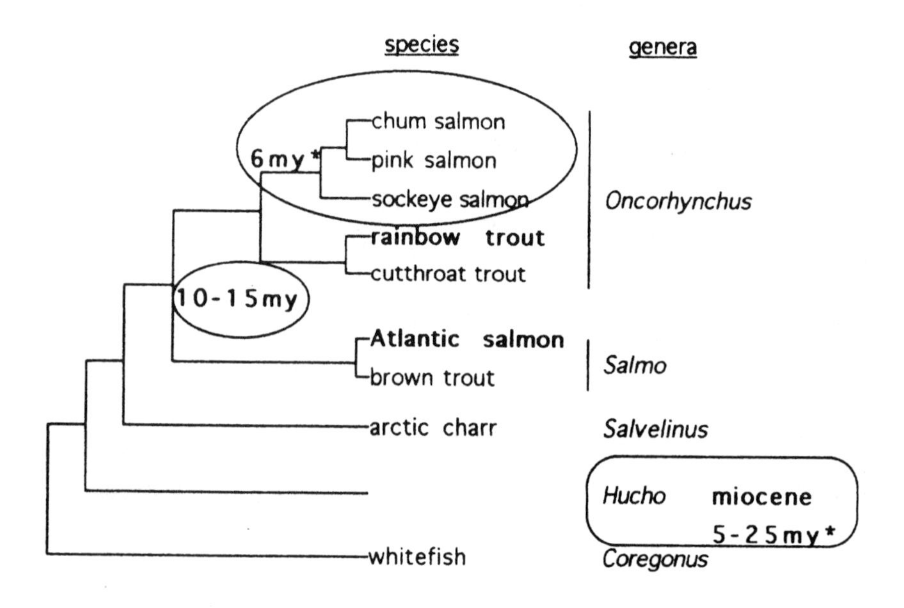

FIGURE 4. Phylogenetic relation of several salmonid fish. Fossil records show that divergence for some *Oncorhynchus* species (chum salmon, pink salmon, and sockeye salmon) occurred before 6 million years ago. Our molecular estimate suggests a divergence time for *Oncorhynchus-Salmo* of 10–15 million years ago.

clone of rainbow trout that we have isolated encodes all CH domains and the secretory peptide. However, we have also isolated by PCR a cDNA clone that encodes CH3 and membrane peptide from rainbow trout spleen mRNA.[26] The sequencing of this cDNA shows a complete deletion of the CH4 domain, indicating that rainbow trout too has unusual IgM splicing.

Although the mechanism for splicing secretory-type mRNA over membrane-type mRNA is not known, specific enzymatic recognition of the splicing boundary seems to be a crucial event in this mechanism.[36] One such sequence is located between CH4 and the COOH-terminal peptide of the secreted IgM proteins of vertebrates.[36] Within this conserved eight-nucleotide sequence, the two codons (GGT AAA) are absolutely conserved throughout the evolution of mammals, birds, amphibians and sharks. This conservation breaks down, however, in all teleost IgM genes known (rainbow trout, AAC CAA, Atlantic salmon, AAC CAA, catfish, AAA ACG, Atlan-

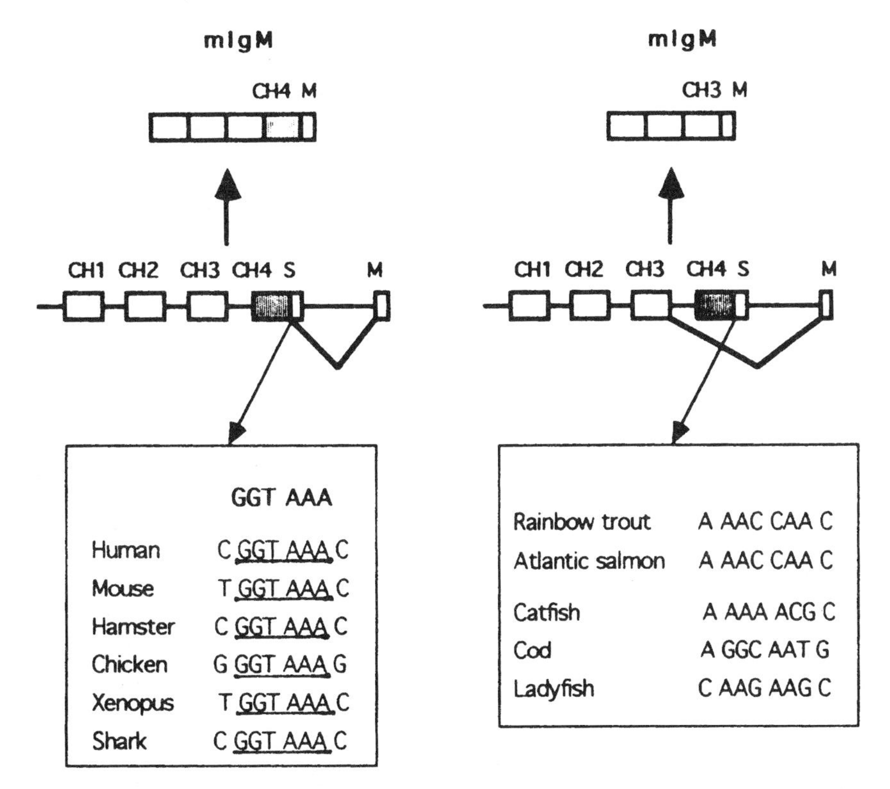

FIGURE 5. Unusual RNA splicing in teleost fish. Membrane IgM of several teleost species have no CH4 domain due to the fact that the mRNA splicing deletes CH4 domain *(right)*. Six nucleotides, GGT, AAA, in the CH4-secretory boundary region is thought important for correct splicing, and this sequencing is absolutely conserved from mammals to the elasmobranch *(left box)*. The same region in teleost fish does not have a conserved sequence *(right box)*, which may be responsible for the unusual splicing and generation of the small-membrane IgM in teleosts.

tic cod, GGC AAT, ladyfish, AAG AAG). Though the unusual RNA splicing has not yet been shown in ladyfish *(Elops),* it seems that this primitive teleost too has membrane IgM with the CH4 domain deleted.

The functional significance of the CH4 domain deletion of membrane IgM is unknown. It may be that the CH4 domain is dispensable in whatever function it might have in the B-cell membrane. Such an evolutionary event, if it happened once, could be "frozen" and inherited in the monophyletic lineage of teleost evolution. This is shown in FIGURE 5.

SELECTION FOR V REGION SPECIFICITIES

Due to the many orders of magnitude difference in generation time between vertebrate hosts and microorganisms, the specificities of germ-line antibody V genes cannot effectively cope with infections. Conservative epitopes or epitopes that are slow to change (e.g. carbohydrate antigens) in microorganisms might be involved in the selection for germ-line antibody V genes. If the early antibody repertoire evolved abruptly, as discussed in the first section of this article, the V genes that encode binding specificities for conservative epitopes on microorganisms would confer selective advantage to host animals (fish). The molecular study of fish antibody genes should also address the question on the major epitopes involved in the antibody responses of fish.

REFERENCES

1. MATSUNAGA, T. & V. TÖRMÄNEN. 1990. Evolution of antibody and T-cell receptor V genes: The antibody repertoire might have evolved abruptly. Dev. Comp. Immunol. **14:** 1–8.
2. ISHIGURO, H., K. KOBAYASHI, M. SUZUKI, K. TITANI, S. TOMONAGA & Y. KUROSAWA. 1992. Isolation of a hagfish gene that encodes a complement component. EMBO J. **11:** 829–837.
3. TONEGAWA, S. 1983. Somatic generation of antibody diversity. Nature **302:** 575–581.
4. MCKEAN, D., K. HUPPI, M. BELL, L. STAUDT, W. GERHARD & M. WEIGERT. 1984. Generation of antibody diversity in the immune response of BALB/c mice to influenza virus hemagglutinin. Proc. Natl. Acad. Sci. USA **81:** 3180–3184.
5. REYNAUD, C. A., V. ANQUEZ, A. DAHAN & J. C. WEIL. 1985. A single rearrangement event generates most of the chicken immunoglobulin light chain diversity. Cell **40:** 283–291.
6. WILSON, M., E. HSU, A. MARCUZ, M. COURTET, L. DU PASQUIER & C. STEINBERG. 1992. What limits affinity maturation of antibodies in *Xenopus*—the rate of somatic mutation or the ability to select mutants? EMBO J. **11:** 4337–4347.
7. BRANDT, D. C., M. GRIESSEN, L. DU PASQUIER & J. C. JATON. 1980. Antibody diversity in amphibians: Evidence for the inheritance pf idiotypic specificities in isogenic *Xenopus.* Eur. J. Immunol. **10:** 731–736.
8. WITZEL, M. C. & J. CHARLEMAGNE. 1985. Antibody diversity in fish. Isoelectrofocalization study of individually purified specific antibodies in the teleost fish species: Tench, carp and goldfish. Dev. Comp. Immunol. **9:** 261–270.
9. DU PASQUIRE, L. 1982. Antibody diversity in low vertebrates—why is it so restricted? Nature **296:** 311–313.
10. MATSUNAGA, T. 1985. Evolution of antibody repertoire. Somatic mutation as a latecomer. Dev. Comp. Immunol. **9:** 585–596.

11. Dover, G. 1982. Molecular drive: A cohesive mode of species evolution. Nature **299:** 111–117.
12. Arnheim, N., M. Kristal, R. Schmickel, G. Wilson, O. Ryder & E. Zimmer. 1980. Molecular evidence for genetic exchange among ribosome genes on non-homologous chromosomes in man and apes. Proc. Natl. Acad. Sci. USA **77:** 7323–7327.
13. Kabat, E. 1967. The paucity of species-specific residues in the variable regions of human and mouse Bence-Jones proteins and its evolutionary and genetic implications. 1967. Proc. Natl. Acad. Sci. USA **57:** 1345–1349.
14. Matsunaga, T., T. Chen & V. Törmänen. 1990. Characterization of a complete immunoglobulin heavy-chain variable region germ-like gene of rainbow trout. Proc. Natl. Acad. Sci. USA **87:** 7767–7771.
15. Andersson, E. & T. Matsunaga. 1991. Evolution of a VH family in low vertebrates. Internat. Immunol. **3:** 527–533.
16. Ohno, S., N. Mori & T. Matsunaga. 1985. Antigen-binding specificities of antibodies are primarily determined by seven residues of VH. Proc. Natl. Acad. Sci. USA **82:** 2945–2949.
17. Matsunaga, T. & S. Ohno. 1987. The evolution of antibody and antibody repertoire. *In* Evolution and vertebrate immunity: The antigen–receptor and MHC gene families. G. Kelsoe & D. H. Schulze, Eds.: 429–451. The University of Texas Press. Austin, TX.
18. Cothia, C., A. M. Lesk, A. Tramontano, M. Levitt, S-J. Smith-Gill, G. Air, S. Sheriff, E. A. Padlan, D. Davies, W. R. Tulip, P. M. Colman, S. Spinelli, P. M. Alzari & R. J. Poljak. 1989. Nature **342:** 877–883.
19. Brodeur, P. & R. Riblet. 1984. The immunoglobulin heavy-chain variable region (Igh-V) locus in the mouse. I. 100 Igh-V genes comprise seven families of homologous genes. Eur. J. Immunol. **14:** 922–940.
20. Kokubu, F., R. Litman, M. J. Shamblott, K. Hinds & G. W. Litman. 1988. Diverse organization of immunoglobulin VH gene loci in a primitive vertebrate. EMBO J. **7:** 3413–3422.
21. Ghaffari, S. H. & C. J. Lobb. 1991. Heavy-chain variable region gene families evolved early in phylogeny. Ig complexity in fish. J. Immunol. **146:** 1037–1046.
22. Wilson, M. R., D. Middleton & G. W. Warr. 1991. Immunoglobulin VH genes of the goldfish, *Carassius auratus:* Reexamination. Mol. Immunol. **28:** 449–457.
23. Bengten, E., T. Leandersson & L. Pilström. 1991. Immunoglobulin heavy-chain cDNA from the teleost Atlantic cod *(Gadus morhua L)*: Nucleotide sequences of secretory and membrane form show an unusual splicing pattern. Eur. J. Immunol. **21:** 3027–3033.
24. Schwager, J., N. Burckert, M. Courtet & L. Du Pasquier. 1989. Genetic basis of the antibody repertoire in *Xenopus:* Analysis of the VH diversity. EMBO J. **8:** 2989–3001.
25. Tanaka, T. & M. Nei. 1989. Positive Darwinian selection observed at the variable region gene of immunoglobulins. Mol. Biol. Evol. **6:** 447–459.
26. Andersson, E. & T. Matsunaga. 1993. Complete sequence of cDNA gene of rainbow trout and evolution of IgM constant domains of vertebrates. Immunogenetics. **38:** 243–250.
27. Amemiya, C. T. & G. Litman. 1990. Complete nucleotide sequence of an immunoglobulin heavy-chain gene and analysis of immunoglobulin gene organization in a primitive teleost species. Proc. Natl. Acad. Sci. USA **87:** 811–815.
28. Kimura, M. 1983. *In* The Neutral Theory of Molecular Evolution. Cambridge University Press. Cambridge, England. pp. 55–116.
29. Hordvik, I., A. M. Voie, J. Glette, R. Male & C. Endersson. 1992. Cloning and sequencing analysis of two isotypic IgM heavy-chain genes from Atlantic salmon, *Salmo salar L.* Eur. J. Immunol. **22:** 2957–2962.
30. Thomas, W. K. & A. T. Beckenbach. 1989. Variation in salmonid mitochondrial DNA: Evolutionary constraints and mechanisms of substitution. J. Mol. Evol. **29:** 233–245.
31. Kido, Y., M. Aono, T. Yamaki, K. Matsumoto, S. Murata, M. Saneyoshi & N. Okada. 1991. Shaping and reshaping of salmonid genomes by amplification of tRNA-derived retroposons during evolution. Proc. Natl. Acad. Sci. USA **88:** 2326–2330.
32. Smith, R. G., K. Swirydczuk, P. G. Zimmel & B. H. Wilkinson. 1982. Fish biostratig-

raphy of late Miocene to Pleiocene sediments of the western Snake River plain, Idaho. Idaho Bur. Mines Geol. Bull. **26:** 519–541.

33. SMITH, G. R. & R. R. MILLER. 1985. Taxonomy of fishes from Miocence Clarkia beds, Idaho. *In* Late Cenozoic History of the Pacific Northwest. C. J. Smiley, Ed.: 75–83. American Association for the Advance of Science. San Francisco..
34. CLEM, L. W., D. MIDDLETON & G. W. WARR. 1990. The immunoglobulin M heavy chain constant region gene of the channel catfish, *Ictalurus punctatus*: An unusual mRNA splicing pattern produces the membrane form of the molecules. Nucleic Acid Res. **18:** 5227–5233.
35. WARR, G. W., D. DELUCA & B. R. GRIFFIN. 1979. Membrane immunoglobulin is present on thymic and splenic lymphocytes of the trout *Salmo gairdneri.* J. Immunol. **123:** 910–917.
36. PETERSON, M. L. & R. P. PERRY. 1989. The regulated production of μ4-to-M1 splice. Mol. Cell. Biol. **9:** 726–738.

Animal Lectins as Self/Non-Self Recognition Molecules

Biochemical and Genetic Approaches to Understanding Their Biological Roles and Evolution[a]

G. R. VASTA, H. AHMED, N. E. FINK,[b] M. T. ELOLA,
A. G. MARSH, A. SNOWDEN, AND E. W. ODOM

Center of Marine Biotechnology
University of Maryland Biotechnology Institute
Baltimore, Maryland 21202

INTRODUCTION

Why include lectins when discussing primordial immunity? It has become clear in recent years that a variety of protein–carbohydrate interactions are directly and indirectly involved in host defense mechanisms[1]—directly as opsonins, complement activating factors, LPS-binding molecules, and toxins and indirectly, as homing receptors, cell adhesion molecules, and inducing cell activation as mitogens or cytokines. Many of these recognition and regulatory functions are part of the acute-phase response to infection, and it is our view that this is the most ancient form of non-self recognition/defense function that may have been conserved through the invertebrate lineages that gave origin to the chordates, and erroneously characterized as lacking specificity. The experimental evidence has suggested that distinct, and probably unrelated, groups of molecules are included under the term "lectin."[2–6]

Animal lectins do not express a recombinatorial diversity like that of antibodies. However, biochemical studies on the carbohydrate specificity and structural analyses of the binding sites of lectins suggest that a limited diversity in recognition capabilities would be accomplished by the occurrence of multiple lectins with distinct specificities; the presence of more than one binding site per polypeptide subunit, specific for different carbohydrates in a single molecule; and by certain "flexibility" of the binding sites that would allow the recognition of a range of structurally related carbohydrates.[7]

Our interest focuses on understanding how protein–carbohydrate interactions may have evolved both in structure and function, in self/non-self recognition processes that mediate host defense against infection. We will discuss here how the valuable information from biochemical and molecular approaches has contributed to the understanding of the structure, function, and evolution of animal lectins. Results

[a]Supported in part by National Science Foundation Award MCB-91-05875-02 and Sea Grant Award NA90AA-D-SG063 to GRV, and National Institutes of Health Award 5 F31 GM14903-02 to EWO/GRV.

[b]On temporary leave from Departamento de Ciencias Biológicas, Cátedra de Hematología, Universidad Nacional de La Plata, Argentina; supported by the Consejo Nacional de Investigaciones Científicas y Técnicas de la República Argentina.

from biochemical studies carried out in a number of laboratories, including our own, that have provided information on the structural and functional properties of invertebrate and vertebrate lectins and their relationships with other vertebrate recognition molecules are discussed in the following sections.

MOLECULAR STRUCTURE AND CLASSIFICATION OF ANIMAL LECTINS

Unlike the immunoglobulin superfamily, a vast diversity in the molecular structures of animal lectins has made virtually impossible the definition of a general pattern. This is possibly due to the fact that carbohydrate binding molecules may represent products of very diverse evolutionary histories. However, certain common motifs have recently emerged from the analysis of the carbohydrate binding sites,[3] suggesting that certain lectin molecules, or at least the amino acid sequences relevant to their biological role, have been conserved in evolution along lineages that gave origin to the vertebrate phyla. Animal lectins are oligomers of equal or distinct subunits and their molecular organization depends on whether they are soluble or integral membrane proteins. The subunits of membrane lectins from vertebrates vary in size, but their molecular structure is common to most integral membrane proteins: a carboxy-terminal domain, variable in length and located in the extracellular side of the membrane, that carries the carbohydrate binding site and eventually the glycosylation sites, followed by a short transmembrane hydrophobic region that anchors the protein to the membrane, and finally by an intracytoplasmic amino-terminal short segment. Although some true integral membrane lectins have been identified and characterized in invertebrate species, none has been sequenced so far. Hence, it is not possible to assess if they follow a domain pattern similar to that established for their vertebrate counterparts. The molecular structure of soluble lectin subunits differ from the integral membrane lectins in that hydrophobic residues are distributed throughout the length of the sequence without forming a transmembrane domain. The COOH-terminal region usually contains the carbohydrate-binding site while the amino-terminal region is variable. The amino-terminal region may consist of a fibrillar collagen-like structure that in one example interacts with complement components, a structure similar to the core protein of the cartilage proteoglycan that exhibits regions that interact with glycosaminoglycans domains similar to epidermal growth factors or even structures similar to those proteins that bind RNA.[3] Among soluble lectins it has been shown that one group, common in vertebrate tissues, exhibits similar properties such as a specificity for β-galactosides and a requirement for free thiol groups, but not divalent cations, for *in vitro* agglutinating activity. This group constitutes a family of molecules that has been substantially conserved in evolution.[2] Although the members of this family exhibit considerable homology, they differ in subunit size, architecture, and oligomeric structure.[5]

The analysis of primary structures of animal lectins has suggested that distinct, and probably unrelated, groups of molecules are included under the term "lectin." Drickamer,[3] Harrison,[4] and Hirabayashi and Kasai[5] have successfully organized and classified the available information on carbohydrate-binding molecules from vertebrates. Within the invertebrate taxa, major groups of lectins can also be identified:

One group would include lectins such as those from the flesh fly *Sarcophaga peregrina* larva,[8] the acorn barnacle *Megabalanus rosa,*[9] and the sea urchin *Anthocidaris crassispina*[10] that show significant homology to membrane-integrated or soluble vertebrate C-type lectins.[3] The second would include those β-galactosyl-specific lectins such as those from the nematode *Caenorhabditis elegans,*[11] the sponge *Geodia cydonium,*[12] and the tunicate *Clavelina picta*[13] (see FIG. 1) homologous to the S-type vertebrate lectins.[5] The third group would be constituted by lectins such as those from the horseshoe crab *Limulus polyphemus*[14] and the tunicate *Didemnum candidum*[15] that show homology to vertebrate pentraxins that exhibit lectin-like properties, such as C-reactive protein and serum amyloid P.[6] Finally, there are examples that do not exhibit similarities to any of the aforementioned categories. Moreover, except for a small number of invertebrate lectins, most of those described so far cannot yet be placed in one or another group because of the lack of information regarding their primary structure (TABLE 1).

Drickamer[3] has classified animal lectins in two major groups, C- and S-type, based on amino acid sequence similarities, particularly in the carbohydrate binding domain, and other properties such as divalent cation dependence and requirement of free thiols. The C-type lectins comprise a wide variety of soluble as well as integral membrane proteins that require Ca^{2+} for binding and exhibit a set of conserved residues, approximately 15% of the recognition domain, among those, cysteine residues probably involved in disulfide bonds that are required to remain intact for binding activity. This constitutes a polypeptide framework in which stretches of sequence with variable residues would confer the carbohydrate specificity. This group includes integral membrane lectins such as mammalian asialoglycoprotein receptors and the chicken hepatic lectin, and soluble lectins such as mammalian liver

TABLE 1. Classification of Animal Lectins

	S-Type	C-Type	Pentraxin-like	Other
Property				
Cysteines	Free thiols	Disulfides	Disulfides	Disulfides
Ca^{2+} requirement	No	Yes	Yes	Variable
Location	Intra- and extracellular	Extracellular	Extracellular	Extracellular
Solubility	Buffer soluble	Variable	Buffer soluble	Variable
Carbohydrate specificity	Mostly β-gal	Variable	Phosphocholine, sialic acids, galactans	Variable
Examples				
Invertebrates	*C. elegans* *C. picta* *G. cydonium*	*S. peregrina* *A. crassispina* *M. rosa* *C. picta*	*L. polyphemus* *D. cnadidum* *C. picta*	*H. pomatia* *L. flavus*
Vertebrates	Galaptins: 14, 16–22, 29–35, and 67-kDa lectins	Selectins Mannose-binding proteins Asialo-glycoprotein receptors	SAP, CRP, HFP	Eel serum anti-O(H) IL-1, IL-2 TNF

and serum mannose-binding proteins, invertebrate lectins, and other proteins such as the 28- to 32-kDa apoprotein of the pulmonary surfactant, the core protein of cartilage and fibroblast proteoglycans, the lymphocyte receptor for the Fc portion of IgE, the mannose receptor from pulmonary macrophages and a fucose-specific hepatic membrane lectin. Lectins of the C-type exhibit little or no homology at all in domains other than the carbohydrate recognition site. Those domains confer specific functional properties to each particular lectin, such as hydrophobic domains, that anchor the protein to the plasma membrane, and complement binding sites, areas for covalent binding of glycosaminoglycans, and so forth, resulting in mosaic or chimeric molecules both from the structural and functional standpoints. The second group, S-type animal lectins, is constituted by β-galactosidic-specific lectins that require the presence of free thiols for binding activity but not Ca^{2+} or other divalent cations. Their location is mostly intracellular, in the cytoplasmic compartment. No detergents are required for their extraction, and hence they are considered soluble lectins, although they may remain membrane-associated through their carbohydrate-binding sites. These lectins constitute the family of related molecules that exhibit considerable similarities in the primary structure and, as the C-type lectins, exhibit an invariant residue pattern in the carbohydrate recognition domains. The conserved residues are different from the C-type lectins, and no cysteines are among them.

The C-type lectins have been further subdivided into four categories based on their primary structure and gene organization of the carbohydrate recognition domains (CRDs), particularly the presence or absence of introns.[16] The first group would include the cartilage and fibroblast proteoglycan core proteins, with specificities for Gal and Fuc. The second group comprises the type II receptors that include integral membrane proteins such as the hepatic asialoglycoprotein receptors, the macrophage lectin, and the IgE-Fc receptor. These members exhibit specificity

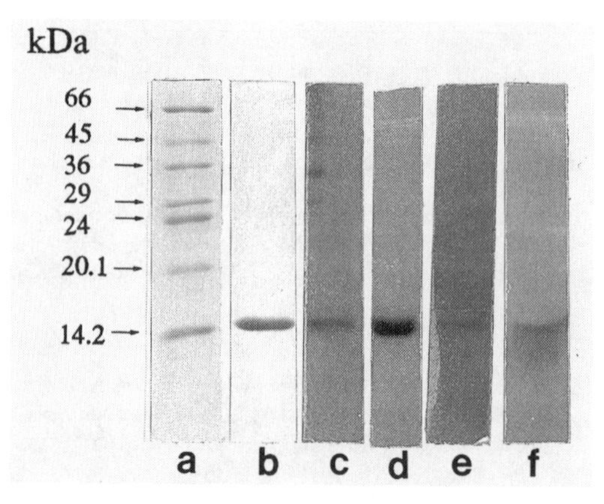

FIGURE 1. S-Type lectins (galaptins). SDS-PAGE of soluble S-type lectins purified by affinity chromatography on lactosyl-Sepharose columns. Samples were run on 15% polyacrylamide gel in presence of 0.5% 2-mercaptoethanol and stained with Coomassie blue. **Lane a:** protein MW markers; **lane b:** bovine spleen; **lane c:** toad ovary; **lane d:** striped bass muscle; **lane e:** striped bass ovary; and **lane f:** tunicate *Clavelina picta.*

for Gal/Fuc, Gal/GalNAc, or GlcNac residues. Most of the CRD coding regions characterized for lectins in these first two groups contain introns. The third group includes the so-called "collectins," lectins that are constituted by a CRD connected to collagen-like sequences, and here are included the mannose-binding proteins from serum and liver, the pulmonary surfactant protein SP-A, and the bovine conglutinin. Sugar specificities in members of this group include mannose, fucose, GlcNac, and galactose. Group IV is constituted by the membrane cell adhesion molecules known as LEC-CAMs (lectin-epidermal growth factor-complement homology-containing cell adhesion molecules), mosaic molecules specific for sialic acids and fucose, among which Mel-14 and LSM-1 (lymphocyte homing receptors), GMP-140 (platelet granule membrane protein), and ELAM-1 (endothelial cell–leukocyte adhesion molecule) are the best characterized. Genes for lectins included in groups III and IV do not have introns in the CRD-coding regions.

Based on the primary structure and polypeptide architecture of the subunits, S-type lectins have been classified into three main subgroups[5]: the "proto," "chimera," and "tandem-repeat" types. The first type includes the well-characterized 14-kDa subunit galaptins, widely distributed in animal taxa, from sponges to mammals. (See FIG. 1.) These lectins form noncovalently bound dimers and carry one binding site per subunit. The chimera type comprises multidomain lectins, so far found only in mammals and previously classified as the 29- to 35-kDa β-galactoside lectin group,[4] that are constituted by a COOH-terminal lectin domain and NH_2-terminal domains homologous to hnRNP components and the L-myc gene product. A well-characterized member of this group is the CBP35, identical to a number of receptors characterized in other contexts, such as the elastin/laminin-binding protein,[17] Mac2 from activated macrophages,[18] and the IgE-binding protein.[19] The tandem-repeat type has only two members described so far: the *C. elegans* 32-kDa lectin[11] and the 36-kDa lectin from rat intestine.[20] These lectins have two repeated domains, most likely products of gene duplication, that carry identical or slightly different carbohydrate-binding sites, separated by a region susceptible to proteolytic activity. Harrison[4] classified the β-galactoside lectins or galaptins according to their molecular masses, in four main groups: 14×10^3 M_r, $16–22 \times 10^3$ M_r, $29–35 \times 10^3$ M_r , and 67×10^3 M_r. The first and third groups correspond to the proto and chimera types from Hirabayashi and Kasai,[5] respectively. Some members of the second, such as the rat intestine lectin, may belong to the tandem-repeat type, and the 67×10^3 M_r group consists of an only member, the elastin/laminin receptor.[17,21]

Although the vertebrate pentraxins constitute a well-defined family of proteins, their lectin-like properties such as binding to carbohydrates and related structures, divalent cation dependence, opsonization functions, and overall molecular structure[22–24] have suggested that these molecules may well be considered a third group of animal lectins.[6] From the information about primary structures available at the present time, invertebrate lectins, including those from *Limulus polyphemus*[14] and *Didemnum candidum*[15] that show homology to the vertebrate pentraxins C-reactive protein and serum amyloid protein can be identified as members of this group.

Certain lectins, such as lectin of the sponge *Axinella polypoides,*[26] do not exhibit similarities to any of the aforementioned categories. Actually, because there is a lack of information regarding their primary structures, the vast majority of invertebrate lectins described so far cannot yet be placed in one or another group. Likewise, other molecules characterized in different contexts such as echinonectin, a galactose-specific lectin from sea urchin embryos that may be analogous to fibronectin,[27,28] and the cytokines IL-Iα, IL-2 and TNF, which exhibit lectin properties[29–31] cannot be classified within the categories indicated above.

CARBOHYDRATE SPECIFICITY AND ARCHITECTURE OF THE BINDING SITE

Why is it important to define the fine specificity and determine in detail the structure of the binding site? Because the biological functions of lectins are presumed to be accomplished, at least in part, through the sugar binding sites, it is evident that the characterization of its architecture and the recognized carbohydrate structures will allow the identification of the "natural" ligands and hence, the elucidation of their biological role. The carbohydrate specificity of animal lectins has been characterized through conventional methods of inhibition of hemagglutination or binding in solid phase (See Ahmed *et al.,* this volume) and quantitative precipitation of glycoconjugates with simple or complex carbohydrates. Equilibrium dialysis, fluorescence quenching, and other related methods have provided information about the affinity and number of binding sites present in the active molecule. Finally, the analysis of the molecular interactions between the polypeptide and the sugar comprises two aspects: the identification and required orientation of the free or substituted hydroxyls on the carbohydrate that actually interact with the protein and the identification of the amino acid residues that are directly involved in that interaction as well as those residues that may not interact with the ligand but determine the architecture of the binding site. The characterization of the molecular interactions in S-type lectins will serve to illustrate this analysis.

The identification of the carbohydrate structures recognized by the S-type lectins have been accomplished in a number of laboratories by inhibition of lectin binding with a variety of mono- and oligosaccharides that carry substitutions in selected positions of the nonreducing or subterminal units. Our results on the bovine spleen galaptin by the use of a novel solid-phase method we developed for that purpose indicated that the lectin has a combining site at least as large as disaccharide.[32] The disaccharides, having nonreducing terminal β-galactosyl residues linked (1,3), (1,4), and (1,6) to Glc, GlcNAc, Ara, Man, or Fru_f were better inhibitors than free Gal or its glycosides. The relevance of 4-OH of the nonreducing sugar was indicated by the fact that Glc β1,4Glc had no activity as compared to Galβ 1,4Glc. Substitution of 6-OH at the Gal residue of Gal β1,4Glc disrupts the binding since NeuAc α2,6Gal β1,4Glc and GlcNAc β1,6Gal β1,4Glc were poor inhibitors. The 4-OH group on the subterminal Glc is also very important for lectin–sugar interaction, inasmuch as Gal β1,3GalNAc and Fuc α1,3[Gal β1,4] Glc (3-fucosyllactose) do not inhibit the binding. The hydroxyls at 2′ and 3′ of Gal may not be very critical as substitutions of these hydroxyls. (Fuc α1,2Gal β1,4Glc and NeuAc α2,3Gal β1,4Glc do not affect binding significantly.) The data clearly demonstrated the preferential binding of bovine spleen S-type lectin towards Gal β[1,3(4)]GlcNAc sequences and are in general agreement with the specificity of other S-type lectins previously characterized.

The identification of the binding site region in the polypeptide subunits and the participation of particular amino acid residues in the interactions with the ligand have been attempted through a variety of approaches. These include amino acid and nucleotide sequence alignment comparisons,[5,33] specific chemical modification of amino acid residues,[34,32] ligand-mediated protection from proteolytic digestion,[32] site-directed mutagenesis,[5,33] and the resolution of the three-dimensional structure of the lectin cocrystallized with the ligand.[35] Despite the fact that the first four are indirect and to a certain extent speculative, these methods have provided substantial preliminary information that to a considerable extent has been confirmed by the resolution of the crystal structure. The comparison of the aligned sequences and identification of conserved residues is based on the notion that these will correlate with the conservation of carbohydrate specificity and biological function. The specific chemical modification of amino acids allows the identification of the nature of the residues involved in the interaction, provided that no drastic conformational changes are detected upon chemical treatment of the protein. Site-directed mutagenesis studies allow the identification of the amino acid residues that, if substituted, will result in loss of binding activity. It fails to discriminate, however, if the loss is due to direct binding to the ligand or by conformational changes produced in the mutated recombinant protein. Finally, the crystallographic approach allows the rigorous identification of the amino acid residues involved in ligand binding and the nature of the interactions established by their direct "visualization" in the molecular models of the lectin–ligand complex constructed with X-ray diffraction data.

We have applied the chemical modification approach to the carboxamidomethylated bovine spleen galaptin and have identified histidine, tryptophan, carboxylic acids, and arginine as the amino acid residues critical for the binding to *N*-acetyl lactosamine.[32] The methods included the chemical modification in the presence and absence of the excess ligand or free amino acid as controls. Circular dichroism studies were conducted in order to monitor for putative conformational changes induced by the chemical treatments, confirming that the losses in binding activity of the lectin upon modification of histidine and tryptophan were not due to changes in quarternary structure, as the circular dichroism spectra of the modified protein remained almost unchanged when compared to the native lectin.[32]

Studies carried out on S-type lectin mutants have identified stretches of amino acid sequences of single residues likely to participate in ligand binding. Alignment of galaptin sequences have shown that a number of conserved hydrophilic residues occur on the central and COOH-terminal region of these molecules. Site-directed mutagenesis carried out by Hirabayashi and Kasai[5] on these regions indicated that even homologous substitutions on His44, Asn46, Arg48, Asn61, Glu71, or Arg73 resulted in loss of binding activity, suggesting that these are residues important for carbohydrate binding, whereas the conserved residues Lys 63, Arg111, or Asp125 are not directly involved. Analysis of deletion, point, and frame shift mutants and size comparisons with C-type recognition domains by Abbott and Feizi[33] suggested that almost the whole polypeptide is required for a fully functional carbohydrate-binding, 14-kDa protein and the presence of Trp on the binding site was predicted by exchanging tryptophan for phenylalanine or leucine.

The resolution of the structure of the bovine spleen galaptin-*N*-acetyl lactosamine complex clearly indicated that a number of residues interact directly with the ligand while others interact with each other to stabilize the structure of the binding site.[35]

It also confirmed the position and orientation of the hydroxyls on the disaccharide that interact with the protein-binding site. The OH on C4 of galactose interacts with Arg48 and His44, whereas the OH on C6 interacts with Asn61. In addition, both OH are restricted by the position of Trp68, and hence no modifications on those positions are allowed, as opposed to OH on C2 and C3 that are almost unrestricted. The interactions with the OH on C3 of the GlcNAc residue occur through Arg73, Arg48, and Glu71, confirming our earlier observations. Other residues, such as Arg48, Asp54, Arg73, and Glu71 interact with each other through a network that provides the correct architecture of the binding site. All residues interacting directly with the ligand or indirectly stabilizing the conformation of the binding site are conserved. S-type lectins from different vertebrate species are now crystallized in our laboratory, and the resolution of these structures will allow a full understanding of the architecture, function, and evolution of their carbohydrate-binding sites.[35]

THE BIOLOGICAL ROLES OF ANIMAL LECTINS

At the present time strong experimental evidence has been obtained in support of a wide variety of physiological roles for lectins present in prokaryotes, invertebrates, and vertebrates. Nevertheless, the biological functions of animal lectins are still far from being elucidated. It becomes clear, however, that within a considerable functional diversity self/non-self recognition is a biological property inherent to animal lectins. Their potential ligands, simple or complex carbohydrates, occur in all living cells and in biological fluids, suggesting that protein–carbohydrate interactions constitute a basic phenomenon common to all organisms. In this regard, evidence has accumulated in support of the current idea that different biological functions, exogenous, and endogenous, are to be found for the different lectin–carbohydrate ligand systems. Animal lectins would participate in humoral and cellular defense, somatic cell adhesion, fertilization, tissue reorganization, larval settlement, induction of the metamorphosis, and so forth, suggesting a multiplicity of biological roles that correlates with the increasing complexity of the organizational levels. S-Type lectins are developmentally regulated and endogenous cell–cell and cell–intercellular matrix interactions as well as a variety of regulatory functions have been proposed for this family of molecules. C-Type lectins represent a higher level of functional complexity due to the presence of multiple domains with distinct recognition and effector roles in these mosaic molecules.

Endogenous Functions of Animal Lectins

Zhou and Cummings[36] have shown that S-type 14-kDa lectins bind preferentially to poly-*N*-acetyllactosamine sequences. The natural ligand for these lectins appears to be laminin, as shown by the binding of the lectin to glycopeptides and glycoproteins from mammalian origin, and its function would consist in promoting the cell adhesion. In stimulated macrophages the Mac-2 antigen is a 29- to 35-kDa, laminin-binding lectin.[37] Hinek *et al.*[17] showed that the elastin receptor complex, present on the surface of chondroblasts producing elastin *in vitro,* contains a 67-kDa galactose-specific lectin that binds specifically to elastin and is important in the elastic fiber assembly. Binding studies performed on glycolipids and neoglycolipids suggest that in addition to glycoproteins, the sugar chains of glycolipids may be potential natural ligands for galaptins outside the intracellular environment.[38]

Studies on myoblast cell cultures indicated that the lectin is located on the cell periphery and ectocytosed,[39] suggesting an extracellular function for the lectin. Immunofluorescence studies using confocal microscopy on a myogenic cell line reveal a patchy distribution in differentiating myoblasts and peripheral distribution in myotubes that suggests that 14-kDa β-galactoside lectins may be involved in myoblast organization, leading to the myotube formation.[40] The mouse embryo fibroblast galaptin functions both as a reversible cytostatic factor and as a cell growth regulator, exerting control on the exit from G0 and at G2, in analogous fashion to interferon. Both functions are not related, however, to the sugar-binding properties and are related to other domains of the lectin molecule.[41] A considerable number of endogenous lectins have been described within the nervous system. The *Xenopus laevis* embryo endogenous galactoside-binding lectins and their carbohydrate receptors have been implicated in the control of neural morphogenesis.[42] Raz and Lotan[43] have postulated that tumor cell endogenous lectins, together with other cell surface adhesive molecules, such as fibronectin, laminin, and collagens, are involved in tumor dissemination in various ways. Interactions with glycoconjugates present on the host cells and the extracellular matrix would promote tumor metastasis mediated by homotypic and heterotypic aggregation, with emboli formation. Reciprocally, lectins on normal host cells would promote similar effects by binding to tumor cell surface carbohydrates. The adhesion of human ovarian carcinoma cells to extracellular matrix would be mediated by galaptins that interact with specific cell surface receptors, lamp-1 and lamp-2, containing galactosyl residues.[44] The addition of epidermal growth factor to fetal rat brain cell cultures causes great stimulation of the cerebellar-soluble lectin CSL biosynthesis, suggesting that EGF-related factors in the brain regulate the expression of the endogenous lectin and thus modulate brain ontogeny.[45]

The presence of glycoproteins in the nucleus compartment has been known for almost two decades, although the mechanism for their delivery to that compartment is unknown. S-type lectins have been discovered recently in the animal cell nucleus,[46] and glycoprotein–lectin interactions have been hypothesized to modulate numerous functions such as DNA duplication and transcription as well as post-transcriptional processes.[47]

Cell adhesion properties of C-type lectins that are independent of the sugar-binding site have been shown for the C-type lectin from the sea urchin *Anthocidaris crassispina* that binds to mucin type oligosaccharides.[10] Echinoidin contains the tripeptide Arg-Gly-Asp, common to fibronectin, vitronectin, and other cell adhesion molecules, and this stretch of amino acid sequence mediates specifically the *in vitro* adhesion of cancer cells. Parallel studies on the frog egg β-galactoside-specific S-type lectin from the frog *Rana catesbeiana* indicates that this lectin promotes cell adhesion through the sugar-binding site.[48]

Non-Self Recognition Functions of Animal Lectins

Regarding the participation of animal lectins as mediators of internal defense mechanisms, the main focus of this review, there is ample evidence for their functions as agglutinins, opsonins, complement activators, and cell-associated recognition molecules.

Soluble C-type lectins that exhibit collagenous regions linked to the COOH-terminal sugar-binding domain, which are found in body fluids of mammals, have been recently classified as "collectins"[49] and include the mannose-binding protein[50,3] and conglutinin[51] from serum and the pulmonary surfactant.[52] The serum mannose-binding protein, an acute-phase response reactant[53] synthesized in the liver, is the best characterized member of this group. These lectins are multimeric proteins (~350 kDa) with a subunit size of about 32 kDa, and their function is related to internal defense against viruses[53,54] and bacteria.[55] They not only bind to mannose but also to other monosaccharides as well, including fucose, ManNAc, GlcNAc, glucose, galactose, and others.[49] The gene structure of its subunit indicates that it is a mosaic polypeptide constituted by a signal peptide, a cysteine-rich domain, a collagen-like domain, a neck region, and a carbohydrate-binding domain; and the presence of stress–response promoter regions suggest that the serum levels are increased upon contact with pathogens.[56]

Several humoral lectins such as the mannose-binding protein,[57] C-reactive protein[25] and the LPS serum-binding protein[58] not only exhibit opsonic activity for microorganisms but also kill bacteria by activating complement. The mouse serum bactericidal factor RaRF binds GlcNAc and LPS in a Ca-dependent manner and is active against *Salmonella* species.[58] The mannose-binding protein activates complement through the classical pathway and hence is similar in function and structure to the C1q, a constituent of complement that binds to immunocomplexes to initiate the activation of the classical pathway.[57] Hence, the mannose-binding protein produces complement-dependent lysis of bacteria and also has complement independent direct opsonic effect.[59] In children, mutational defects of this lectin, resulting in low serum levels, are the cause for predisposition to severe recurrent infections, diarrhea, and failure to thrive.[60]

Serum amyloid P component is a Ca-dependent mammalian plasma lectin specific for the cyclic, 4,6-pyruvate acetal of galactose.[24] SAP is a member of the pentraxin family, a highly conserved group of molecules that includes C-reactive protein and has been shown to bind specifically, through its carbohydrate-binding site, to a variety of pathogenic bacteria including *Klebsiella,* group A *Streptococcus,* and *Xanthomonas.*

IL-2 binds with high affinity to the glycoproteins uromodulin and OVA and to yeast mannans.[30] It is specifically inhibited by diacetyl chitobiose and high-mannose glycopeptides and recognizes the core structure of N-linked oligosaccharides and together with IL-1α[29] and TNF,[31] that have been shown to exhibit carbohydrate binding activity towards N-linked oligosaccharides, should be considered mammalian lectins.[30] IL-2 shares 27% of the primary structure of the carbohydrate recognition domain with the mannose binding receptor, a humoral lectin that is an acute-phase component that functions as an opsonin.[30]

Within the invertebrates and protochordates the participation of lectins as mediators of immune mechanisms is supported by extensive experimental evidence for their roles as agglutinins, opsonins, and cell-associated recognition molecules. The mechanism through which the plasma lectins would function as defense molecules has not been experimentally demonstrated, but it appears reasonable to think that once the pathogenic bacteria have reached the internal cavities in the invertebrate they would be immediately agglutinated by the humoral lectins. Agglutination would immobilize the pathogen and would prevent its dispersion and penetration of other tissues. Numerous reports indicate that invertebrate lectins actually agglutinate *in vitro* potentially pathogenic bacteria. Purified plasma lectins from the blue crab *Callinectes sapidus* agglutinated 10 of 11 isolates of *Vibrio parahaemolyticus* from several sources. Interestingly enough, the nonrecognized *Vibrio* strain has been isolated from *Callinectes* hemolymph (Cassels *et al.,* this volume). The *in vitro* results of bacterial agglutination have been reflected *in vivo* studies of clearance of bacteria or model ligands such as erythrocytes and glycoconjugates. Agglutination probably induces the release of bacterial products that may act as chemotactic factors increasing migration and degranulation of hemocytes, which in turn stimulates phagocytosis or killing of pathogens by the effector molecules released by the hemocytes. The agglutinated and immobilized pathogens may be more susceptible to the local action of those effector molecules, which would make the pathogens "more foreign" to the hemocyte and facilitate their phagocytosis. Furthermore, it is clear that agglutination of the pathogens increases the size of the foreign particle, favoring its phagocytosis by the hemocytes (free phagocytes) and results in multiple pathogens being endocytosed, and thus cleared from circulation, in a single phagocytosis event. It also increases the chances of the particle being randomly trapped by the sessile phagocytes lining the hemal sinuses. Opsonic functions have been described for invertebrate humoral lectins. Although hemocytes from the oyster *Crassostrea gigas* can phagocytose in saline in the absence of lectin, suggesting that the hemocytes have the appropriate recognition molecules on the surface, it has been shown that previous exposure of the particle to the lectin enhances its phagocytosis and clearance rates.[56] Similar results have been obtained by Renwrantz[62] on the mussel *Mytilus.* With respect to the molecular mechanisms by which lectins could act as opsonins, the binding of the lectin to the non-self substance would occur through its carbohydrate-binding sites, as discussed above in relation to their agglutination properties, but how the lectin–ligand complex binds to the surface of the phagocyte has not been elucidated yet. A distinct possibility with considerable support from the experimental evidence is that the binding to ligand (and/or divalent cations) produces conformational changes in the lectin molecule[63] that may expose cryptic carbohydrate-binding sites with affinity for sugars on the hemocyte surface, or on certain regions of the lectin molecule that are detected as "foreign" by hemocyte receptors, this "foreignness" being a reflection of changes in charge or more likely by increased hydrophobicity.

Are invertebrate lectins inducible? The occurrence of adaptive defense responses in invertebrates has been demonstrated. (See Karp, this volume.) Results obtained in the attempt of inducing humoral lectins (agglutinins/opsonins) by specific challenges such as injections of bacteria or erythrocytes, or nonspecific stimuli such as trauma, have been controversial. In general, the responses have been weak and short-lived. Two models, however, have yielded relatively clear-cut results: the induction of lectins in the larvae of the flesh fly *Sarcophaga peregrina* by injury or injection of erythrocytes[8] and in the oyster *Crassostrea gigas,* a filter feeder, by exposing the molluscs to an environment with a high concentration of bacteria.[61]

Different populations of invertebrate hemocytes participate in a series of defense-related reactions, including phagocytosis, encapsulation, hemolymph clotting, and release of phenoloxidase and melanization in response to challenge with "non-self" components. However, the molecular basis of those cellular mechanisms of recognition are not yet understood. The presence of carbohydrate-binding proteins on the cell membrane of invertebrate hemocytes has been demonstrated by several groups including our laboratory. Our research on *Crassostrea virginica* hemocyte and serum lectins showed that the hemocyte membrane exhibited a lectin that shared the carbohydrate specificity of a fraction of the oyster plasma lectins.[64] *In vivo* evidence for the participation of serum lectins and possibly hemocyte-associated lectins in invertebrate cellular responses to non-self was obtained in the cockroach *Periplaneta americana*: Glycoproteins that were the most effective inhibitors for the purified plasma lectins, when injected in solution or bound to Sepharose beads in the hemocoele of the insect, produced a cellular response, nodulation, and encapsulation significantly higher than for glycoproteins that behaved as poor inhibitors for the purified lectins.[65]

Self/Non-Self Recognition by **Clavelina picta** *Lectins*

Clavelina picta lectins are present in plasma but also on the surface of the hemocytes. Some of these lectins are mosaic molecules: CPL-I exhibits a peptide that shows high homology to human proclotting factor VIIIc, an acute-phase reactant synthesized by the liver[66]; CPL-III is highly cross-reactive with mammalian CRP and SAP and binds to similar ligands (see TABLE 2).[67] *Clavelina* lectins recognize a number of bacteria from the environment and those associated with the colony, but also bind to intact sulfated glycans from the colony tunic.[67] On the basis of our experimental evidence, we have proposed that these lectins participate in non-self recognition and defense by agglutinating and opsonizing potentially pathogenic

TABLE 2. Amino Acid Sequences of Selected Peptides from Tunicate and Horseshoe Crab Lectins and Mammalian Acute-Phase Reactants[a]

Protein	Amino Acid Residue/Number												
	7											18	
DCL-I	*S*	*R*	V	G	N	*V*	*F*	F	*K*	L	F	*D*	
	5											16	
CRP	*S*	*R*	*K*	A	*F*	*V*	*F*	*P*	*K*	*E*	*S*	*D*	
SAP	*S*	G	*K*	*V*	*F*	*V*	*F*	*P*	R	*E*	*S*	V	
	7											18	
LIM	T	S	*K*	*V*	K	F	P	*P*	S	S	*S*	P	
CPL-I	X	*F*	X	*D*	*D*	*N*	*S*	*P*	*S*	*F*	*I*	*Q*	*I*
F VIIIc	R	*F*	D	*D*	*D*	*N*	*S*	*P*	*S*	*F*	*I*	*Q*	*I*

[a]DCL-I: *Didemnum candidum* lectin I; CRP: C-reactive protein; SAP: serum amyloid P; LIM: *Limulus polyphemus* lectin; CPL-I: *Clavelina picta* lectin I; FVIIIc: human proclotting factor VIIIc.

bacteria from the environment, as well as in self recognition and wound repair by binding to L-fucose or L-galactose from the nonreducing terminal residues from the tunic glycan exposed to the plasma and hemocytes upon damage of the body wall (Vasta, unpublished). In this aspect *Clavelina* lectins are similar to tetranectin, a mammalian lectin that binds to sulfated glycans and to components of the clotting cascade.[68]

EVOLUTION OF ANIMAL LECTINS

Sequences homologous to C-type lectin domains have been identified in a wide variety of organisms, from mammals to prokaryotes. Like many pathogenic bacterial species, *Bordetella pertussis* attaches to cells through adhesins, large nonfimbrial multimeric agglutinins with multiple binding domains. *B. pertussis* toxin is one of the two main adhesins and a major virulence factor, constituted by two types of subunit A and B. The subunit B carries the recognition domains, is responsible for bacterial adhesion and delivery of the toxic subunit S1, and causes mitogenic activation of T lymphocytes. The recognition domains carry two distinct specificities, for sialylated and nonsialylated carbohydrate ligands. The most interesting aspect of these *B. pertussis* toxin recognition domains, as well as those from the *E. coli* fimbrial protein PapG, is that they are homologous to eukaryotic carbohydrate recognition lectin C-type lectin domains. These, however, are not related to any lectin from plant origin.[69] The anticoagulant protein factor IX/factor X-binding protein consists of two chains that share 25 to 37% of the sequence of the C-type carbohydrate recognition domain at the COOH-terminal region of several C-type lectins, including the proteoglycan core protein, asialoglycoprotein liver receptors, tetranectin, lymphocyte Fcε receptor for IgE, and pancreatic stone protein. It shows considerable homology to the b subunit of factor XIII.[70] Tetranectin binds to the kringle 4 part of plasminogen[71] as well as to sulfated polysaccharides.[68] The cartilage proteoglycan core protein has lectin-like activity with fucose and galactose being the preferred ligands. The primary structure shows that a segment of approximately 130 residues near the COOH-terminus is highly homologous to the carbohydrate-binding domain from the chicken hepatic lectin and other C-type vertebrate lectins.

Drickamer and collaborators[16] have proposed an order of events that may have occurred during the evolution of lectins containing the C-type CRDs: A primordial CRD gene would have suffered duplication, insertion, and deletion events giving origin to two types of divergent precursor CRD genes, containing or lacking introns. Subsequent events of exon shuffling would have resulted in associations with various domains, generating the precursors to the four CRD groups mentioned above. Through gene duplication and divergence within each of the four groups, the carbohydrate specificities and effector functions particular to each lectin would have evolved. A final process of convergent evolution would have resulted in similar binding properties for some CRDs.

The earliest ancestor of the S-type lectins has been found in the sponge *Geodia cydonium*.[12] The lectins LECT-1 (subunit M_r 15,000) and -2 (subunit M_r 13,000 and 16,000) have been classified in the L-14 subgroup, and the former may consist of a disulfide-bonded dimer. Although with a common origin, these two lectins may have evolved separately with a high conservation of the binding domain.[12] A 32-kDa, S-type lectin from the nematode *Caenorhabditis elegans*[11] was shown to share 25 to 30% sequence identity with the vertebrate members of that group. Its primary structure reveals two tandem repeated domains homologous to the 14-kDa galaptin from vertebrates and suggests the presence of two binding sites per polypeptide. A smaller galaptin (16 kDa) was also isolated from *C. elegans* with an amino acid composition very similar to the 32-kDa species, and it was proposed that it may originate in limited proteolysis of the 32-kDa product.[11]

In addition to several Ca^{2+}-dependent lectins specific for L-fuc, we have identified and isolated three S-type lectins from the protochordate *Clavelina picta* of 14.5, 16, and 35-kDa.[13]

Hirabayashi and Kasai[5] have analyzed sequence alignments for S-type lectins from several vertebrate and invertebrate species and calculated the divergence times. These agree with the divergence times for the species under consideration, nematode, sponge, conger eel, *Xenopus,* chicken, rat, and human. These authors propose a common ancestor to all S-type lectins and calculate that sponge and nematode S-type lectins may have diverged 800 and 680 million years ago. Chimera type lectins may have diverged between 100 and 300 million years ago.

Some lectins appear as intermediate forms between S- and C-type lectins. The lectin from the rattlesnake *Crotalus atrox* venom is related to the C-type lectins but exhibits a rather simple structure and was proposed as a primitive member of this superfamily that may constitute an intermediate form between this and the 14-kDa S-type lectins.[72] The primary structure indicates that this lectin is related to both C-type lectins from vertebrates and invertebrates.

For some examples, it is likely that convergent evolution is to be responsible for the similarities in structure and/or function. CD22 is a B lymphocyte cell surface glycoprotein and has been defined as a sialic acid–binding lectin.[73] CD22 exhibits high homology to adhesion molecules such as V-CAM and N-CAM, and hence is considered a member of the immunoglobulin gene superfamily and shows no homology even to those selectins that exhibit similar carbohydrate specificity. Two isoforms have been identified: CD22α is the smaller and is constituted by five Ig-like extracellular domains, whereas CD22β exhibits two additional extracellular domains and recognizes multiple sialoglycoproteins, preferentially with sialyl-oligosaccharides with α2–6 linkages, on T cells. Side chains of sialic acids are required for binding of CD22, in contrast to E- and L-selectins for which mild periodate oxidation of the sialoglycoprotein ligand enhances binding. The three-dimensional structure of the bovine S-type lectin overlaps surprisingly well with those from plant lectins such as ConA, pea lectin, and *Erythrina corallodendron* lectin.[35] These proteins, however, share a very small number of identities in their primary structure, and their similarities in folding are most likely to result from structural constraints for the binding to carbohydrate ligands.

Finally, it appears that throughout evolution, some lectins may have conserved structural features but may have lost or changed their functional properties. Some molecules carrying C-type lectin domains have lost the ability to bind to carbohydrates: The type II anti-freeze protein from herring, smelt, and sea raven are C-type lectins, sharing related primary structure and folding structure and Ca-binding sites. They bind to ice crystals and require Ca for activity but do not bind carbohydrates any longer.[74] The Charcot-Leyden crystal protein, a lysophospholipase associated with inflammatory processes shows considerable similarity in primary structure to S-type lectins,[75] and its function consists in preventing the effect of cytotoxic activity of lysophospholipids resulting from the activity of phospholipase A2.

CONCLUSIONS

As knowledge on protein and gene sequences and the three-dimensional structure of animal lectins accumulated in recent years, we have begun to understand not only their complex molecular and gene organization but also their evolution and multiple biological roles. The analysis of the primary structure of a number of lectins has shown that various functional domains can be identified, including those related to collagens, growth factors, calmodulin, complement regulatory proteins, cell-adhesion proteins and others, and common to all, the carbohydrate-binding domains comprising a set of highly conserved residues that constitute a "framework" supporting stretches of variable residues.

The participation of lectins as direct and indirect mediators of immune mechanisms include their functions as agglutinins, opsonins, complement activators, and cell-associated recognition molecules, among others. Most of these functions are associated with the acute-phase response, a complex network of recognition and effector mechanisms, apparently lacking the exquisite specificity of the immunoglobulin-mediated immune response. However, the acute-phase response may be the most ancient internal defense system, widespread from the invertebrate taxa to the mammals and despite its relative lack of specificity and memory, it appears to have "done the job" throughout their evolution.

SUMMARY

In recent years, the significant contributions from molecular research studies on animal lectins have elucidated structural aspects and provided clues not only to their evolution but also to their multiple biological functions. The experimental evidence has suggested that distinct, and probably unrelated, groups of molecules are included under the term "lectin." Within the invertebrate taxa, major groups of lectins can be identified: One group would include lectins that show significant homology to membrane-integrated or soluble vertebrate C-type lectins. The second would include those β-galactosyl-specific lectins homologous to the S-type vertebrate lectins. The third group would be constituted by lectins that show homology to vertebrate pentraxins that exhibit lectin-like properties, such as C-reactive protein and serum amyloid P. Finally, there are examples that do not exhibit similarities to any of the aforementioned categories. Moreover, the vast majority of invertebrate lectins described so far cannot yet be placed in one or another group because of the lack of information regarding their primary structure. (See TABLE 1.) Animal lectins do not express a recombinatorial diversity like that of antibodies, but a limited diversity in recognition capabilities would be accomplished by the occurrence of multiple lectins with distinct specificities, the presence of more than one binding site, specific for different carbohydrates in a single molecule, and by certain "flexibility" of the binding sites that would allow the recognition of a range of structurally related carbohydrates. In order to identify the lectins' "natural" ligands, we have investigated the interactions between those proteins and the putative endogenous or exogenous glycosylated substances or cells that may be relevant to their biological function. Results from these studies, together with information on the biochemical properties of invertebrate and vertebrate lectins, including their structural relationships with other vertebrate recognition molecules, are discussed.

REFERENCES

1. VASTA, G. R. 1991. The multiple biological roles of invertebrate lectins: Their participation in non-self recognition mechanisms. *In* Phylogenesis of Immune Functions. N. Cohen & G. W. Warr, Eds.: 73–101. CRC Press. Boca Raton, FL.
2. BARONDES, S. H. 1984. Soluble lectins: A new class of extracellular proteins. Science **223:** 1259–1264.
3. DRICKAMER, K. 1988. Two distinct classes of carbohydrate recognition domains in animal lectins. J. Biol. Chem. **263:** 9557–9560.
4. HARRISON, F. L. 1991. Soluble vertebrate lectins: Ubiquitous but inscrutable proteins. J. Cell Sci. **100:** 9–14.
5. HIRABAYASHI, J. & K.-i. KASAI. 1993. The family of metazoan metal-independent β-galactoside-binding lectins: Structure, function and molecular evolution. Glycobiology **3:** 297–304.
6. VASTA, G. R. 1990. Invertebrate lectins, C-reactive and serum amyloid. Structural relationships and evolution. *In* Defense Molecules. J. J. Marchalonis & C. L. Reinisch, Eds.: 183–199. Wiley-Liss. New York.
7. VASTA, G. R. 1992. Invertebrate lectins: Distribution, synthesis, molecular biology and function. *In* Glycoconjugates: Composition, Structure and Function. H. J. Allen & E. C. Kisailus, Eds.: 593–634. Marcel Dekker. New York.
8. TAKAHASHI, H., H. KOMANO, N. KAWAGUCHI, N. KITAMURA, S. NAKANISHI & S. NATORI. 1985. Cloning and sequencing of cDNA of *Sarcophaga peregrina* humoral lectin induced on injury of the body wall. J. Biol. Chem. **260:** 12228–12233.
9. MURAMOTO, K., K. OGATA & H. KAMIYA. 1985. Comparison of the multiple agglutinins of the acorn barnacle *Megabalanus rosa.* Agr. Biol. Chem. **49:** 85–93.
10. GIGA, Y., A. IKAI & K. TAKAHASHI. 1987. The complete amino acid sequence of echinoidin, a lectin from the coelomic fluid of the sea urchin *Anthocidaris crassispina.* J. Biol. Chem. **262:** 6197–6203.
11. HIRABAYASHI, J., M. SATOH & K.-i. KASAI. 1992. Evidence that *Caenorhabditis elegans* 32-kDa β-galactoside-binding protein is homologous to vertebrate β-galactoside-binding lectins. J. Biol. Chem. **267:** 15485–15490.
12. PFEIFER, K., M. HAASEMAN, V. GAMULIN, H. BRETTING, F. FAHRENHOLZ & W. E. G. MÜLLER. 1993. S-Type lectins occur also in invertebrates: High conservation of the carbohydrate recognition domain in the lectin genes from the marine sponge *Geodia cydonium.* Glycobiology **3(2):** 179–184.
13. AHMED, H. & G. R. VASTA. 1993. Unpublished observation.
14. NGUYEN, N. Y., A. SUZUKI, K. A. BOYKINS & T.-Y. LIU. 1986. The amino acid sequence of *Limulus* C-reactive protein. Evidence of polymorphism. J. Biol. Chem. **261:** 10456–10465.
15. VASTA, G. R., J. HUNT, J. J. MARCHALONIS & W. W. FISH. 1986. Galactosyl-binding lectins from the tunicate *Didemnum candidum.* Purification and physiocochemical characterization. J. Biol. Chem. **261:** 9174–9181.
16. BEZOUSKA, K., G. V. CRICHLOW, J. M. ROSE, M. E. TAYLOR & K. DRICKAMER. 1991. Evolutionary conservation of intron position in a subfamily of genes encoding carbohydrate-recognition domains. J. Biol. Chem. **1991:** 11604–11609.

17. Hinek, A., D. S. Wrenn, R. P. Mecham & S. H. Barondes. 1988. The elastin receptor: A galactoside-binding protein. Science **239:** 1539–1541.
18. Cherayl, B. J., S. J. Weiner & S. Pillai. 1989. The Mac-2 antigen is a galactoside specific lectin that binds IgE. J. Exp. Med. **170:** 1959–1972.
19. Albrandt, K., N. K. Orida & F.-T. Liu. 1987. An IgE-binding protein with a distinctive sequence and homology with an IgG receptor. Proc. Natl. Acad. Sci. USA **84:** 6859–6863.
20. Oda, Y., J. Herrman, M. A. Gitt, C. W. Turck, A. L. Burlingame, S. H. Barondes & H. Leffler. 1993. Soluble lactose-binding lectin from rat intestine with two different carbohydrate-binding domains in the same peptide chain. J. Biol. Chem. **268:** 5929–5939.
21. Mecham, R. P., L. Whitehouse, M. Hay & M. Sheetz. 1991. Ligand affinity of the 67-kD elastin/laminin binding protein is modulated by the protein's lectin domain: Visualization of elastin/laminin-receptor complexes with gold-tagged ligands. J. Cell Biol. **113:** 187–194.
22. Uhlenbruck, G., J. Solter, E. Janssen & H. Haupt. 1982. Anti-galactan and anti-haemocyanin specificity of CRP. Ann. N.Y. Acad. Sci. **389:** 476–479.
23. Baltz, M. L., F. C. de Beer, A. Feinstein, E. A. Munn, C. P. Milstein, T. C. Fletcher, J. F. March, J. Taylor, C. Bruton, J. R. Clarfnp, A. J. S. Davies & M. B. Pepys. 1982. Phylogenetic aspects of C-reactive protein and related proteins. Ann. N.Y. Acad. Sci. **389:** 49–75.
24. Hind, C. R. K., P. M. Collins, M. L. Baltz & M. B. Pepsy. 1985. Human serum amyloid P component, a circulating lectin with specificity for the cyclic 4,6-pyruvate acetal of galactose. Biochem. J. **225:** 107–111.
25. Kilpatrik, J. M. & J. E. Volanakis. 1985. Opsonic properties of C-reactive protein. Stimulation by phorbol myristate acetate enables human neutrophils to phagocytize C-reactive protein coated cells. J. Immunol. **134:** 3364–3370.
26. Buck, F., C. Luth, K. Strupat & H. Bretting. 1992. Comparative investigations on the amino-acid sequence of different isolectins from the sponge *Axinella polypoides.* Glycobiology **3:** 179–184.
27. Alliegro, M. C., C. A. Burdsal & D. R. McClay. 1990. *In vitro* biological activities of echinonectin. Biochemistry **29:** 2135–2141.
28. Veno, P. A., M. A. Strumski & W. H. Kinsey. 1990. Purification and characterization of echinonectin, a carbohydrate-binding protein from sea urchin eggs. Dev. Growth Differ. **32:** 315–319.
29. Muchmore, A. V. & J. M. Decker. 1987. Evidence that recombinant IL-1a exhibits lectin-like specificity and binds homogenous uromodulin via N-linked oligosaccharides. J. Immunol. **138:** 2541.
30. Sherblom, A. P., N. Sathyamoorthy, J. M. Decker & A. V. Muchmore. 1989. IL-2, a lectin with specificity for high mannose glycopeptides. J. Immunol. **143:** 939–944.
31. Sherblom, A. P., J. M. Decker & A. V. Muchmore. 1988. The lectin-like interaction between recombinant tumor necrosis factor and uromodulin. J. Biol. Chem. **263:** 5418–5424.
32. Ahmed H. & G. R. Vasta. Unpublished observation.
33. Abott, M. W. & T. Feizi. 1991. Soluble 14-kDa β-galactoside-specific bovine lectin. J. Biol. Chem. **266:** 5552–5557.
34. Powell, J. T. 1985. Chemical modification of arginine residues of lung galaptin and fibronectin. Biochem. J. **232:** 919–922.
35. Liao, D.-I., G. Kapadia, H. Ahmed , G. R. Vasta & O. Herzberg. 1994. Structure of S-lectin, a developmentally regulated vertebrate β-galactoside binding protein. Proc. Natl. Acad. Sci. USA. In press.
36. Zhou, Q. & R. D. Cummings. 1993. L-14 Lectin recognition of laminin and its promotion on *in vitro* cell adhesion. Arch. Biochem. Biophys. **300:** 6–17.
37. Woo, H.-J., L. M. Shaw, J. M. Messier & A. M. Mercurio. 1990. The major non-integrin laminin binding protein to macrophages is identical to carbohydrate binding protein 35 (Mac-2). J. Biol. Chem. **265:** 7097–7099.
38. Solomon, J. C., M. S. Stoll, P. Penfold, W. M. Abbott, R. A. Childs, P. Hanfland &

T. FEIZI. 1991. Studies of the binding specificity of the soluble 14,000 dalton bovine heart muscle lectin using immobilised glycolipids and neoglycolypids. Carbchyd. Res. **213:** 293–307.
39. COOPER, D. N. W. & S. H. BARONDES. 1990. Evidence for export of a muscle lectin from cytosol to extracellular matrix and for a novel secretory mechanism. J. Cell Biol. **110:** 1681–1691.
40. HARRISON, F. L. & T. J. G. WILSON. 1992. The 14-kDa β-galactoside binding lectin in myoblast and myotube cultures: Localization by confocal microscopy. J. Cell Sci. **101:** 635–646.
41. WELLS, B. & L. MALLUCCI. 1991. Identification of an autocrine negative growth factor: Mouse β-galactoside-binding protein is a cytostatic factor and cell growth regulator. Cell **64:** 91–97.
42. MILOS, N. C., Y. MA & Y. N. FRUNCHAK. 1989. Involvement of endogenous galactoside-binding lectin of *Xenopus laevis* in pattern formation to *Xenopus* neurites in vitro. Cell Diff. Dev. **28:** 203–210.
43. RAZ, A. & R. LOTAN. 1987. Endogenous galactoside-binding lectins: A new class of functional tumor cell surface molecules related to metastasis. Cancer Metastasis Rev. **6:** 433–452.
44. SKRINCOSKY, D. M., H. J. ALLEN & R. J. BERNACKI. 1993. Galaptin-mediated adhesion of human ovarian carcinoma A121 cells and detection of cellular galaptin-binding glycoproteins. Cancer Res. **53:** 2667–2675.
45. TENOT, M., S. KUCHLER, J.-P. ZANETTA, G. VINCENDON & P. HONEGGER. 1989. Epidermal growth factor enhances the expression of an endogenous lectin in aggregating fetal brain cell cultures. J. Neurochem. **53:** 1435–1441.
46. SEVE, A.-P., J. HUBERT, D. BOUVIER, M. BOUTIELLE, C. MAINTIER & M. MONSIGNY. 1985. Detection of sugar-binding proteins in membrane-depleted nuclei. Exp. Cell Res. **157:** 533–538.
47. HUBERT, J., A.-P. SEVE, P. FACY & M. MONSIGNY. 1989. Are nuclear lectins and nuclear glycoproteins involved in the modulation of nuclear functions? Cell Diff. Dev. **27:** 69–81.
48. OZEKI, Y., T. MATSUI & K. TITANI. 1991. Cell adhesive activity of two animal lectins through different recognition mechanisms. FEBS **289:** 145–147.
49. HOLMSKOV, U., P. HOLT, K. B. M. REID, A. C. WILLIS, B. TEISNER & J. C. JENSENIUS. 1993. Purification and characterization of bovine mannan-binding protein. Glycobiology **3(2):** 147–153.
50. DRICKAMER, K., M. S. DORDAL & L. REYNOLDS. 1986. Mannose-binding proteins isolated from rat liver contain carbohydrate-recognition domains linked to collagenous tails. Complete primary structures and homology with pulmonary surfactant apoproteins. J. Biol. Chem. **261:** 6878–6887.
51. FRIIS-CHRISTIANSEN, P., S. THIEL, S. SVEHAG, R. DESSAU, P. SVENDSEN, O. ANDERSEN, S. B. LAURSEN & J. C. JENSENIUS. 1990. *In vivo* and *in vitro* antibacterial activity of conglutinin, a mammalian plasma lectin. Scan. J. Immun. **31:** 453–460.
52. LU, J., A. C. WILLIS & K. B. M. REID. 1992. Purification, characterization and cDNA cloning of human lung surfactant protein D. Biochem. J. **284:** 795–802.
53. EZEKOWITZ, R. A. B., L. E. DAY & G. A. HERMAN. 1988. A human mannose-binding protein is an acute-phase reactant that shares sequence homology with other vertebrate lectins. J. Exp. Med. **167:** 1034–1046.
54. ANDERS, W. M., C. A. HARTLEY & D. C. JACKSON. 1990. Bovine and mouse serum inhibitors of influenza A viruses are mannose-binding lectins. Proc. Natl. Acad. Sci. USA **87:** 4485–4489.
55. KAWASAKI, N., T. KAWASAKI & I. YAMASHINA. 1989. A serum lectin (mannan-binding protein) has complement-dependent bactericidal activity. J. Biochem. **106:** 483–489.
56. TAYLOR, M. E., P. M. BRICKELL, R. K. CRAIG & J. A. SUMMERFIELD. 1989. Structure and evolutionary origin of the gene encoding a human serum mannose-binding protein. Biochem. J. **262:** 763–771.
57. IKEDA, K., T. SANNOH, N. KAWASAKI, T. KAWASAKI & I. YAMASHINA. 1987. Serum lectin

with known structure activates complement through the classical pathway. J. Biol. Chem. **262:** 7451–7454.
58. KAWAKAMI, M., I. IHARA, A. SUZUKI & Y. HARADA. 1982. Properties of a new complement-dependent bactericidal factor for Ra chemotype *Salmonella* in sera of conventional and germ-free mice. J. Immunol. **129(5):** 2198–2201.
59. KUHLMAN, M., K. JOINER & A. B. EZEKOWITZ. 1989. The human mannose-binding protein functions as an opsonin. J. Exp. Med. **169:** 1733–1740.
60. SUPER, M., S. THIEL, J. LU, R. J. LEVINSKY & M. W. TURNER. 1989. Association of low levels of mannan-binding protein with a common defect of opsonisation. Lancet **25:** 1236–1239.
61. OLAFSEN, J. A. 1988. Role of lectins in invertebrate humoral defense. *In* Disease Processes in Marine Bivalve Molluscs. W. S. Fisher, Ed. Am. Fish. Soc. Vol. **18:** 189–205.
62. RENWRANTZ, L. & A. STAHMER. 1983. Opsonizing properties of an isolated hemolymph agglutinin and demonstration of lectin-like recognition molecules at the surface of hemocytes from *Mytilus edulis.* J. Comp. Physiol. **149:** 535–546.
63. MOHAN, S., D. T. DORAI, S. SRIMAL, B. K. BACHHAWAT & M. K. DAS. 1984. Circular dichroism studies on carcinoscorpin, the sialic acid binding lectin of horseshoe crab *Carcinoscorpius rotunda cauda.* Indian J. Biochem. Biophys. **21:** 151–154.
64. VASTA, G. R., T. C. CHENG & J. J. MARCHALONIS. 1984. A lectin on the hemocyte membrane of the oyster (*Crassostrea virginica*). Cell Immunol. **88:** 475–488.
65. LACKIE, A. M. & G. R. VASTA. 1988. The role of a glactosyl-binding lectin in the cellular immune response of the cockroach *Periplaneta americana* (Dictyoptera). Immunology **64:** 353–357.
66. VASTA, G. R. & J. POHL. Unpublished observation.
67. ELOLA, M. T. & G. R. VASTA. Unpublished observation.
68. CLEMMENSEN, I. 1989. Interaction of tetranectin with sulphated polysaccharides and trypan blue. Scan. J. Clin. Lab. Invest. **49:** 719–725.
69. SAUKKONEN, K., W. N. BURNETTE, V. L. MAR & H. R. MASURE. 1992. Pertussis toxin has eukaryotic-like carbohydrate recognition domain. Proc. Natl. Acad. Sci. **89:** 118–122.
70. ATODA, H., M. HYUGA & T. MORITA. 1991. The primary structure of coagulation factor IX/factor X-binding protein isolated from the venum of *Trimeresurus flavoviridis.* J. Biol. Chem. **266:** 14903–14911.
71. CLEMMENSEN, I., L. C. PETERSEN & C. KLUFT. 1986. Purification and characterization of a novel, oligomeric, plasminogen kringle 4 binding protein from human plasma: Tetranectin. Eur. J. Biochem. **156:** 327–333.
72. HIRABAYASHI, J., T. KUSUNOKI & K.-I. KASAI. 1991. Complete primary structure of a galactose-specific lectin from the venom of the rattlesnake *Crotalus atrox.* J. Biol. Chem. **266:** 2320–2326.
73. SGROI, D., A. VARKI, S. BRAESCH-ANDERSEN & I. STAMENKOVIC. 1993. CD22, a B cell-specific immunoglobulin superfamily member, is a sialic acid-binding lectin. J. Biol. Chem. **268:** 7011–7018.
74. EWART, K. V. & G. L. FLETCHER. 1993. Herring antifreeze protein: Primary structure and evidence for a C-type lectin evolutionary origin. Mol. Marine Biol. Biotechnol. **2:** 20–27.
75. ACKERMAN, S. J., S. E. CORRETTE, H. F. ROSENBERG, J. C. BENNETT, D. M. MASTRIANNI, A. NICHOLSON-WELLER, P. F. WELLER, D. T. CHIN & D. G. TENEN. 1993. Molecular cloning and characterization of human eosinophil Charcot-Leyden crystal protein (lysophospholipase). J. Immunol. **150:** 456–468.

Recognition Molecules and Immunoglobulin Domains in Invertebrates[a]

S. F. SCHLUTER,[b] J. SCHROEDER, E. WANG,
AND J. J. MARCHALONIS

Microbiology and Immunology
University of Arizona
College of Medicine
Tucson, Arizona 85724

INTRODUCTION

Many proteins are classified as belonging to the immunoglobulin (Ig) superfamily on the basis of containing one or more Ig-like domains.[1] The Ig domain is a segmental motif consisting of 55–75 amino acids contained within a disulfide bond and adopting a common structure comprising a fold forming a sandwich of two β-sheets. Ig domains are recognized by the presence of the disulfide bond, certain highly conserved amino acid residues, and by analysis of sequence similarity or homology[1] using the Dayoff scoring matrix.[2] Numerous molecules of the immune system, in addition to immunoglobulins, contain Ig domains. These include, for example, Thy-1, MHC, CD4, CD8, and poly Ig receptor.[1] Several proteins quite separate from the immune system, however, also contain Ig domains. These include such molecules as neural adhesion molecule (NCAM), myelin-associated glycoprotein, and carcinoembryonic antigen.[1] Most of the members of the Ig superfamily appear to be involved in recognition and signaling functions, with the Ig domains mediating either homophilic (self) recognition or non-self recognition (antigen binding by antibody and T-cell receptor).

Several molecules have been identified in invertebrates and shown by sequence analysis to contain Ig domains. However, these are not immunoglobulin proteins. For instance, the fasciclin II[3] and amalgam[4] molecules isolated from insects are homologs to N-CAM and function essentially the same way as they do in vertebrates in the development of the nervous system. Hemolin can be induced in the hemolymph of cecropia larvae by injection of a killed bacterial vaccine and also appears to be homologous to N-CAMs.[5] Proteins that are putatively homologs of Ig superfamily members Thy-1 and β2-microglobulin have been identified in various invertebrates[6–8] by serological cross-reactions, but definitive sequence evidence is not yet available.

The evolutionary relationship of the Ig superfamily members is unclear. The

[a]This work was supported in part by grants from the National Science Foundation (DCB 9106934) and the National Institutes of Health (GM 42437).

[b]Address for correspondence: Samuel F. Schluter, Ph.D., Microbiology and Immunology, College of Medicine, University of Arizona, Tucson, AZ 85724.

simplest idea is that these molecules evolved by gene duplication from a primordial gene coding for about 100 amino acids containing a single Ig domain. The N-CAMs certainly share a common ancestor that must have arisen before the divergence of the protostomes and deuterostomes. Therefore, the Ig domain must be equally ancient. All vertebrates from sharks (and probably cyclostomes) to man possess immunoglobulins that are structurally essentially the same.[9] In contrast, immunoglobulin molecules have not been found in invertebrates. Thus, the rearranging immunoglobulin system seems to have rapidly arisen at the time of the emergence of the chordates from the protochordates.[10] This may have involved the utilization of an unrelated molecule containing an Ig domain(s) such as, for instance, an N-CAM-like molecule.[11] Alternatively, the Ig system may have arisen independently and the similarity of the Ig domain is the result of convergent evolution imposed by the functional properties of the Ig fold.

The protochordates are a group of extant invertebrates whose ancestors are considered most likely to have given rise to the vertebrates.[12] To better understand immunoglobulin evolution, we have been trying to identify immunoglobulin-related molecules (i.e., possible proto-immunoglobulins) in this group of deuterostome invertebrates.

SEARCHING FOR IMMUNOGLOBULIN-RELATED MOLECULES IN PROTOCHORDATES

One of the conclusions resulting from our work on the Igs of lower vertebrates is that their antigenic and 3-D structures have been highly conserved.[13–15] Serological experiments showing antigenic cross-reactions between immunoglobulins of sharks and man directly demonstrate that certain antigenic determinants (structural motifs) have been phylogenetically conserved over more than 400 million years.[13] Such determinants would also be expected to be conserved on immunoglobulin-related molecules found in species phylogenetically ancestral to sharks. Our strategy, therefore, has been to screen the hemolymph of various tunicate species using antibody probes to these derminants.

We have developed two rabbit antisera reagents that best define conserved immunoglobulin determinants. One was raised against shark IgM immunoglobulin heavy chain and reacts with immunoglobulin of shark, amphibian, bird, rodent, and human origin.[13,15] The other was raised against a synthetic peptide corresponding to the J or joining segment of a T-cell receptor β chain. This reagent reacts not only with T-cell receptor β chains, but also with immunoglobulins from shark to man,[16] a result consistent with the conservation of peptide sequence in this region.

These reagents were used in immunoblot transfer analyses (Western blots) to analyze the hemolymph of various tunicate species. FIGURE 1 shows the two positive results obtained using the anti-μ chain sera. A single, strongly cross-reactive band of approximately 30 kDa is apparent in hemolymph from *Pyura hausteria* and *Boltenia ovipera.* Prebleed sera and sera raised against shark light chain is negative in these assays. The Boltenia molecule also reacts with the anti-Jβ peptide reagent.

The Boltenia shark μ-chain cross-reactive molecule (μCRM) was purified using Western blotting to monitor the purification at each step. Briefly, the supernatant

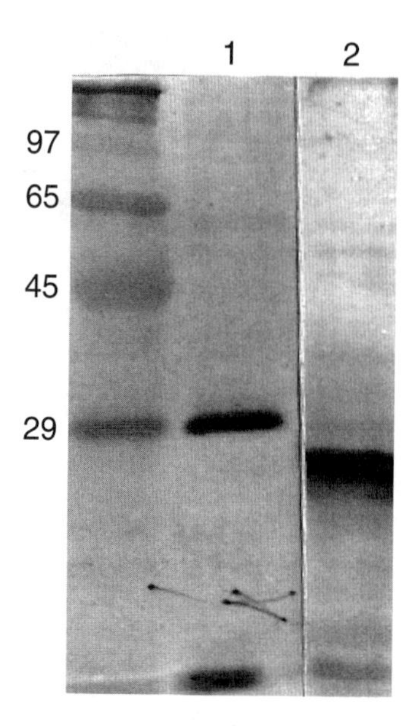

FIGURE 1. Western blot analysis of proteins in tunicate hemolymph reactive with rabbit antibodies to Galapagos shark IgM heavy chain. Western blotting was performed as described in Rosenshein *et al.*[15] MW standards in kDa are indicated. *Lanes:* (1) *Pyura hausteria* hemolymph; (2) *Boltenia ovipera* hemolymph.

from a 40% saturated ammonium sulfate solution was made to 60% saturated ammonium sulfate. The precipitate was resuspended in Tris-buffered saline (TBS) and dialyzed against 25 mM Tris, pH 8.0. This sample was subjected to HPLC ion-exchange chromatography on a MONO Q column (Pharmacia, Piscataway, NJ). The μCRMs eluted at approximately 50 mM NaCl, and the fractions were concentrated using Centricon 10 ultrafiltration concentrators (Amicon, Danvers, MA). The sample was pure at this point, and 200 to 300 μg of Boltenia μCRM was obtained from 300 ml of hemolymph. Analysis of the purified protein is shown in FIGURE 2. It is composed of nondisulfide-linked monomeric subunits of 28 kDa. The nonreduced protein has a slightly higher mobility than the reduced protein, indicating that it probably has internal disulfide bonds. The native μCRM appears to be a monomer, since it elutes as an approximately 30-kDa protein during chromatography on Sephadex G-200 (Pharmacia, Piscataway, NJ) in TBS (not shown).

Nearly all of the protein forms a single band at pH 7.5 in isoelectric focusing. Two to three minor bands are apparent at pH 4.5 to 5.5. This apparent microheterogeneity does not appear to be an artifact of the purification procedures, since 2-D analysis of whole hemolymph (visualized by Western blot) indicates the same result.

The peptide sequence was obtained using an Applied Biosystems Pulsed Liquid Phase Sequencer at the University of Arizona Biotechnology Center. The first 15 NH_2-terminal residues are I P G F/G Q L A P P A P I G/P F Q A (the shaded residues were minor components). To obtain internal sequences, the protein was fully reduced and alkylated in 6M GuHCl. After dialysis against digestion buffer (50 mM Tris, pH 8, 1 M urea, 20 mM methylamine), the protein was digested with sequencing-grade endoprotinease Lys-C (Boehringer Mannheim). The peptides were separated and manually collected by reversed-phase HPLC on a C4 column (Vydac, Hespera, CA). One peak yielded a reliable sequence: Q K D T N I M L S. Interestingly, the first

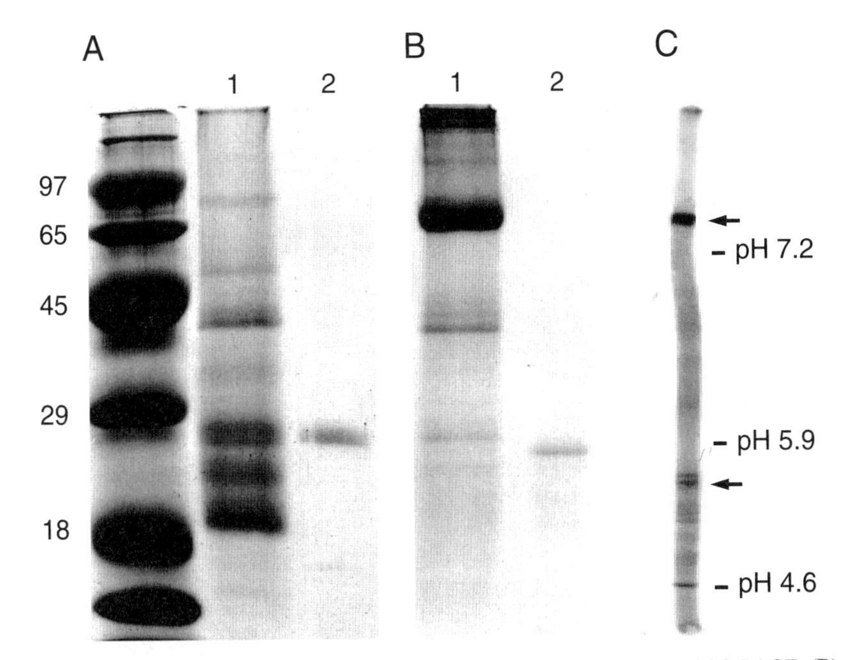

FIGURE 2. Analysis of purified Boltenia μCRM protein. **(A)** Nonreducing SDS-PAGE. **(B)** Reducing SDS-PAGE. *Lanes:* (1) *Boltenia* hemolymph; (2) 0.5 μg μCRM. **(C)** Isoelectric focusing of 1 μg of μCRM. The pI positions of IEF standards are indicated.

cycle for this peptide yielded five peaks, with Y being the major residue. The small amount of sequence heterogeneity seen with both peptides may be an explanation for the minor heterogeneity seen in IEF. Comparison of the NH_2-terminal peptide with shark μ chain sequence using the GCG (Sequence Analysis Software Package by Genetics Computer, Inc.) program BESTFIT produced an alignment that was 42% identical and 71% similar to a segment of one of the internal domains of the μ chain, *viz:*

```
peptide   P G F Q L A P P A I G . . . F Q A
          | . : . |   | | . | :       : : |
shark μ   P V I K L L P P S I E Q V L L E A
```

Although we cannot draw a definitive conclusion from such a small amount of sequence information, this result is encouraging.

Whether the tunicate molecules are homologous to Igs, as is suggested by the immunochemical and sequence results, will only be determined from complete sequence information. We are presently attempting to clone the gene using the polymerase chain reaction (PCR) with degenerate primers derived from the peptide sequences. The complete characterization of the μCRMs and the elucidation of their relationship to the Ig superfamily will likely prove important to our understanding of the origins of the vertebrate-type immune system.

CHARACTERIZATION OF LAMPREY EGG LECTIN

In addition to Ig-related molecules, we have been studying non-Ig-related molecules in primitive vertebrates and invertebrates. An example of such a molecule is a lectin isolated from the eggs of a primitive cyclostome vertebrate, the sea lamprey *(Petromyzon marinus)*. Lectins are carbohydrate-binding proteins whose functions are largely unknown, though it has been shown that they can play a role in self/non-self recognition in some invertebrates[17] and as cell homing and differentiation markers in vertebrates.[18] We have focused on the egg lectin since preliminary data suggests it is probably related to molecules containing C-type lectin domains and is found in vertebrates as well as in a protochordate.

The lectin is simply extracted from lamprey eggs by homogenization in TBS containing 10 mM $CaCl_2$ using a polytron tissue homogenizer (Brinkmann Instruments, Westbury, NY). The supernatant is cleared by centrifugation at 15,000 *g* for 30 min. Activity is measured by agglutination of 1% horse red blood cells (Cleveland Sci. Co., Bath, OH) in microtiter trays. The agglutinating activity is very high, being greater than 1/100,000. Ca^{2+} ions are required for activity. Thus, EDTA completely abrogates agglutinating activity, but this is restored by the addition of excess Ca^{2+} or dialysis against Ca^{2+} buffers. Agglutinating activity is inhibited by glycoproteins and simple sugars. Fetuin was the most potent inhibitor tested, but asialofetuin was negative, indicating that sialic acid was involved in binding activity. Mannose and sucrose also were inhibitory, but lactose and galactose were negative. The lectin binds to Sepharose, Superose, and Sephacryl (Pharmacia, Piscataway, NJ) as measured by adsorption of agglutinating activity from egg extracts. Sepharose and Superose are composed of cross-linked agarose, which is a linear polysaccharide composed of galactopyranose residues. Sephacryl is made from cross-linked dextrans (anhydroglucose polymer). Thus, the binding specificity of this lectin appears quite complex. Preliminary indications are that the native lectin molecule is a very large polymer. Over 50% of the agglutinating activity is pelleted by centrifugation for 30 minutes at 50,000 *g*. The purified lectin is excluded from a Zorbax GF250 gel filtration column (Dupont, DEL), indicating a size of at least 500,000 kDa.

Superose had the highest binding capacity for the lectin, and we used this property to affinity purify the lectin. One hundred milliliters of extract (50 g of eggs) were applied to a column packed with 70 ml of Superose 12 in TBS plus 10 mM $CaCl_2$. The agglutinating activity was completely adsorbed. Because EDTA inhibited activity, the column was eluted with 20 mM EDTA in TBS, and 100% of the activity was recovered. Total protein recovery was approximately 40 mg. On SDS-PAGE, the reduced protein shows several bands, but the majority of the protein comprises two bands of approximately 30 kDa and 65 kDa. The subunits were purified by SDS-PAGE preparative electrophoresis using a BioRad (Richmond, CA) Model 491 Prep Cell. Proteins are isolated by continuous elution as they migrate off the bottom of the gel. These were used to (a) prepare antisera in rabbits, (b) obtain NH_2-terminal sequences, (c) to prepare peptides by digestion with endoproteins Lys-C or trypsin to obtain internal sequences as described above, and (d) obtain amino acid composition data.

The 30-kDa and 65-kDa subunits appear to be very similar to each other. First, there is extensive serological cross-reaction between the two bands in Western blots

(Fig. 3A). Second, the amino acid composition (expressed as mole percent values) are very similar. Third, the NH_2-terminal sequences show 8 out of 10 identities. However, the NH_2-terminal sequences are not completely identical. These sequencing reactions have been repeated several times, and the results confirmed. Thus the subunits are not identical, though they must share extensive homology.

Some interesting homologies are suggested from the peptide sequence data we have. Table 1 shows an alignment of a peptide from the 65-kDa subunit with a group of C-type lectins. Like the C-type lectins,[19] the lamprey agglutinin requires Ca^{2+} for activity and has specificity for galactopyranose residues. The human hepatic lectin versus the set gives seven matches while the fly lectin gives six matches. In comparison, the lamprey lectin gives six matches, which suggests that this peptide is part of a C-type domain. The human leukocyte adhesion molecule-1 (LEM1) contains such a C-type domain. Interestingly, the NH_2-terminal 30-kDa subunit shows significantly homology with a segment of this molecule, *viz:*

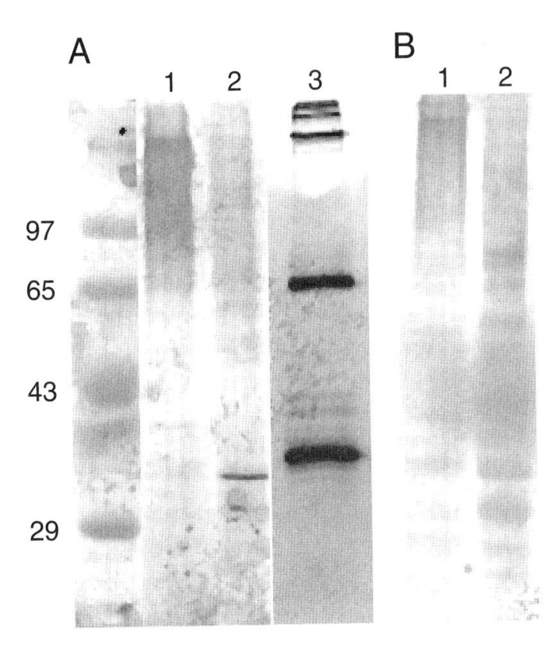

FIGURE 3. Western blots showing reactivity of **(A)** rabbit antiserum raised against the 35-kDa subunit of lamprey egg lectin and **(B)** normal prebleed serum. *Lanes:* (1) *Didemnum candidum* hemolymph; (2) *Halocynthia pyriformis* hemolymph; (3) purified lamprey egg lectin.

TABLE 1. 65-kDa Subunit Peptide 2 versus a Homologous Region of C-Type Lectins

Lam-Pep	S	W	A	N	L	N	T	F	G	R
Tun	N	Y	A	D	A	G	T	Y	C	Q
S.U.	T	W	A	E	G	E	Q	F	C	Q
CHL	S	W	H	K	A	K	A	E	C	E
FLY	N	W	H	Q	A	W	H	E	C	A
CPG	T	W	V	D	A	E	R	R	C	R
HuHL	A	W	A	D	A	D	N	Y	C	R
Bar	R	W	N	D	A	Q	L	A	C	Q

NOTE: Tun, tunicate; S.U., sea urchin; CHL, chicken hepatic lectin; FLY, fly lectin; CPG, cartilage proteoglycan core protein; HuHL, human hepatic lectin; Bar, barnacle. Adapted from Drickamer.[19]

A relationship of the lamprey egg lectin with other proteins can be directly demonstrated by immunochemical techniques. FIGURE 3 shows the reactivity in a Western blot of antisera raised to the 30-kDa subunit. A band of approximately 35 kDa is seen in the hemolymph from the tunicate *Halocynthia pyriformis*. Hemolymph from *Didemnum candidum* and several other tunicates are negative, as is the normal prebleed serum. A triton detergent lysate of a human T-cell line also yields a positive reaction with a band, again at approximately 35 kDa. The higher molecular weight bands are present in the negative control. The sequence and immunochemical data suggest that this lectin from a primitive vertebrate contains a C-type lectin domain and is related to molecules found in both protochordate invertebrates and higher vertebrates. Thus, research on this molecule should provide an interesting case study of the evolution of recognition molecules.

SUMMARY

We have used specific antibody probes to conserved antigenic motifs to identify and characterize immunoglobulin-related molecules in tunicates and a C-type lectin found in lamprey that is related to molecules found in tunicates and mammals. The tunicate immunoglobulin cross-reactive molecule (μCRM) reacts with antibodies raised to shark IgM heavy chains. Intact tunicate μCRM is a monomer of Ig light-chain-sized subunits and is oligoclonal by IEF. That this molecule is related to Ig is indicated both by immunochemical data and by peptide sequence homologies. The lamprey lectin is a large polymer (>500,000 kDa) of 35-kDa and 60-kDa subunits. It appears to be related to C-type lectins as shown by peptide sequence homology and the requirement of Ca^{2+} for activity. Related molecules appear to be present in tunicates and mammals as shown by cross-reactivity of antibodies in Western blots with single bands from hemolymph and T-cell extracts.

ACKNOWLEDGEMENT

We would like to thank Ms. Diana Humphreys for her assistance in the preparation of this manuscript.

REFERENCES

1. Williams, A. F. & A. N. Barclay. 1988. Ann. Rev. Immunol. **6:** 381–385.
2. Dayhoff, M. O. 1978. Atlas of Protein Sequences and Structure. Vol. 5, Suppl. 3. National Biomedical Research Foundation. Washington. DC.
3. Harrelson, A. L. & C. S. Goodman. 1988. Science **242:** 700–708.
4. Seeger, M. A., L. Haffley & T. C. Kaufman. 1988. Cell **55:** 489–600.
5. Sun, S. C., I. Lindstrom, H. G. Boman, I. Faye & O. Schmidt. 1990. Science **250:** 1729–1732.
6. Williams, A. F. & J. Gagnon. 1982. Science **216:** 696–670.
7. Mansour, M. H., H. I. Negm & E. L. Cooper. 1987. Dev. Comp. Immunol. **11:** 3–15.
8. Shalev, A., A. H. Greenberg, L. Logdberg & L. Bjorck. 1981. J. Immunol. **127:** 1186–1191.
9. Hohman, V. S., S. F. Schluter & J. J. Marchalonis. 1992. Proc. Natl. Acad. Sci. USA **89:** 276–280.
10. Marchalonis, J. J. & S. F. Schluter. 1990. BioScience **40:** 758–768.
11. Edelman, Gerald M. 1987. Immunol. Rev. **100:** 12–45.
12. Berril, N. J. 1955. The Origin of Vertebrates. Oxford University Press. New York and London.
13. Marchalonis, J. J., S. F. Schluter, H.-Y. Yang, V. S. Hohman, K. McGee & L. Yeaton. 1992. Comp. Biochem. Physiol. **101A**(No. 4): 675–687.
14. Marchalonis, J. J., V. S. Hohman, H. Kaymaz & S. F. Schluter. 1993. Comp. Biochem. Physiol. **105B:** 423–444.
15. Rosenshein, I. L., S. F. Schluter, G. R. Vasta & J. J. Marchalonis. 1985. Dev. Comp. Immunol. **9:** 783–795.
16. Schluter, S. F., I. L. Rosenshein, R. A. Hubbard & J. J. Marchalonis. 1987. Biochem. Biophys. Res. Commun. **145:** 699–705.
17. Marchalonis, J. J. & S. F. Schluter. 1989. Defense Molecules. J. J. Marchalonis & C. L. Reinisch, Eds.: 265–280. Alan R. Liss. New York..
18. Sharon, N. & H. Lis. 1989. Lectins. Chapmen and Hall. New York.
19. Drickamer, K. 1988. J. Biol. Chem. **263:** 9557–9560.

Evolution of Adaptive Immunity: Inducible Responses in the American Cockroach[a]

RICHARD D. KARP, LAURA E. DUWEL,
LENORE M. FAULHABER, AND MELLISA A. HARJU

Department of Biological Sciences
University of Cincinnati
Cincinnati, Ohio 45221

INTRODUCTION

Our extensive studies on the American cockroach *(Periplaneta americana)* have indicated that this insect has developed an impressive array of immune defense mechanisms that have undoubtedly played a large role in the ability of this species to survive over the last 350 to 400 million years. We have focused on three major aspects of the roach immune response, which include the humoral response to soluble proteins, the response to bacteria, and the response to foreign tissue transplants. Our current efforts are concentrated on determining the nature of the mediators of these mechanisms, and whether or not these mediators play multiple roles in the roach's defense system. An example of an approach taken in each of these categories is outlined in this paper as follows: (1) the successful use of Western blotting and autoradiography to identify hemolymph proteins in immunized animals that specifically bind to the original antigen; (2) the use of an *in vitro* assay that demonstrates that cell-free hemolymph from roaches injected with killed *Pseudomonas aeruginosa* is bactericidal at 48 hr post injection; and (3) roaches that have rejected integumentary xenografts develop enhanced levels of a 102-kDa protein that has proven to be associated with the humoral response to soluble proteins.

MATERIALS AND METHODS

Cockroaches

Adult American cockroaches (females in some studies and males in others as outlined below) were obtained from our own inbred colonies. The animals were housed in 2-gallon clear plastic containers (Tristate Plastics, Dixon, KY) with vented lids. The top inside rims of the containers were lined with petrolatum to prevent escape. Corrugated cardboard rolls were employed to increase surface area, and sawdust was added to the bottom of the containers to control moisture. The animals were maintained at 25 to 28°C with 50 to 60% relative humidity, exposed to a 12-hr

[a]The work presented here was supported by National Institutes of Health research Grant GM 39398 and the University of Cincinnati Foundation Fund for Comparative Immunology.

light/12-hr dark illumination cycle, and fed Purina laboratory canine chow and water *ad libitum*.

Immunization and Cell-Free Hemolymph Preparation

Roaches were anesthetized with CO_2 and injected with either soluble protein toxoids or killed *Pseudomonas aeruginosa* in the studies described here. The various preparations were loaded into a Hamilton microsyringe and injected into the hemocoel by sliding the needle horizontally through the intersegmental membrane between abdominal sternites. After various periods of time, the roaches were anesthetized again and bled by severing the legs just proximal to the femoral/tibial joint and expressing hemolymph by applying gentle pressure to the abdomen. The hemolymph from the animals in each group was collected in a common sterile tube and stored on ice during the procedure. The hemolymph was separated from the hemocytes by centrifugation at 600 × *g* for 10 min (100 × *g* in the bacterial studies), and the supernatant was subjected to a second centrifugation at 3200 × *g* for 20 min.

Reducing and Nonreducing Polyacrylamide Gel Electrophoresis

SDS-PAGE analysis was performed using a modification of Laemmli's discontinuous method.[1] The upper "stacking" portion of the gel was prepared at an acrylamide concentration of 4.5% and the lower "resolving" portion of the gel was prepared at a concentration of 7.5%. All gels (both SDS and NR) contained a 2.7% concentration of bisacrylamide. The actual analysis was carried out using SDS-PAGE minigels, employing a Hoefer "Mighty Small" miniature slab gel electrophoresis unit at a constant voltage of 150 V.

Nondissociating discontinuous NR-PAGE gels were cast with an acrylamide concentration of 3.75% in the "stacking" portion of the gel and 6.0% in the "resolving" portion of the gel. Standard size (14 cm × 16 cm) NR-PAGE gels were run at 4°C, using constant currents of 30 mA for the stacking gel and 60 mA for the resolving gel.

Protein concentrations of samples to be analyzed were adjusted to 2 mg/ml using the Bradford protein assay[2] before electrophoresis. Protein bands were visualized using standard Coomassie blue staining.

Western Blot Binding Assay

Cell-free hemolymph samples from immune or control adult female cockroaches were subfractionated using a CM Affi-Gel Blue ion-exchange column (Bio-Rad). Unbound proteins were collected in the starting buffer (Buffer A, 0.1 M K_2HPO_4, pH 7.25, containing 0.15M NaCl and 0.02% NaN_3). Bound proteins were eluted using buffer A containing 1.4 M NaCl. We have found that CM Affi-Gel Blue concentrates NR-PAGE bands 2 and 4 in the salt wash (for unknown reasons), such that they are in nearly equal concentrations with NR-PAGE bands 1 and 3 (which are the major hemolymph protein bands observed in unfractionated hemolymph after NR-PAGE analysis). The buffer A and salt washes were concentrated to 2 mg/ml and electrophoresed on 6% NR-PAGE gels. The separated proteins were electroblotted onto Immunobilon nitrocellulose paper. Duplicate blots were incubated with either ^{125}I-labeled immunizing or heterologous antigen for 2 hr at room temperature. Binding of the antigen was then visualized using autoradiography.

In Vitro *Antibacterial Assay*

Adult male roaches were injected with either pyrogen-free Burns-Tracey saline[3] (BTS) or 10^6 killed *Pseudomonas aeruginosa.* The animals were bled 48 hr post injection, cell-free hemolymph was obtained as above, and the protein concentration adjusted to 10 to 13 mg/ml. *P. aeruginosa* in log-phase growth were washed 2 × with 0.9% NaCl and diluted to 2.5×10^7 colony-forming units (CFU)/ml. Fifty microliters of this dilution was incubated in 950 μl of hemolymph protein to give a final concentration of 1.25×10^6 CFU/ml. The bacteria and hemolymph were incubated at 37°C, and the number of viable bacteria was determined at various times during the incubation by serial diluting and plating of samples on trypticase soy (TSA) agar plates (Difco).

Transplantation of Integumentary Xenografts

Adult female *Periplaneta americana* from our colony stock served as recipients in this study, and adult male *Blatta orientalis,* also from our colonies, served as the xenogeneic donors. Grafting was performed according to our established methods,[4] which briefly involve the following: Xenografts (approximately 3 × 4 mm in size) were removed from donor roaches and placed epidermal side down in sterile BTS. Recipient roaches were anesthetized with CO_2 for 10 sec and placed on ice for 10 min. Recipients were surface sterilized in 4% formalin, followed by 70% ethanol, placed on a sterile surface, wings clipped, and swabbed with ethanol again. The intersegmental membranes were loosened under the second and fourth tergites, and the integumental pieces fully removed, floated in sterile BTS for 10 sec, and replaced in the graft beds. Animals in the experimental groups received a piece of recipient integument at one site to serve as an autograft control, whereas a second site was grafted with a piece of donor xenogeneic integument of comparable size. The grafted area was then covered with antibiotic salve (20 g white petrolatum, 100 mg penicillin, 100 mg chloramphenicol) and a parafilm bandage.

RESULTS

In Vitro *Binding of Antigen by Immune Hemolymph Proteins*

Our previous studies established that the American cockroach responds to injections of soluble protein toxoids, such as honey bee venom, by generating an adaptive humoral response demonstrating both specificity and memory.[5,6] The inducible humoral factor (IHF) mediating this response has been found to be protein in nature and capable of precipitating the inducing antigen.[7] Sodium dodecyl sulfate–polyacrylamide gel electrophoresis (SDS-PAGE) analysis of immune hemolymph indicated that certain proteins were enhanced in immune animals as compared to saline-injected controls.[8] Most notable was the appearance of a 102-kDa protein that was induced no matter what antigen was injected. Nonreducing-PAGE (NR-PAGE) analysis revealed that bands 2 and 4 were greatly enhanced in immune versus control animals.[9] Interestingly, the reduced 102-kDa protein is a major component of both of the enhanced NR-PAGE bands.[9] Our current studies have been focused on investigating whether the enhanced NR-PAGE bands contain the immune proteins that are biologically active, so that we know where to concentrate our efforts in isolating the IHF. For this purpose, we devised a combination Western blot/autoradiography assay to determine if any of the bands would bind radiolabeled antigen. FIGURE 1 shows the results of incubating ^{125}I-labeled honey bee phospholipase A_2 (BPA2) with Western blots of buffer wash and salt wash proteins run on NR-PAGE gels from animals injected with BPA2 toxoid or saline. There was significant uptake of labeled BPA2 in bands 2 and 4 of immune hemolymph, with evidence that not all binding was limited to salt wash proteins. Although there was binding present in band 2 in the control hemolymph in both buffer A and salt wash proteins, there was slight (as in FIG. 1) or no binding (as in other trials not shown) of labeled antigen to band 4. In specificity experiments, blots containing protein bands from hemolymph of animals immunized against BPA2 toxoid were incubated with antigenically distinct ^{125}I-labeled cytochrome *c* (which is similar in size and charge to BPA2). The results indicated that although some binding to bands 2 and 4 could be detected, it was at very low levels (FIG. 2). Thus, binding to band 4 was far more intense when the immune proteins were incubated with the correct antigen.

Monitoring the Antibacterial Response to Pseudomonas aeruginosa

We have previously established that a biphasic defense response is generated when the American cockroach is injected with killed *Pseudomonas aeruginosa.*[10] The first phase of the response, which occurs between 1 and 3 days post injection, is very acute and nonspecific. After day 4 post injection, the characteristics of the response begin to change, with protection being longer lasting (up to 14 days) and more selective. We set out to develop a simple *in vitro* assay to quantitate activity in immunized animals not only for use in future biological studies, but also to employ in planned purification protocols. Our first use of the assay was to determine if there was a cell-free antibacterial factor in the hemolymph of immunized animals during the first phase of the response. Cell-free immune and control hemolymph samples obtained 2 days post injection, were incubated with 1.25×10^6 CFU/ml of log-phase *P. aeruginosa* in a shaker bath at 37°C. Aliquots of the incubation mixture were removed at various times and plated on TSA agar. The preliminary results indicated that by 24 hr of incubation, the immune hemolymph had killed all of the incubated bacteria, whereas in control hemolymph, the bacteria were actually increasing in number (FIG. 3). Thus, there appears to be a cell-free bactericidal factor that participates in the early response to injected bacteria.

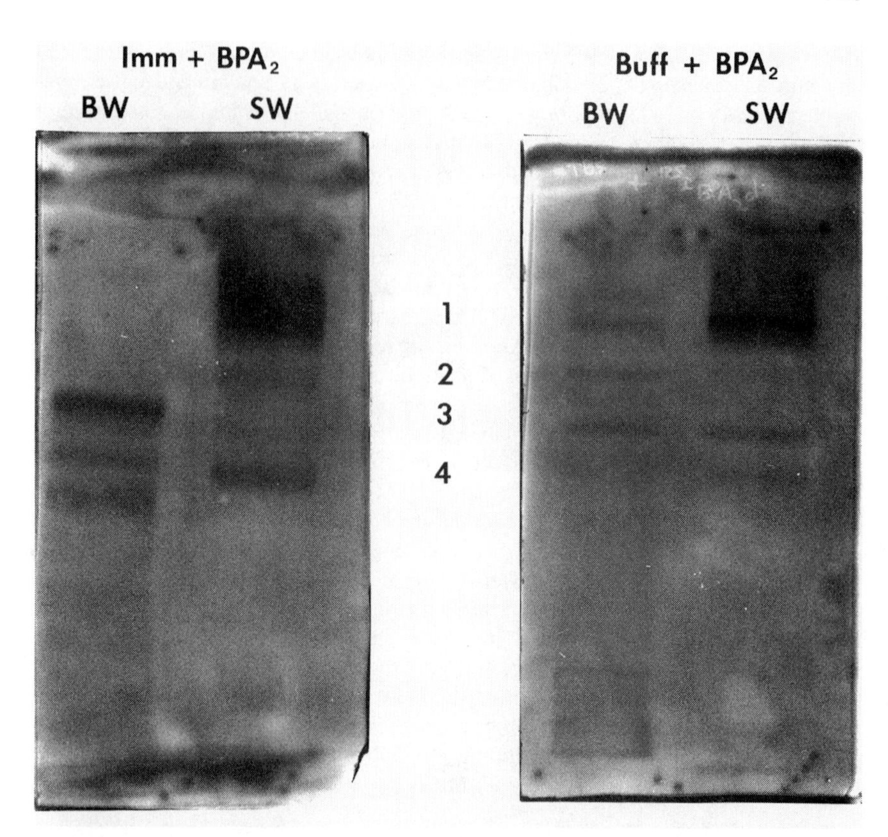

FIGURE 1. Western blot/autoradiographic analysis of binding proteins in immunized American cockroaches. Buffer A and salt wash proteins from CM Affi-gel blue-subfractionated BPA2 immune and control hemolymph samples run on 6% NR-PAGE gels electroblotted onto Immobilon and incubated with ^{125}I-BPA2. BPA2: ^{125}I-labeled honey bee phospholipase A_2; Imm: hemolymph from immunized animals; Buff: hemolymph from animals injected with saline buffer; BW: proteins obtained by washing the Affi-gel blue column with buffer A; SW: proteins obtained by washing the Affi-gel blue column with buffer containing 1.4M NaCl; numbers 1-4 indicate the major protein bands that are visualized on NR-PAGE analysis of cockroach hemolymph. Note the selective binding of band 4 by ^{125}I-BPA2 in the SW proteins of immune hemolymph versus control hemolymph.

Induction of Immune Proteins during the Rejection of Integumentary Xenografts

We have conducted several studies establishing that *Periplaneta americana* is capable of recognizing and rejecting integumentary xenografts and allografts.[4,11–13] Thus, the cockroach possesses both well-developed cell-mediated defense mechanisms as well as humoral defense mechanisms. The natural progression of the grafting studies had led us to design experiments that would not only identify the cell

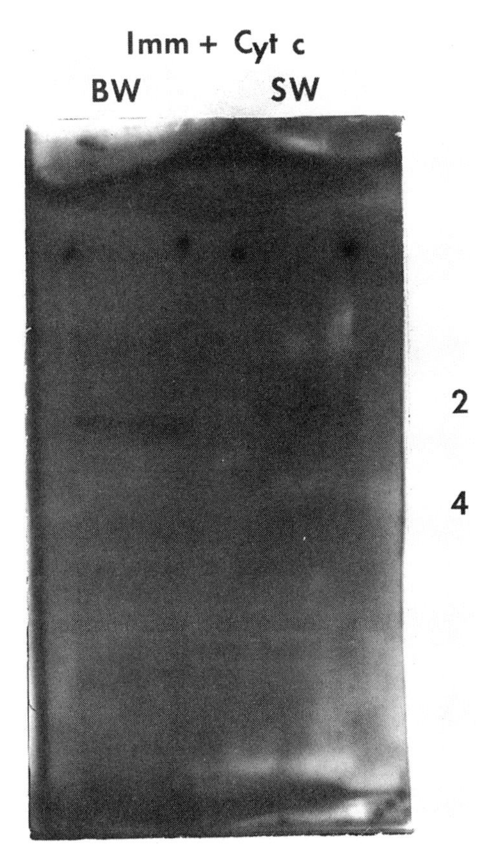

FIGURE 2. Western blot/autoradiographic analysis of the specificity of the binding proteins in immunized American cockroaches. Buffer A and salt wash proteins from CM Affi-gel blue-subfractionated BPA2 immune hemolymph samples run on 6% NR-PAGE gels electroblotted onto Immobilon and incubated with ^{125}I-cytochrome *c*. Imm: hemolymph from animals immunized with BPA2; Cyt *c*: ^{125}I-labeled cytochrome *c*; BW: proteins obtained by washing the Affi-gel blue column with buffer A; SW: proteins obtained by washing the Affi-gel blue column with buffer containing 1.4 M NaCl; numbers 2 and 4 indicate the NR-PAGE protein bands that are typically enhanced in the hemolymph of immunized animals. Note that incubation with ^{125}I-Cyt *c* results in only background levels of binding to SW band 4.

type mediating the response, but also the nature of the cell receptor that recognizes non-self in the roach. Because the humoral response to soluble proteins is strongly associated with the consistent induction of increased levels of hemolymph proteins, particularly the reduced 102-kDa species, we investigated the possibility that graft rejection might also induce these proteins. For this purpose, adult female *Periplaneta* received either a xenograft from *Blatta orientalis* and an autograft, or two autografts (to serve as a trauma control). The hemolymph from grafted animals was removed

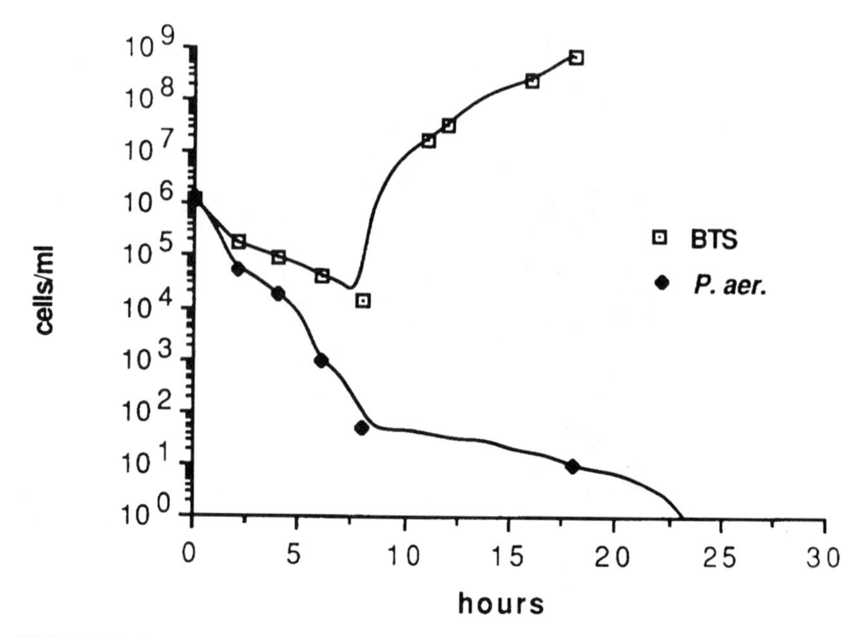

FIGURE 3. Bactericidal activity of cell-free hemolymph from roaches injected with killed *Pseudomonas aeruginosa* 48 hr post injection. BTS: hemolymph from animals injected with Burns-Tracey saline; *P. aer.*: hemolymph from animals injected with killed *Pseudomonas aeruginosa.*

at days 7, 10, 14, and 21 posttransplant and analyzed by SDS-PAGE for increases in protein band concentration. At the same time, the grafts from these animals were processed for histological examination. The results from three trials (representative data in FIG. 4) indicated that at day 14 posttransplant, the same proteins found to be enhanced in animals immunized with soluble proteins were also enhanced in xenografted animals, including a very prominent induction of the 102-kDa band. However, this activity was found to be absent in the control animals receiving two autografts. The increased production of the hemolymph proteins was not found at other time points tested in the grafted animals. Histological examination of the xenografts indicated that 100% of the transplants had been rejected at each time point. Thus, we can safely assume that antigenic stimulation was taking place in the xenografted animals, and that this provided the stimulus for the observed induction of increased hemolymph protein levels.

DISCUSSION

Our studies on the American cockroach have revealed that this animal possesses an impressive array of immune defense mechanisms. The studies we have completed that have defined the nature of the humoral response to soluble proteins, the biphasic

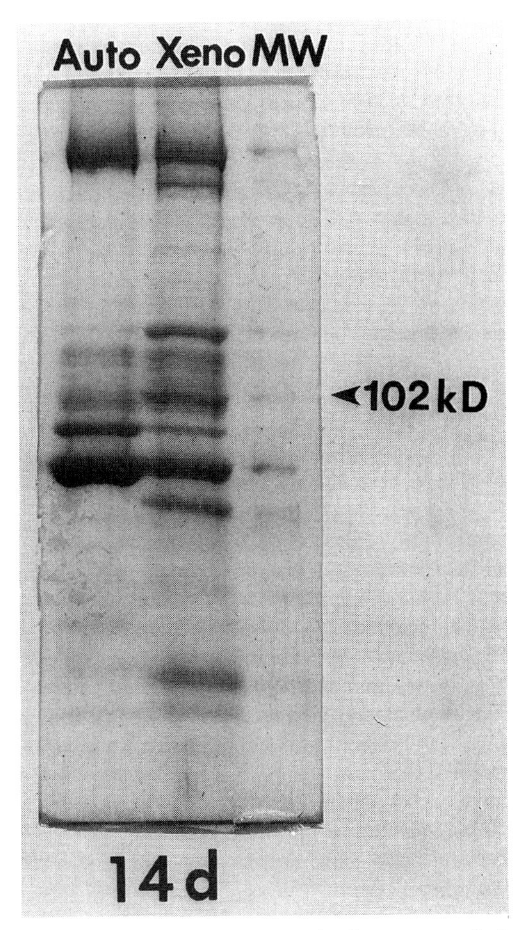

FIGURE 4. SDS-PAGE analysis of hemolymph proteins from xenografted animals 14 days posttransplant. Auto: hemolymph from control animals that had received two autografts; Xeno: hemolymph from animals that had received a xenograft from *Blatta orientalis* and an autograft; MW: molecular weight markers. Note the strong induction of the 102-kDa band as compared to autografted controls.

response to injected bacteria, and the rejection of integumentary xenografts and allografts indicate that there are some common themes associated with these different types of responses. First, all of these mechanisms, including the bacterial response, eventually recognize antigen in a specific manner. The humoral response to soluble proteins can discriminate between similar toxins, since animals are not protected against venoms that they have not been immunized against.[4] The Western blot studies reported here show that the proteins induced by immunizing animals to BPA2 toxoid will bind ^{125}I-BPA2 quite well, but bind little or no ^{125}I-cytochrome *c*, which is of similar size and charge. Thus, there is now both *in vivo* and *in vitro* evidence for specific recognition of antigen by the induced immune protein(s). Our previous studies on the antibacterial response in the roach indicated that after day 4 postinjection, a more specific and long-term response begins to develop.[10] The specificity gets more refined as the response continues to develop with time. Our ability to follow immune reactivity with the *in vitro* assay reported here will allow us to determine whether there actually is a specific factor that is induced immediately, but is masked, perhaps by nonspecific cellular mechanisms such as encapsulation, or whether two different factors are generated, one nonspecific and the other specific, during the development of the protective response. The cell-mediated response to transplants is by definition specific, since each animal has an autograft control that is not rejected. This has been demonstrated very directly by injecting allogeneic hemocytes underneath an autograft, and finding that although the host reacted to the foreign cells with a heavy hemocyte infiltration, the autograft remained intact.[11] The finding of specificity necessitates envisioning a cell surface receptor that is capable of recognizing foreign epitopes. The findings reported here that rejection of xenografts induces the enhanced production of the same protein species that are associated with specific binding in the humoral response, particularly the 102-kDa protein, raises the possibility that these molecules may also be acting as specific recognition molecules in roach cell-mediated immunity.

The second major theme that is common among roach immune responses would be the existence of immunologic memory. Although we have not investigated this possibility yet in the antibacterial response, we have established the existence of long-term memory in the inducible humoral response to soluble proteins[6] and have found at least short-term memory in the response to integumentary allografts.[14] Because specificity is so closely tied to memory in vertebrate responses, it should come as no surprise that we should find memory as a component of roach immunity. It must be remembered that the cockroach is a relatively long-lived animal that has to deal with repeated exposure to noxious agents as well as pathogens. Thus, it is very logical that this animal developed the capacity to react in an enhanced fashion upon re-exposure to harmful substances in order to assure its survival.

We have spent a great deal of time characterizing the various types of immune responses present in the American cockroach. The work presented in this paper is the start of the second generation of research, which will concentrate on identifying and characterizing the mediators of the different forms of adaptive immunity present in this fascinating animal. The results of those studies should provide some very interesting data that will help us to understand the evolutionary significance of immune responses in whatever form they are found throughout the animal kingdom.

REFERENCES

1. LAEMMLI, U. K. 1970. Cleavage of structural proteins during the assembly of the head of bacteriophage T4. Nature **227:** 680–685.
2. BRADFORD, H. M. 1976. A rapid and sensitive method for the quantification of microgram quantities of protein utilizing the principle of protein-dye binding. Anal. Biochem. **72:** 248–254.
3. LUDWIG, D., K. M. TRACEY & M. BURNS. 1957. Ratios of ions required to maintain the heart beat of the American cockroach, *Periplaneta americana*. Ann. Entomol. Soc. Am. **50:** 244–246.

4. GEORGE, J. F., R. D. KARP & L. A. RHEINS. 1984. Primary integumentary xenograft reactivity in the American cockroach, *Periplaneta americana.* Transplantation **37:** 478–484.
5. RHEINS, L. A., R. D. KARP & A. BUTZ. 1980. Induction of specific humoral immunity to soluble proteins in the American cockroach *(Periplaneta americana).* I. Nature of the primary response. Dev. Comp. Immunol. **4:** 447–458.
6. KARP, R. D. & L. A. RHEINS. 1980. Induction of specific humoral immunity to soluble proteins in the American cockroach *(Periplaneta americana).* II. Nature of the secondary response. Dev. Comp. Immunol. **4:** 629–639.
7. RHEINS, L. A. & R. D. KARP. An inducible humoral factor in the American cockroach *(Periplaneta americana)*: Precipitin activity that is sensitive to a proteolytic enzyme. J. Invert. Pathol. **40:** 190–196.
8. GEORGE, J. F., R. D. KARP, B. L. RELLAHAN & J. L. LESSARD. 1987. Alteration of the protein composition in the haemolymph of American cockroaches immunized with soluble proteins. Immunology **62:** 505–509.
9. DUWEL-EBY, L. E., L. M. FAULHABER & R. D. KARP. 1991. The inducible humoral response in the cockroach. *In* Immunology of Insects and Other Arthropods. A. P. Gupta, Ed.: 385–402. Academic Press. New York.
10. FAULHABER, L. M. & R. D. KARP. 1992. A diphasic immune response against bacteria in the American cockroach. Immunology **75:** 378–381.
11. GEORGE, J. F., T. K. HOWCROFT & R. D. KARP. 1987. Primary integumentary allograft reactivity in the American cockroach, *Periplaneta americana.* Transplantation **43:** 514–519.
12. HOWCROFT, T. K. & R. D. KARP. 1987. Demonstration of cell-mediated cytotoxicity to allogeneic and xenogeneic tissue in the American cockroach, *Periplaneta americana,* using a combination *in vivo/in vitro* assay. Transplantation **44:** 129–135.
13. KARP, R. D. & C. C. MEADE. 1993. Transplantation immunity in the American cockroach, *Periplaneta americana:* The rejection of integumentary grafts from *Blatta orientalis.* Dev. Comp. Immunol. **17:** 301–307.
14.. HARTMAN, R. S. & R. D. KARP. 1989. Short-term immunologic memory in the allograft response of the American cockroach. Transplantation **47:** 920–922.

Autoimmunoregulation and the Importance of Opioid Peptides

G. B. STEFANO,[a] M. K. LEUNG,[a] A. SZÛCS,[b] K. S.-RÒZSA,[b]
E. M. SMITH,[c] T. K. HUGHES,[d] E. OTTAVIANI,[e]
C. FRANCESCHI,[f] O. DUVAUX-MIRET,[g] A. CAPRON,[g]
AND B. SCHARRER[h]

In recent years, there has been growing interest in documenting the presence and significance of neuropeptides in the immune system. The same is true for the presence of cytokines and other immunoactive signal molecules in nervous tissue. Thus, some of the boundaries defining academic disciplines and subdisciplines have been eroding. The present review focuses on information gained mainly in invertebrates demonstrating that opioid and other neuropeptides are important in autoimmunoregulation.

IS METHIONINE ENKEPHALIN PRESENT IN INVERTEBRATE HEMOLYMPH?

The demonstration of an endogenous Met-enkephalin-like material in the hemolymph was carried out by means of HPLC and RIA.[3] The results of HPLC fractionation of concentrated extracts of mollusc and insect cell-free hemolymph and hemocytes, respectively, show the highest activities in fractions 5, 6, and 11 (FIGS. 1, 2). Authentic Met-enkephalin under our chromatographic conditions has a retention time of 10.5 min and would have eluted in fraction 11. Moderate activity was detected in fractions 7 and 13 of the cell-free hemolymph, and very little activity in 7 and 13 of the cellular material. Because of the presence of Met-enkephalin-like material in a relatively large number of HPLC fractions showing more than one peak, it is within reason to assume that the hemolymph may contain more than one opioid-like substance. Moreover, the peak values for hemolymph fluid should be

[a]Multidisciplinary Center for the Study of Aging, Old Westbury Neuroscience Research Institute, State University of New York, College at Old Westbury, Old Westbury, N.Y. 11568-0210.
[b]Balaton Limnological Research Institute of the Hungarian Academy of Sciences, Tihany, Hungary.
[c]Department of Psychiatry, University of Texas Medical Branch at Galveston, Galveston, Texas 77550.
[d]Department of Behavioral Sciences and Microbiology, University of Texas Medical Branch at Galveston, Galveston, Texas 77550.
[e]Department of Animal Biology, University of Modena, 41100 Modena, Italy.
[f]Institute of General Pathology, University of Modena, 41100 Modena, Italy.
[g]Center for Immunology and Parasitology, Institut Pasteur, 59019 Lille Cedex, France.
[h]Department of Anatomy and Structural Biology and Department of Neuroscience, Albert Einstein College of Medicine, Bronx, New York 10461.

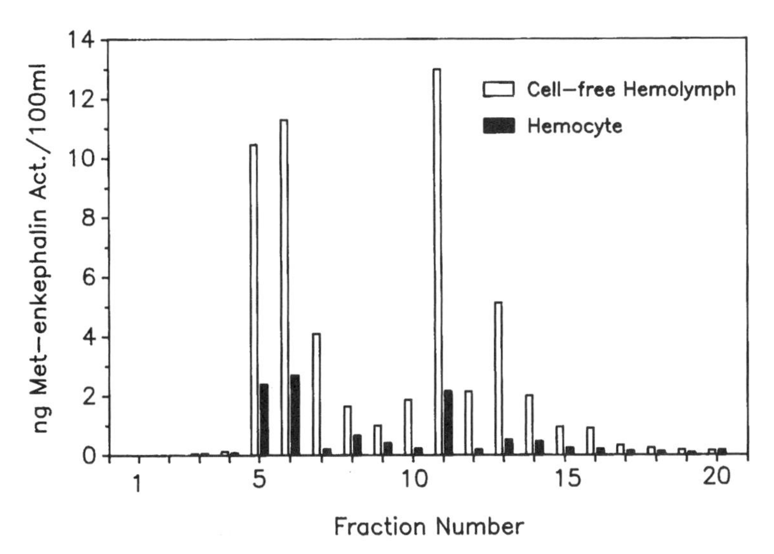

FIGURE 1. Met-enkephalin-like activity (Me Act.) determined by RIA in selected HPLC fractions of cell-free hemolymph and hemocytes of *Mytilus.* (Taken from Stefano *et al.*[3] with permission from the *Proceedings of the National Academy of Sciences USA.*)

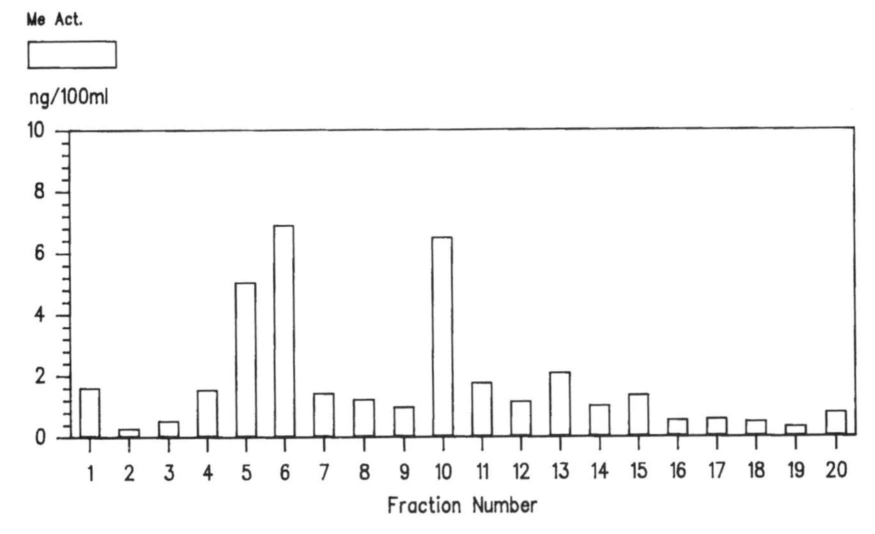

FIGURE 2. Met-enkephalin-like activity (Me Act.) determined by RIA in selected HPLC fractions of cell-free hemolymph and hemocytes of *Leucophaea maderae.* Fraction 10 contains Met-enkephalin-like activity, whereas fractions 5 and 6 contain Met-enkephalin sulfoxide.

taken to represent minimum rather than actual levels, since they may have surpassed the upper limit of detection by the RIA system used (2500 pg/ml). Nevertheless, the results show a pronounced difference in the amounts of opioid content of hemolymph fluid and hemocytes.

HOW ARE NEUROPEPTIDE LEVELS REGULATED IN INVERTEBRATE HEMOLYMPH?

CD10 (CALLA, common acute lymphoblastic leukemia antigen/ neutral endopeptidase 24.11 (NEP, "enkephalinase") hydrolyzes a number of naturally occurring peptides including the endogenous opioid pentapeptides Met- and Leu-enkephalin.[4] Hence, the enzyme in the brain has been termed "enkephalinase." In invertebrate organisms such as the mollusc *M. edulis,* Met-enkephalin triggers inflammatory responses by inducing morphological changes, directed migration, and aggregation of hemocytes.[3–6] We recently found that *M. edulis* hemocytes express a CD10/NEP-related structure and that abrogation of CD10/NEP enzymatic activity reduces the amount of Met-enkephalin required for hemocyte activation by five orders of magnitude.[4] Human $CD10^+$ polymorphonuclear leukocytes are similarly responsive.

Immunocyte cell surface CD10/NEP enzymatic activity is also shown to be regulated by the activation state of the cell during the time period in which the enzyme has its most pronounced effects.[7–9] These results suggest that in immunocytes, CD10/NEP functions to control responsiveness to Met-enkephalin and other peptides. Furthermore, exposure to cytokine downregulates NEP toward neuropeptide stimulation.

In a more recent study we found that Met-enkephalin is degraded by peptidases present in the hemolymph fluid and hemocyte membrane suspension of *Mytilus edulis.*[10] Degradation of Met-enkephalin is rapid in the fluid and slower in the membrane preparation. Aminopeptidase activity is bestatin sensitive in hemocyte membrane and highest in the fluid of the hemolymph, which appears to have a component that is insensitive to inhibition. ACE activity is found only in the fluid of the hemolymph. Carboxypeptidase and NEP (CD10: "enkephalinase") are membrane bound, and the former appears to predominate. Phosphoramidon not only inhibits NEP, as expected, but the invertebrate carboxypeptidase as well.

ARE THERE OTHER ENDOGENOUS OPIOID MOLECULES IN THE HEMOLYMPH?

A possible immunoregulatory function for Met-enkephalin-Arg^6-Phe^7 involving human and invertebrate immunocytes has been reported.[11] This study demonstrates that Met-enkephalin-Arg^6-Phe^7 exhibits stimulatory effects comparable to those of Met-enkephalin. Furthermore, since NEP exists in invertebrate immunocyte membranes, we demonstrated that its specific inhibitor, phosphoramidon, potentiates the effects of the heptapeptide in inducing conformational change in both human and invertebrate granulocytes.[4] The major metabolic products of NEP activity, Phe-Met-

Arg Phe and Tyr-Gly-Gly, appear to be potent antagonists of this enzymes activity, especially the tetrapeptide.[11]

ARE THERE HIGH-AFFINITY OPIOID-BINDING SITES ON IMMUNOCYTES?

The effects of the opioid neuropeptide deltorphin I, isolated from amphibian skin, on immunoregulatory activities were studied in representatives of vertebrates and invertebrates. (D-Ala^2)-Deltorphin I binding and pharmacological evidence for a special subtype of delta opioid receptor on human and invertebrate immune cells was found.[12] The high potency of this compound parallels that of Met-enkephalin previously demonstrated in vertebrate plasma and invertebrate hemolymph. The addition of deltorphin I at a concentration of 10^{-11} M to human granulocytes or immunocytes of the mollusc *Mytilus edulis* resulted in cellular adherence and conformational changes indicative of cellular activation. This value is in line with those obtained with Met-enkephalin, tested in the presence of the special NEP inhibitor phosphoramidon and this opioid's synthetic analogue DAMA which, like deltorphin, is resistant to proteolytic degradation. Both ligands appear to be acting on the same population of immunocytes.[12] The same relationship was estimated to exist in the insect *Leucophaea maderae* in which the high viscosity of the hemolymph makes the quantification of reactive cells more difficult than in *Mytilus.* In addition, deltorphin is as potent as β-endorphin in affecting the proliferation of lymphocytes in response to mitogen.[12] Cold saturation experiments with the radioligand [^{3}H](D-Ala^2)-deltorphin I and [^{3}H]DAMA revealed the presence of two high-affinity binding sites on human and invertebrate immunocytes, one sensitive to the nonequilibrium delta opioid antagonist DALCE and the other relatively insensitive[12] (FIG. 3). The results obtained with deltorphin I support the view that the special role played by endogenous Met-enkephalin in immunobiological activities of vertebrates and invertebrates is mediated by a special subtype of delta opioid receptor.

IS THERE AN OPIOID–CYTOKINE LINK THAT HAS THE POTENTIAL TO INFLUENCE AUTOIMMUNOREGULATION?

Opioid induction of an interleukin-1-like substance in *Mytilus edulis* pedal ganglia and immunocytes exists.[13–15] We demonstrate that recombinant human interleukin-1 (IL-1) can induce the formation of a TNF-like substance as well as initiate specific immunocyte conformational changes in immunocytes.[13] Both the immune and nervous system of *Mytilus* contain an IL-1-like molecule.[15–16] Opioid challenge can induce the formation of an endogenous IL-1-type molecule that stimulates immunocytes as does authentic IL-1.[15] This immunocyte stimulation can be blocked by specific IL-1 antibody. DAMA-induced stimulation of the production of an IL-1-like substance is shared by both the immune and nervous system. These data also imply that immune signal molecules may have functions that transcend immunomodulation.

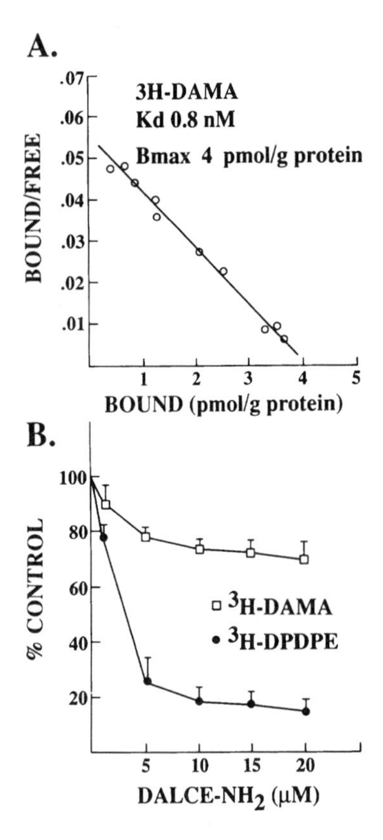

FIGURE 3. Binding analysis of [^{3}H]DAMA to *Mytilus* immunocyte membrane suspensions. **(A)** Scatchard analysis. **(B)** Displacement analysis. (Taken from Stefano *et al.*[12] with permission from the *Proceedings of the National Academy of Sciences USA.*)

CAN CYTOKINES ACT AS NEUROMODULATORS/NEUROTRANSMITTERS?

This question implies that if cytokines can influence select neurophysiological activities, immunocytes can communicate with neurons and/or glial-like cells. Mammalian cytokines have been shown to alter highly specific neurophysiological activities in invertebrates.[9,17] The effect of interleukin-1 was studied on the voltage-activated ion currents of the identified central neurons of *Helix pomatia* L. by use of two microelectrode voltage clamps. The voltage-activated inward current (I_{Ca}) was decreased, whereas the outward current ($I_{net\ K}$) was increased by IL-1. IL-1 affects both the transient and the delayed rectifying potassium currents. The IL-1 modulatory effect on the voltage-activated ion currents is voltage and dose dependent. The threshold concentration for IL-1 was 2 U/ml. The proposed modulatory effect of IL-1 appears to have more than one site of action on the neuron membrane ion channels. Rabbit anti-human IL-1 polyclonal antiserum eliminated the IL-1 effects on the voltage-activated inward and outward currents.[17] Clearly, as recent data indicate, the role of neuropeptides has been broadened to include immunoregulatory mechanisms. Now we demonstrate that immunoregulatory molecules can alter specific neuronal ionophoric domains. Thus, the signal molecules used for autoimmunoregulation appear to have the potential for neuroimmune communication in invertebrates as well.

ARE INVERTEBRATE CORTICOTROPIN-RELEASING FACTOR, ADRENOCORTICOTROPIN, AND MELANOTROPIN-LIKE SUBSTANCES INVOLVED IN AUTOIMMUNOREGULATION?

The elucidation of the immunomodulatory activities of adrenocorticotropin (ACTH) and other neuropeptides has made considerable progress by recent investigations in higher invertebrates.[18] ACTH-like immunoreactivity is present in representatives of most phyla (see Smith *et al.*[18]). This compound is found in neuroendocrine as well as immunoreactive structures. A material resembling vertebrate ACTH and MSH was found in the neuroendocrine system of the insect *Leucophaea maderae.*[19] Ottaviani *et al.*[20] demonstrated ACTH and β-endorphin in the hemolymph and hemocytes of the freshwater snail *Planorbarius corneus.*

As in vertebrates, the function of ACTH in the nervous system is virtually unknown. Rafaeli *et al.*[21] demonstrated an immunological cross-reactivity between ACTH and the diuretic hormone of locusts. It may be that ACTH has endocrine and neurotransmitter functions in many organisms. Therefore, it is not too surprising to find ACTH and other POMC products in invertebrate immune cells.[18,22]

We have detected irACTH by radioimmunoassay in hemolymph of *Mytilus edulis* and have preliminarily shown it to be biologically active and antigenically identical with ACTH standards.[18] A possible source of this irACTH is the immunocyte, as seen in vertebrates (Smith *et al.*[18] for review) and in the snail *Planorbarius.*[20] The other obvious possibility is that some or all of the irACTH in hemolymph comes from the nervous system. The role of ACTH in invertebrate hemolymph may be immunomodulation in addition to some other signal functions as suggested above. When ACTH and MSH were assessed for the ability to alter immunocyte conformation and motility in *M. edulis,* results similar to those in human granulocytes were obtained.[23,24] ACTH itself did not appear to alter the activity, that is, the conformational state of inactive or active hemocytes in short-term (20-min) assays, whereas both α- and β-MSH induced active-ameboid hemocytes to become rounded-inactive in a dose-dependent fashion. α-MSH also inactivated ameboid hemocytes that had been stimulated with β-endorphin or tumor necrosis factor.[23] As in human granulocytes, ACTH inactivated the immunocytes after four hours of incubation, whereas the effect of α-MSH occurred in minutes.[24] The effect of ACTH was blocked by phosphoramidon. Therefore, it is likely that the major inhibitory activity on *Mytilus* immunocytes is due to MSH following ACTH cleavage by NEP.[24]

Furthermore, at extremely low doses ACTH has been reported to enhance immunocyte migration and phagocytic activity of immunocytes of the snail *Planorbarius corneus.*[22] The kinetics of the phagocytosis shows a dramatic jump in activity after two hours which, based on the discussion above, suggests that processing may also occur in this system and that MSH may be a mediator of the effect. Another effect of ACTH on *Planorbarius* immunocytes reported by Ottaviani *et al.*[22] is the release of biogenic amines. Exposure of the hemocytes to ACTH caused a reduction in the cellular contents of norepinephrine, epinephrine, and dopamine with a concomitant rise in the hemolymph level of these amines. The most interesting finding is that only glass-adherent phagocytic spreading hemocytes (SH), but not round hemocytes (RH), contain these neuropeptides. However, the presence of irACTH both in the cytoplasm and on the cell membrane of SH is intriguing. IrACTH may be produced by hemocytes.[20] Because the major characteristic of SH is their ability to phagocytize foreign material, a role of ACTH in phagocytosis could be predicted. Accordingly, the effect of ACTH on the phagocytosis of bacteria was studied on *P. corneus* hemocytes.[20] The effect of ACTH was evident at physiological concentrations, that is, down to 10^{-13} M.

The result of the studies by Ottaviani and colleagues is that antigenic challenge may represent the equivalent of the immune system stress response of the nervous system. They feel that the difference between antigens and stressors is often only quantitative and semantic. The invasion of pathogenic bacteria triggers a complex set of responses that, besides chemotaxis and phagocytosis, include the antigenic-stress response.

Corticotropin-releasing factor-like (CRF) molecules were first identified both in hemocytes and serum by RIA of *P. corneus.*[20] Recently, CRF was shown to mimic the effects of α-MSH while exhibiting a longer duration of action.[25] Alpha-helical CRF, a specific inhibitor of CRF, antagonized CRF-induced cellular immunosuppression but was ineffective in altering MSH-induced immunosuppression. Both human and *Mytilus* immunocytes appear to have specific CRF receptors. In another experiment, both CRF and MSH antagonize tumor necrosis factor stimulation of immunocytes. Again, α-helical CRF only antagonized the CRF activity, suggesting the presence of a separate CRF receptor on these cells.[25] The results suggest that invertebrates also have numerous molecules associated with inhibiting immunocyte activation and responsiveness to stimulatory signal molecules. We can only surmise that this immunocyte inhibitory activity was essential during the course of evolution to halt the process of excessive immune stimulation and/or at the end of a successful immune defense. In other words, given the potent cascading stimulatory signal molecule pathways involving different types of signal molecules, it became important to devise a mechanism(s) to prevent immune-hyperstimulation. Reports from our laboratory indicate that several types of molecules fulfill this role, for example, MSH, NPY (neuropeptide Y[26]), CRF, and morphine.[31]

CAN PARASITES ESCAPE IMMUNE SURVEILLANCE BY INFLUENCING AUTOIMMUNOREGULATION?

Evidence supporting the concept that the parasite trematode, *Schistosoma mansoni,* may escape immune reactions from its vertebrate host (man) by using signal molecules common to both host and parasite was obtained by the following experiments. The presence of immunoactive proopiomelanocortin (POMC)-derived peptides and their release from *S. mansoni* was demonstrated.[27] Coincubation of adult worms with human polymorphonuclear leukocytes or *B. glabrata* immunocytes led to the appearance of α-MSH in the medium. The conclusion is that this α-MSH resulted from conversion of the parasite's ACTH by NEP. Because it is active at lower concentrations, it may be used for distant signaling. Recent data indicates that *S. mansoni* also can synthesize a Met-enkephalin-like molecule that it can "secrete" into an *in vitro* incubation medium.[28]

CONCLUSION

Thus, it would appear that opioid peptides have not only been conserved during the course of evolution but also that their activities in immunoregulation have been conserved as well.[29,30] Stereoselect, degradation, and dynamic receptor mechanisms also exist to ensure that opioid signals exhibit the highest fidelity. The potential for immunocytes to communicate with neural elements using opioid as well as cytokine signal molecules also exists. Furthermore, opioid molecules may be used by parasites to escape immune surveillance. Thus, the comparative study of opioid-associated immunoregulation represents a scientific area of timely interest that will be the subject of our future endeavors.[31]

REFERENCES

1. HUGHES, T. K., E. M. SMITH, J. A. BARNETT, R. CHARLES & G. B. STEFANO. 1991. LPS-stimulated invertebrate hemocytes: A role for immunoreactive TNF and IL-1. Dev. Comp. Immunol. **15:** 117–122.
2. HUGHES, T. K., E. M. SMITH, E. M. BARNETT, R. CHARLES & G. B. STEFANO. 1991. LPSand opioids activate distinct populations of *Mytilus edulis* immunocytes. Cell Tiss. Res. **264:** 317–320.
3. STEFANO, G. B., M. K. LEUNG, X. ZHAO & B. SCHARRER. 1989. Evidence for the involvement of opioid neuropeptides in the adherence and migration of immunocompetent invertebrate hemocytes. Proc. Natl. Acad. Sci. USA **86:** 626–630.
4. SHIPP, M. A., G. B. STEFANO, L. DADAMIO, S. N. SWITZER, F. D. HOWARD, J. SINISTERRA, B. SCHARRER & E. REINHERZ. 1990. Downregulation of enkephalin-mediated inflammatory responses by CD10/neutral endopeptidase 24.11. Nature **347:** 394–396.
5. STEFANO, G. B., P. CADET & B. SCHARRER. 1989. Stimulatory effects of opioid neuropeptides on locomotory activity and conformational changes in invertebrate and human immunocytes: Evidence for a subtype of delta receptor. Proc. Natl. Acad. Sci. USA **86:** 6307–6311.
6. STEFANO, G. B. 1989. Role of opioid neuropeptides in immunoregulation. Prog. Neurobiol. **33:** 149–159.
7. SHIPP, M. A., G. B. STEFANO, S. N. SWITZER, J. D. GRIFFIN & E. L. REINHERZ. 1991. CD10 (CALLA) neutral endopeptidase 24.11 modulates inflammatory peptide-induced changes in neutrophil morphology, migration, and adhesion proteins and is itself regulated by neutrophil activation. Blood **78:** 1834–1841.
8. STEFANO, G. B., L. R. PAEMEN & T. K. HUGHES. 1992. Autoimmunoregulation: Differential modulation of CD10/Neutral endopeptidase 24.11 by tumor necrosis factor and neuropeptides J. Neuroimmunol. **41:** 9–14.
9. STEFANO, G. B. 1992. Invertebrate and vertebrate immune and nervous system signal molecule commonalties. Cell. Mol. Neurobiol. **12:** 357–366.
10. LEUNG, M. K., S. LE, S. HOUSTON & G. B. STEFANO. 1992. Degradation of Met-enkephalin by hemolymph peptidases in *Mytilus edulis.* Cell. Mol. Neurobiol. **12:** 367–378.
11. STEFANO, G. B., M. A. SHIPP & B. SCHARRER. 1991. A possible immunoregulatory function for Met-enkephalin-Arg^6-Phe^7 involving human and invertebrate granulocytes. J. Neuroimmunol. **31:** 97–103.

12. STEFANO, G. B., P. MELCHIORRI, L. NEGRI, T. K. HUGHES & B. SCHARRER. 1992. (D-Ala^{2})-Deltorphin I binding and pharmacological evidence for a special subtype of delta opioid receptor on human and invertebrate immune cells. Proc. Natl. Acad. Sci. USA **89:** 9316–9320.
13. HUGHES, T. K. JR., E. M. SMITH, P. CADET, J. SINISTERRA, M. K. LEUNG, M. A. SHIPP, B. SCHARRER & G. B. STEFANO. 1990. Interaction of immunoactive monokines (IL-1 and TNF) in the bivalve mollusc *Mytilus edulis.* Proc. Natl. Acad. Sci. USA **87:** 4426–4429.
14. HUGHES, T. K., R. CHIN, E. M. SMITH, M. K. LEUNG & G. B. STEFANO. 1991. Similarities of signal systems in vertebrates and invertebrates: Detection, action, and interactions of immunoreactive monokines in the mussel, *Mytilus edulis* Adv. Neuroimmunol. **1:** 59–70.
15. STEFANO, G. B., E. R. SMITH & T. K. HUGHES. 1991. Opioid induction of immunoreactive interleukin-1 in *Mytilus edulis* and human immunocytes: An interleukin-1-like substance in invertebrate neural tissue. J. Neuroimmunol. **32:** 29–34.
16. PAEMEN, L. R., E. PORCHET-HENNERE, M. MASSON, M. K. LEUNG, T. K. HUGHES & G. B. STEFANO. 1992. Glial localization of interleukin-1α in invertebrate ganglia cell. Mol. Neurobiol. **12:** 463–472.
17. SZÛCS, A., B. G. STEFANO, T. K. HUGHES & K. S.-RÓZSA. 1992. Modulation of voltage-activated ion currents on identified neurons of *Helix pomatia* L. by interleukin-1. Cell Mol. Neurobiol. **12:** 429–438
18. SMITH, E. M., T. K. HUGHES, M. K. LEUNG & B. G. STEFANO. 1991. The production and action of ACTH-related peptides in invertebrate hemocytes. Adv. Neuroimmunol. **1:** 7–16.
19. HANSEN, B. L., B. N. HANSEN & B. SCHARRER. 1986. Immunocytochemical demonstration of a material resembling vertebrate ACTH and MSH in the corpus cardiacum—corpus allatum complex of the insect *Leucophaea maderae. In* Handbook of Comparative Aspects of Opioid and Related Neuropeptide Mechanisms, Vol. 1. G. B. Stefano, Ed.: 213–222. CRC Press. Boca Raton, FL.
20. OTTAVIANI, E., F. PETRAGLIA, G. MONTAGNANI, A. COSSARIZZA, D. MONTI & C. FRANCESCHI. 1990. Presence of ACTH and β-endorphin immunoreactive molecules in the freshwater snail *Planorbarius corneus* (L.) (Gastropoda, Pulmonata) and their possible role in phagocytosis. Reg. Peptides **27:** 1–9.
21. RAFAELI, A., P. MOSHITZKY & S. W. APPLEBAUM. 1987. Diuretic action and immunological cross-reactivity of corticotropin and locust diuretic hormone. Gen. Comp. Endocrinol. **67:** 1–6.
22. OTTAVIANI, E., E. CASELGRANDI, M. BONDI, A. COSSARIZZA, D. MONTI & C. FRANCESCHI. 1991. The "immune-mobile brain": Evolutionary evidence. Adv. Neuroimmunol. **1:** 27–39.
23. STEFANO, G. B., D. M. SMITH, E. M. SMITH & T. K. HUGHES. 1991. MSH can deactivate both TNF stimulated and spontaneously active immunocytes. *In* Molluscan Neurobiology. K. S. Kits, H. H. Boer & J. Joosse, Eds.: 206–209. North Holland Publishing Co. Amsterdam.
24. SMITH, E. M. T. K. HUGHES, F. HASHEMI & G. B. STEFANO. 1992. Immunosuppressive effects of ACTH and MSH and their possible significance in human immunodeficiency virus infection. Proc. Natl. Acad. Sci. USA **89:** 782–786.
25. SMITH, E. M., T. K. HUGHES, P. CADET & G. B. STEFANO. 1992. CRF-induced immunosuppression in human and invertebrate immunocytes cell. Mol. Neurobiol. **12:** 473–482.
26. DUREUS, P., D. M. LOUIS, A. V. GRANT, T. V. BILFINGER & G. B. STEFANO. 1993. Neuropeptide Y inhibits human and invertebrate immunocyte chemotaxis, chemokinesis and spontaneous activation. Cell. Mol. Neurobiol. In press.
27. DUVAUX-MIRET, O., G. B. STEFANO, E. M. SMITH, C. DISSOUS & A. CAPRON. 1992. Immunosuppression in the definitive and intermediate hosts of the human parasite *Schistosoma mansoni* by release of immunoactive neuropeptides. Proc. Natl. Acad. Sci. USA **89:** 778–781.
28. DUVAUX-MIRET, O., MICHAEL K. LEUNG, ANDRE CAPRON & GEORGE B. STEFANO. 1993.

An enkephalinergic system in *Schistosoma mansoni* that may participate in intra- and extra-organismic signaling. Exp. Parasit. **76:** 76–84.
29. STEFANO, G. B. 1991. Conformational matching a stabilizing signal system factor during evolution: Additional evidence in comparative neuroimmunology. Adv. Neuroimmunol. **1:** 71–82.
30. SCHARRER, B. 1991. Neuroimmunology: The importance and role of the comparative approach. Adv. Neuroimmunol. **1:** 1–6.
31. STEFANO, G. B., A. DIGENIS, S. SPECTOR, M. K. LEUNG, T. V. BILFINGER, M. H. MAKMAN, B. SCHARRER & N. N. ABUMRAD. 1993. Opiatelike substances in an invertebrate, a novel opiate receptor on invertebrate and human immunocytes, and a role in immunosuppression. Proc. Natl. Acad. Sci. USA, in press.

Role of Hemocyte-Derived Granular Components in Invertebrate Defense[a]

SADAAKI IWANAGA,[b] TATSUSHI MUTA,
TAKESHI SHIGENAGA, YOSHIKI MIURA, NORIAKI SEKI,
TETSU SAITO, AND SHUN-ICHIRO KAWABATA

Department of Biology
Faculty of Science
Kyushu University 33
Fukuoka 812, Japan

INTRODUCTION

The horseshoe crab hemolymph contains mainly one type of cells, called amebocytes or granulocytes, and these cells are extremely sensitive to bacterial endotoxins (lipopolysaccharides, LPS).[1-3] The cytoplasm of this hemocyte is filled with two types of granules, larger (L) and less dense and smaller (S) but dense.[4,5] The hemocytes release both these granular components into hemolymph plasma via an exocytosis mediated with LPS. The released components apparently participate in self-defense against invading microbes.[6-8] Recent immunocytochemical studies of these granules suggest that at least three clotting factors, including factor C, proclotting enzyme, and a clottable protein, coagulogen, all of which constitute the *Limulus* coagulation cascade, are localized in the L granules, and that the antimicrobial peptide, tachyplesin, is located exclusively in the S granules.[5] Therefore, these L and S granules seem to play an important role not only in the storage of biologically active substances, but also as a host defense mechanism against invading microbes.

This short review will focus on a role of the hemocyte-derived granular protein and peptide components participating in biological defense of *Limulus*. The more detailed review has been recently published.[9]

BIOLOGICAL FUNCTION OF HEMOCYTES

On the basis of cell morphology, the *Limulus* hemocyte is an oval, plate-shaped structure, 15 to 20 μm in its longest dimension. FIGURE 1 shows an electron micrograph of the hemocyte separated from the Japanese horseshoe crab, *Tachypleus tridentatus*.[5] The cell contains numerous dense granules classed into two major types of L and S granules. The L granules are larger (up to 1.5 μm in diameter) and less dense than the S granules (less than 0.6 μm in diameter).

[a]This work was supported by a Grant-in-Aid for Scientific Research from the Ministry of Education, Science and Culture of Japan.

[b]Address for correspondence: Sadaaki Iwanaga, Ph. D., Department of Biology, Faculty of Science, Kyushu University 33, Higashi-ku, Fukuoka 812, Japan.

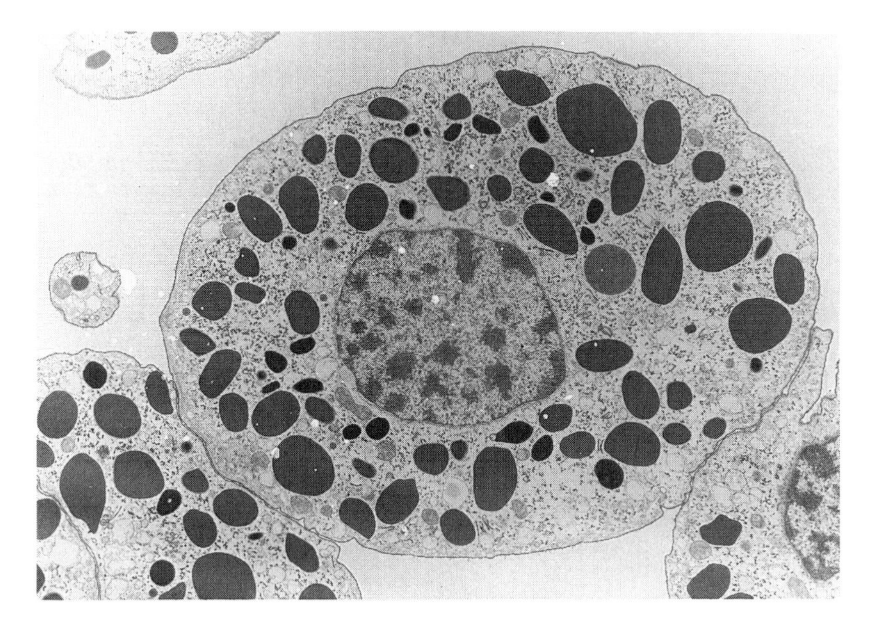

FIGURE 1. An electron micrograph of a cross section of a granular hemocyte showing the distribution of various organelles.

FIGURE 2 illustrates how *Limulus* hemocyte functions against invading microbes. When *Limulus* is infected by gram-negative bacteria, such as *Vibrio cholerae,* hemocytes recognize the microbes and bring about cell adhesion and aggregation.[4,5] Through this contact, some cellular reactions via LPS receptor/LPS-binding protein induce the fusion of granules and the plasma membrane, resulting in dispersal of the granular components into the hemolymph plasma. This exocytosis activates the intracellular clotting system,[9] and the clot generated during activation immobilizes the invaders. The peptide tachyplesins and anti-LPS factor released from small and large granules, respectively, act as the bactericidal substances.

HEMOCYTE-DERIVED LARGE AND SMALL GRANULAR COMPONENTS

The separation of L granules from the postnuclear supernatant was first done by Mürer *et al.* in 1975 using *Limulus polyphemus* hemocytes.[4] There has been no information, however, on the distribution of S granules determined in a centrifugal analysis. We have recently designed a method for separating L and S granules from the postnuclear supernatant of *T. tridentatus* hemocytes mechanically disrupted in the presence of phenylmethane sulfonyl fluoride (PMSF) and propranolol.[10] The additions of PMSF to the Tris-HCl buffer used for preparation of the postnuclear supernatant and the presence of heparin during the sucrose density gradient cen-

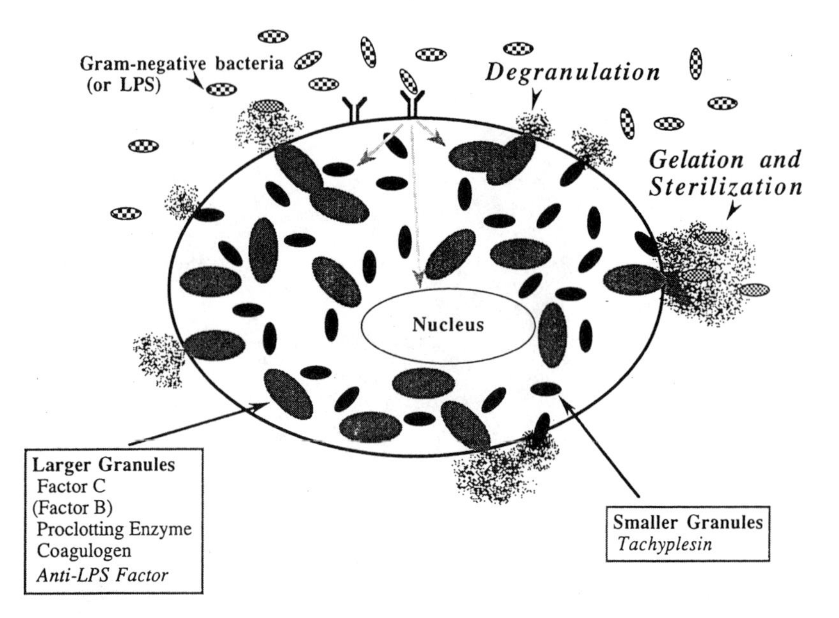

FIGURE 2. The hypothetical processes of the coagulogen-based clotting system and containment of invading microbes. On contact with bacterial LPS, the L granules are released more rapidly than the S granules, although almost all granules are finally exocytosed. This exocytosis is associated with clot formation, the process being complete within 90 seconds.[5]

trifugation makes it feasible to significantly reduce some aggregation of the granules and to avoid their destruction. Through these procedures, the two types of L and S granules could be clearly separated into the fractions, at around 2 M sucrose density. The method is reproducible, and each separated granule seems to be a relatively pure form.[10] Comparisons of granular components thus subfractionated reveal that proteins and peptides in L and S granules are quantitatively unique. Protein profiles obtained from different batches of each of the granules are always similar, showing the coexistence of four major components of coagulogen, factor C, proclotting enzyme and anti-LPS factor in the L-granules, and a major tachyplesin I peptide in the S granules (FIG. 2). In particular, coagulogen and tachyplesin are highest in L and S granules so that there is a coagulogen granule for the former and a tachyplesin one for the latter. Therefore, these proteins and peptides can serve as molecular markers for studies on LPS-mediated exocytosis and for the intracellular formation of granules.

TABLE 1 summarizes the biochemical properties of the hemocyte-derived granular components so far characterized in the author's laboratory.[9] Many components are closely associated with hemolymph coagulation cascade and biological defense systems in *Limulus*. In addition to the exclusive coexistence of tachyplesin and its analogues, S granules contain six other protein components with molecular masses less than 30 kDa.[10] The amino acid compositions of these six proteins show a characteristic property,[10] indicating the presence of Arg-rich (S1), Asx-rich (S2),

TABLE 1. Hemocyte Components Closely Associated with Biological Defense System in *Limulus (Tachypleus tridentatus)*

Component	Mol. Wt. (residues)	Carbohydrate Content (%)	Active Form	Function	Intracellular Localization
Coagulogen	19,700 (175)	None	Coagulin (two chains)	Gelation	L-granule[a]
Factor C	123,000 (994)	9%	Factor $\bar{C}$ (three chains)	Serine protease (LPS-sensitive)	L-granule
Factor B	64,000 (400)	Presence	Factor $\bar{B}$ (two chains)	Serine protease	Unknown
Proclotting enzyme	54,000 (346)	11%	Clotting enzyme (two chains)	Serine protease	L-granule
Factor G	110,000 (a=673) (b=278)	Presence	Factor $\bar{G}$ (two subunits)	Serine protease	Unknown
Factor D	42,000 (386)	—	Unknown	Pseudo-serine protease	Unknown
Transglutaminase (TGase)	86,000 (764)	None	TGase (Ca^{2+}-dependent)	Cross-link	Cytosol
Anti-LPS factor	11,600 (102)	None		Antimicrobial	L-granule
Tachyplesin precursor	9,335 (77)	None	Tachyplesin	—	ER
Tachyplesin	2,263 (17)	None		Antimicrobial	S-granule[a]
Polyphemusin	2,453 (18)	None		Antimicrobial	S-granule
Factor C inhibitor	50,000 (394)	Presence		Anticoagulant	L-granule
Trypsin inhibitor	6,800 (61)	None		Protease inhibitor (Aprotinin-type)	Unknown
L-6	27,000 (221)	None		LPS-binding	L-granule
8.6-kDa protein	8,671 (81)	None		TGase substrate	L-granule
Pro-rich protein	80,000	Presence		TGase substrate	L-granule

[a]L, large; S, small.

Gly/Arg-rich (S3), Leu/Asx-rich (S4), and Ala-rich (S5) proteins and one glycoprotein (S6). Although biological functions of these components remain to be determined, evidence for the existence of an antimicrobial tachyplesin peptide in the S granules is pertinent, since this peptide is stored in the mature form, like mammalian defensins in azurophil granules.[11]

On the other hand, L granules contain at least 20 proteins, the majority of which have molecular masses between 8 kDa and 94 kDa. The major component is a clottable protein, coagulogen, which plays a central role in the *Limulus* clotting system. As suggested by Mürer *et al.*,[4] all the clotting factors required for the LPS-mediated coagulin gel formation could exist in L granules. In fact, both activities of LPS-sensitive serine protease zymogens, factor C, and the proclotting enzyme, except for clotting factor B (not examined), have been detected in this granule, in addition to coagulogen (TABLE 1). Also, L granules obtained after sucrose gradient centrifugation form a gel during dialysis to remove sucrose, under conditions of no sterilization. These results are in good agreement with those in Mürer's report[4] and the subcellular localizations of these clotting factors identified histocytochemically.[5] Furthermore, chemical characterizations of other protein components in L granules indicate the coexistence of the trypsin inhibitor, LTI,[45] which has been recently isolated from the hemocyte lysate of *L. polyphemus.* The most interesting finding is that two natural substrates, the 8.6-kDa protein and the Pro-rich protein, rapidly cross-linked enzymatically by *Limulus* transglutaminase,[12] exist in L granules. Since these protein substrates, in addition to several clotting factors, are secreted from the hemocyte into the external milieu during LPS-mediated degranulation, they may participate in immobilization against invading microbes, as suggested in cellular fibronectin cross-linked by mammalian transglutaminase, factor XIII.[13]

BIOCHEMICAL PROPERTIES OF *LIMULUS* CLOTTING FACTORS

Since the 1980s, we have directed attention to molecular mechanisms involved in hemolymph coagulation in *Limulus,* and we defined the protease cascade, shown in FIGURE 3.[8,9] This cascade is based on three kinds of serine protease zymogens, factor C, factor B, proclotting enzyme, and coagulogen. LPS activates the zymogen factor C to the active factor $\bar{C}$. Factor $\bar{C}$ then activates factor B to factor $\bar{B}$, which in turn converts the proclotting enzyme to the clotting enzyme. Each activation proceeds by limited proteolysis. The resulting clotting enzyme cleaves two bonds in coagulogen, a fibrinogen-like molecule in arthropods, to yield an insoluble coagulin gel. The coagulation cascade is also activated by (1,3)-β-D-glucan; the serine protease zymogen, factor G, which is initially activated leads to activation of the proclotting enzyme. The biochemical properties of these clotting factors are briefly described below.

Coagulogen and Coagulin

Among clotting factors so far identified in hemocytes, coagulogen plays a central role in the *Limulus* clotting system (FIG. 3). In the intact animal, the soluble precursor

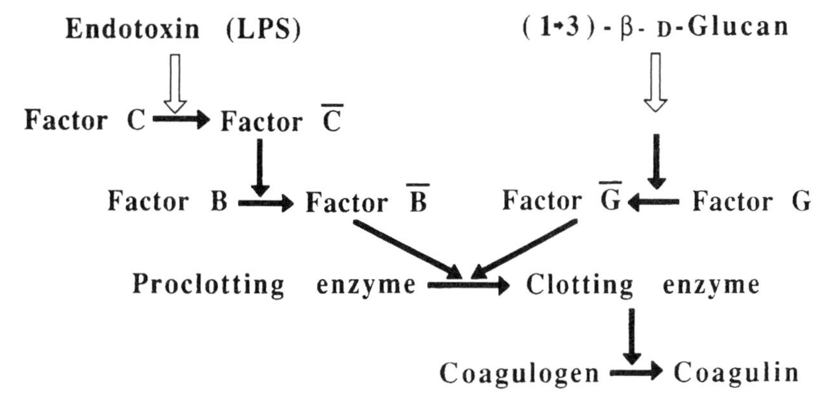

FIGURE 3. Hemolymph coagulation cascade system in *Limulus (Tachypleus tridentatus).*

of this protein is absent in the plasma but is sequestered in the cytoplasmic L granules of hemocytes. Exocytosis from this granule releases coagulogen into the external milieu. The extracellular clot is composed of fibers of the protein named coagulin. Coagulogen consists of a single basic polypeptide chain with a molecular weight of 19,700 (±50).[14] The clotting enzyme cleaves the Arg-Gly and Arg-Thr linkages, both located at the amino-terminal region. The amino acid sequences of four coagulogens isolated from the hemocytes of American *(L. polyphemus),* Japanese *(T. tridentatus),* and two Southeast Asian *(T. gigas* and *Carcinoscorpius rotundicauda)* horseshoe crabs have now been elucidated.[8] In addition, the cDNAs for two of them *(L. polyphemus* and *T. tridentatus)* have been cloned.[15,16] Upon gelation of all the coagulogens by a *Limulus* clotting enzyme, a large peptide, named peptide C (28 residues in length), is released in common from the inner portion of the parent molecules. The resulting gel consists of two chains of A (18 residues) and B (129 residues), bridged by two disulfide linkages.[8] Among *Limulus* clotting factors, the coagulogen molecule essential for gelation of hemolymph is the smallest. This is in contrast to vertebrate fibrinogen, which is known to be one of the largest macromolecules. On activation, however, coagulogen molecules like fibrinogen polymerize by forming coagulin gel and convert to a protofibril observed electron microscopically. *Limulus* TGase may also participate in formation of the coagulin gel network and lead to development of covalently cross-linked heterooligomers with other proteins, such as 8.6-kDa protein and Pro-rich protein coexisting in L granules in hemocytes.[12] These processes seem important for immobilization of invading microbes. At present, however, the biochemical mechanisms related to polymerization and gelation of coagulogen are poorly understood.

Proclotting Enzyme and Clotting Enzyme

A clotting enzyme that catalyzes the transformation of coagulogen to coagulin exists as its zymogen form, in hemocytes.[17,18] This proclotting enzyme is activated by both factor $\overline{B}$ and factor $\overline{G}$ (FIG. 3), and it is a single chain glycoprotein with an apparent molecular mass of 54 kDa.[18] Upon activation by factor $\overline{B}$, it is converted into a two-chain active form of clotting enzyme, composed of light (L, 25-kDa) and heavy (H, 31-kDa) chains. The proclotting enzyme consists of 346 amino acid residues, and the cleavage site associated with the zymogen activation is Arg_{98}-Ile_{99}, the sequence of which connects the L and H chains.[18] The H chain of the proclotting enzyme shows a typical structure of serine protease. It contains the His_{143}-Asp_{199}-Ser_{297} triad known to be the catalytic triad of serine proteases. The substrate binding site corresponding to Ser_{189} in chymotrypsin numbering is aspartic acid, thereby indicating that this clotting enzyme has a typical trypsin-like specificity (FIG. 4). The amino-terminal portion of the L chain consists of a novel disulfide-knotted structure. This structure forms a "clip-like" shape consisting of a discrete domain, which has a sequence similarity in its amino-terminal light-chain portion with *Drosophila* protein called serine protease easter precursor.[9] Although the significance is unknown, these results do suggest that the "clip-like" domain found in *Limulus* proclotting enzyme and in factor B (described below) is one of the common structural elements in the serine protease zymogen of invertebrate animals.

Factor B and Factor $\overline{B}$

The zymogen factor B is activated by factor $\overline{C}$, and it in turn activates proclotting enzyme (FIG. 3). Factor B is a single-chain glycoprotein with a molecular mass of 64 kDa (TABLE 1). Upon activation of the zymogen factor B by active factor $\overline{C}$, it is converted into factor $\overline{B}$ with heavy (32-kDa) and light (25-kDa) chains, releasing an activation peptide.[19] We have determined the cDNA sequence for zymogen factor B.[20] It consists of 400 amino acid residues with 23 residues of signal sequence. The mature protein has 377 amino acids, and its entire sequence is similar to that of the *Limulus* proclotting enzyme.[21] The sequence identity of the carboxy-terminal serine protease domains between factor B and proclotting enzyme is 43.9% (FIG. 4). Moreover, the amino-terminal regions up to the 60th residue of both proteins share a similar structure with six cysteines, suggesting that these proteins arose from gene duplication.

Factor C and Factor $\overline{C}$

As shown in FIGURE 3, factor C is an initial activator of the clotting cascade triggered by LPS.[22] The zymogen factor C is a glycoprotein with a molecular mass of 123 kDa, consisting of a heavy chain (80 kDa) and a light chain (43 kDa).[23] Factor C is converted autocatalytically to an activated form, designated factor $\overline{C}$ (123 kDa), in the presence of LPS or lipid A.[24] Upon activation, a single cleavage of the Phe-Ile bond in the light chain occurs, resulting in accumulation of two new fragments, a B chain (34 kDa) and an A chain (8.5 kDa).[25] The zymogen factor C consists of 994 amino acid residues[26] (TABLE 1). In the H and L chains, there are several interesting amino acid sequences, including five repeating units ("sushi" domain or short consensus repeat) of about 60 amino acids each, an epidermal growth factor (EGF)-like

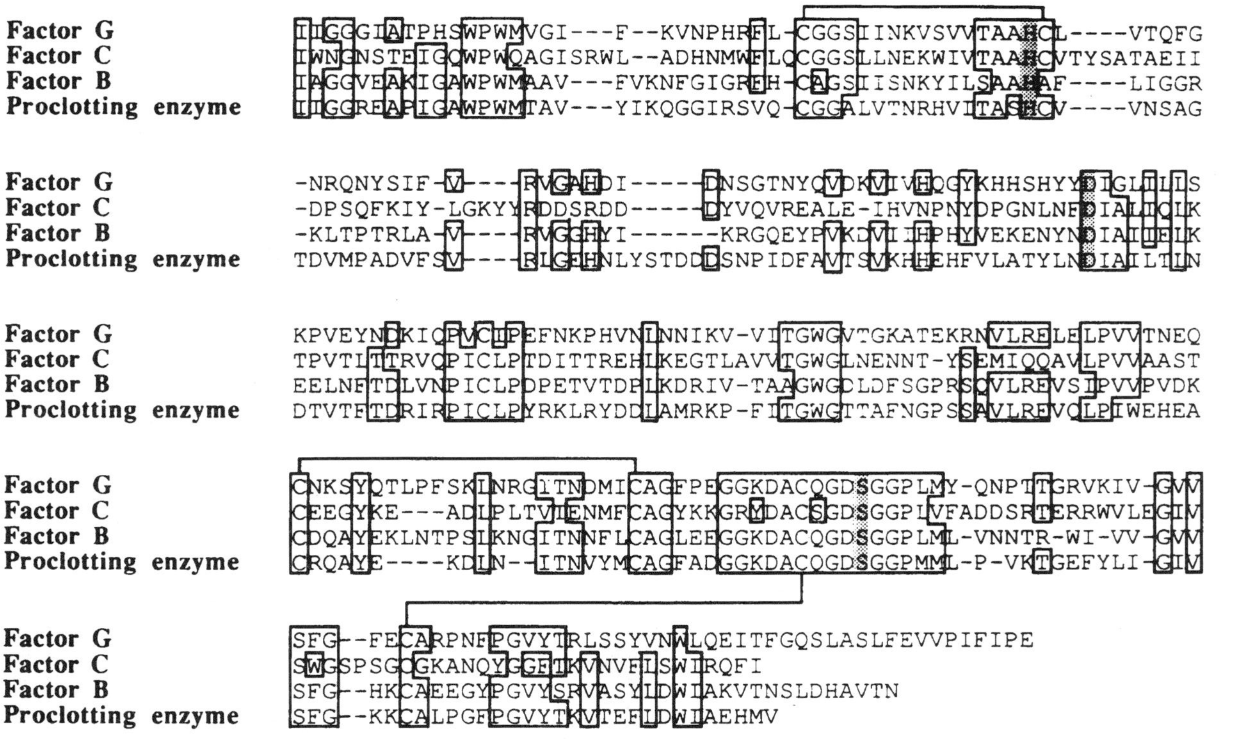

FIGURE 4. Sequence alignment of serine protease domains in *Limulus* clotting factors. The disulfide locations are taken from data on the *Limulus* proclotting enzyme.[18] A serine active-site triad, Asp-102, His-57, Ser-195, in chymotrypsin numbering is shadowed.

domain, and a C-type lectin-like domain. The B chain portion of factor C is composed of a typical serine protease with an active-site triad[26,27] (FIG. 4), which shows the strongest homology with human α-thrombin (36.7% identity).

Factor G and Factor $\bar{G}$

In 1981, we and others obtained evidence for a (1,3)-β-D-glucan-mediated clotting pathway in the hemocyte lysate of *T. tridentatus.*[28,29] An initiator of this pathway was named factor G. This (1,3)-β-D-glucan-sensitive factor G consists of two subunits, a (72 kDa) and b (34 kDa). In the presence of (1,3)-β-D-glucan, factor G is autocatalytically converted to active factor $\bar{G}$ (FIG. 3). We have just cloned and sequenced cDNA, encoding for the a and b subunits of factor G, respectively (TABLE 1). Subunit b is a typical serine protease zymogen with a calculated molecular mass of 30.8 (FIG. 4) and shows the strongest sequence similarity with the serine protease domain of *Limulus* factor B (40.5% identity). Subunit a is a protein with intriguing features. The amino-terminal amino acid sequence of this protein shows similarity with the carboxy-terminal sequence of β-1,3-glucanase A1 (EC 3.2.1.39) from *Bacillus circulans,* suggesting that this region may interact with (1,3)-β-D-glucan polymer and implying glucanase activity. Furthermore, the carboxy-terminal sequence of subunit a has a long direct repeat (126 amino acids in length), which shows a sequence similarity to the amino-terminal portion of xylanase Z (EC 3.2.1.8) from *Clostridium thermocellum.* Between the amino-terminal glucanase and the carboxy-terminal xylanase homologous regions, there are five "QQWSY-like" motifs, the functions of which are unknown. Therefore, factor G is a newly identified type of serine protease zymogen (subunit b), noncovalently associated with the (1,3)-β-D-glucan recognition subunit a, which plays an important role in the (1,3)-β-D-glucan-mediated clotting pathway.

Transglutaminase (TGase)

In the mammalian clotting system, the fibrin clot generated via the cascade reaction is cross-linked to form a huge fibrin network as well as being cross-linked to other plasma proteins, such as soluble fibronectin, α_2-macroglobulin, and α_2-plasmin inhibitor. This final step is catalyzed by plasma TGase factor XIIIa, and the cross-linking of fibrin with itself and with other proteins is essential for normal hemostasis and wound healing. It is, therefore, expected that in the *Limulus* clotting system a TGase might participate not only in cross-linking of the coagulin gel, but also in immobilization against invading microorganisms. Indeed, in 1973, Wilson *et al.* discovered the existence of a Ca^{2+}-dependent TGase in *L. polyphemus* amebocytes and showed the TGase-catalyzed incorporation of [^{14}C]putrescine into β-lactoglobulin and α-casein.[30] Recently, we purified and characterized a cellular TGase from the hemocytes of *T. tridentatus*[12] and examined the cDNA sequence for TGase mRNA.[31] The *Limulus* TGase has a molecular mass of 86 kDa and properties of the mammalian type II TGase-like enzyme. One of the cloned cDNAs for TGase consists of 2,884 base pairs. An open reading frame of 2,292 base pairs encodes a sequence comprising 764 residues of the mature protein with a molecular mass of 87,021 Da. *Limulus* TGase shows significant sequence similarity with the mammalian TGase family, as follows: guinea pig liver TGase (32.7%), human factor XIIIa subunit (34.7%), human keratinocyte TGase (37.6%), and human erythrocyte band 4.2 (23.0%). *Limulus* TGase has an unique amino-terminal cationic extension of 60 residues with no homology to the amino-termini of mammalian TGases.[31] The TGase activity is Ca^{2+}-dependent and is inhibited by primary amines, EDTA, and SH-reagents. Moreover, two major potential substrates for TGase have been identified in the hemocyte L granules, using dansylcadaverine incorporation in the presence of 10 mM $CaCl_2$ and 10 mM dithiothreitol. Of these protein substrates, an 80-kDa protein contains a large number of proline residues, amounting to about 22% of the total amino acids (TABLE 1). On the other hand, an 8.6-kDa protein abundantly present in L granules is a Cys-rich protein consisting of 81 amino acid residues and a calculated molecular mass of 8,671. Also, the 8.6-kDa protein is readily cross-linked intermolecularly by TGase, forming multimers as large as pentamers.[12] We speculate that, like plasma factor XIIIa, *Limulus* TGase and its two protein substrates in hemocytes may play an important role in the defense of the animal against invading microbes.

FIGURE 5 summarizes the gross structures of *Limulus* clotting factors so far described. Like mammalian clotting factors, factor C, factor B, and proclotting enzyme, which constitute the LPS-mediated clotting pathway (FIG. 3), are typical serine protease zymogens related to the trypsin family. They have structural domains in their amino-terminal portions that are different from those of mammalian clotting factors. There are no γ-Gla, kringle, finger, and apple domains.[32] Moreover, mammalian clotting factors do not have the "sushi" domain, whereas *Limulus* factor C contains several "sushi" domains, as do complement factors, and the proclotting enzyme and factor B contain a "clip-like" domain in the amino-terminal portions. The zymogen factor G, which is an initiator for (1,3)-β-D-glucan-mediated clotting pathway (FIG. 3), is an intriguing molecule, as described above. When the zymogen factor G is exposed to (1,3)-β-D-glucan polymer, it is autocatalytically activated to active factor $\bar{G}$. Although structural changes of subunits a and b during the β-glucan-mediated activation are unknown, a heretofore undefined mechanism may be operative in activation of the zymogen factor G.

ANTIMICROBIAL SUBSTANCES FOUND IN LARGE AND SMALL GRANULES

Anti-LPS Factor

In 1982, a protein component that inhibits the coagulation cascade was detected in the hemocyte lysate from Japanese and American horseshoe crabs and was named anti-LPS factor. The purified protein, which consists of 102 amino acid residues, specifically inhibits the LPS-mediated activation of zymogen factor C and has a strong antibacterial effect, especially on the growth of gram-negative, R-type bacteria.[33,34] Furthermore, it has hemolytic activity on red blood cells sensitized with LPS and cytolytic activity on LPS-sensitized polymorphonuclear leukocytes, mono-

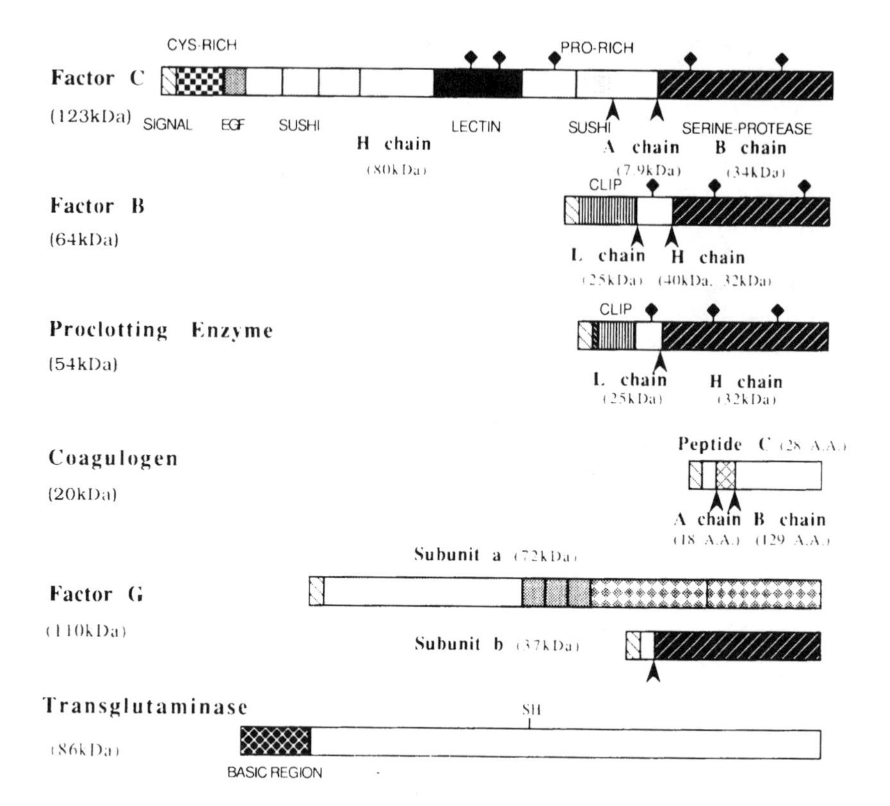

FIGURE 5. Clotting factors. Domain structures of *Limulus* clotting factors. The arrowheads indicate cleavage sites for zymogen activation. The potential carbohydrate attachment sites are indicated by closed diamonds.

nuclear cells, and human leukemia cells.[35] This anti-LPS factor is now known to be localized in the L granules.[5] In our continued study of the *Limulus* clotting system, we also found tachyplesin, which binds with LPS and has antimicrobial activity against both gram-negative and gram-positive bacteria, and also against fungi.[36]

Tachyplesin Family and Its Precursors

Tachyplesin I is a cationic peptide first isolated from acid extracts of Japanese horseshoe crab *(T. tridentatus)* hemocytes.[36] Later, two peptide analogues of tachypresin I, named tachyplesins II and III, were isolated from hemocytes of two species of Southeast Asian horseshoe crabs, *T. gigas* and *C. rotundicauda.*[37,38] In the hemocyte debris of *L. polyphemus,* two tachyplesin analogues, named polyphemusins I and II, were also found.[37] The primary structures of the tachyplesin family are shown in FIGURE 6. They are composed of 17 or 18 residues with arginine α-amide at the carboxy-terminal end. Tachyplesin has a characteristic structure with three tandem

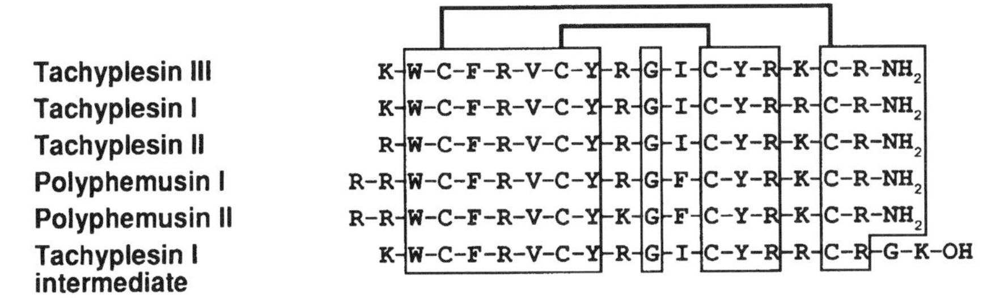

FIGURE 6. Primary structures of six kinds of tachyplesins and its analogues isolated from four species of horseshoe crab. Tachyplesin III was isolated from *T. gigas;* I from *T. tridentatus, T. gigas,* and *C. rotundicauda;* and II from *T. tridentatus,* respectively. Polyphemusins I and II were isolated from *L. polyphemus.*

repeats of a tetrapeptide sequence, namely hydrophobic amino acid Cys, hydrophobic amino acid Arg.[36] The ^{1}H-NMR studies indicate that tachyplesin I takes on a fairly rigid conformation constrained by two disulfide bridges and adopts a conformation consisting of an anti-parallel β-sheet (residues 3–8 and 11–16) connected by a β turn (residues 8–11).[39] In this planar conformation, five bulky hydrophobic side groups are localized on one side of the plane, and six cationic side groups are distributed at the "tail" part of the molecule (residues 1–5 and 14–17). This conformation may be closely associated with the bactericidal activity.

The tachyplesin precursors consist of 77 amino acids with 23 residues in a presegment,[40] and there are two types of mRNAs corresponding to the isopeptides of tachyplesins I and II. Both precursors contain a putative signal peptide, a processing peptide sequence, and a carboxy-terminal amidation signal, Gly-Lys-Arg, connected to the mature tachyplesin peptide. Moreover, an unusual acidic amino acid cluster with Asp-Glu-Asp-Glu-Asp-Asp-Asp-Glu-Glu-COOH is present in the carboxy-terminal portions of both precursors. This acidic region might interact with a cationic part of the tachyplesin peptide to stabilize a conformational structure of the precursor for proteolytic degradation. Tachyplesin displays potent antimicrobial activity against several strains of microorganisms.[37] In the presence of tachyplesin at 3.5 μg/ml, *Salmonella* strains irreversibly lose viability. Therefore, tachyplesin seems likely to act as an arthropod antibiotic for the defense of horseshoe crabs against microbial infections.[9]

SUMMARY

FIGURE 2 illustrates an outline of the cellular and humoral defense systems in limulus. On the basis of the knowledge described above, it is suggested that granular components present in L and S granules in the hemocytes play a decisive role in the biological defense for this animal. The isolated L granules contain at least three clotting factors plus coagulogen as the major component. The known anti-LPS factor and a number of additional unknown protein components are also present in the L granules. On the other hand, the isolated S granules contain antimicrobial tachyplesins as the major component, in addition to six unidentified proteins. We speculate that the L-granule-derived protein components, which probably contain all the factors essential for the *Limulus* clotting system participate, in immobilizing invading microbes, and that the S-granule-derived tachyplesins contribute to a self-defense system against invaders. Although we have not mentioned hemolymph plasma components, there are many humoral factors,[3] such as proteinase inhibitors, α_2-macroglobulin, various lectins, C-reactive protein,[41] and polyphemin, all of which are important for antimicrobial defense. Furthermore, Liu and colleagues have reported several endotoxin-binding proteins[42] and a cell-adhesion protein found in the *Limulus* hemocytes.[43] Although the exact functions of these substances are unknown, they may act in concert with other components to provide biological defense for the animal.[44] Nevertheless, compared to our knowledge of mammalian blood cells, much less remains to be learned of biological/physiological events in horseshoe crab hemocytes.

ACKNOWLEDGMENTS

We would like to thank Drs. T. Morita, Y. Toh, M. Niwa, T. Miyata, T. Nakamura, F. Tokunaga, Y. Kawano, R. Hashimoto, and T. Oda for collaboration in an earlier stage of this work. We also thank Mrs. Itsuko Edamitsu for her expert secretarial assistance.

REFERENCES

1. Bang, F. B. 1956. A bacterial disease of *Limulus polyphemus.* Bull. Johns Hopkins Hosp. **98:** 325–350.
2. Levin, J. & F. B. Bang. 1964. The role of endotoxin in the extracellular coagulation of *Limulus* blood. Bull. Johns Hopkins Hosp. **115:** 265–274.
3. Armstrong, P. B. 1991. Cellular and humoral immunity in the horseshoe crab, *Limulus polyphemus. In* Immunology of Insects and Other Arthropods. A. P. Gupta, Ed.: 3–17. CRC Press. Boca Raton, FL.
4. Mürer, E. H., J. Levin & R. Holm. 1975. Isolation and studies of the granules of the amebocytes of *Limulus polyphemus,* the horseshoe crab. J. Cell Physiol. **86:** 533–542.
5. Toh, Y., A. Mizutani, F. Tokunaga, T. Muta & S. Iwanaga. 1991. Morphology of the granular hemocytes of the Japanese horseshoe crab *Tachypleus tridentatus* and immunocytochemical localization of clotting factors and antimicrobial substances. Cell Tissue Res. **266:** 137–147.
6. Armstrong, P. B. & F. R. Rickles. 1982. Endotoxin-induced degranulation of the *Limulus* amebocyte. Exp. Cell Res. **140:** 15–24.
7. Ornberg, R. L. 1985. Exocytosis in *Limulus* amebocytes. *In* Blood Cells of Marine Invertebrates: Experimental Systems in Cell Biology and Comparative Physiology. MBL Lectures in Biology. Vol. 6: 127–142. W. D. Cohen, Ed. Alan R. Liss, Inc. New York.
8. Iwanaga, S., T. Morita, T. Miyata, T. Nakamura & J. Aketagawa. 1986. The hemolymph coagulation system in invertebrate animals. J. Protein Chem. **5:** 225–268.
9. Iwanaga, S., T. Miyata, F. Tokunaga & T. Muta. 1992. Molecular mechanism of hemolymph clotting system in limulus. Thrombos. Res. **68:** 1–32.
10. Shigenaga, T., Y. Takayenoki, S. Kawasaki, N. Seki, T. Muta, Y. Toh, A. Ito & S. Iwanaga. 1993. Separation of large and small granules from horseshoe crab *(Tachy-*

pleus tridentatus) hemocytes and characterization of their components. J. Biochem. (Tokyo) **114:** 307–316.
11. Gabay, J. E. & R. P. Almeida. 1993. Antibiotic peptides and serine protease homologs in human polymorphonuclear leukcocytes: Defensins and azurocidin. Curr. Opinion Immunol. **5:** 97–102.
12. Tokunaga, F., M. Yamada, T. Miyata, Y. L. Ding, M. Hiranaga-Kawabata, T. Muta, S. Iwanaga, A. Ichinose & E. W. Davie. 1993. Limulus transglutaminase: Its purification, characterization and identification of the intracellular substrates. J. Biol. Chem. **268:** 252–261.
13. Mosher, D. F. & R. A. Proctor. 1980. Binding and factor XIIIa-mediated cross-linking of a 27-kilodalton fragment of fibronectin to *Staphylococuss aureus.* Science **209:** 927–929.
14. Miyata, T., M. Hiranaga, M. Umezu & S. Iwanaga. 1984. Amino acid sequence of the coagulogen from *Limulus polyphemus* hemocytes. J. Biol. Chem. **259:** 8924–8933.
15. Miyata, T., H. Matsumoto, M. Hattori, Y. Sakaki & S. Iwanaga. 1986. Two types of coagulogen mRNAs found in horseshoe crab *(Tachypleus tridentatus)* hemocytes: Molecular cloning and nucleotide sequences. J. Biochem. (Tokyo) **100:** 213–220.
16. Cheng, S. M., A. Suzuki, G. Zon & T.-Y. Liu. 1986. Characterization of a complementary deoxyribonucleic acid for the coagulogen of *Limulus polyphemus.* Biochim. Biophys. Acta **868:** 1–8.
17. Nakamura, T., T. Morita & S. Iwanaga. 1985. Intracellular proclotting enzyme in limulus *(Tachypleus tridentatus)* hemocytes: Its purification and properties. J. Biochem. (Tokyo) **97:** 1561–1574.
18. Muta, T., R. Hashimoto, T. Miyata, H. Nishimura, Y. Toh & S. Iwanaga. 1990. Proclotting enzyme from horseshoe crab hemocytes: cDNA cloning, disulfide locations, and subcellular localization. J. Biol. Chem. **265:** 22426–22433.
19. Nakamura, T., T. Horiuchi, T. Morita & S. Iwanaga. 1986. Purification and properties of intracellular clotting factor, factor B, from horseshoe crab *(Tachypleus tridentatus)* hemocytes. J. Biochem. (Tokyo) **99:** 847–857.
20. Muta, T., R. Hashimoto, T. Oda, T. Miyata & S. Iwanaga. 1991. *Limulus* clotting enzyme and factor B associated with enodotoxin-sensitive coagulation cascade: Novel serine protease zymogens with a new type of "disulfide-knotted domain." Thrombos. Haemostas. **65:** 935 (abstract).
21. Muta, T., T. Oda & S. Iwanaga. 1993. Horseshoe crab coagulation factor B: A serine protease zymogen activated by a cleavage between ILe-ILe bond. J. Biol. Chem. **268:** 21384–21388.
22. Nakamura, T., T. Morita & S. Iwanaga. 1986. Lipopolysaccharide-sensitive serine-protease zymogen (factor C) found in *Limulus* hemocytes: Isolation and characterization. Eur. J. Biochem. **154:** 511–521.
23. Tokunaga, F., T. Miyata, T. Nakamura, T. Morita, K. Kuma, T. Miyata & S. Iwanaga. 1987. Lipopolysaccharide-sensitive serine-protease zymogen (factor C) of horseshoe crab hemocytes: Identification and alignment of proteolytic fragment produced during the activation show that it is a novel type of serine protease. Eur. J. Biochem. **167:** 405–416.
24. Nakamura, T., F. Tokunaga, T. Morita, S. Iwanaga, S. Kusumoto, T. Shiba, T. Kobayashi & K. Inoue. 1988. Intracellular serine-protease zymogen, factor C, from horseshoe crab hemocytes: Its activation by synthetic lipid A analogues and acidic phospholipids. Eur. J. Biochem. **176:** 89–94.
25. Nakamura, T., F. Tokunaga, T. Morita & S. Iwanaga. 1988. Interaction between lipopolysaccharide and intracellular serine protease zymogen, factor C, from horseshoe crab *(Tachypleus tridentatus)* hemocytes. J. Biochem. (Tokyo) **103:** 370–374.
26. Muta, T., T. Miyata, Y. Misumi, F. Tokunaga, T. Nakamura, Y. Toh, Y. Ikehara & S. Iwanaga. 1991. *Limulus* factor C: An endotoxin-sensitive serine protease zymogen with a mosaic structure of complement-like, epidermal growth factor–like, and lectin-like domains. J. Biol. Chem. **266:** 6554–6561.
27. Tokunaga, F., H. Nakajima & S. Iwanaga. 1991. Purification and characterization of lipopolysaccharide-sensitive serine protease zymogen (factor C) isolated from *Limulus*

polyphemus hemocytes: A newly identified intracellular zymogen activated by α-chymotrypsin, not by trypsin. J. Biochem. (Tokyo) **109:** 150–157.
28. Morita, T., S. Tanaka, T. Nakamura & S. Iwanaga. 1981. A new (1-3)-β-D-glucan-mediated coagulation pathway found in *Limulus* amebocytes. FEBS Lett. **129:** 318–321.
29. Kakinuma, A., T. Asano, H. Torii & Y. Sugino. 1981. Gelation of *Limulus* amebocyte lysate by an antitumor (1-3)-β-D-glucan. Biochem. Biophys. Res. Commun. **101:** 433–439.
30. Wilson, J., F. R. Rickles, P. B. Armstrong & L. Lorand. 1992. N^{ε} (γ-glutamyl) lysine crosslinks in the blood clot of the horseshoe crab, *Limulus polyphemus.* Biochem. Biophys. Res. Communs. **188:** 655–661.
31. Tokunaga, F., T. Muta, S. Iwanaga, A. Ichinose, E. W. Davie, K. Kuma & T. Miyata. 1993. *Limulus* hemocyte transglutaminase: cDNA cloning, amino acid sequence, and tissue localization. J. Biol. Chem. **268:** 262–268.
32. Davie, E. W., K. Fujikawa & W. Kisiel. 1991. The coagulation cascade: Initiation, maintenance and regulation. Biochemistry **30:** 10363–10370.
33. Aketagawa, J., T. Miyata, S. Ohtsubo, T. Nakamura, H. Hayashida, T. Miyata & S. Iwanaga. 1986. Primary structure of *Limulus* anticoagulant anti-lipopolysaccharide factor. J. Biol. Chem. **261:** 7375–7365.
34. Muta, T., T. Miyata, F. Tokunaga, T. Nakamura & S. Iwanaga. 1987. Primary structure of anti-lipopolysaccharide factor from American horseshoe crab, *Limulus polyphemus.* J. Biochem. (Tokyo) **101:** 1321–1330.
35. Ohashi, K., M. Niwa, T. Nakamura, T. Morita & S. Iwanaga. 1984. Anti-LPS factor in the horseshoe crab, *Tachypleus tridentatus*: Its hemolytic activity on the red blood cell sensitized with lipopolysaccharide. FEBS Lett. **176:** 207–210.
36. Nakamura, T., H. Furunaka, T. Miyata, F. Tokunaga, T. Muta, S. Iwanaga, M. Niwa, T. Takao & Y. Shimonishi. 1988. Tachyplesin, a class of antimicrobial peptide from hemocytes of the horseshoe crab, *(Tachypleus tridentatus).* J. Biol. Chem. **263:** 16709–16713.
37. Miyata, T., F. Tokunaga, Y. Yoneya, K. Yoshikawa, S. Iwanaga, M. Niwa, T. Takao & Y. Shimonishi. 1989. Antimicrobial peptides isolated from horseshoe crab hemocytes, tachyplesin II, and polyphemusins I and II: Chemical structures and biological activity. J. Biochem. (Tokyo) **106:** 663–668.
38. Muta, T., T. Fujimoto, H. Nakajima & S. Iwanaga. 1990. Tachyplesins isolated from hemocytes of Southeast Asian horseshoe crabs *(Carcinoscorpius rotundicauda* and *Tachypleus gigas)*: Identification of a new tachyplesin, tachyplesin III, and a processing intermediate of its precursor. J. Biochem. (Tokyo) **108:** 261–266.
39. Kawano, K., T. Yoneya, T. Miyata, K. Yoshikawa, F. Tokunaga, Y. Terada & S. Iwanaga. 1990. Antimicrobial peptide, tachyplesin I, isolated from hemocytes of the horseshoe crab *(Tachypleus tridentatus)*: NMR determination of the β-sheet structure. J. Biol. Chem. **265:** 15365–15367.
40. Shigenaga, T., T. Muta, Y. Toh, F. Tokunaga & S. Iwanaga. 1990. Antimicrobial tachyplesin peptide precursor: cDNA cloning and cellular localization in horseshoe crab *(Tachypleus tridentatus).* J. Biol. Chem. **265:** 21350–21354.
41. Nguyen, N. Y., A. Suzuki, S-M. Cheng, G. Zon & T-Y. Liu. 1986. The amino acid sequence of *Limulus* C-reactive protein. J. Biol. Chem. **261:** 10450–10455.
42. Minetti, C. A. S. A., Y. Lin, T. Cislo & T-Y. Liu. 1991. Purification and characterization of an endotoxin-binding protein with protease inhibitory activity from *Limulus* amebocytes. J. Biol. Chem. **266:** 20773–20780.
43. Liu, T., Y. Lin, T. Cislo, C. A. S. A. Minetti, J. M. K. Baba & T-Y. Liu. 1991. Limunectin: A phosphocholine-binding protein from *Limulus* amebocytes with adhesion-promoting properties. J. Biol. Chem. **266:** 14813–14821.
44. Iwanaga, S. 1993. The limulus clotting reaction. Curr. Opinion Immunol. **5:** 74–82.
45. Donovan, M. A. & T. M. Laue. 1991. A novel trypsin inhibitor from the hemolymph of the horseshoe crab, *Limulus polyphemus.* J. Biol. Chem. 266. **2121:** 2125.

Regulation of Antibacterial Protein Synthesis Following Infection and During Metamorphosis of *Manduca sexta*[a]

PETER E. DUNN,[b] TONY J. BOHNERT, AND VIRGINIA RUSSELL

Department of Entomology
Purdue University
West Lafayette, Indiana 47907-1158

INTRODUCTION

Larvae of the tobacco hornworn, *Manduca sexta,* respond to the detection of bacteria within the hemocoel with a series of active defenses. Initial antibacterial defenses are mediated by circulating hemocytes that sequester invading bacteria near their point of entry into the hemocoel via phagocytosis and nodule formation.[1] In phagocytosis, individual hemocytes adhere to and endocytose bacteria, which are subsequently killed and digested. Nodule formation is the simultaneous entrapment of many bacteria within large hemocytic aggregates, which leave circulation by adhering to tissues and are melanized as a result of activation of the enzyme phenoloxidase. The fate of bacteria within hemocytic nodules is not well characterized. The formation of hemocytic nodules following infection by large numbers of bacteria may severely deplete the population of circulating hemocytes.

Hemocyte-mediated antibacterial responses are supplemented by the induced synthesis of a suite of bacteria-elicited hemolymph proteins.[2] The suite of bacteria-elicited proteins synthesized by *M. sexta* includes the bacteriolytic enzyme lysozyme[3]; families of homologous cecropin-like bactericidal peptides[4]; bacteriostatic attacin-like proteins[5]; hemolin, which is a member of the immunoglobulin superfamily[6]; and protein M13,[7] whose function has not been clearly defined.

In *M. sexta,* fat body is a major source of the antibacterial proteins found in the hemolymph of bacterially challenged insects,[8] but several other tissues have been shown to synthesize and secrete lysozyme, including pericardial cells,[9] hemocytes,[10] Malpighian tubules,[5] and midgut epithelium.[11] Several larval tissues have also been shown to contain mRNA for cecropin-like peptides, including the midgut epithelium.[4]

Previous studies of the regulation of the synthesis of bacteria-elicited proteins by *M. sexta*[8,12,13] have demonstrated that bacterial cell walls and cell wall peptidoglycan

[a]Research described was supported by funds from the Indiana Agricultural Experiment Station and from National Institutes of Health Grant 1R01-GM41753 to PED.

[b]Address for correspondence: Peter E. Dunn, Department of Entomology, 1158 Entomology Hall, Purdue University, West Lafayette, IN 47907-1158.

are sufficient signals to induce the dose-dependent synthesis of these proteins both *in vivo* and *in vitro*. Soluble fragments of depolymerized peptidoglycan generated by extensive digestion of *Micrococcus luteus* cell walls[12] elicited synthesis of lysozyme, cecropin-like peptides, and attacin-like protein when injected into naive larvae[4,5,12] or when added to primary explants[8,12] of fat body from naive larvae. In both of these systems, the elevated synthesis of antibacterial proteins was preceded by the accumulation of elevated levels of specific RNA transcripts encoding the antibacterial proteins. When added to medium supporting growth of MRRL-CH-1, an established cell line derived from embryos of *M. sexta*,[13] peptidoglycan fragments induced an enhanced abundance of lysozyme, cecropin-like peptide, and attacin-like protein-specific transcripts in the cells and accumulation of lysozyme and attacin-like protein in the growth medium.

While performing experiments to characterize the regulation of antibacterial protein synthesis in *M. sexta*, it was noted that, in addition to exhibiting enhanced synthesis of hemolymph antibacterial proteins, larvae receiving injections of bacteria, bacterial cell walls, or peptidoglycan fragments also exhibited a "malaise syndrome" reminiscent of human disease states. Larvae exhibiting this "malaise syndrome" seemed to eat less, accumulate mass more slowly, and develop more slowly. Because of the consistency of these observations, we felt it necessary to quantitate these physiological and developmental responses associated with the induction of antibacterial protein synthesis and to determine if they were actually peptidoglycan elicited. In this report, we describe the initial physiological and developmental characterization of this "malaise syndrome."

During the process of metamorphosis in lepidoptera, the midgut epithelium is restructured, the larval epithelium is sloughed off and replaced by an epithelium derived from "regenerative cells," and the protective peritrophic membrane lining the gut is lost.[11,14,15] These processes have the potential to expose internal tissues to direct contact with the bacterial flora present in the gut lumen. Antibacterial agents appearing in the midgut lumen at this time would confer needed protection at a susceptible site. In an earlier study, we examined the synthesis of lysozyme by the midgut of naive fifth instar *M. sexta* during the beginning of metamorphosis to the pupa.[11] These experiments demonstrated accumulation of lysozyme in apical vacuoles of differentiating pupal midgut epithelial cells and the release of this bacteriolytic enzyme into the midgut lumen immediately after the larval epithelium degenerated and was sloughed off. As a continuation of this study, we have examined the fluid present in the lumen of the metamorphosing midgut to determine if other antibacterial factors may also be present. This paper also describes the spectrum of such antibacterial factors, which are present, in addition to lysozyme, in the midgut of *M. sexta* during metamorphosis.

MATERIALS AND METHODS

Insects and Bacteria

M. sexta larvae were derived from a laboratory colony reared from surface-sterilized eggs on an artificial diet under stringent conditions that minimize exposure to microorganisms.[16]

For studies of the peptidoglycan-elicited "malaise syndrome," larvae were staged during the fourth to fifth larval molt and provided with artificial diet *ad libitum* before injection.

For studies of the metamorphosing midgut, larvae were selected at the molt to the fifth instar and staged until two days after the pupal molt (day 10.5) by external indicators as well as chronological observations of days post emergence. "Wandering" occurred at day 5, proleg retraction occurred at day 6.5, mouth part retraction at day 7, light dorsal tanning at day 8, head capsule slippage at day 9, and lateral tanning at day 10, just before the molt.[17]

Escherichia coli D-31 was cultured, and viable bacteria quantitated as described.[16]

Injection of Insects

Insects were washed free of diet materials and frass (fecal material), immobilized by chilling on ice for 5 minutes, and surface sterilized by immersion in 70% (wt/vol) ethanol. Materials to be injected were delivered in a 5-μl volume by a calibrated syringe driven by an ISCO Model M microapplicator. The 30-gauge needle was inserted between the third and fourth abdominal prolegs and advanced anteriorly to the second proleg before the sample was deposited. Peptidoglycan fragments (D50A1) were prepared as described by Kanost *et al.*[12] and dissolved in sterile, endotoxin-free water (Sigma Chemical Co., St. Louis, MO) to a final concentration of 20 μg/μl (stock solution). *N*-acetylglucosamine and *N*-acetylmuramyl-L-alanyl-D-isoglutamine (adjuvant peptide) were obtained from Sigma Chemical Co. (St. Louis, MO) and dissolved in sterile, endotoxin-free water to a final concentration of 20 μg/μl.

Effects of Peptidoglycan on Development and Feeding

To determine the effects of peptidoglycan on larval development and pupation, a population of 90 larvae in the first day of the fifth stadium was obtained from the laboratory colony and weighed. On day 2 of the experiment, 30 larvae received injections of either D50A1 or sterile, endotoxin-free water. Weights of these larvae were recorded on each day of the fifth larval stadium until all had pupated. The remaining 30 larvae were maintained as an untreated control group (naive), and weights were recorded as in the treated populations. The date on which each larva pupated was recorded. The mean pupation dates and the mean weights of larvae were compared using the general linear model (GLM) because of unequal sample sizes, and statistically different means were ranked using the Student-Newman-Keuls (SNK) test.

To determine the effects of peptidoglycan on feeding, individual untreated, naive larvae and larvae receiving injections of D50A1 or sterile, endotoxin-free water were placed in individual rearing units. All experiments were initiated with larvae in the third day of the fifth larval stadium. At times after treatment, weights of larvae, food, and frass were recorded. The means of the changes in weight of larvae, food consumed, and frass produced were compared by analysis of variance (ANOVA, $p < 0.05$). Statistically different means were ranked using the SNK test.

Tissue Samples

Dissections were performed on chilled larvae, and midgut tissue was rinsed twice in saline [0.85% (wt/vol) NaCl]. Tissue was homogenized in a ground glass homogenizer containing 1 ml AMS buffer [50 mM sodium acetate, pH 5.0, 5 mM $MgCl_2$, and 0.1% (wt/vol) SDS], centrifuged for 3 min at 12,000 *g*, and the supernatant was assayed for bactericidal activity.

Midgut contents were collected in a 1.5-ml microcentrifuge tube from larval midguts that had been ligated at the junctions of the midgut to the foregut and hindgut and excised from the animal. After pupation, midgut contents were collected from intact animals by removing the dorsal sclerites and puncturing the distended midgut. A few crystals of phenylthiourea (Sigma Chemical Co., St. Louis, MO) were added to inhibit phenoloxidase activity. Midgut contents were then centrifuged at 12,000 *g* for three minutes to remove food particles. The supernatant volume was measured, and the contents were assayed for bactericidal activity.

Hemolymph was collected from a wounded proleg into a 1.5-ml microcentrifuge tube containing a few crystals of phenylthiourea. Samples were centrifuged at 12,000 *g* for three minutes to remove hemocytes and assayed for bactericidal activity.

Tissue from five animals was used for each experimental data point unless otherwise indicated.

Bactericidal Activity

Bactericidal activity against *E. coli* was measured by radial diffusion assay.[18] Activity was quantitated as the area (in mm^2) of clearance in a lawn of *E. coli* D-31 which had been grown in Luria agar at 37°C.

Gel Electrophoresis

Proteins from hemolymph and gut lumen contents were resolved by electrophoresis in the presence of sodium dodecyl sulfate in gels containing 15% (wt/vol) acrylamide (SDS-PAGE).[19] Protein standards for SDS-PAGE were obtained from Bethesda Research Laboratories (Gaithersburg, MD). *M. sexta* lysozyme,[3] attacin-like protein (ALP),[5] rabbit anti-lysozyme serum,[8] and rabbit anti-ALP serum[5] were isolated or generated in this laboratory as described. *M. sexta* hemolin and rabbit anti-hemolin serum were generously provided by M. Kanost, Department of Biochemistry, Kansas State University (Manhattan, KS). Proteins separated by SDS-PAGE were transferred to nitrocellulose as described,[8] and Western blots were probed using polyclonal antibodies prepared in rabbits to lysozyme, ALP, and hemolin. The antibody–antigen complex was visualized with horseradish peroxidase conjugated goat anti-rabbit immunoglobulin according to the procedure outlined in the BioRad Immuno Blot (GAR-HRP) Assay Kit (BioRad Laboratories, Richmond, CA).

RESULTS

Effects of Peptidoglycan Elicitors on Larval Growth and Pupation

To quantitate differences in larval growth and time to pupation in *M. sexta* after injection of peptidoglycan elicitors, populations of insects received injections of sterile, endotoxin-free water or D50A1 (100 μg) on day 2 of the fifth larval stadium or were left untreated. Profiles of daily weights throughout the fifth stadium for D50A1-treated, endotoxin-free water–treated, and untreated insects are compared in FIGURE 1. On day 3 of the stadium, one day after injection of D50A1, there was a significant reduction in the mean weight of D50A1-treated larvae, which persisted until day 7 of the stadium. As shown in TABLE 1, significant changes in the weight of D50A1-treated larvae were observed as soon as 6 hours after injection, at which time a decrease in weight from time 0 was observed.

Maximum weight during the fifth stadium was attained for the insects receiving an injection of endotoxin-free water and for the naive insects on day 6, while the D50A1-treated group achieved maximum weight one day later (FIG. 1). The maximum weight achieved was greatest for the endotoxin-free water and the naive groups, which were not significantly different, and was significantly lower for the D50A1-treated population.

Larvae treated with D50A1 exhibited a significant delay in the date of pupation when compared to either the naive or water-treated populations (FIG. 1, TABLE 2).

Effects on Food Consumption after Injection of Peptidoglycan Elicitors

To measure changes in feeding behavior, the consumption of food and production of frass were recorded from insects after injection of sterile, endotoxin-free water or D50A1 (100 μg). The net consumption of food after the various treatments is compared in TABLE 3. From the time of injection until 18 hours postinjection, a trend similar to that observed with the weight gain data (TABLE 2) was also observed for food consumption. Naive larvae consumed the most diet, endotoxin-free water larvae consumed an intermediate quantity, and D50A1-injected insects consumed the least amount of food. The decrease in food consumption exhibited by D50A1-treated insects was observed as early as 6 hours after treatment. At 24 and 36 hours after treatment, there was no significant difference between the naive group and the water group, but the D50A1-injected larvae were still consuming less food during these intervals. After 2 days, there was no significant difference in food consumption among the treatment groups.

Net frass production after the same treatments is compared in TABLE 4. At 6 hours after injection, there was no significant difference among the treatments. From 12 to 36 hours postinjection, there was no significant difference between the naive larvae and the larvae receiving an injection of water. At these latter time points, however, there was significantly less frass produced by the D50A1-injected larvae, and the frass produced was notably more watery. By 48 hours postinjection, all groups of insects were again producing the same quantities of frass.

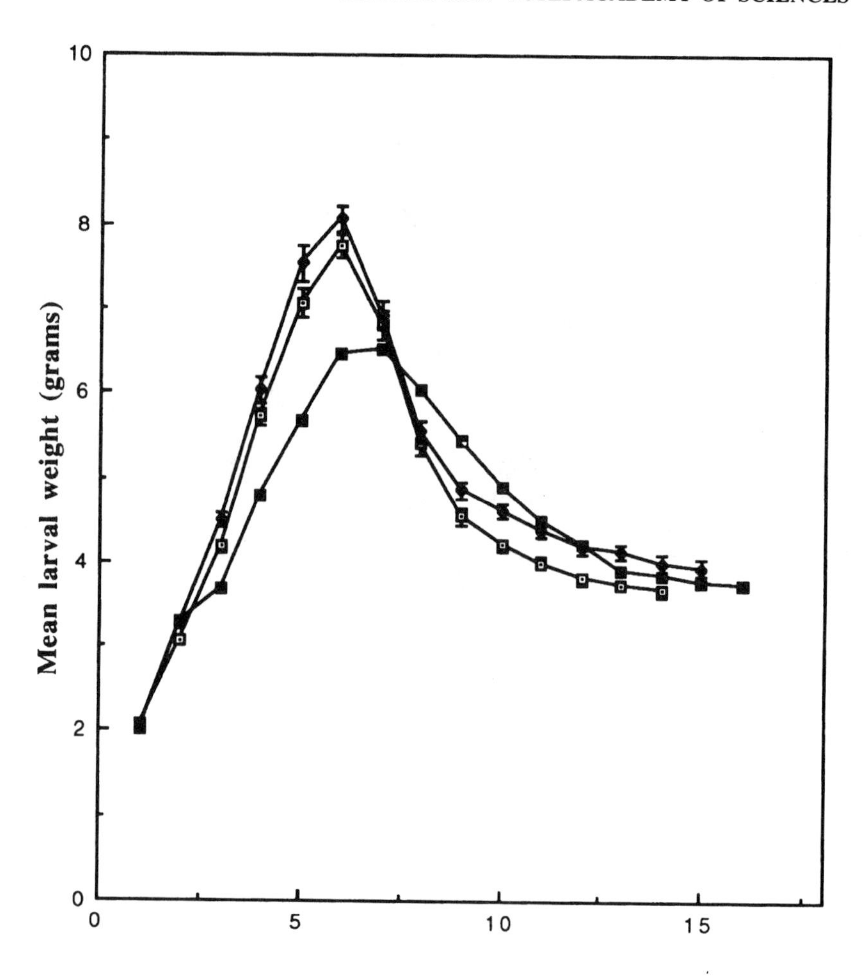

FIGURE 1. Mean larval weight of fifth instar *Manduca sexta.* Data represent the mean (± SD) larval weight of 30 insects. When error bars are not shown, they are smaller than the symbol size. *Open square,* naive control insects; *solid circles,* larvae injected with sterile, endotoxin-free water on day 2 of the fifth stadium; *solid squares,* larvae injected with 100 μg D50A1 on day 2 of the fifth stadium.

To investigate further the specificity of peptidoglycan-elicited alterations in excretory behavior, fifth instar larvae were administered doses of water, *N*-acetylglucosamine (100 μg), *N*-acetylmuramyl-L-alanyl-D-isoglutamine (adjuvant peptide, 100 μg), or D50A1 (100 μg) by injection; and frass production was monitored at intervals for 72 hours (Fig. 2). Only the population receiving the D50A1 mixture of peptidoglycan fragments exhibited decreased production of frass and an increase in the fluidity of the frass produced.

TABLE 1. Net Weight Gain of Tobacco Hornworm Larvae after Injection of D50A1 (100 μg) or Endotoxin-free Water[a]

Hours after Injection	Net Larval Weight Gain (grams)		
	Naive	H_2O	D50A1
6	0.37 ± 0.02(A)	0.22 ± 0.03(B)	−0.11 ± 0.03(C)
12	0.61 ± 0.03(A)	0.46 ± 0.03(B)	0.40 ± 0.04(B)
18	0.60 ± 0.03(A)	0.50 ± 0.04(B)	0.36 ± 0.03(C)
24	0.65 ± 0.03(A)	0.56 ± 0.02(B)	0.40 ± 0.04(C)
36	0.93 ± 0.05(A)	1.04 ± 0.05(A)	0.92 ± 0.06(A)
48	0.87 ± 0.06(A)	0.99 ± 0.06(A)	1.03 ± 0.04(A)

[a]Data are the mean ± SEM of three replicates of 15 larvae. Numbers at a specific time point followed by the same letter are not significantly different (SNK test, $p \leq 0.05$).

TABLE 2. Duration of Fifth Stadium in Days following Treatment of Larvae with D50A1 (100 μg) or Endotoxin-free Water[a]

Naive	H_2O	D50A1
12.30 ± 0.13(B)	11.83 ± 0.11(C)	13.33 ± 0.16(A)

[a]Data are the mean ± SEM for 30 larvae, except D50A1 treatment where $n = 27$. Numbers followed by the same letter are not significantly different (SNK test, $p \leq 0.05$).

TABLE 3. Net Consumption of Food (in grams) by Tobacco Hornworm Larvae after Injection of D50A1 (100 μg) or Endotoxin-free Water[a]

Hours after Injection	Net Consumption of Food		
	Naive	H_2O	D50A1
6	1.53 ± 0.04(A)	1.26 ± 0.06(B)	0.85 ± 0.03(C)
12	1.90 ± 0.05(A)	1.51 ± 0.06(B)	1.21 ± 0.07(C)
18	1.91 ± 0.05(A)	1.73 ± 0.07(B)	1.32 ± 0.08(C)
24	1.88 ± 0.04(A)	1.88 ± 0.06(A)	1.39 ± 0.07(B)
36	3.28 ± 0.08(A)	3.17 ± 0.08(A)	2.85 ± 0.11(B)
48	3.23 ± 0.11(A)	3.15 ± 0.08(A)	3.14 ± 0.08(A)

[a]Data are the mean ± SEM of three replicates of 15 larvae. Numbers at a specific time point followed by the same letter are not significantly different (SNK test, $p \leq 0.05$).

TABLE 4. Net Production of Frass by Tobacco Hornworm Larvae after Injection of D50A1 (100 μg) or Endotoxin-free Water[a]

Hours after Injection	Net Production of Frass (wet weight, grams)		
	Naive	H_2O	D50A1
6	0.40 ± 0.02(A)	0.43 ± 0.02(A)	0.36 ± 0.02(A)
12	0.53 ± 0.02(A)	0.48 ± 0.04(A)	0.25 ± 0.03(B)
18	0.57 ± 0.02(A)	0.53 ± 0.04(A)	0.36 ± 0.04(B)
24	0.59 ± 0.02(A)	0.64 ± 0.04(A)	0.38 ± 0.04(B)
36	1.40 ± 0.07(A)	1.37 ± 0.05(A)	1.18 ± 0.06(B)
48	1.45 ± 0.06(A)	1.37 ± 0.05(A)	1.34 ± 0.06(A)

[a]Data are the mean ± SEM of three replicates of 15 larvae. Numbers at a specific time point followed by the same letter are not significantly different (SNK test, $p \leq 0.05$).

Results from previous experiments have demonstrated that the magnitude of antibacterial protein synthesis elicited by peptidoglycan fragments in intact larvae, in primary fat body explants, and in cultured embryonic cells increased with increasing D50A1 dose. To investigate the dose dependence of peptidoglycan-elicited altered excretion, fifth instar larvae received injections containing increasing amounts of the D50A1 mixture (FIG. 3). A clear dose-dependent trend in the decrease in the amount of frass produced with increasing dose of D50A1 over the range 1 to 50 μg per insect was observed.

ANTIBACTERIAL FACTORS IN THE METAMORPHOSING MIDGUT

No antibacterial activity against *E. coli* was detected in either the midgut homogenate or in the gut lumenal contents until the period immediately before pupation (day 10). At the same time the larval gut was sloughed off, bactericidal activity was detected in the midgut lumen by radial diffusion assay, although none could be detected in the midgut epithelium homogenate at any time by this procedure. Lumenal bactericidal activity appeared to be high at this time and persisted at elevated levels for at least 3 days post pupation (FIG. 4). When hemolymph samples were examined for antibacterial activity, which might indicate a systemic response to the presence of a bacterial inducer, only a very few animals (10%) demonstrated detectable hemolymph bactericidal activity during any stage.

Midgut lumen samples from pupae were subjected to denaturing gel electrophoresis (SDS-PAGE) to detect other antibacterial proteins that might be present. When Western blots of the electrophoresed samples were probed with antibody prepared against *M. sexta* attacin-like protein, no reaction with the antibody was observed. However, Western blots of electrophoresed midgut lumen samples probed with an antibody prepared against *M. sexta* hemolin did exhibit a reactive antigen that co-electrophoresed with authentic *M. sexta* hemolin (48×10^3 daltons) (data not shown).

The presence of prophenoloxidase was inferred by preliminary observations of melanization of gut content samples collected without the addition of phenylthiourea

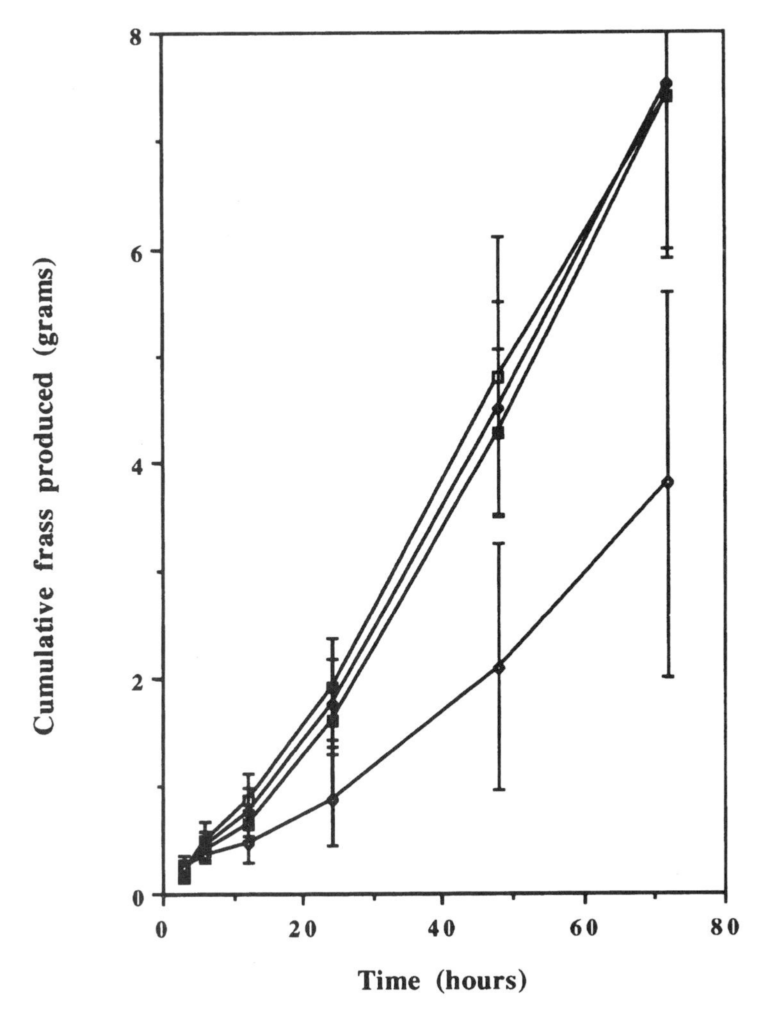

FIGURE 2. Cumulative weight of frass excreted by fifth instar *Manduca sexta* injected with various potential elicitors. Data represent the mean (± SD) cumulative weight of frass excreted by 10 insects. When error bars are not shown, they are smaller than the symbol size. *Open square,* naive control insects; *solid circle,* larvae injected with 100 μg *N*-acetylglucosamine on day 2 of the fifth stadium; *solid square,* larvae injected with 100 μg of adjuvant peptide on day 2 of the fifth stadium; *open circle,* larvae injected with 100 μg D50A1 on day 2 of the fifth stadium.

(PTU). Darkening of samples was prevented by the addition of PTU, a well-characterized inhibitor of phenoloxidase. Furthermore, samples collected at increasing times after pupation showed progressive darkening of luminal contents, indicative of melanization due to phenoloxidase activity. Spectrophotometric quantitative analysis of the levels of phenoloxidase activity was prevented by the presence of an interfering pigment in the lumen contents.

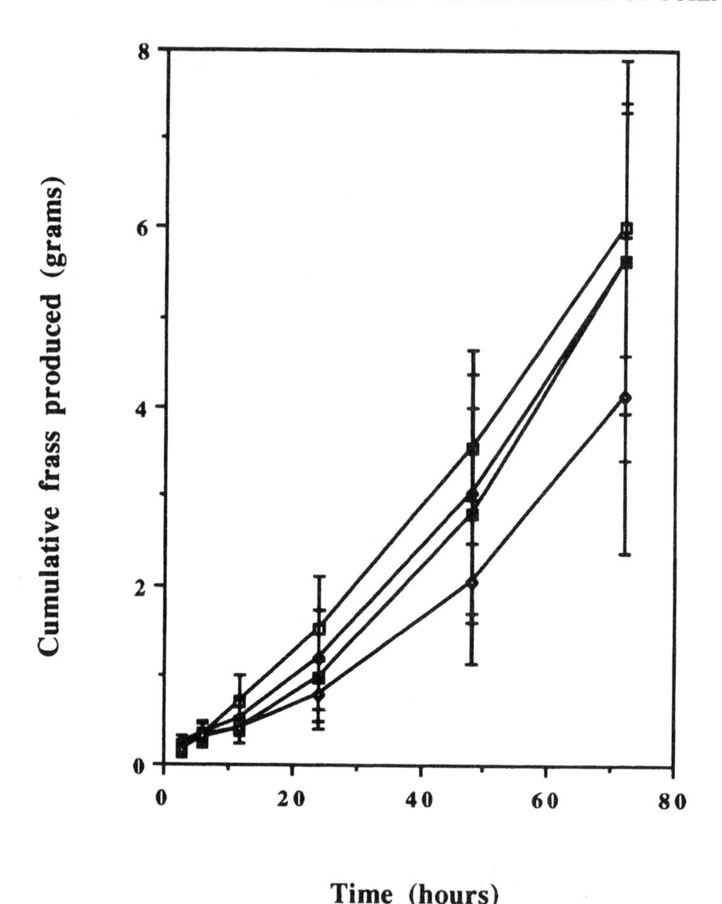

FIGURE 3. Cumulative weight of frass excreted by fifth instar *Manduca sexta* injected with varying doses of D50A1. Data represent the mean (± SD) cumulative weight of frass excreted by 10 insects. When error bars are not shown, they are smaller than the symbol size. *Open square,* naive control insects; *solid circle,* larvae injected with 1 μg D50A1 on day 2 of the fifth stadium; *solid square,* larvae injected with 10 μg D50A1 on day 2 of the fifth stadium; *open circle,* larvae injected with 50 μg D50A1 on day 2 of the fifth stadium.

DISCUSSION

Peptidoglycan-Elicited "Malaise Syndrome"

Data reported confirm quantitatively previous observations of altered growth, development, feeding, and excretory behavior in larvae receiving injected doses of bacterial peptidoglycan as compared to naive controls and to experimental insects injected with sterile, endotoxin-free water. Alterations in growth, development, feeding, and excretion in peptidoglycan-treated larvae were manifest rapidly. De-

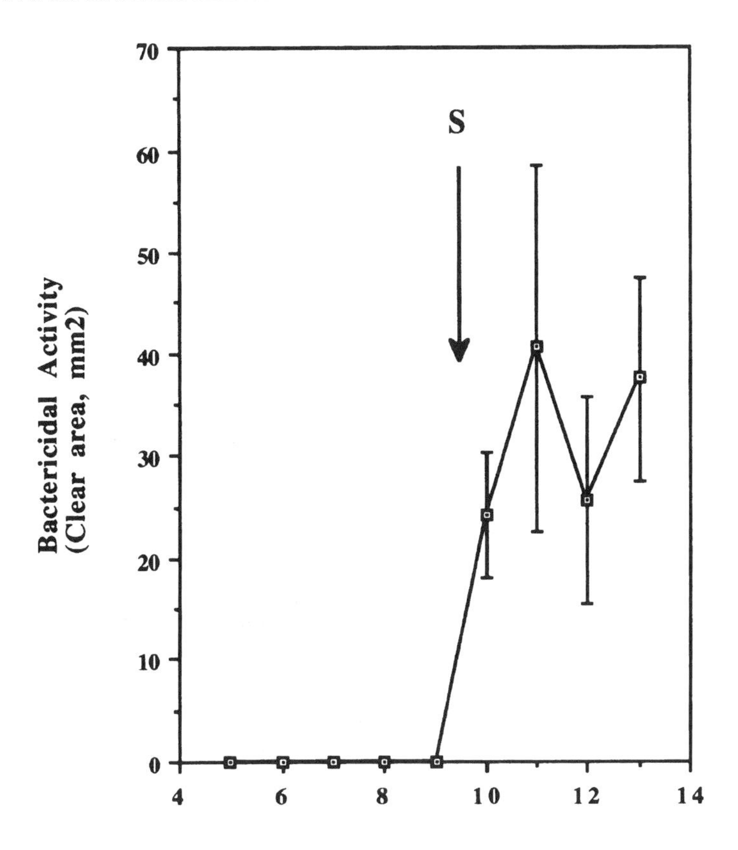

FIGURE 4. Bactericidal activity in the lumen of the metamorphosing midgut of *Manduca sexta.* Data represent the mean (± SD) bactericidal activity of midgut contents from five insects. S = sloughing off of larval midgut epithelium.

creased consumption of food and decreased weight were evident as soon as 6 hours post injection, and decreased excretion of frass was evident by 12 hours. Decreased feeding persisted for 24 to 36 hours, and resulted in a significant decrease in maximal weight achieved and a delay in metamorphosis. In addition to the perhaps not surprising observation of decreased excretion of frass accompanying the decrease in ingestion of food, a consistent and marked increase in the fluidity of the frass excreted by peptidoglycan-treated larvae was noted. Using frass production as an index, the structural requirements for elicitation of the syndrome were investigated. Neither *N*-acetylglucosamine nor adjuvant peptide, which are both components of peptidoglycan, were sufficient to elicit the syndrome. However, these components of peptidoglycan were also not elicitors of enhanced synthesis of antibacterial hemolymph proteins in a previous study with *M. sexta* larvae.[12] Thus, there may be a relationship between peptidoglycan regulation of antibacterial protein synthesis and the peptidoglycan elicitation of this "malaise syndrome" in larvae.

The involvement of polymeric peptidoglycan and of peptidoglycan fragments in the regulation of inflammation and immunity is not unique to insects. Peptidoglycan is known to act as an adjuvant,[20] a mitogen,[21] an activator of the complement pathway,[22] and as a causative agent of arthritis in mammals.[23] It should not be surprising that an agent with such diverse biological activity in mammals would also exert profound physiological effects on insects.

It is interesting to speculate on the potential significance of this peptidoglycan-elicited "malaise syndrome" to the insect's antibacterial defense mechanisms. The primary physiological response underlying the symptoms described would appear to be a dramatic decrease in feeding. The net effect of decreased ingestion may be a temporary emptying of the gut. The observed production of decreasing amounts of frass approximately 6 hours after the decrease in food consumption would be consistent with this putative emptying of the gut. The apparent increase in the fluidity of frass excreted during the period of decreased total excretion suggests further that gut emptying may be coupled with increased fluid secretion into the gut to produce the watery frass. Because the two most likely sources of bacterial infection for an insect are wounds to the integument and wounds to the midgut, it is tempting to speculate that the observed symptoms are manifestations of a mechanism to flush out and sterilize the midgut at times when a massive infection of the hemocoel has occurred as signaled by the presence of peptidoglycan fragments in hemolymph. Another set of observations, which are consistent with this hypothesis, are the previously described peptidoglycan-induced increase in the relative abundance of specific RNA transcripts encoding lysozyme,[24] cecropin-like bactericidins,[4] and attacin-like proteins[5] in the midgut and/or Malpighian tubules of peptidoglycan-treated tobacco hornworm larvae.

Antibacterial Factors in the Metamorphosing Midgut

Earlier studies had shown that the midgut epithelium of fifth instar *M. sexta* accumulates lysozyme as the insect prepares for pupation. This lysozyme is present in large vacuoles in the presumptive pupal epithelium and is secreted into the gut lumen at pupation.[11] Lysozyme is an antibacterial agent that is most active against gram-positive bacteria. From the data presented here, it appears that the antibacterial activity that develops in the lumen of the metamorphosing gut also includes a bactericidal factor active against *E. coli.* A protein that reacted with hemolin antibody was also observed, and prophenoloxidase/phenoloxidase activity was inferred by persistent phenylthiourea-inhibited melanization of the gut contents.

The small proportion (10%) of the "naive" larval population used in this study that did demonstrate bactericidal activity in their hemolymph most probably represents a segment of the population that has been exposed to bacterial or fungal inducers despite efforts to minimize contamination. In later periods of the larval stadium, after "wandering," yeast infestation in the diet could occasionally be observed, and minimal, undetected bacterial contaminants could give rise to an induced state in individuals. Of greater note is the observation that 90% of the insects studied showed no elevated hemolymph antibactericidal activity but still demonstrated the presence of such activity in the gut lumen. Although lysozyme appears to accumulate in the midgut epithelium according to a developmental program, since it appears at a very specific time and in animals which have not been exposed experimentally to bacteria,[11] the occurrence of bactericidal proteins in the gut lumen is evident only after the dissolution of the larval epithelium. This is a time when either intact bacteria or cell wall fragments generated by lysozyme digestion would potentially come into contact with the new pupal epithelium and possibly induce synthesis and secretion of defense compounds.

Although lysozyme has been observed in large apical vacuoles of the pupal epithelium and can readily be detected in homogenates of the gut epithelium, we did not detect other antibacterial proteins and peptides or bactericidal activity in homogenates. This most probably is a reflection of the fact that they are rapidly secreted after synthesis and never stored in sufficient amounts for ready detection.

Results reported herein demonstrate clearly that the active antibacterial responses of insects are not limited to the dramatic and effective clearance responses mediated by circulating hemocytes and the strikingly enhanced synthesis of potent hemolymph antibacterial proteins. These classic responses focused on bacteria in circulating hemolymph are clearly augmented by broader physiological and developmental responses targeted against perennial bacterial populations associated with the gut, which may gain access to the hemocoel when the insect host is weakened by prior infection or vulnerable because of metamorphosis-associated restructuring of tissues.

SUMMARY

Larvae of the tobacco hornworm, *Manduca sexta,* respond to intrahemocoelic injection of bacteria or bacterial cell wall peptidoglycan with induced synthesis of a suite of antibacterial proteins. Previous studies have demonstrated peptidoglycan regulation of the synthesis of these antibacterial proteins. In addition to eliciting enhanced synthesis of antibacterial proteins, peptidoglycan fragments also elicit a "malaise syndrome" characterized by decreased feeding and growth, delayed metamorphosis, and altered excretion. We speculate that these symptoms may be components of a mechanism to flush out and sterilize the midgut lumen, one of the primary sources of bacterial infection in insects. Studies of naive larvae have demonstrated the accumulation of lysozyme in the differentiating pupal midgut epithelium and release of lysozyme into the pupal midgut lumen after the larval midgut epithelium has been sloughed off. These observations have been extended by the identification of potent bactericidal activity against *E. coli* and immunoreactive hemolin, together with lysozyme, in the lumen of the newly differentiated pupal midgut.

REFERENCES

1. DUNN, P. E. 1986. Biochemical aspects of insect immunology. Ann. Rev. Entomol. **31:**321–339.
2. DUNN, P. E. 1991. Insect antibacterial proteins. *In* Phylogenesis of Immune Functions. G. W. Warr & N. Cohen, Eds.: 19–44. CRC Press. Boca Raton, FL.

3. Dunn, P. E., M. R. Kanost & D. R. Drake. 1987. Increase in serum lysozyme following injection of bacteria into larvae of *Manduca sexta*. *In* Molecular Entomology. J. H. Law, Ed.: 381–390. Alan R. Liss. New York.
4. Dickinson, L., V. Russell & P. E. Dunn. 1988. A family of bacteria-regulated, cecropin-like peptides from *Manduca sexta*. J. Biol. Chem. **263:** 19424–19429.
5. Dai, W. 1988. Structure and regulation of the synthesis and secretion of a attacin-like protein from *Manduca sexta*. Ph.D. thesis, Purdue University, West Lafayette, IN.
6. Ladendorff, N. & M. R. Kanost. 1991. Bacteria-induced protein P4 (hemolin) from *Manduca sexta,* a member of the immunoglobulin superfamily which can inhibit hemocyte aggregation. Arch. Insect Biochem. Physiol. **18:** 285–300.
7. Minnick, M. F., R. A. Rupp & K. D. Spence. 1986. A bacteria-induced lectin which triggers hemocyte coagulation in *Manduca sexta*. Biochem. Biophys. Res. Commun. **137:** 729–735.
8. Dunn, P. E., W. Dai, M. R. Kanost & C. Geng. 1985. Soluble peptidoglycan fragments stimulate antibacterial protein synthesis by fat body from larvae of *Manduca sexta*. Dev. Comp. Immunol. **9:** 559–568.
9. Russell, V. W. & P. E. Dunn. 1990. Lysozyme in the pericardial complex of *Manduca sexta*. Insect Biochem. **20:** 501–509.
10. Geng, C. 1986. Studies of hemocytes from the tobacco hornworm, *Manduca sexta* (Lepidoptera: Sphingidae). M. S. thesis, Purdue University, West Lafayette, IN.
11. Russell, V. W. & P. E. Dunn. 1991. Lysozyme in the midgut of *Manduca sexta* during metamorphosis. Arch. Insect Biochem. Physiol. **17:** 67–80.
12. Kanost, M. R., W. Dai & P. E. Dunn. 1988. Peptidoglycan fragments elicit antibacterial protein synthesis in larvae of *Manduca sexta*. Arch. Insect Biochem. Physiol. **8:** 147–164.
13. Liu, K. 1992. Peptidoglycan regulation of antibacterial gene expression in MRRL-CH-1, an embryonic cell line derived from *Manduca sexta*. M.S. thesis, Purdue University, West Lafayette, IN.
14. Judy, K. J. & L. I. Gilbert. 1970. Histology of the alimentary canal during the metamorphosis of *Hyalophora cecropia (L.)*. J. Morph. **131:** 277–299.
15. Waku, Y. & K. Sumimoto. 1971. Metamorphosis of midgut epithelial cells in the silkworm *Bombyx mori (L)* with special regard to the calcium salt deposits in the cytoplasm. I. Light microscopy. Tiss. Cell **3:** 27–136.
16. Dunn, P. E. & D. R. Drake. 1983. Fate of bacteria injected into naive and immunized larvae of the tobacco hornworm, *Manduca sexta*. J. Invertebr. Pathol. **41:** 77–85.
17. Goodman, W. J., R. O. Carlson & K. L. Nelson. 1985. Analysis of larval and pupal development in the tobacco hornworm (Lepidoptera: Sphingidae) *Manduca sexta*. Ann. Entomol. Soc. Am. **78:** 70–80.
18. Hultmark, D., A. Engstrom, K. Andersson, H. Bennich, R. Kapur & H. G. Boman. 1982. Insect immunity. Isolation and structure of cecropin D and four minor antibacterial components for cecropia pupae. Eur. J. Biochem. **127:** 207–217.
19. Ryrie, I. J. & A. Gallagher. 1979. The yeast mitochondrial ATPase complex: Subunit composition and evidence for a latent protease contaminant. Biochem. Biophys. Acta. **545:** 1–14.
20. Adam, A., R. Ciorbara, F. Ellouz, J. F. Petit & E. Lederer. 1974. Adjuvant activity of monomeric bacterial cell wall peptidoglycans. Biochem. Biophys. Res. Commun. **56:** 561–567.
21. Dziarski, R. & A. Dziarski. 1979. Mitogenic activity of staphylococcal peptidoglycan. Infect. Immun. **23:** 706–710.
22. Greenblatt, J., R. J. Boackle & J. H. Schwab. 1978. Activation of the alternate complement pathway by peptidoglycan from streptococcal cell wall. Infect. Immun. **19:** 296–303.
23. Cromartie, W. J., J. G. Craddock, J. H. Schwab, S. K. Anderle & C. H. Yang. 1977. Arthritis in rats after systemic injection of streptococcal cells or cell walls. J. Exp. Med. **146:** 1585–1602.
24. Mulnix, A. B. & P. E. Dunn. 1994. Structure and induction of a lysozyme gene from the tobacco hornworm, *Manduca sexta*. Insect Biochem. Molec. Biol. In press.

Invertebrate α_2-Macroglobulin: Structure–Function and the Ancient Thiol Ester Bond[a]

JAMES P. QUIGLEY[b,c] AND
PETER B. ARMSTRONG[b,d,e]

[b]*Marine Biological Laboratory*
Woods Hole, Massachusetts 02543

[c]*Department of Pathology*
Health Sciences Center
State University of New York
Stony Brook, New York 11794-8691

[d]*Department of Molecular and Cellular Biology*
University of California
Davis, California 95616-8755

INTRODUCTION

Individual metazoans are threatened by frequent challenge from microbial and metazoan pathogens. The life cycle time for most microbial pathogens is significantly shorter than that of the metazoan host, making it essentially impossible for the host population to outgrow populations of the pathogens. Instead, metazoans have evolved a varied array of defense strategies that limit the susceptibility to attack by potential pathogens. These include the presence of surface tissues that restrict the penetration of pathogens into the interior and a variety of specialized cells and humoral factors that fight those pathogens that have succeeded in gaining access into the body. In higher animals, a majority of the internal defense systems involve cells and soluble factors of the blood. One of the most ancient of these is embodied in the plasma protein α_2-macroglobulin (α_2M), which is present in mammals (for review see Sottrup-Jensen[1]) and several phyla of invertebrates (for review see Armstrong & Quigley[2]). α_2M is a high-molecular-mass protein that binds endopeptidases and several other proteins, including peptide mitogens[3,4] and various basic proteins.[5–7]

α_2-MACROGLOBULIN, AN ANCIENT IMMUNE EFFECTOR

The best-studied species of α_2M, human plasma α_2M, is a homotetramer[8] organized as a noncovalently linked dimer of disulfide-linked homodimers of a 180-kDa

[a]This report was supported by Grant No. MCB 9218460 from the National Science Foundation.
[e]Address for correspondence: Peter B. Armstrong, Department of Molecular and Cellular Biology, Storer Hall, University of California, Davis, CA 95616-8755.

subunit.[9–11] The disulfide-linked homodimer is thought to be the basic functional unit,[12] but dimers of the noncovalently linked subunits can also bind proteases.[13,14] Certain other members of the α_2-macroglobulin family of protease-binding proteins exist as disulfide-linked homodimers[15–19] or as monomers.[20]

The mechanism of protease binding by α_2M is unique. In contrast to the active-site inhibitors of proteases, α_2M selectively inhibits the interaction of protease with macromolecules without affecting the enzymatically active site. α_2M-bound proteases are still capable of hydrolyzing low-molecular-mass substrates. Binding to tetrameric human α_2M can involve both covalent[21–27] and noncovalent[23] interaction. The latter involves a physical folding of the α_2M molecule around the protease molecule, so as to "trap" it,[7] with the arms of the α_2M forming a stearic barrier to prevent contact between the protease and large molecules. We know of no other enzyme inhibitor that works by enfolding the target enzyme. Covalent binding involves the establishment of an isopeptide bond with the γ-carboxyl of the glutamyl residue of a reactive intrachain thiol ester.[21–24,26–33] Members of the α_2M superfamily, which includes C3, C4, and C5 of the complement cascade, are the only proteins that have been described with an internal thiol ester bond[f] (for review, see Tack[34]). The unique features of α_2M have excited the interest of structural biochemists, while the possible importance of α_2M in immunity has interested the physiological biochemists.

Because a broad spectrum of proteolytic enzymes can be bound by α_2M, the molecule has been envisioned to function as a carrier molecule both for endogenous proteases involved in coagulation, fibrinolysis, and inflammation and for foreign proteases introduced by invading pathogens.[1] Although unreacted α_2M has a long lifetime in the plasma, α_2M–protease complexes are rapidly cleared from the circulation.[35] Clearance involves binding to a cell-surface receptor that specifically recognizes reacted forms of α_2M.[36] Because of the abundance of other protease inhibitors in mammalian plasma, it has been difficult to identify the specific physiological roles for α_2M and to determine which specific target proteases are uniquely interactive with mammalian α_2M *in vivo.* α_2-Macroglobulin is presumed to be essential to survival because, in contrast to many other plasma proteins, full genetic deficiency of α_2M in humans has not yet been found, presumably because deficiency is lethal. Additionally, the structure and function of α_2M appears to be conserved throughout the evolution of higher animals.[2] It is our expectation that investigation of α_2M homologues from species evolutionarily distant from mammals and possessing a less complex plasma will contribute to the identification of specific immune functional roles for α_2M.

DISCOVERY OF α_2-MACROGLOBULIN IN THE PLASMA OF INVERTEBRATES

In 1982, we reported the presence of a protease inhibitor in the plasma of the American horseshoe crab, *Limulus polyphemus,* that we proposed was a functional

[f]The reactive internal thiol ester bond is absent in C5 and certain forms of α_2M.[92]

and molecular homologue to mammalian α_2M.[37,38] Homologues of vertebrate α_2M have now been described from a variety of invertebrates, including chelicerate[38] and mandibulate[18,39–42] arthropods and mollusks.[43,44] These molecules share numerous functional properties with mammalian α_2M and have remarkable identity at the level of peptide sequence in key functional domains.[45–48] The present review describes our investigations of the structural and functional features of the forms of α_2M found in invertebrates. The presence of homologues of α_2M in an ancient invertebrate such as *Limulus,* whose plasma protein and circulating hemocyte composition is relatively simple, recommends *Limulus* as a model for the elucidation of physiological function of α_2M.

Characterization of the protease inhibitor of the plasma of *Limulus* was facilitated by our ability to prepare large quantities of plasma uncontaminated by products released from the blood cells following their activation and resultant degranulation (for techniques, see Armstrong[48,49]). This latter feature was important because the only protease inhibitor detectable in the plasma is the α_2M homologue,[38] whereas the activated blood cells of *Limulus* undergo degranulation to release proteases[50–54] and also several active-site protease inhibitors[55–57] that complicate the analysis of serum from this animal, containing as it does the secretory products of the blood cells. Our initial characterization of the plasma α_2M included the demonstration that *Limulus* plasma contained an activity that abolished the ability of a diverse array of proteases to digest [^{14}C]casein, a protein substrate, but failed to affect the amidolytic activity of trypsin against the low-molecular-mass substrate BAPNA. The protease inhibitory activity was abolished if the plasma was treated with the primary amine, methylamine, consistent with the possibility that inhibition is dependent on an internal thiol ester bond (see discussion below). Finally, we demonstrated that *Limulus* plasma protected the amidolytic activity of trypsin against the macromolecular active-site inhibitor, soybean trypsin inhibitor.[39] The only protease inhibitor that had been described with these characteristics was α_2M, which was known principally from mammals, but that had been shown also to be present in representatives from all classes of vertebrates.[19,58–59] Subsequent studies have shown that a protease inhibitor with most of these same characteristics is present in the blood of crustaceans[18,39,40] and bivalve, gastropod, and cephalopod mollusks[43,44] and the eggs of certain echinoderms (Armstrong, unpublished).

Purification of *Limulus* α_2M was facilitated by the relative simplicity of the pattern of proteins in the plasma: There are only three abundant proteins—hemocyanin (about 95% of the total protein), C-reactive protein (CRP),[1] and α_2M. A number of other, less abundant proteins are also present in the plasma. The α_2M from *Limulus* has been purified by ultracentrifugation and treatment with zinc acetate to remove hemocyanin, differential precipitation with polyethylene glycol (PEG),[1] treatment with phosphorylethanolamine-Sepharose to remove CRP, and gel filtration (FIG. 1)[60,61] to yield a product that exhibits a single band by SDS-polyacrylamide gel electrophoresis and a single amino-terminal amino acid sequence.[47] The α_2M, isolated from *Limulus,* has a subunit molecular weight, estimated by SDS-polyacrylamide gel electrophoresis conducted under reducing conditions, that is very similar to that of human α_2M (*viz.* 180–185 kDa).[61] Al-

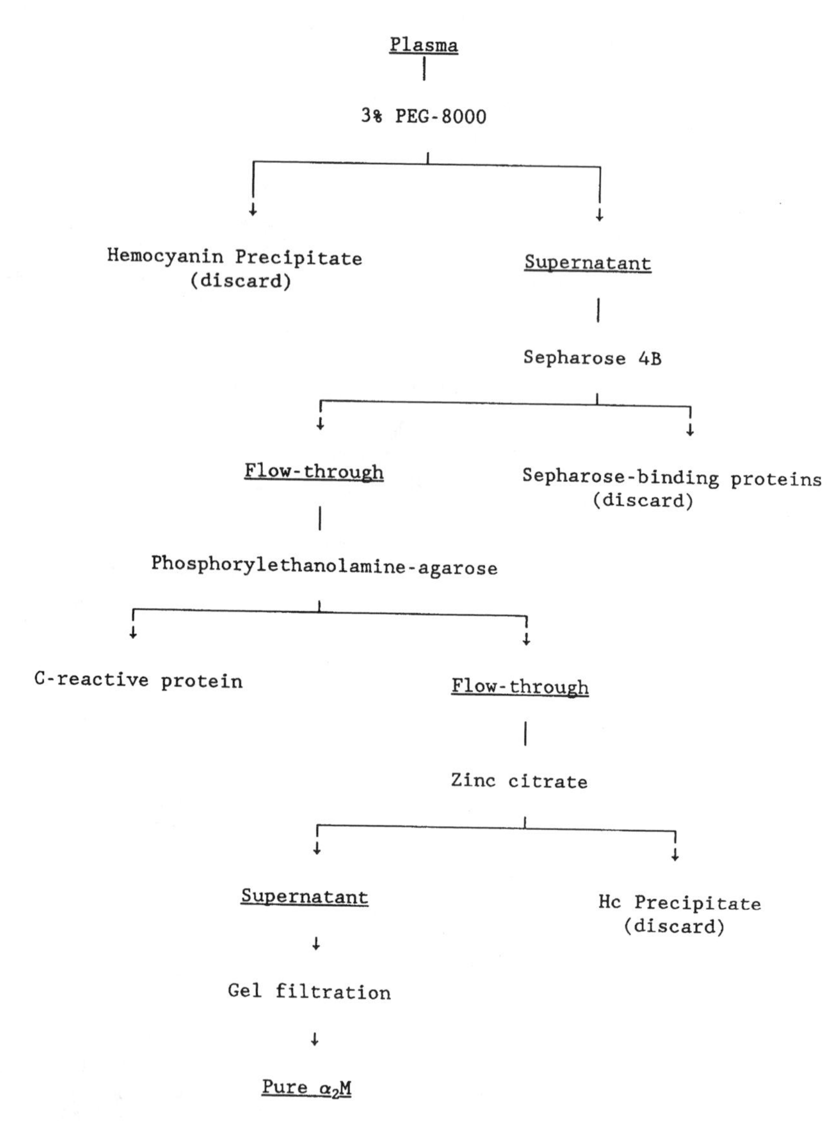

FIGURE 1. Scheme for purifying α_2M from the plasma of the American horseshoe crab, *Limulus polyphemus*.

though molecular exclusion chromatography suggested a native molecular mass of ~ 550 kDa,[61] recent studies involving ultracentrifugation, SDS-polyacrylamide gel electrophoresis, transmission electron microscopy, scanning transmission electron microscopy, and peptide sequencing have shown that *Limulus* α_2M is a disulfide-linked homodimer with a native molecular mass of 370 kDa and a subunit molecular mass of 185 kDa.[60]

THE INTERNAL THIOL ESTER OF α_2-MACROGLOBULIN

One unique feature of human α_2M is the presence of a reactive internal thiol ester bond.[62,63] This bond is broken when the α_2M reacts with proteases, resulting in the exposure of a new thiol group on the cysteine and a highly reactive γ-carbonyl group on the glutamine (FIG. 2).[21,63–67] This latter group reacts rapidly with nucleophiles, including amino groups of the entrapped protease molecule.[23–28,30–33,68–71] Significant fractions of entrapped trypsin become covalently linked to α_2M by this reaction. The thiol ester bond in native α_2M is reactive with low-molecular-mass primary amines, such as methylamine, resulting in loss of the ability to react subsequently with proteases.[12,24,28,30,59,63,72–76] The thiol ester is also responsible for the susceptibility of α_2M to a specific fragmentation of the peptide chain in response to mild denaturing conditions.[12,22,76–79] This apparently results from a nucleophilic attack by the amino group of the glutamic acid residue on the γ-carbonyl group, resulting in rupture of the peptide chain at this site. The presence of thiol ester bonds in candidate proteins can be established by the reactivity with methylamine, the exposure of new thiols during reaction with proteases, and the specific fragmentation pattern during mild denaturation. Interestingly, only certain of the members of the

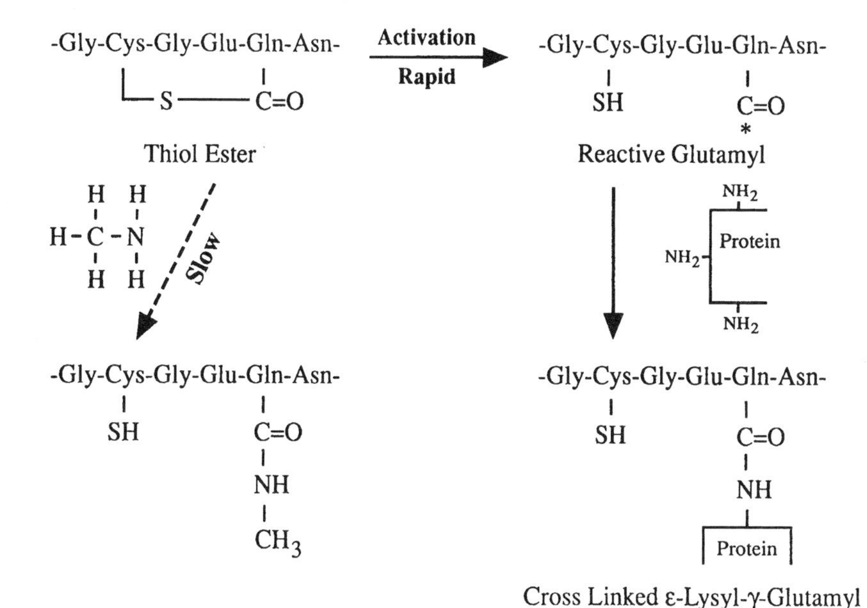

FIGURE 2. Thiol ester reactivity in α_2-macroglobulin. Activation of the internal thiol ester of α_2M exposes a new thiol group on the cysteine and a reactive γ carbonyl on the glutamyl residue. The reactive internal thiol ester of members of the α_2M protein family is cleaved following proteolysis at the distantly located protease bait region of the protein. Thiol ester cleavage generates an activated γ-carbonyl at the glutamyl residue and a free thiol at the cysteinyl residue. The reactive glutamyl can form amide linkages with proteins (*right arm of the diagram*). The thiol ester can also react with small primary amines, such as methylamine (*left arm of the diagram*).

α_2M family of proteins contain the thiol ester bond. For example, the bond is present in C3 and C4 but is absent in C5[34] and ovostatin, an α_2M homologue from the egg white of chickens.[80]

Application of these tests to the α_2M homologues from crustaceans and *Limulus* indicate that these proteins do contain the internal thiol ester bond and that this bond is necessary for function.[18,39,40,45,47,81] The protease-binding activity of these forms of α_2M is inactivated by exposure to low-molecular-mass primary amines, such as methylamine.[18,38,39] *Limulus* and *Homarus* α_2M also experience autolytic fragmentation during mild thermal denaturation, yielding fragments of approximate molecular masses of 125 and 55 kDa, whereas the methylamine-treated protein is stable under these conditions. Reaction with protease results in the rapid generation of new thiols.[18,81] In addition, the amino acid sequence of the thiol ester of *Limulus* α_2M is identical to that of human α_2M (FIG. 3).[47] Because it is unlikely that so unique a functional group as the intrachain thiol ester could have evolved independently in the α_2M of arthropods and mammals, it is concluded that it is an ancient feature of the molecule, one that must have been present in the α_2M of the common ancestor of mammals and arthropods. Because the lineages leading to the vertebrates and the arthropods diverged approximately 0.55 billion years ago,[82–84] the α_2M molecule and the intrachain thiol ester bond are truly ancient.

REACTION OF *LIMULUS* α_2-MACROGLOBULIN WITH PROTEASES

α_2-Macroglobulin binds proteases[1,85] and various other proteins, including peptide mitogens[3,4,29] and basic proteins.[5–7] Both covalent[21,23,24,27,28,31,33,86–89] and noncovalent[23–26,90–93] binding operate in the α_2-macroglobulin–protease interaction. The latter involves a physical folding of α_2-macroglobulin around the protease molecule to "trap" it in the folds of an α_2-macroglobulin "cage."[7] The dimeric and monomeric homologues of α_2-macroglobulin from mammals are incapable of binding proteases by noncovalent means[20,94] and, thus, rely entirely on covalent bonding. Covalent binding involves the internal thiol ester bond that is activated following proteolysis of α_2-macroglobulin by the reacting protease.[22,28,95] Isopeptide bonds are established between the γ-carbonyl group of the glutamyl residue and nucleophilic groups, primarily ε-amino groups of lysine, of the protease.[69] Proteases that react well with mammalian α_2-macroglobulin, such as trypsin, usually form the protease–α_2-macroglobulin isopeptide bond with high efficiency.[27,33,88,93,96–98] Covalent binding results

Human α_2M :	-	M	P	Y	G	C	G	E	Q	N	M	V	L^{956}	-
		•	*		*	*	*	*	*	*	*	•		
<u>Limulus</u> α_2M:	-	L	P	T	G	C	G	E	Q	N	M	I	K	-
			*	•	*	*	*	*	*	*	*	*		
Human C3:	-	T	P	S	G	C	G	E	Q	N	M	I	G	-
Human C4:	-	L	P	R	G	C	G	E	Q	T	M	I	Y	-

FIGURE 3. Peptide sequence similarity among members of the α_2M family of proteins.

in the formation of high-molecular-mass complexes that can be demonstrated by SDS-polyacrylamide gel electrophoresis as protein bands with retarded migration from that of unreacted α_2-macroglobulin.[33,88,89,96] Although the exact composition of the different high-molecular-mass products remains controversial,[33,69] it has been established that one or more moles of protease are covalently bound in each mole of product. It has been proposed that, in the larger products, the bound protease molecules form molecular bridges between two or more proteolytic fragments of α_2-macroglobulin.[33,69]

As mentioned above, unreacted *Limulus* α_2-macroglobulin has a subunit molecular mass of 185 kDa. Trypsin-reacted samples contained two prominent peptides smaller (85 and 100 kDa) than the intact subunit, but also three peptides that were larger (200, 250, and 300–350 kDa). Formation of the larger peptides clearly involved isopeptide bond formation mediated by the thiol ester, since they failed to form when *Limulus* α_2M that had been pretreated with methylamine was reacted with trypsin. Only the 85- and 100-kDa fragments formed. Because the α_2-macroglobulin species from *Limulus* has proven to be similar to mammalian α_2-macroglobulin, especially to the dimeric forms of α_2-macroglobulin found in many mammals and because the latter require covalent bonding to the reacting proteases for α_2-macroglobulin–protease complex formation,[20,94] we were surprised to observe that *Limulus* α_2-macroglobulin–protease interaction involves solely noncovalent bonding and the reaction products larger than the intact subunit do not include bound trypsin. When *Limulus* α_2-macroglobulin was reacted with biotinylated trypsin, almost all of the biotin label detected in Western blots of reducing SDS-polyacrylamide gel electrophoretic separations was found at the molecular mass value of trypsin. Small amounts of trypsin had become covalently linked to α_2M fragments smaller than the native subunit (e.g., 100 and 120 kDa), but all of the large products were free of label. As expected, biotinylated trypsin did form covalent complexes with human α_2M. Thus, *Limulus* α_2M departs from the behavior shown by mammalian α_2M in regard to the relative importance of covalent and noncovalent bonding of the reacting proteases.

CHARACTERIZATION OF THE PLASMA-BASED CYTOLYTIC SYSTEM OF *LIMULUS*

One of the important immune defense strategies employed by animals is to kill invading pathogenic organisms by inducing their cytolysis in the blood. In higher vertebrates, cytolysis is mediated by the complement system with its associated regulators and receptors.[99] The key factor in the mammalian complement system is the protein C3, which binds to the surfaces of target cells, marking them for destruction by cytolysis and phagocytosis. Binding involves the covalent bonding of a reactive internal thiol ester of the C3 molecule with hydroxyl and amino residues at the surface of the target particle.[100] C3 is a member of the α_2 macroglobulin family of proteins, based on peptide sequence homology and the presence of the reactive thiol ester.[34] A recently emergent topic in the complement field is the evolution of this complex defense system.[101,102]

Lower vertebrates show many but not all of the elements of the mammalian system.[101] Invertebrates also have plasma- or hemocyte-based cytolytic systems,[103–111] but none had been convincingly demonstrated to be related to the vertebrate complement system until Enghild *et al.*[45] observed that the form of α_2-macroglobulin found in the blood of the American horseshoe crab *Limulus polyphemus*[38,61,112] is a component of the plasma-based cytolytic system of that animal. Based on the molecular similarity between α_2-macroglobulin and C3 and the involvement of the thiol ester in the activities of both molecules, these results are consistent with the possibility that in *Limulus,* α_2-macroglobulin serves a function like that of C3 in vertebrates.

We have confirmed the basic observations of Enghild *et al.*[45] Hemocyanin-depleted *Limulus* plasma showed a dose-dependent, Ca^{2+}-dependent hemolytic activity (FIG. 4). The red cells employed for this assay were not sensitized by pretreatment with anti-erythrocyte antisera. An involvement of α_2-macroglobulin in hemolysis was demonstrated by the reduction in the hemolytic titer of plasma following treatment with methylamine under conditions that eliminated the protease-binding capacities of the plasma (FIG. 4). Hemolytic activity was restored to the methylamine-treated plasma by the addition of purified *Limulus* α_2-macroglobulin (FIG. 4). Methylamine-treated *Limulus* α_2-macroglobulin was unable to restore the hemolytic activity of methylamine-treated plasma.

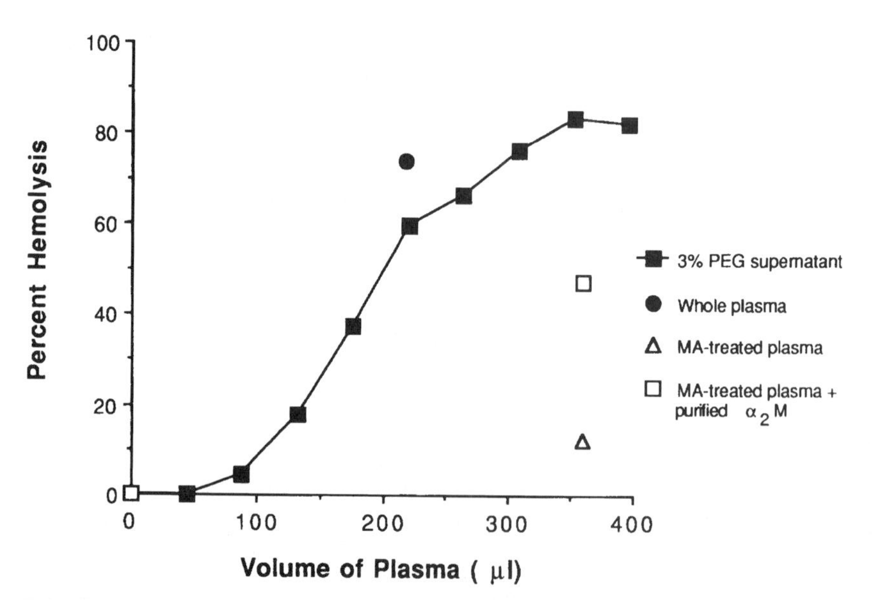

FIGURE 4. Hemolytic activity of *Limulus* plasma. Plasma was depleted in hemocyanin by addition of polyethylene glycol to 3% (final concentration). Hemocyanin-depleted plasma (■) shows a similar hemolytic activity to that of whole plasma (●). Control samples, containing equivalent volumes of 3% PEG in sea water (to replicate the ionic composition of *Limulus* plasma), were nonhemolytic. Methylamine-treated plasma (Δ) has lost hemolytic activity, which can be partially restored by the addition of 0.38 mg/ml purified *Limulus* α_2-macroglobulin (□).

The three most abundant proteins of *Limulus* plasma are α_2-macroglobulin (185 kDa),[61] hemocyanin (70 kDa), and C-reactive protein (24 kDa).[113–115] α_2-macroglobulin is present in *Limulus* plasma at 1 to 2 mg/ml[45] and C-reactive protein at 1 to 5 mg/ml.[113] Hemocyanin is present at 30 to 50 mg/ml. The hemolytic activity of *Limulus* plasma is unaffected by the nearly complete removal of hemocyanin by ultracentrifugation, indicating that this molecule plays no essential role in the process. The third abundant plasma protein, C-reactive protein, is an essential component of the hemolytic reaction. Treatment of plasma with the C-reactive protein reactant, phosphorylethanolamine, resulted in dose-dependent elevation of hemolytic activity (FIG. 5). Removal of C-reactive protein with phosphorylethanolamine-agarose (FIG. 6) significantly reduced activity (FIG. 7). Activity was restored to the phosphorylethanolamine-agarose-treated plasma by the addition of purified C-reactive protein. Purified C-reactive protein and a minor subfraction of C-reactive protein that was purified by affinity chromatography with phosphorylethanolamine-agarose and fetuin agarose were hemolytic in the absence of other components (FIG. 8). The fetuin-binding subfraction of C-reactive protein comprised less than 1% of the total C-reactive protein and had a sialic acid–dependent hemagglutinating activity. The

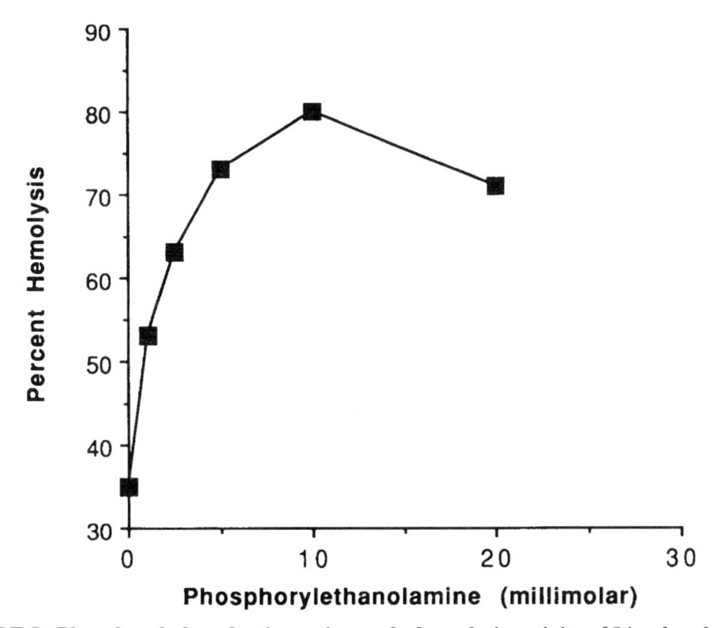

FIGURE 5. Phosphorylethanolamine activates the hemolytic activity of *Limulus* plasma. A 1 M solution of phosphorylethanolamine buffered to pH 6.8 with 1 M Tris base was added to an 0.8-ml reaction mixture to give the indicated final concentration of phosphorylethanolamine and incubated for 5 min with plasma before red blood cells were added. All samples contained 100 μl of *Limulus* plasma that had been cleared of hemocyanin by ultracentrifugation. An equivalent amount of 1 M Tris, pH 7.5, failed to activate plasma. Phosphorylethanolamine applied in the absence of plasma failed to elicit hemolysis.

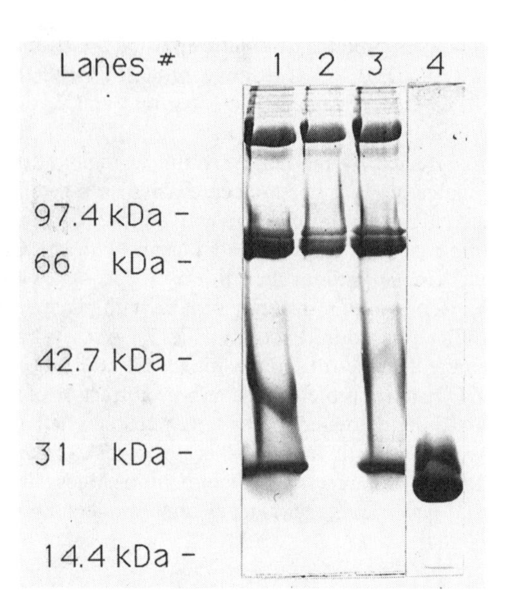

FIGURE 6. SDS-polyacrylamide gel electrophoresis (reducing conditions: samples treated with 5% β-mercaptoethanol) of 3% of polyethylene glycol supernatant of *Limulus* plasma (lane 1 = untreated plasma; lane 2 = plasma exposed to phosphorylethanolamine-agarose; lane 3 = plasma exposed to plain agarose) and *Limulus* C-reactive protein (lane 4).

hemolytic activity of purified C-reactive protein was unaffected by treatment with methylamine.

The possibility that invertebrates possess isolated elements of the vertebrate complement system has been suggested previously,[103,107,116,117] but the demonstration that α_2-macroglobulin and a subfraction of C-reactive protein with lectin activity participate in hemolysis in *Limulus* is the first direct identification of an involvement of components related to elements of the vertebrate complement system in an invertebrate cytolytic system. However, important differences between complement-mediated hemolysis and hemolysis in *Limulus* are readily apparent. In the complement system, C3 is absolutely required, whereas in *Limulus* the C3 homologue, α_2-macroglobulin, is dispensable. Although inactivation of α_2-macroglobulin reduced the hemolytic activity of unpooled plasma samples, for reasons that are not clear, the hemolytic activity of two independent, pooled plasma samples was unaffected by methylamine treatments that inactivated the α_2-macroglobulin present. α_2-Macroglobulin has been reported to be the only methylamine-binding protein in the plasma of *Limulus*[45] (Dodds, personal communication), and the hemolytic activity of methylamine-inactivated plasma was restored by addition of purified *Limulus* α_2-macroglobulin (FIG. 1), so an involvement of α_2-macroglobulin is indicated. Our results, however, suggest that the role is not the obligatory role that is exercised by C3 in the mammalian complement system. C-reactive protein is an activator of the complement system of mammals, but not a direct participant in the

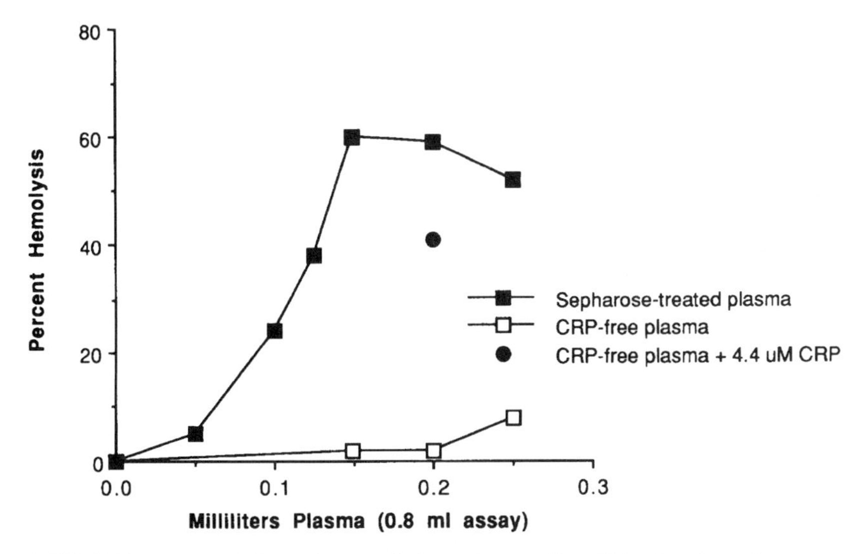

FIGURE 7. Removal of C-reactive protein from ultracentrifuged *Limulus* plasma by passage over phosphorylethanolamine–Sepharose reduces its hemolytic activity. Ultracentrifuged plasma passed over unconjugated Sepharose 4B (■) was strongly hemolytic, whereas phosphorylethanolamine–Sepharose-treated plasma (□) showed reduced activity. Phosphorylethanolamine-agarose removes C-reactive protein. Addition of 4.4 μM purified C-reactive protein (●) restored activity.

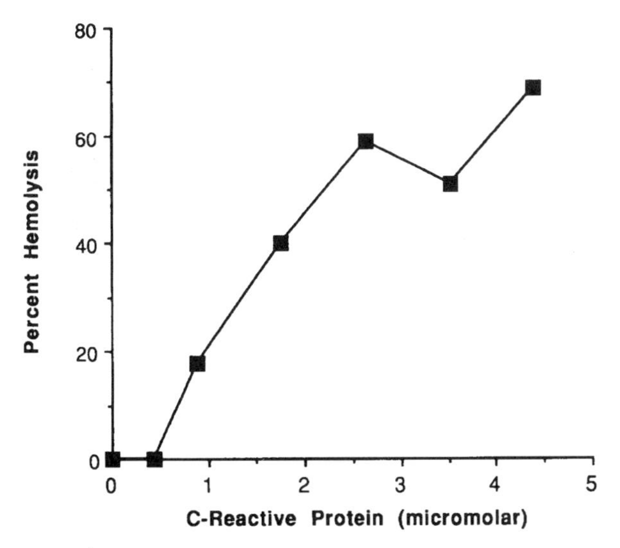

FIGURE 8. Hemolytic activity of purified *Limulus* C-reactive protein. Hemolysis was conducted in the presence of 0.19 M $NaCl_2$ and in the absence of other plasma components. The added protein was purified *Limulus* C-reactive protein.

process of complement-mediated cytolysis.[118–120] In *Limulus,* the subfraction of C-reactive protein with lectin activity appears to play a more central role in cytolysis because removal of C-reactive protein from plasma profoundly depressed its hemolytic activity and the purified subfraction of *Limulus* C-reactive protein that has sialyl lectin activity could produce hemolysis in the absence of other plasma proteins.

In summary, invertebrates possess complex and elaborate internal defenses against invading pathogenic organisms.[121] The precise role in immunity that is served by α_2M is only partially clear, in either invertebrates or vertebrates. This molecule probably plays an important role in the removal of endopeptidases from the body fluids, with the exceptionally broad spectrum of activity facilitating its ability to interact with proteases from endogenous and exogenous sources. In *Limulus,* α_2M may play an additional role in the cytolysis and destruction of foreign cells. In vertebrates, C3, which is the most important component in the complement-mediated destruction of foreign cells is a member of the α_2M family of proteins.

REFERENCES

1. Sottrup-Jensen, L. 1987. *In* The Plasma Proteins. Structure, Function, and Genetic Control, 2nd edit. F.W. Putnam, Ed. Vol. V: 191–291. Academic Press. Orlando, FL.
2. Armstrong, P. B. & J. P. Quigley. 1991. *In* Immunology of Insects and Other Arthropods. A. P. Gupta, Ed.: 291–310. CRC Press. Boca Raton, FL.
3. Huang, J. S. 1989. Am. J. Respir. Cell Mol. Biol. **1:** 169–170.
4. James, K. 1990. Immunol. Today **11:** 163–167.
5. Peterson, C. G. B. & P. Venge. 1987. Biochem. J. **245:** 781–787.
6. Santambrogio, P. & W. H. Massover. 1989 Br. J. Haematol. **71:** 281–290.
7. Barrett, A. J. & P. M. Starkey. 1973. Biochem. J. **133:** 709–724.
8. Swenson, R. P. & J. B. Howard. 1979. J. Biol. Chem. **254:** 4452–4456.
9. Harpel, P. C. 1973. J. Exp. Med. **138:** 508–521.
10. Sottrup-Jensen, L., T. M. Stepanik, T. Kristensen, D. M. Wierzbicki, C. M. Jones, P. B. Lonblad, S. Magnusson & T. E. Petersen. 1984. J. Biol. Chem. **259:** 8318–8327.
11. Jensen, P. E. H. & L. Sottrup-Jensen. 1986. J. Biol. Chem. **261:** 15863–15869.
12. Barrett, A. J., M. A. Brown & C. A. Sayers. 1979. Biochem. J. **181:** 401–418.
13. Gonias, S. L. & S. V. Pizzo. 1983. Biochemistry **22:** 536–546.
14. Gonias, S. L. & S. V. Pizzo. 1983. Biochemistry **22:** 4933–4940.
15. Feldman, S. R. & S. V. Pizzo. 1984. Biochim. Biophys. Acta **235:** 267–275.
16. Feldman, S. R. & S. V. Pizzo. 1986. Biochemistry **25:** 721–727.
17. Sand, O., J. Folkersen, J. G. Westergaard & L. Sottrup-Jensen. 1985. J. Biol. Chem. **260:** 15723–15735.
18. Spycher, S. E., S. Arya, D. E. Isenman & R. H. Painter. 1987. J. Biol. Chem. **262:** 14606–14611.
19. Starkey, P. M. & A. J. Barrett. 1982. Biochem. J. **205:** 91–95.
20. Enghild, J. J., G. Salvesen, I. B. Thøgersen & S. V. Pizzo. 1989. J. Biol. Chem. **264:** 11428–114345.
21. Feinman, R. D., A. I. Yuan, S. R. Windwer & D. Wang. 1985. Biochem. J. **231:** 417–423.
22. Howard, J. B. 1981. Proc. Natl. Acad. Sci. USA **78:** 2235–2239.
23. Salvensen, G. S. & A. J. Barrett. 1980. Biochem. J. **187:** 695–701.
24. Salvensen, G. S., C. A. Sayers & A. J. Barrett. 1981. Biochem. J. **195:** 453–461.
25. Sottrup-Jensen, L., T. E. Petersen & L. Magnusson. 1981. FEBS Lett. **128:** 123–126.
26. Van Leuven, F., J. J. Cassiman & H. Van Den Berghe. 1981. J. Biol. Chem. **256:** 9023–9027.

27. Wu, K., D. Wang & R. D. Feinman. 1981. J. Biol. Chem. **256:** 10409–10414.
28. Sottrup-Jensen, L., T. E. Petersen & S. Magnusson. 1980. FEBS Lett. **121:** 275–279.
29. Sottrup-Jensen, L., P. B. Lonblad, T. M. Stephanik, T. E. Petersen, S. Magnusson & H. Jornvall. 1981. FEBS Lett. **127:** 167–173.
30. Swenson, R. P. & J. B. Howard. 1979. Proc. Natl. Acad. Sci. USA **76:** 4313–4316.
31. Wang, D., K. Wu & R. D. Feinman. 1981. Arch. Biochem. Biophys. **211:** 500–506.
32. Wang, D., K. Wu & R. D. Feinman. 1983. Arch. Biochem. Biophys. **222:** 117–122.
33. Wang, D., A. I. Yuan & R. D. Feinman. 1984. Biochemistry **23:** 2807–2811.
34. Tack, B. F. 1983. Springer Semin. Immunopathol. **6:** 259–282.
35. Van Leuven, F. 1984. Mol. Cell. Biochem. **58:** 121–167.
36. Strickland, D. K., J. D. Ashcom, S. Williams, W. H. Burgess, M. Migliorini & W. S. Argraves. 1990. J. Biol. Chem. **265:** 17401–17404.
37. Quigley, J. P., P. B. Armstrong, P. Gallant, F. R. Rickles & W. Troll. 1982. Biol. Bull. (Woods Hole) **163:** 402.
38. Quigley, J. P. & P. B. Armstrong. 1983. J. Biol. Chem. **258:** 7903–7906.
39. Armstrong, P. B., M. Rossner & J. P. Quigley. 1985. J. Exp. Zool. **236:** 1–9.
40. Hergenhahn, H. -G. & K. Söderhäll. 1985. Comp. Biochem. Physiol. **81B:** 833–935.
41. Hergenhahn, H. -G., M. Hall & K. S214derhäll. 1988. Biochem. J. **255:** 801–806.
42. Stöcker, W. & R. Zwilling. 1986. Verh. Deutschen Zool. Ges. **79:** 190–191.
43. Armstrong, P. B. & J. P. Quigley. 1992. Veliger **35:** 161–164.
44. Thøgersen, I. B., G. Salvesen, F. H. Brucato, S. V. Pizzo & J. J. Enghild. 1992. Biochem. J. **285:** 521–527.
45. Enghild, J. J., I. B. Thorgersin, G. Salvesen, G. H. Fey, N. L. Figler, S. L. Gonias & S. V. Pizzo. 1990. Biochemistry **29:** 10070–10080.
46. Hall, M., K. Söderhäll & L. Sottrup-Jensen. 1989. FEBS Lett. **254:** 111–114.
47. Sottrup-Jensen, L., W. Borth, M. Hall, J. P. Quigley & P. B. Armstrong. 1990. Comp. Biochem. Physiol. **96B:** 621–625.
48. Armstrong, P. B. 1985. *In* Blood Cells of Marine Invertebrates. W. Cohen, Ed.: 77–124. A.R. Liss. New York.
49. Armstrong, P. B. 1985. *In* Blood Cells of Marine Invertebrates. W. Cohen, Ed.: 77–124. A.R. Liss. New York.
50. Muta, T., R. Hashimoto, T. Miyata, H. Nishimura, Y. Toh & S. Iwanaga. 1990. J. Biol. Chem. **265:** 22426–22433.
51. Muta, T., T. Miyata, Y. Misumi, F. Tokunaga, T. Nakamura, Y. Toh, Y. Ikehara & S. Iwanaga. 1991. J. Biol. Chem. **266:** 6554–6561.
52. Roth, R. I. & J. Levin. 1992. J. Biol. Chem. **267:** 24097–24102.
53. Tai, J. Y. & T. -Y. Liu. 1977. J. Biol. Chem. **252:** 2178–2181.
54. Tokunaga, F., H. Nakajima & S. Iwanaga. 1991. J. Biochem. **109:** 150–157.
55. Armstrong, P. B. & J. P. Quigley. 1985. Biochem. Biophys. Acta **827:** 453–459.
56. Donovan, M. A. & T. M. Lane. 1991. J. Biol. Chem. **266:** 2121–2125.
57. Nakamura, T., T. Hirai, F. Tokunaga, S. Kawabata & S. Iwanaga. 1987. J. Biochem. **101:** 1297.
58. Starkey, P. M. & A. J. Barrett. 1982. Biochem. J. **205:** 105–115.
59. Starkey, P. M., T. C. Fletcher & A. J. Barrett. 1982. Biochem. J. **205:** 97–104.
60. Armstrong, P. B., W. F. Mangel, J. S. Wall, J. F. Hainfield, K. E. Van Holde, A. Ikai & J. P. Quigley. 1991. J. Biol. Chem. **266:** 2526–2530.
61. Quigley, J. P. & P. B. Armstrong. 1985. J. Biol. Chem. **260:** 12715–12719.
62. Swenson, R. P. & J. B. Howard. 1980. J. Biol. Chem. **225:** 8087–8091.
63. Tack, B. F., R. A. Harrison, J. Janatova, M. L. Thomas & J. M. Prahl. 1980. Proc. Natl. Acad. Sci. USA **77:** 5764–5768.
64. Larsson, L. -J. & I. Bjork. 1984. Biochemistry **23:** 2802–2807.
65. Straight, D. L. & P. A. Mc Kee. 1984. J. Biol. Chem. **259:** 1272–1278.
66. Strickland, D. K. & P. Bhattacharya. 1984. Biochemistry **23:** 3115–3124.
67. Larsson, L. -J., S. T. Olson & I. Bjork. 1985. Biochemistry **24:** 1585–1593.
68. Sottrup-Jensen, L., T. E. Petersen & L. Magnusson. 1981. FEBS Lett. **128:** 127–132.
69. Sottrup-Jensen, L., J. F. Hansen, H. S. Petersen & L. Kristensen. 1990. J. Biol. Chem. **265:** 17727–17737.

70. HOWARD, J. B. 1981. Proc. Natl. Acad. Sci. USA **78:** 2235–2239.
71. HOWARD, J. B., R. SWENSON & E. ECCLESTON. 1983. Ann. N.Y. Acad. Sci. **421:** 160–166.
72. HOWARD, J. B., M. VERMEULEN & R. B. SWENSON. 1980. J. Biol. Chem. **255:** 3820–3823.
73. DANGOTT, L. J. & L. W. CUNNINGHAM. 1982. Biochem. Biophys. Res. Commun. **107:** 1243–1251.
74. GONIAS, S. L., A. E. BALBER, W. J. HUBBARD & S. V. PIZZO. 1983. Biochem. J. **209:** 99–105.
75. FELDMAN, S. R., S. L. GONIAS, K. A. NEY, C. W. PRATT & S. V. PIZZO. 1984. J. Biol. Chem. **259:** 4458–4462.
76. ESNARD, F., N. GUTMAN, A. EL MORIJAHED & F. GARTHIER. 1985. FEBS Lett. **182:** 125–129.
77. HARPEL, P. C., M. B. HAYES & T. E. HUGLI. 1979. J. Biol. Chem. **254:** 8669–8678.
78. SEYA, T. & S. NAGASAWA. 1981. J. Biochem. (Tokyo) **89:** 659–664.
79. NELLES, L. P. & H. P. SCHNEBLI. 1982. Hoppe-Zeyler's Z. Physiol. Chem. **363:** 677–682.
80. NAGASE, H., H. HARRIS, J. F. WOESSNER & K. BREW. 1983. J. Biol. Chem. **258:** 7481–7489.
81. ARMSTRONG, P. B. & J. P. QUIGLEY. 1987. Biochem. J. **248:** 703–707.
82. DURHAM, J. W. 1978. Annu. Rev. Earth Planet. Sci. **6:** 21–24.
83. RUNNIGAR, B. 1982. J. Geol. Soc. Australia **29:** 395–411.
84. SEPKOSKI, J. J. 1978. Paleobiology **4:** 223–251.
85. FEINMAN, R. D., Ed. 1983. Chemistry and Biology of α_2-Macroglobulin. Ann. N.Y. Acad. Sci. **421:** 478 pp.
86. HARPEL, P. C. & M. S. BROWER. 1983. Ann. N.Y. Acad. Sci. **421:** 1–9.
87. LONBERG-HALL, K., D. L. REED, D. C. ROBERTS & D. DAMATO-MCCABE. 1987. J. Biol. Chem. **262:** 4844–4853.
88. VAN DER GRAAF, F., A. RIETVELD, F. J. A. KEUS & B. N. BOUMA. 1984. Biochemistry **23:** 1760–1766.
89. WALLER, E. K., W. -D. SCHLEUING & E. REICH. 1983. Biochem. J. **215:** 123–131.
90. CREWS, B. C., M. W. JAMES, A. H. BETH, P. GETLINS & L. W. CUNNINGHAM. 1987. Biochemistry **26:** 5963–5967.
91. GONIAS, S. L. & S. V. PIZZO. 1983. Biochemistry **22:** 4933–4940.
92. NAGASE, H. & H. HARRIS. 1983. J. Biol. Chem. **258:** 7490–7498.
93. POCHON, F. 1987. Biochim. Biophys. Acta **915:** 37–45.
94. CHRISTENSEN, U. & L. SOTTRUP-JENSEN. 1984. Biochemistry **23:** 6619–6626.
95. SALVESEN, G. S., C. A. SAYERS & A. J. BARRETT. 1981. Biochem. J. **195:** 453–461.
96. MEIJERS, J. C. M., P. N. M. TIJBURG & B. N. BORMA. 1987. Biochemistry **26:** 5932–5937.
97. PIZZO, S. V., S. RAJAGOPALON, P. A. ROCHE, H. E. FUCHS, S. R. FELDMAN & S. L. GONIAS. 1986. Biol. Chem. Hoppe-Seyler **367:** 1177–1182.
98. SOTTRUP-JENSEN, L. & H. BIRKENDAL-HANSEN. 1989. J. Biol. Chem. **264:** 393–401.
99. LAW, S. K. A. & K. B. M. REID. 1988. Complement. IRL Press. Oxford. 72 pp.
100. LAW, L. K. & R. P. LEVINE. 1977. Proc. Natl. Acad. Sci. USA **74:** 2701–2705.
101. DODDS, A. W. & A. J. DAY. 1993. *In* The Molecular Biology and Biochemistry of Complement. R. B. Sim, Ed. MTP Press. Lancaster, UK. In press.
102. FARRIES, T. C. & J. P. ATKINSON. 1991. Immunol. Today **12:** 295–300.
103. BERTHEUSSEN, K. 1983. Dev. Comp. Immunol. **7:** 637–640.
104. CANICATTI, C. & D. CUILLA. 1987. Dev. Comp. Immunol. **11:** 705–712.
105. CANICATTI, C. & D. CUILLA. 1988. Dev. Comp. Immunol. **12:** 55–64.
106. CENINI, P. 1983. Dev. Comp. Immunol. **7:** 637–640.
107. DAY, N. K. B., H. GEWURZ, R. JOHANNSEN, J. FINSTAD & R. A. GOOD. 1970. J. Exp. Med. **132:** 941–950.
108. KOMANO, H. & S. NATORI. 1985. Dev. Comp. Immunol. **9:** 31–40.
109. NOGUCHI, H. 1903. Zentr. Bakter. Parasitenk. Infekt. **33:** 353–362.
110. PHIPPS, K., J. S. CHADWICK, R. G. LEEDER & W. P. ASTON. 1989. Dev. Comp. Immunol. **13:** 103–111.
111. TUCKOVÁ, L., J. REJNEK, P. ŠÍMA & R. ONDREJOVÁ. 1986. Dev. Comp. Immunol. **10:** 181–189.

112. QUIGLEY, J. P., A. IKAI, H. ARAKAWA, T. OSADA & P. B. ARMSTRONG. 1991. J. Biol. Chem. **266:** 19426–19431.
113. NGUYEN, N. Y., A. SUZUKI, S. -M. CHENG, G. ZON & T. -Y. LIU. 1986. J. Biol. Chem. **261:** 10450–10455.
114. NGUYEN, N. Y., A. SUZUKI, R. A. BOYKINS & T. -Y. LIU. 1986. J. Biol. Chem. **261:** 10456–10465.
115. ROBEY, F. A. & T. -Y. LIU. 1981. J. Biol. Chem. **256:** 969–975.
116. BERTHEUSSEN, K. 1982. Dev. Comp. Immunol. **6:** 635–642.
117. DAY, N., H. GEIGER, J. FINSTAD & R. A. GOOD. 1972. J. Immunol. **109:** 164–167.
118. JIANG, H., J. N. SIEGEL & H. GEWURZ. 1991. J. Immunol. **146:** 2324–2330.
119. MIYAZAWA, K. & K. INOUE. 1990. J. Immunol. **145:** 650–654.
120. VOLANAKIS, J. E. 1982. Ann. N.Y. Acad. Sci. **389:** 235–250.
121. ARMSTRONG, P. B. 1991. *In* Immunology of Insects and Other Arthropods. A. P. Gupta, Ed.: 1–17. CRC Press. Boca Raton, FL.

C-Reactive Proteins, Limunectin, Lipopolysaccharide-Binding Protein, and Coagulin

Molecules with Lectin and Agglutinin Activities from *Limulus polyphemus*

TEH-YUNG LIU,[a] CONCEICAO A. S. MINETTI, CONSUELO L. FORTES-DIAS, TERESA LIU, LEEWEN LIN, AND YUAN LIN

Division of Allergenic Products and Parasitology
Office of Vaccines Research and Review
Center for Biologics Evaluation and Research
Food and Drug Administration
Rockville, Maryland 20852-1448

In the invertebrate *Limulus polyphemus,* a number of molecules with lectin-like activities exists in the hemolymphs and in the hemocytes that may work in concert to serve the role of immunoglobulins in higher animals. This article summarizes our biochemical, molecular, and functional characterization of the C-reactive protein (*Limulin),* the cell-adhesion molecule (limunectin), the endotoxin-binding protein (LEBP-PI), and the coagulin (a proteolytic fragment of coagulogen). Each of these molecules are endowed with strong cell agglutinin and/or adhesion activities and are found as components of extra cellular fluid during exocytosis of amebocytes. Binding of these molecules to invading organisms could target them for the phagocytic activities of amebocytes.

LIMULIN–*LIMULUS* C-REACTIVE PROTEIN

C-reactive protein (CRP), the prototypic acute-phase protein in humans, was first discovered in patients with pneumonia and identified as a precipitin for the C-polysaccharide of the pneumococcal cell wall.[1] Calcium-dependent precipitation of this protein with the phosphocholine ligand of the C-polysaccharide provides the functional definition of CRPs.[2] The role CRP plays in helping to survive inflammation and to repair damage is still a matter of conjecture. *In vitro,* CRP has been found to bind phosphocholine[2] and chromatin.[3] Robey *et al.*[3] suggested that CRP may act as a scavenger for chromatin released from damaged cells. *In vitro* it, like the immunoglobulins, possesses the ability to promote agglutination and phagocytosis of bacteria and complement fixation.[4–7] In rodents, in conjunction with platelets, CRP is able to mediate protection against immature schistosomes or directly inhibit

[a]Address for correspondence: Division of Allergenic Products and Parasitology, Office of Vaccines Research and Review, Center for Biologics Evaluation and Research, Food and Drug Administration, 1401 Rockville Pike (HFM-410), Rockville, MD 20852-1448.

schizont development of malaria sporozoites.[8,9] The various biological functions attributed to CRP have been related to the host nonspecific defense mechanism against infectious agents and wound healing. It is possible that CRP is multifunctional, and there is truth in all of these proposals.

Over the last two decades, our laboratory has been engaged in the study of humans,[10] rabbits,[11] *Xenopus*,[12] and *Limulus* CRPs.[13] The results of these studies are summarized in TABLE 1. It has now been well established that liver is the primary site of CRP synthesis in mammals and that IL-6 is the essential and sufficient factor that is required for the induction of CRP synthesis[10] (FIG. 1). At the site of tissue injury or inflammation, IL-1 is generated, which stimulates the production of IL-6 by T cells. A number of "acute-phase" reactants, including CRP, fibrinonectin, α_1-antitrypsin, and α_1-acid glycoprotein, produced in liver under the influence of IL-6, migrate to the sites of inflammation or tissue injury.

In mammals, the concentration of serum CRP may increase 100 to 1,000-fold under inflammatory conditions.[14] *In vitro,* it has been demonstrated that the human CRP gene expressions is induced by IL-6.[10] Woo *et al.*[15] noted that in the 5′ end of the human CRP gene, there are heatshock consensus sequences. In both the *Xenopus* and the *Limulus* CRP genes, neither the IL-6 responsive element, nor the heatshock consensus sequences are found at the 5′-upstream region (TABLE 1). By protein and mRNA analyses, it has been shown that *Xenopus* CRP gene is not inducible by either heatshock or an inflammatory agent that induces the production of IL-6 in mammals. Although similar experiments have not been done with the *Limulus,* the fact that these *cis*-acting elements are missing in the 5′ upstream region of *Limulus* and *Xenopus* CRP genes, reinforces the suggestion that these elements are intimately involved in the induction of CRPs in human and rabbit under acute-phase conditions (TABLE 1).

CRP is a primitive pattern recognition, lectin-like molecule. It is evolutionarily conserved and is found from the invertebrate *L. polyphemus* to mammals. A number of functional capabilities have been demonstrated for CRP; however, the importance of these functions may vary from species to species. The marked differences in the mode of expression of CRP in various species again suggests functional difference.

TABLE 1. Comparison of CRPs

	Human	Rabbit	*Xenopus*	*Limulus*
Protein subunit	20,000	20,000	20,000	20,000
Structure	Pentamer	Pentamer	—	Hexamer
Homology (%)	100	76	45	35
mg/ml	<0.02–>2.0	<0.02–>2.0	<0.001	2–5
Gene copy	1	2 Alleles	1	>3 Polymorphic
Intron	$-(GT)_{15}-$	$-(GT)_{15}-$ $-(GT)_{12}-$	2	None
Heat shock elem.	2	2	None	None
Inducer	IL-6	IL-6	—	—
Immunoglobulin	Yes	Yes	Yes	No
Molecular weight (under physiological conditions)	~500,000	~500,000	50,000×X	$\sim 1 \times 10^6$

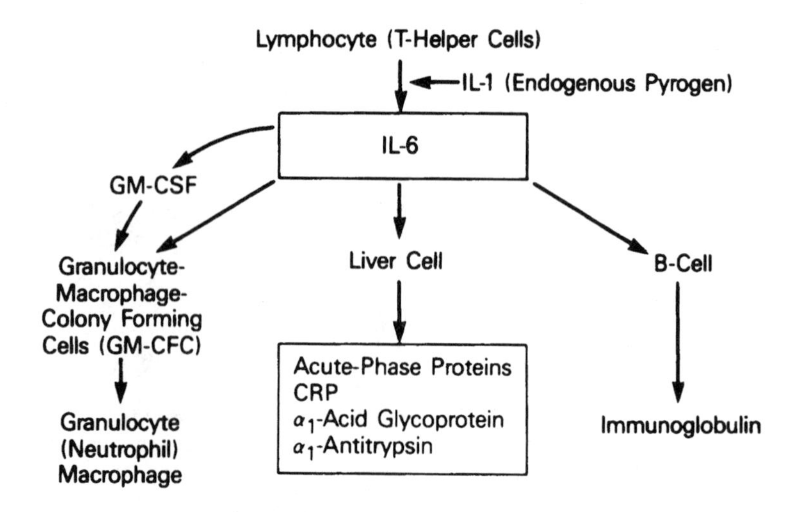

FIGURE 1. Induction of C-reactive protein.

On the other hand, the antiquity and remarkable conservation of structure and phosphocholine-binding property of CRP in evolution indicates the importance of at least certain functions. It is possible that the biological significance of CRP switched along with its mode of expression.

It has been suggested that CRP might serve the role of immunoglobulins in *Limulus*.[13,16] The absence of immunoglobulins, the presence of multiple CRP genes, the polymorphic nature of CRP, and the lectin-like property of CRP in *Limulus* make this an attractive proposal. It is possible that with the development of a more advanced immune system and other pattern recognition proteins or specialized defense agents in animals, for example, lipopolysaccharide-binding protein[17] and mannose-binding protein,[18] the biological functions of CRP gradually became less essential.

LIMUNECTIN: A CELL-ADHESION MOLECULE

Adhesion proteins are a group of complex glycoproteins that play critical roles in the regulation of the immune response, in wound healing, cellular development, morphogenesis, and metastasis. A substantial body of literature dealing with tissue injury and repair indicates that adhesion molecules, soluble or membrane bound, are involved. These proteins serve as mediators for cell-to-cell adhesion and cell to extracellular matrix interaction through different mechanisms. Fibrinonectin and von Willebrand factor contain several types of internal repeat, whereas vitronectin contains no internal repeat.[19–21] Adhesion proteins in general share little sequence homology but often contain several separate domains, each of which exhibits specific physiological functions.

By using phosphocholine affinity column chromatography similar to that used in the purification of CRPs, a 54-kDa intracellular protein from *Limulus* amebocytes that binds to PC independent of Ca^{2+} has been isolated and shown to possess cell-adhesion properties.[22]

The most striking feature of the molecule, as revealed by sequence analysis, is the repeated homologous segment. With the exception of 25 amino acids at the NH_2 terminus, the rest of the molecule consists of 10 continuous segments of about 45 amino acids that share extensive homology (FIG. 2). The alignment of repeated internally homologous segments is reminiscent of the Ig superfamily of proteins including the N-cell adhesion molecule,[23] where five internally homologous segments are found to be aligned contiguously in the molecule. These proteins serve, among other functions, as mediators of cell–cell interaction. Another group of adhesion proteins, regulators of complement activities, are also known to contain a variable number of 4 to 30 internal repeats, corresponding to the consensus sequence

```
                                                            Residue#   %Homology
                      EVSQTDKTEL HSTGMEILQS IFPS            (1-24)
(1)  IDAVFKWSNG VTYIFKGSCY FRYEDKTNEI SNCRRLSAWG GLTGP      (25-69)      91
(2)  VDAVFRWRNG VTYFFQGDCY YRYEDKTDEI SKCSPVTAWG GMTGP      (70-114)     93
(3)  VDAVFRWSNG ITYFFKEDCY YRYEDKDNKI SKCTPITAWG KMTGP      (115-159)    96
(4)  IDAVFRWSNG VTYFFKRDCY FRYEDKPNEI SKCRAIALWG ASSYQP     (160-205)    85
(5)  LDAVFRWNDG VTYFFKGFCY YHNDLK---- -KCKPISAWG GISKP      (206-245)    76
(6)  VSAVLLWNNK ETYFFEGKCY HSYEAKNNSI SKCIPISTWA KKIRV      (246-290)    56
(7)  VDAVFRWSNG ITYFFKGNCY YRYEDKTNKL SQCSPVTEWG GMTGP      (291-335)    98
(8)  VDAVFRWSNG ATYFFQGNCY YRYDDKNNKL SQCSPVTAWG GMTGP      (336-380)    98
(9)  VDAVFRWSNG ATYFFQGNCY YRYDDKNNKL SQCSPVTAWG GMTGP      (381-425)    98
(10) VDAVFRWSNG ATYFFKEDCY MKYEDKPQKL SGCNPISAWG GGIY       (426-469)    82
----------------------------------------------------
(c)  vdAVfrWsng vTYfFkgdCY yryedKtnki skCspitaWg gmtgp
```

FIGURE 2. Internal homology of limunectin indicating 10 consecutive repeating segments of 45 amino acids 1 through 10. (c) is a consensus sequence constructed from the repeating units by selecting amino acids that appear most frequently (used three to nine times) at each position; *capital letters* designate those amino acids that appear in all 10 segments. The homology is calculated by comparing amino acids in each of the 10 segments with those of the consensus sequence, allowing the substitution of similar amino acids. The presumptive disulfide bonds connecting cysteines are indicated. (Taken from Liu *et al.*[22])

```
                                                          %Identity   %Homology

    Limunectin  VDAVFRW--SNGITYFFKGDCYYRYED
(A)             ·||·|    ··| ||·|||  |·|·||                  52          80
    Vitronectin IDAAFTRINCQGKTYLFKGNQYWRFED
                207                      233

    Limunectin  VDAVFRWSNG-ITYFFKGDCYYRYEDKTNKISKCSP---ITAW
(B)             |||·|·||·   ||·| || ··||  ·  ·|·   |   · ||  44          69
    Gelatinase  VDAAFNWSKNKKTYIFAGDKFWRYNEVKKKMDPGFPKLIADAW
                539                                       582

    Limunectin  VDAVFRWSNGVTYFFKGDCYYRYEDKTNEISKCSPITAW
(C)             |||||   ·|· |||·|   |···  ||· |      ·  |      39          59
    Collagenase VDAVFM-KDGFFYFFHGTRQYKFDPKTKRILTLQKANSW
                427                                   464
```

FIGURE 3. Sequence comparison between limunectin *versus* vitronectin (human) (A), limunectin *versus* gelatinase (human) (B), and limunectin *versus* collagenase (human) (C). The symbol | indicates identical sequences; • indicates similar sequences. The consensus sequence (see FIG. 2) of limunectin was used for comparison. (Taken from Liu *et al.*[22] with permission from the *Journal of Biological Chemistry*.)

of complement regulatory proteins.[24] A granular membrane protein isolated from platelets and endothelial cells belongs to this family and contains nine homologous repeats and six functional domains.[25]

Some limited sequence homologies were found between the consensus sequence of limunectin and segments of vitronectin, gelatinase, and collagenase (FIG. 3). Such homologies become more significant when one considers the fact that these sequences are repeated 10 times in the *Limulus* protein. This finding was the reason for naming the *Limulus* 54-kDa protein limunectin and was the basis of our investigation into the possible functional similarities between limunectin and cell-adhesion molecules. Limunectin was found to bind collagen, gelatin, and fibrinonectin, the natural substrates for collagenase, gelatinase, and vitronectin (TABLE 2). The finding that limunectin shares a common sequence with vitronectin, gelatinase, and collagenase suggests that this common sequence might be responsible for the binding of limunectin to gelatin and collagen and fibrinonectin.

In addition to binding to a number of extracellular matrix molecules, limunectin binds to bacterial cells and fixed amebocytes, with an ID_{50} of about 10^{-7} M. Heparin, glucosamine, and lipopolysaccharide were ineffective in inhibiting this binding. When phosphocholine was used to compete for the binding to formalin-fixed *S. aureus* cells, a high concentration was needed, with an ID_{50} of 18 mM.

ENDOTOXIN-BINDING PROTEIN WITH PROTEASE INHIBITORY ACTIVITY: LEBP-PI

Protease inhibitors are often found to coexist with proteases in biological systems. There is significant evidence to suggest that protease inhibitors play a key role in preventing unwanted proteolysis as seen in mammalian blood plasma, where

TABLE 2. Binding of ^{125}I-Labeled Limunectin to Collagen, Gelatin, Heparin, and Fibronectin

	^{125}I-Limunectin Bound (CPM)	
Compound	Total	Specific[a]
Collagen		
Type I (native)	5506	1959
Type I (denatured)	4120	2708
Type III (native)	9310	5749
Type III (denatured)	8160	3721
Type IV (native)	7505	3506
Type IV (denatured)	6855	3783
Type V (native)	5719	2915
Type V (denatured)	6736	3129
Gelatin	4949	3652
Heparin	3503	3367
Fibronectin	4196	2289
Control[b]	1595	0

NOTE: The binding was carried out with 100 μl of the test compound (10 μg/ml) and 10,000 cpm of ^{125}I-labeled limunectin in the presence and in the absence of unlabeled limunectin.

[a]Specific binding was calculated by subtracting the nonspecific binding (with 100-fold molar excess of unlabeled limunectin) from the total binding. The values are the average of three determinations.

[b]As a control, the wells received only the binding buffer.

proteolytic processes are tightly regulated in part by the presence of a number of inhibitors. Activities inhibitory to proteases have been observed in both the plasma and amebocytes of *Limulus.* An α_2-macroglobulin-like inhibitor was identified in plasma and a panel of protease inhibitors,[26] including a 16-kDa serine protease inhibitor,[27] have been purified from *Limulus* amebocytes.

Using a lipopolysaccharide affinity column and ion-exchange chromatography, a 12-kDa protein (LEBP-PI) with inhibitory activity toward trypsin but not chymotrypsin was purified from the acid extract of *Limulus* amebocytes.[28] The protein is a major component of the cytoplasmic proteins (1%). In solid-phase binding assays, the radio-labeled protein binds specifically to LPS with a K_d value on the order of 10^{-7} M. The binding to LPS can be displaced by the unlabeled 12-kDa protein, polymyxin B, lipid A, and to a lesser extent by D-glucosamine. In whole-cell binding assays, the 12-kDa protein has also been shown to bind to *Escherichia coli.*

Using both [^{14}C]casein and a synthetic substrate, the protein has been shown to inhibit the proteolytic activity of trypsin, with an IC_{50} of approximately 10^{-7} M. In the presence of LPS, the antitryptic activity of LEBP-PI remains unaffected.

A cDNA coding for LEBP-PI has been isolated and sequenced. The amino acid sequence deduced from the cDNA indicates that this protein shares no sequence homology with LPS-binding proteins isolated from different species of vertebrates[29] and invertebrates.[30] Although LEPB-PI does not share any homology with other LPS-binding proteins, it is a basic protein (pI = 9.6) similar to most LPS-binding proteins. The basic nature (rich in Arg and Lys) of LEBP-PI could be an important factor in its interaction with the negatively charged LPS molecule and as inhibitor to trypsin.

When the purified LEBP-PI protein was subjected to chromatography on a Superose 12 column using a buffer containing 0.5 M KCl (the physiological ionic strength of *Limulus*), the protein eluted at a position following vitamin B_{12} (M_r = 1350). This observation indicates that LEBP-PI interacts strongly with the polysaccharide gel matrix of the agarose column. The lectin-like property of LEBP-PI is most likely responsible for its binding to bacterial cells.

Although the exact function of LEBP-PI is not known, intracellularly it probably acts as an inhibitor of trypsin-like enzymes to prevent the onset of the coagulation process of *Limulus* lysate. In the absence of LPS, spontaneous degranulation and gelation occurs *in vitro* in isolated amebocytes, albeit very slowly due to the presence of trypsin-like activity in the *Limulus* amebocytes. The presence of the relatively high concentration of LEBP-PI in the granules may serve to ensure that the proteolytic enzymes remain in their inactive form under normal physiological conditions. At the time of cell lysis or exocytosis, LEBP-PI is expelled from the amebocytes together with other cellular components, including the proteins in the coagulation cascade system. Binding of LEBP-PI to bacterial cells could target these microorganisms for the phagocytic activities of amebocytes. In this manner, LEBP-PI could serve its dual functions.

COAGULIN: A PROTEOLYTIC FRAGMENT OF COAGULOGEN

Coagulogen is a major protein in the hemocytes of many invertebrates, and its role as a substrate in clot formation has been studied extensively. In *Limulus polyphemus,* coagulogen is a 175-amino-acid protein containing eight disulfide bonds.[31] During the last step in the clotting process, the endotoxin-induced clotting enzyme cleaves coagulogen between amino acids 18 and 19 and between 46 and 47 to release a peptide fragment of 28 amino acids. The 18-amino-acid peptide at the NH_2-terminal end of coagulogen remains linked to the rest of the molecule through a disulfide bond to form the coagulin clot.[32]

During the course of our studies of proteins with agglutines and/or lectin activities from *Limulus* amebocytes, we have observed that the ability of amebocyte to promote cell agglutination of rabbit erythrocytes or human leukocytes is dependent on the activation by endotoxin of the clotting cascades. A 24-kDa protein with strong cell agglutination properties was purified from endotoxin-activated *Limulus* amebocytes. Characterization of this protein revealed that it corresponds to a proteolytic fragment of the coagulogen. Similar cell agglutination activity can be generated, *in vitro,* by proteolytic cleavage of this coagulogen with either trypsin or a preparation of endogenous protease from *Limulus* amebocytes. Only the proteolytically cleaved coagulin, not the intact coagulogen or the partially cleaved product at position 17 or 46, binds to rabbit erythrocytes and formalin-fixed *Limulus* amebocytes. These bindings were not inhibited by various mono-and oligosaccharides, and were not Ca^{2+}-dependent. Thus, coagulin, like its mammalian counterpart fibrin, is an agglutinin with adhesion properties. In addition to rabbit erythrocytes and human leukocytes, coagulin binds to other cells such as bacteria and *Leishmania* in a concentration-dependent manner. The consequence of this binding is the entrapping and sequestering of cell particles for their eventual removal from the circulation.

SUMMARY

In 1964, Levin and Bang discovered that gram-negative bacterial endotoxin could rapidly induce gelation of *Limulus* amebocyte lysate.[33] This observation has led to the development of the most sensitive and specific method for the detection of bacterial endotoxin in pharmaceuticals and drugs intended for human use. Over 10 years ago, Bang[34] injected endotoxin into young horseshoe crabs and observed a time and dose-dependent coagulation of the whole hemolymph. Limunectin, LEBP-PI, and *Limulus* CRP are found together with coagulin as part of the hemolymph clot at the time of endotoxin-induced exocytosis of amebocytes. In this manner, these molecules with agglutinin/lectin activities could work in concert to assist in the recognition and eventual removal of invading microorganisms from the circulating system. Although the mechanism of endotoxin-induced clot formation is to a large extent understood, the mechanism of clot dissolution and removal in the *Limulus* hemolymph remains to be clarified.

REFERENCES

1. TILLET, W. S. & T. FRANCESES. 1930. J. Exp. Med. **52:** 561–571.
2. KAPLAN, M. H. &. J. E. VOLANAKIS. 1974. J. Immunol. **112:** 2135–2147.
3. ROBEY, F. A., K. D. JONES, T. TANAKA & T. -Y. LIU. 1984. J. Biol. Chem. **259:** 7311–7316.
4. GAL, K. & M. MILTENYI. 1955. Acad. Sci. Hung. **3:** 41–51.
5. HOKAMA, Y., M. K. COLEMAN & R. F. RILEY. 1962. J. Bacteriol. **83:** 1017–1024.
6. JAMES, K., L. BAUM, C. ADAMOWSKI & H. GEWURZ. 1983. J. Immunol. **131:** 2930–2934.
7. ROBEY, F. A., K. D. JONES & A. D. STEINBERG. 1985. J. Exp. Med. **161:** 1344–1356.
8. BOUT, D., M. P. JOSEPH, H. VORNG, D. DESLEE & A. CAPRON. 1986. Science **231:** 153–156.
9. PIED, S., A. NUSSLER, M. PONTET, F. MILTGEN, H. MATILE, P. -H. LAMBERT & D. MAZIER. 1989. Infect. Immun. **57:** 278–282.
10. LI, S. -P., T. -Y. LIU & N. D. GOLDMAN. 1990. J. Biol. Chem. **265:** 4136–4142.
11. SYIN, C., E. C. GOTSCHLICH & T. -Y. LIU. 1986. J. Biol. Chem. **261:** 5473–5479.
12. LIN, L. & T. -Y. LIU. 1993. J. Biol. Chem. **268:** 6809–6815.
13. NGUYEN, N. Y., A. SUZUKI, S. -M. CHENG, G. ZON & T. LIU.1986. J. Biol. Chem. **261:** 10450–10455.
14. KUSHNER, I. 1982. Ann. N.Y. Acad. Sci. **389:** 39–48.
15. WOO, P., J. R. KORENBERG & A. S. WHITEHEAD. 1985. J. Biol. Chem. **260:** 13384–13388.
16. MARCHALONIS, J. J. & G. M. EDELMAN. 1968. J. Mol. Biol. **32:** 453–465.
17. TOBIAS, P. S., J. C. MATHISON & R. J. ULEVITCH. 1988. J. Biol. Chem. **263:** 13479–13481.
18. EZEKOWITZ, R. A. B. & P. D. STAHL. 1988. J. Cell Sci. **9:** 121–133.
19. VERWEIJ, C. L., P. J. DIERGAARDE, M. HART & H. PANNEKOEK. 1986. EMBO J. **5:** 1839–1847.
20. PETERSEN, T. E., H. C. THOGERSEN, K. SKORSTENGAARD, K. VIBE-PEDERSEN, P. SAHL, L. SOTTRUP-JENSEN & S. MAGNUSSON. 1983. Proc. Natl. Acad. Sci. USA **80:** 137–141.
21. SUZUKI, S., A. OLDBERG, E. G. HAYMAN, M. D. PIERSCHBACHER & E. RUOSLAHTI. 1985. EMBO J. **4:** 2519–2524.
22. LIU, T., Y. LIN, T. CISLO, C. A. S. A. MINETTI, J. M. K. BABA & T. -Y. LIU. 1991. J. Biol. Chem. **266:** 14813–14821.

23. Cunningham, B. A., J. J. Hemperly, B. A. Murray, E. A. Prediger, R. Brackenbury & G. M. Edelman. 1987. Science **236:** 799–806.
24. Hourdace, D., V. M. Holers & J. P. Atkinson. 1989. Adv. Immunol. **45:** 381–416.
25. Johnston, G. I., R. G. Cook & R. P. McEver. 1989. Cell **56:** 1033–1044.
26. Armstrong, P. B. & J. P. Quigley. 1985. Biochem. Biophys. Acta **827:** 453–459.
27. Donovan, M. A. & T. M. Laue. 1991. J. Biol. Chem. **266:** 2121–2125.
28. Minetti, C. A. S. A., Y. Lin, T. Cislo & T. -Y. Liu. 1991. J. Biol. Chem. **266:** 20773–20780.
29. Schumann, R. R., S. R. Leong, G. W. Flaggs, P. W. Gray, S. D. Wright, J. C. Mathison, P. S. Tobias & R. J. Ulevitch. 1990. Science **249:** 1429–1431.
30. Aketagawa, J., T. Miyata, S. Ohtsubo, T. Nakamura, T. Morita, H. Hayashida, T. Miyata, S. Iwanaga, T. Takao & Y. Shimonishi. 1986. J. Biol. Chem. **261:** 7357–7365.
31. Cheng, S. -M., A. Suzuki, G. Zon & T. -Y. Liu. 1986. Biochem. Biophys. Acta **868:** 1–8.
32. Fortes-Dias, C. L., C. A. S. A. Minetti, Y. Lin & T. -Y. Liu. 1993. Comp. Biochem. Physiol. **105B:** 79–85.
33. Levin, J. &. F. B. Bang. 1964. Bull. Johns Hopkins Hosp. **115:** 265–274.
34. Bang, F. B. 1979. Biomedical Applications of the Horseshoe Crab (*Limulidas*).: 109–123. Alan R. Liss. New York.

The Prophenoloxidase Activating System and Its Role in Invertebrate Defence[a]

KENNETH SÖDERHÄLL,[b] LAGE CERENIUS, AND MATS W. JOHANSSON

Department of Physiological Botany
University of Uppsala
S-752 36 Uppsala, Sweden

INTRODUCTION

Invertebrates have an open circulatory system and must therefore have rapid and immediate noninducible defence and coagulation mechanisms to entrap parasites and prevent blood loss after wounding. As in most animals, these processes are mainly carried out by the blood cells or, as they are called in arthropods, hemocytes. The hemocytes of arthropods and other invertebrates have been shown to be important in defence, since they are responsible for phagocytosis of small foreign particles such as bacteria or fungal spores and form capsules around parasites that are too large to be internalized by individual hemocytes.

Because the hemocytes obviously can react to and remove a foreign particle that has succeeded in gaining entry into the body cavity of an arthropod, it appears as if these animals can differentiate non-self from self, and thus a system that can carry out this process ought to be present. Agglutinins or lectins may be one candidate for such a system.[1–5] Another likely candidate is the so-called prophenoloxidase (proPO) activating system,[6] and recently evidence has accumulated mainly from work done on crustaceans that this may be the case. The prime reason why this system early was proposed to function in recognizing foreign particles was the finding that the enzyme phenoloxidase in crayfish blood was turned into its active form by fungal cell wall β-1,3-glucans.[7,8] This was later also confirmed to be the case in several insect species[9–11] and other invertebrates.[12–14] Other microbial products such as lipopolysaccharides and peptidoglycans from bacterial cell walls are also active as elicitors of the proPO system.[15,16] Thus, regardless of which events follow after the proPO system is activated, it is clear that it can react to foreign microbial polysaccharides and as such can be defined as a recognition system.

Recent research has also provided clear evidence that, upon activation of the proPO system, factors are produced that will aid in the elimination of foreign particles such as parasites within the body cavity. This brief overview will report some of these studies, which have mainly been carried out on arthropods, and where

[a]This research was mainly supported by grants from the Swedish Natural Science Research Council and the Swedish Council for Forestry and Agricultural Research.

[b]Address for correspondence: Professor Kenneth Söderhäll, Department of Physiological Botany, University of Uppsala, Villavägen 6, S-752 36 Uppsala, Sweden.

appropriate make comparisons with the vertebrate immune system as well as with other invertebrates.

THE PROPHENOLOXIDASE SYSTEM: LOCALIZATION AND ACTIVATION

The activity of the terminal component of the proPO system, phenoloxidase, has been detected in several invertebrate groups, such as in crustaceans, insects, millipedes, mollusks, bivalves, brachiopods, echinoderms, and ascidians,[12–14,17,18] but the biochemical mechanism of proPO system activation has been studied in greatest detail in the silkworm *Bombyx mori* and the freshwater crayfish *Pacifastacus leniusculus.* Prophenoloxidase has been isolated in a homogeneous form from the insects *B. mori,*[19] *Manduca sexta,*[20] *Hyalophora cecropia*[21]; *Blaberus discoidalis*[22]; and the freshwater crayfish *P. leniusculus*[23] (TABLE 1). The molecular mass is 76 kDa for the cockroach and crayfish proPO and 80 kDa for the silkworm proenzyme. A prophenoloxidase-activating enzyme (ppA), a serine proteinase with a mass of 36 kDa, was purified from crayfish blood cells,[24] and this proteinase could cleave the proPO into two peptides with masses of 60 and 62 kDa, which both exhibited phenoloxidase activity.[23] In *B. mori,* Ashida and colleagues have shown that a cuticular proteinase,[25] which may not be the endogenous proPO-activating enzyme, could cleave and thus activate proPO from silkworm blood into an active PO by removing a 5-kDa peptide from the zymogen.[26] Nevertheless, it has to be emphasized that the complete mechanism of proPO-system activation is by no means clear, and, for example in freshwater crayfish, yet another hemocyte serine proteinase with a mass of 38 kDa also seems to be able to convert proPO into its active form.[23] In addition, at least two other

TABLE 1. Components of the Prophenoloxidase Activating System Purified from Arthropods

Protein and Species	Molecular Mass (kDa)	Reference
Prophenoloxidase		
Bombyx mori	80	Ashida, 1971[19]
Manduca sexta	71 & 77	Aso *et al.*, 1985[20]
Hyalophora cecropia	76	Andersson *et al.*, 1989[21]
Pacifastacus leniusculus	76	Aspán & Söderhäll, 1991[23]
Blaberus discoidalis	76	Durrant *et al.*, 1993[22]
Prophenoloxidase Activating Enzyme		
Bombyx mori (cuticle)	*ca.* 35	Dohke, 1973[25]
Pacifastacus leniusculus	36	Aspán *et al.*, 1990[24]
Prophenoloxidase Activating Enzyme Inhibitor		
Pacifastacus leniusculus	155	Aspán *et al.*, 1990[24]
Locusta migratoria	3.76 & 3.8	Boigegrain *et al.*, 1992[51]

serine proteinases with masses of 50 and 67 kDa are present in the crayfish blood cells,[23] but their function, if any, in the proPO system is as yet unknown.

Both proPO and ppA are present in the secretory granules of the semigranular and granular blood cells of crayfish[27]; thus, the compartmentalization of the system in these vesicles may be one important way to control and regulate the release of the proPO system at least in crustaceans. Other means by which the activation of the proPO system may be regulated are proteinase inhibitors present in crayfish plasma such as an α-macroglobulin[28] and a high-molecular-mass inhibitor of trypsin-like proteinases.[29] The trypsin inhibitor, which has a mass of 155 kDa, is by far the most efficient endogenous inhibitor of ppA.[30]

Another way to control production and activity of proPO factors is proteolytic degradation, and in crayfish this occurs with the 76-kDa cell-adhesion protein,[31] which is rapidly degraded in a hemolymph sample in which the proPO system is activated into a 30-kDa peptide with lower biological activity.[32] (See TABLE 2 for list of defence proteins purified from crayfish blood.)

BIOLOGICAL FUNCTION OF FACTORS OF THE PROPO SYSTEM

Upon activation of the proPO system by microbial elicitors, the terminal component of the system, phenoloxidase, can oxidize phenols into quinones, and then these will autocatalyze into melanin. Melanin and its precursors have been shown to inhibit growth of fungi and bacteria.[33–35]

More importantly, proteins that are associated with the proPO system have been shown to be directly involved in the communication between different blood cell types and also in the removal of foreign particles within the hemocele of crayfish.

TABLE 2. Defence Proteins Purified from Crayfish Blood

Protein	Molecular Mass (kDa)	Reference
Prophenoloxidase	76	Aspán & Söderhäll, 1991[23]
Prophenoloxidase activating enzyme (ppA)	36	Aspán *et al.*, 1990[24]
Cell adhesion, degranulating, and opsonic protein	76	Johansson & Söderhäll, 1988,[31] 1989[48] Kobayashi *et al.*, 1990[32]
β-1,3-Glucan binding protein (βGBP)	100	Duvic & Söderhäll, 1990[39]
α-Macroglobulin	2 × 190	Hergenhahn *et al.*, 1988[28]
ppA inhibitor	155	Hergenhahn *et al.*, 1987[29]
Subtilisin inhibitor	23	Häll & Söderhäll, 1982[52]
Receptor for βGBP and cell adhesion protein	340[a]	Duvic & Söderhäll, 1992[42] Cammarata *et al.*, unpublished
Clotting protein	2 × 210	Kopacek *et al.*, 1993[53]
Hemagglutinin	420[b]	Kopacek *et al.*, 1993[54]

[a]Multimer of 90-kDa subunits.
[b]Multimer of 65–85-kDa subunits.

Thus, it has been demonstrated that proteins associated with the proPO system have, for example, encapsulation-promoting activities and can act as opsonins.

ELICITORS OF proPO SYSTEM ACTIVATION

The proPO system is induced to its active form by different microbial polysaccharides, such as β-1,3-glucans[6] and peptidoglycans or lipopolysaccharides.[16,36] The activation exerted by β-1,3-glucans is very specific; most other glycans that do not contain β-1,3-D-glucopyranosyl residues fail to elicit any activation of the proPO system.[8] Consequently, the proPO system can specifically recognize different types of microorganisms by the presence of different polysaccharides on their surface; for example, the cell walls of almost all fungi contain β-1,3-glucans. Thus, it was not surprising to see that proteins that can bind to β-1,3-glucans and induce activation of the proPO system are present in the plasma of both insects and crustaceans.

THE β-1,3-GLUCAN-BINDING PROTEIN

A protein that specifically binds β-1,3-glucans has been isolated and characterized from a number of different arthropods such as the insects *B. mori,*[37] and *B. craniifer*[38] and several crustaceans[39,40] (Thörnqvist *et al.,* unpublished) (TABLE 3). This protein, besides inducing activation of the proPO system *in vitro,* also functions as a degranulation factor for crayfish granular cells.[41] Similarly to the 76-kDa factor, this protein can function as a phagocytosis-stimulating opsonin for crustacean hyaline cells (Thörnqvist *et al.,* unpublished).

The most detailed information regarding both the biochemical characteristics and the biological activities of any β-1,3-glucan-binding protein (βGBP) has been obtained from the crayfish *P. leniusculus.*[39,41,42] Recently, the cDNA cloning of the *P. leniusculus* protein was achieved and a sequence established. This protein is synthesized in the hepatopancreas from which it appears to be secreted into the plasma.

Sequence analysis against data banks have not revealed any strong homology to any known protein. Interestingly, however, the sequence of the crayfish βGBP

TABLE 3. β-1,3-Glucan Binding Proteins Purified from Arthropods

Species	Molecular Mass (kDa)	Reference
Insects		
Blaberus craniifer	90	Söderhäll *et al.,* 1988[38]
Bombyx mori	62	Ochiai & Ashida, 1988[37]
Crustaceans		
Pacifastacus leniusculus	100	Duvic & Söderhäll, 1990[39]
Procambarus clarkii	100	Duvic & Söderhäll, 1992[42]
Astacus astacus	95, 105	Duvic & Söderhäll, 1992[42]
Carcinus maenas	110	Thörnqvist *et al.,* unpublished

contains an RGD (arginine-glycine-aspartic acid) sequence. Such sequences are known to be involved in ligand binding to receptors of the integrin family in vertebrates.[43] We have earlier demonstrated that a pentapeptide containing the RGD sequence triggers degranulation and cell-substratum attachment of crayfish granular hemocytes.[44] Whether the degranulation provoked by the binding of β-1,3-glucan-treated βGBP to the hemocytes and the effects obtained by the RGD peptide involve the same receptor has not been established.

The βGBPs from different crustaceans are very similar. For example, an antiserum raised against the *P. leniusculus* βGBP reacts with βGBP from other crayfish and crab species[40] (Thörnqvist *et al.*, unpublished). In addition, crab βGBP exerts the same biological activities on crayfish hemocytes as it does on crab cells and vice versa (Thörnqvist *et al.*, unpublished). Furthermore, the NH_2-terminal amino acid sequence of βGBP from *P. leniusculus* and the crab *Carcinus maenas* are similar (Thörnqvist *et al.*, unpublished). In TABLE 3 β-1,3-glucan binding proteins isolated so far from different arthropods are shown.

THE 76-kDa PROTEIN

This crayfish protein (for reviews, see Johansson and Söderhäll[45,46]), which is synthesized by the blood cells, stored in a biologically inactive form in their secretory granules,[47] and is released from them during degranulation, gains its biological activity concomitant with proPO system activation[31] (TABLE 4). How this protein is converted to its active form is still unknown, although some recent data clearly indicate that a limited proteolysis by a serine proteinase appears to be involved (Johansson & Söderhäll, unpublished). If this turns out to be the case, then only a very small peptide may be cleaved off from the 76-kDa protein to generate the active factor. In its active form this protein functions as a cell-adhesion factor for crayfish semigranular and granular blood cells[31] and as a strong degranulation factor for the same cells.[48] Kobayashi *et al.*[32] could demonstrate that the 76-kDa protein also is an encapsulating promoting factor, since particles coated with the purified protein were more avidly encapsulated by the semigranular cells. The 76-kDa protein also functions as a phagocytosis-stimulating opsonin for crab hyaline cells (Thörnqvist *et al.*, unpublished), and polyclonal antiserum to the 76-kDa protein recognizes two bands

TABLE 4. Cell Communication Proteins in Crayfish Blood

76-kDa Protein	
Biosynthesis:	Hemocytes
Localization of inactive protein:	Hemocyte granules
Activation:	Concomitant with proPO-system activation
β-1,3-Glucan Binding Protein	
Biosynthesis:	Hepatopancreas
Localization of protein:	Plasma
Generation of hemocyte receptor binding activity:	After reacting with β-1,3-glucans

of 81 and 85 kDa in the crab *C. maenas*[46] (Thörnqvist *et al.*, unpublished). From the insect *B. craniifer,* a similar cell-adhesion and degranulating protein with a mass of 90 kDa has been isolated, and it reacts with the anti-crayfish 76-kDa protein antiserum.[49] Furthermore, the crayfish 76-kDa protein cross-reacts with the 75-kDa mammalian cell-adhesion protein vitronectin[46] (Thörnqvist *et al.,* unpublished), which is involved in the coagulation and complement systems.[50]

A MEMBRANE RECEPTOR FOR THE βGBP AND THE 76-kDa PROTEIN

In hemocyte membranes we were able to identify a receptor protein for the βGBP.[42] Interestingly, the βGBP did only bind the receptor if it had previously been reacted with a β-1,3-glucan. The receptor protein could be isolated, purified, and partially characterized. It was shown to be composed of subunits of 90 kDa, and its native molecular mass is around 340 kDa[42] (Cammarata *et al.,* unpublished). Recently, we have also been able to demonstrate that the 76-kDa protein binds to the same receptor with a higher apparent affinity and that βGBP that has been reacted with β-1,3-glucans competes with the 76-kDa protein for this binding (Cammarata *et al.,* unpublished). The membrane receptor is presently being cloned and sequenced to possibly reveal its structural similarity with receptor proteins from other animals.

CELLULAR COMMUNICATION

In crustaceans, two proteins, which are associated with proPO system activation in some as yet unknown way, have been demonstrated to be directly involved in the communication between blood cells (TABLE 4). More structural data, however, need to be obtained about these two molecules, their receptors, and their tentative analogues in other arthropods and invertebrates before we can reveal the fine details of ligand–receptor interaction during these defence processes. Such work is presently under way in our laboratory. We are currently also investigating the intracellular signaling pathways triggered by binding of these ligands to the cells.

REFERENCES

1. YEATON, R. W. 1981. Dev. Comp. Immunol. **5:** 391–402.
2. YEATON, R. W. 1981. Dev. Comp. Immunol. **5:** 535–554.
3. RATCLIFFE, N. A., A. F. ROWLEY, S. W. FITZGERALD & C. P. RHODES. 1985. Int. Rev. Cytol. **97:** 183–350.
4. NATORI, S. 1986. *In* Fundamental and Applied Aspects of Invertebrate Pathology. R. A. Samson, J. M. Vlak & D. Peters, Eds. 411–414. Foundation of the Fourth International Colloquium of Invertebrate Pathology. Wageningen, Holland.
5. OLAFSEN, J. 1986. *In* Immunity in Invertebrates. M. Brehèlin, Ed. 94–111. Springer Verlag. Berlin.
6. SÖDERHÄLL, K. 1982. Dev. Comp. Immunol. **6:** 601–611.
7. UNESTAM, T. & K. SÖDERHÄLL. 1977. Nature **267:** 45–46.
8. SÖDERHÄLL, K. & T. UNESTAM. 1979. Can. J. Microbiol. **25:** 404–416.

9. Ashida, M., Y. Ishizaki & H. Iwahana. 1983. Biochem. Biophys. Res. Commun. **113:** 562–568.
10. Dularay, B. & A. M. Lackie. 1985. Insect Biochem. **15:** 827–834.
11. Leonard, C., K. Söderhäll & N. A. Ratcliffe. 1985. Insect Biochem. **15:** 803–810.
12. Canicatti, C. & P. Götz. J. Invertebr. Pathol. **58:** 305–310.
13. Smith, V. J. & K. Söderhäll. 1991. Dev. Comp. Immunol. **15:** 251–261.
14. Jackson, A. D., V. J. Smith & C. M. Peddie. 1993. Dev. Comp. Immunol. **17:** 97–108.
15. Yoshida, H. & M. Ashida. 1986. Insect Biochem. **16:** 539–545.
16. Söderhäll, K., A. Aspan & B. Duvic. 1990. Res. Immunol. **141:** 896–904.
17. De Aragao, G. A. & M. Bacila. 1976. Comp. Biochem. Physiol. **54B:** 179–182.
18. Krishnan, G. & M. H. Ravindranath. 1989. J. Insect Physiol. **19:** 647–653.
19. Ashida, M. 1971. Arch. Biochem. Biophys. **144:** 749–762.
20. Aso, Y., K. J. Kramer, T. L. Hopkins & G. L. Lookhart. 1985. Insect Biochem. **15:** 9–17.
21. Andersson, K., S. C. Sun, H. G. Boman & H. Steiner. 1989. Insect Biochem. **19:** 629–637.
22. Durrant, H. J., N. A. Ratcliffe, C. R. Hipkin, A. Aspan & K. Söderhäll. 1993. Biochem. J. **289:** 87–91.
23. Aspan, A. & K. Söderhäll. 1991. Insect Biochem. **21:** 363–373.
24. Aspán, A., J. Sturtevant, V. J. Smith & K. Söderhäll. 1990. Insect Biochem. **20:** 709–718.
25. Dohke, K. 1973. Arch. Biochem. Biophys. **157:** 210–221.
26. Ashida, M. 1974. Biochem. Biophys. Res. Commun. **57:** 1089–1095.
27. Johansson, M. W. & K. Söderhäll, 1985. J. Comp. Physiol **156B:** 175–181.
28. Hergenhahn, H.-G., M. Hall & K. Söderhäll. 1988. Biochem. J. **255:** 801–806.
29. Hergenhahn, H.-G., A. Aspan & K. Söderhäll. 1987. Biochem. J. **248:** 223–228.
30. Aspan, A., M. Hall & K. Söderhäll. 1990. Insect Biochem. **20:** 485–492.
31. Johansson, M. W. & K. Söderhäll. 1988. J. Cell Biol. **106:** 1795–1803.
32. Kobayashi, M., M. W. Johansson & K. Söderhäll. 1990. Cell Tissue Res. **260:** 13–18.
33. Söderhäll, K. & R. Ajaxon. 1982. J. Invertebr. Pathol. **39:** 105–109.
34. St. Leger, R. J., R. M. Cooper & A. K. Charnley. 1988. J. Invertebr. Pathol. **52:** 459–470.
35. Rowley, A. F., J. L. Brookman & N. A. Ratcliffe. 1990. J. Invertebr. Pathol. **56:** 31–38.
36. Ashida, M. & H. I. Yamazaki. 1990. *In* Molting and Metamorphosis. E. Ohnishi & H. Ishizaki, Eds.: 239–265. Springer Verlag. Berlin.
37. Ochiai, M. & M. Ashida. 1988. J. Biol. Chem. **263:** 12056–12062.
38. Söderhäll, K., W. Rögener, I. Söderhäll, R. P. Newton & N. A. Ratcliffe. 1988. Insect Biochem. **18:** 323–330.
39. Duvic, B. & K. Söderhäll. 1990. J. Biol. Chem. **265:** 9327–9332.
40. Duvic, B. & K. Söderhäll. 1993. J. Crust. Biol. **13:** 403–408.
41. Barracco, M., B. Duvic & K. Söderhäll. 1991. Cell Tissue Res. **266:** 491–497.
42. Duvic, B. & K. Söderhäll. 1992. Eur. J. Biochem. **207:** 223–228.
43. Ruoslahti, E. 1991. J. Clin. Invest. **87:** 1–5.
44. Johansson, M. W. & K. Söderhäll. 1989. Insect Biochem. **19:** 573–579.
45. Johansson, M. W. & K. Söderhäll. 1989. Parasitol. Today **5:** 171–176.
46. Johansson, M. W. & K. Söderhäll. 1992. Animal Biol. **1:** 97–107.
47. Liang, Z., P. Lindblad, A. Beauvais, M. W. Johansson, J.-P. Latge, M. Hall, L. Cerenius & K. Söderhäll. 1992. J. Insect Physiol. **12:** 987–995.
48. Johansson, M. W. & K. Söderhäll. 1989. Insect Biochem. **19:** 183–190.
49. Rantamäki, J. H. Durrant, Z. Liang, N. A. Ratcliffe, B. Duvic & K. Söderhäll. 1991. J. Insect Physiol. **37:** 627–634.
50. Tomasini, B. R. & D. M. Mosher. 1990. Prog. Thromb. Haemost. **10:** 269–305.
51. Boigegrain, R.-A., H. Mattras, M. Brehélin, P. Paroutaud & M.-A. Colleti-Previero. 1992. Biochem. Biophys. Res. Commun. **189:** 790–793.
52. Häll, L. & K. Söderhäll. 1982. J. Invertebr. Pathol. **39:** 9–37.
53. Kopacek, P., M. Hall & K. Söderhäll. 1993. Eur. J. Biochem. **213:** 591–597.
54. Kopacek, P., L. Grubhofer & K. Söderhäll. 1993. Dev. Comp. Immunol. In press.

Phagocytosis and Invertebrate Opsonins in Relation to Parasitism[a]

CHRISTOPHER J. BAYNE AND SARAH E. FRYER

Department of Zoology
Oregon State University
Corvallis, Oregon 97331-2914

INTRODUCTION

At the Woods Hole Meeting, the group of papers in which this was the first to be presented was challenged to address the question: Was a cellular defense the first defense? Phagocytosis is a very primitive cell behavior. Our protist ancestors relied on it as a means of obtaining food, and this was retained by the earliest (metazoan) animals—the sponges, cnidarians, and free-living flatworms. If we entertain the possibility that the origins of the eukaryote condition depended, in part, on the acquisition of intracellular mutualists that are retained as mitochondria, plastids, and other "organelles," we *de facto* admit the likelihood that members of the Monera (bacteria) were entering bodies long before the animals (Metazoa) appeared. Phagocytic cells were relied on to meet such challenges in early metazoans. Before the evolution of circulatory systems, ameboid phagocytes wandered through the body, phagocytosing foreign materials and damaged self.

These considerations do not predicate the conclusion that a cellular defense was the first defense. Allelochemicals released by protists and metazoans alike provide effective "humoral" defenses, killing or stopping the growth of other species in the habitat. We proffer that it is unimportant to be able to say which came first—cellular or humoral; we must be content in the knowledge that both forms of defense have been present from the very earliest times.

MOLLUSCAN PHAGOCYTIC SYSTEMS

Our contributions to this field have dealt with phagocytes, opsonins, and parasites of molluscs, which are animals with well-developed circulatory systems. In the 1950s, Les Stauber did "Metchnikoff-type" experiments with oysters and learned that they possess the means to rapidly clear injected colloidal carbon.[1] When bacteria were injected into *Helix aspersa* (a land snail), the clearance capacity was found to be enormous.[2] Furthermore, the discriminative ability of molluscan reticulo-endothelial-type functions is impressive: Chitons clear injected hemocyanins at rates that correlate directly with the degree of foreignness of the source species.[3] In *Helix pomatia,* clearance rates for bacteria (*Azotobacter vinelandii*) are strain-specific, and for human erythrocytes are blood group–specific.[4]

[a]This work has been supported by National Institutes of Health Grant AI-16137.

Although the circulating, free phagocytes may not be the earliest agents of particle clearance in mollusks,[4] these cells do phagocytose injected foreign materials and are easily obtained for *in vitro* studies.[5] They constitute remarkably good models of vertebrate macrophages with respect to their complement of lysosomal enzymes,[6] their phagocytic behaviors,[7] their ability to secrete lytic molecules,[8,9] and their oxidative burst potentials.[10,11] The parallelism of functions has limits, of course. For example, despite the existence of such reactivities in other invertebrate taxa (e.g., Bertheussen[12] and Parrinello *et al.*[13]), we have been unable (Haisch and Bayne, unpublished results) to demonstrate any allogeneic, or even xenogeneic, cytotoxic capacity against other gastropod hemocytes (the leukocytes of molluscan "blood," most or all of which are capable of some degree of phagocytosis). This observation is consistent with reports that allogeneic[14] and congeneric xenogeneic hearts (Sullivan, personal communication, 1993) implanted into the snail *Biomphalaria glabrata* are typically accepted, continuing to beat for weeks.

Phagocytes and Parasites

The parasites and pathogens of a given host species are a very special subsample of the potential invaders that share the host habitat. Most foreign agents entering the body of a potential host encounter effective defenses that kill them. Those that succeed in establishing infections have evolved mechanisms to escape, avoid, or disarm the innate aggression of host phagocytes and humoral components.

Haplosporidium nelsoni is a protist parasite of certain oysters. It has decimated populations, for example, in Chesapeake Bay. When the plasmodia of *H. nelsoni* are placed in culture with phagocytes of *Geukensia demissa,* a mussel that is not susceptible to the disease, they are engulfed. However, when cultured with phagocytes of their oyster hosts, *Crassostrea virginica,* they not only escape detection, but actively repel the advances of wandering cells.[15] Killed parasites do not do this. Similarly, when oyster hemocytes are cultured with cercariae larvae of marine trematodes, adhesion of hemocytes and encapsulation are limited to encounters with dead parasites; live cercariae are ignored.[16] Gastropod hemocytes behave in a similar manner. When *Biomphalaria glabrata* hemocytes are given the opportunity to interact with larvae of *Echinostoma paraensei* (a digenetic trematode that normally parasitizes this snail), they fail to adhere—unless the parasites have been killed previously.[17] Larvae of another digenetic trematode, *Schistosoma mansoni*, that parasitizes *B. glabrata* are recognized by the snail's hemocytes. However, those from snail strains with which the trematode is compatible soon detach, whereas those from strains of snail that are resistant to this parasite adhere closely, interact with the tegument, and go on to kill the parasite.[18]

Phagocytes Are the Effectors of Snail Resistance to Trematode Parasites

First indications of an aggressive role for hemocytes in the *S. mansoni–B. glabrata* parasitism are observable *in vivo* as massive encapsulations forming around mother sporocysts within hours of their entry into the body of a resistant snail (reviewed in Bayne[5]). Such responses, however, could be the consequence of alternative (humoral) killing mechanisms; that is, the hemocytes could be attracted to dead or dying sporocysts. Evidence that this is not the case is seen when *in vitro* systems are used to study the interactions between hemocytes and *in vitro*-derived sporocysts.[18] These parasite larvae are killed only when hemocytes are present; the cell-free plasma fails to kill. This holds whether the plasma is from genetically resistant or susceptible strains of the snail. Furthermore, hemocytes from resistant snails kill the parasites even in the absence of plasma. Interestingly, the plasma from resistant snails makes it possible for hemocytes from susceptible snails to kill sporocysts.[19] This observation, made with an *in vitro* system, was independently confirmed when plasma from resistant snails was injected into susceptible snails, with the result that resistance was partially transferred.[20] The effective component(s) of resistant plasma has proven elusive. Additional properties unique to resistant-strain plasma have been found, and one or more of these may play a role in mediating resistance to *S. mansoni.* One such property is the ability to agglutinate sporocysts,[21] and another is the ability to opsonize yeast for phagocytosis.[22]

The process of larval parasite encapsulation has been characterized as "frustrated phagocytosis." The view is that, as phagocytes encounter the surface of a foreign object larger than themselves, they spread pseudopodia over it as though they were in the act of phagocytosing it, only to find that it is too large to be internalized. Instead, the phagocyte becomes spread over the foreign surface. As additional phagocytes meet the same fate, encapsulation occurs. This proposition is important for what is to follow, as we have exploited the phagocytosis of yeast as a model of hemocyte responsiveness to larger biological materials. The phagocytic process is a sequence of distinct stages: attraction, recognition (attachment), engulfment, digestion, and subsequent processing. The key element that validates our model is that recognition precedes encapsulation of a large particle just as it precedes engulfment of a small one, and differences of subsequent fates are irrelevant to common features of recognition of the two particle sizes.

A most important reason for our adoption of yeast phagocytosis as a model for parasite encapsulation stems from frustrations encountered during attempts to isolate all but one variable in a system that includes two living, responding organisms. The use of heat-killed yeast obviates such problems. In addition, yeast have proven in our lab (M. R. Tripp, personal communication) to be more consistently phagocytosed than alternative, often-used test particles, namely, vertebrate erythrocytes, and their merits as test particles in phagocytosis studies are well known.

THE MODEL

Schistosoma mansoni, the trematode parasite otherwise known as human blood fluke, requires passage through a pulmonate snail, *Biomphalaria glabrata,* in which it multiplies by asexual reproduction (Fig. 1). In the field, both susceptible and resistant phenotypes of this snail are found. Laboratory-derived strains of the snail are available that express genetically based differences in compatibility with strains of the trematode. Laboratories exploring the cellular and molecular bases of compatibility in this host–parasite system (see Van der Knaap and Loker[23]) often exploit the

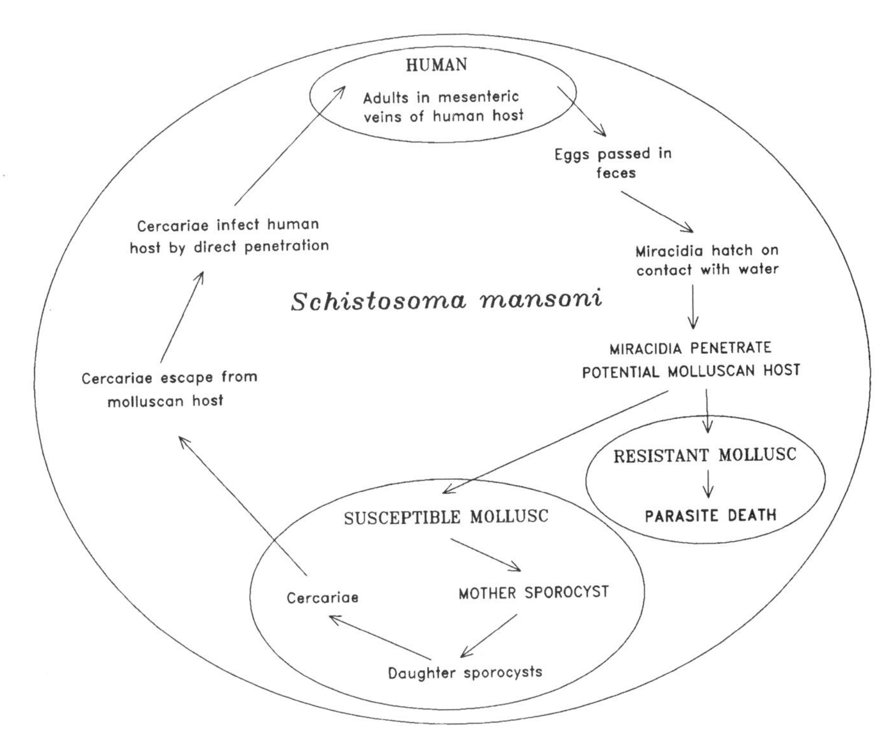

FIGURE 1. The life cycle of *Schistosoma mansoni,* the trematode parasite causing human and molluscan schistosomiasis. When a miracidia larva penetrates a snail, it is either recognized by the internal defense (natural immune) system and killed within hours, or escapes recognition, going on to produce a second generation of sporocysts, and eventually numerous cercariae larvae that are infective for humans.

Puerto Rican PR-1 strain of *S. mansoni,* and three strains of the snail: the susceptible M line and the resistant 13-16-R1 and 10-R2 lines.

When heat-killed *Saccharomyces cerevisiae* are washed in physiological saline, then offered to snail hemocytes *in vitro* in the absence of plasma, they are efficiently phagocytosed.[22] Hemocytes taking in yeast constitute 15 to 25% of the population after 20 minutes. However, when yeast are preincubated in snail plasma (10% in buffer) and washed before being offered to phagocytes in saline, they are subsequently more avidly phagocytosed if the plasma for incubation was taken from resistant strains of snail, but not if taken from individuals of the susceptible strain[22] (FIG. 2). Phagocytes from either susceptible or resistant strains recognize the opsonized yeast better than naive yeast. Because the yeast surface is rich in mannan and glucan,[24] we investigated the effects of a soluble, yeast-derived mannan and two soluble glucans on the apparent opsonic process and on phagocytosis.[25] Phagocytosis of native yeast or plasma-treated yeast was inhibited by laminarin, (a β,1-3-linked glucan), but not by glycogen (an α,1-4-linked glucan) nor by mannan (FIG. 3). This implied a role for lectino-phagocytosis[26] in this system and the existence of phag-

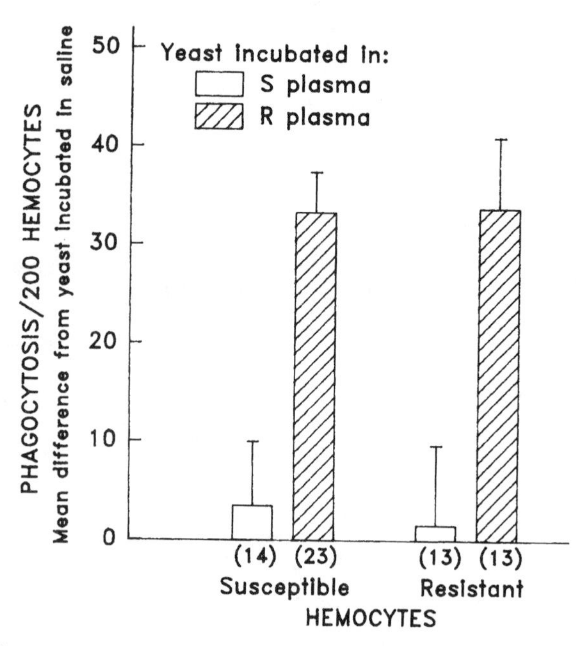

FIGURE 2. Phagocytosis of heat-killed yeast that had been exposed to snail plasma (10% in saline) for an hour before being washed in saline and added to phagocyte monolayers; R plasma was from a strain of snail that is resistant to PR1 *Schistosoma mansoni;* S plasma was from a strain of snail that is susceptible to PR1 *S. mansoni.* Data are expressed as differences from phagocytosis of saline-treated yeast by aliquots of the same hemocytes.

ocyte surface receptors for β,1-3-linked glucan. As all other molluscan lectins thought to be involved in non-self recognition (see below) had so far been reported to have identical specificities in cell-bound or soluble (humoral) form, we were surprised to find that opsonization was inhibited in our system by mannan, and not by either of the glucans! This (FIG. 4) implied that a soluble mannan-binding plasma component (perhaps similar to vertebrate mannan-binding protein[27]) was responsible for opsonization of the yeast. These results have led to the model cartooned in Figure 5[28]: A soluble mannan-binding protein in the snail plasma binds to the yeast surface and undergoes a conformational change that results in the exposure of previously cryptic glucan determinants. As with the plasma factor that bestows cytotoxic capacity in our *in vitro* encapsulation assays, this putative opsonin remains elusive.

Because we followed the kinetics of opsonization in this system, another unusual feature revealed itself, that is, that even strongly opsonic plasma, in the *short* term (a few minutes), makes particles *less* easily recognized by phagocytes (FIG. 6). In the simplest of cases, opsonization (as evidenced by increased associations between yeast and phagocyte) would have proceeded from zero time values to higher values in a linear fashion as opsonic lectin molecules bound to the yeast. Alternative scenarios had to be envisioned, and we hypothesized two. First, we considered the possibility that opsonization involved the binding of plasma proteins that, in their

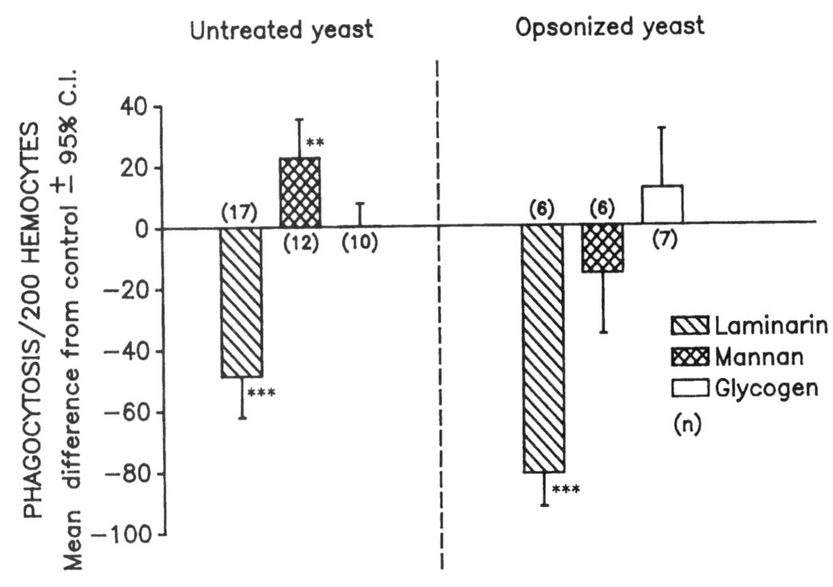

FIGURE 3. The effects of carbohydrates on phagocytosis of unopsonized and opsonized yeast cells. The yeast cells were suspended in 1% solutions of laminarin, mannan, or glycogen in physiological saline; controls lacked added carbohydrates, and levels of phagocytosis are presented relative to those of (*left*) yeast not exposed to plasma, or (*right*) yeast exposed to resistant-strain plasma for one hour. Paired Student's *t*-test compared to control: $^{**}p < 0.01$; $^{***}p < 0.001$. (Adapted from Fryer *et al.*,[25] *Developmental and Comparative Immunology*, with permission.)

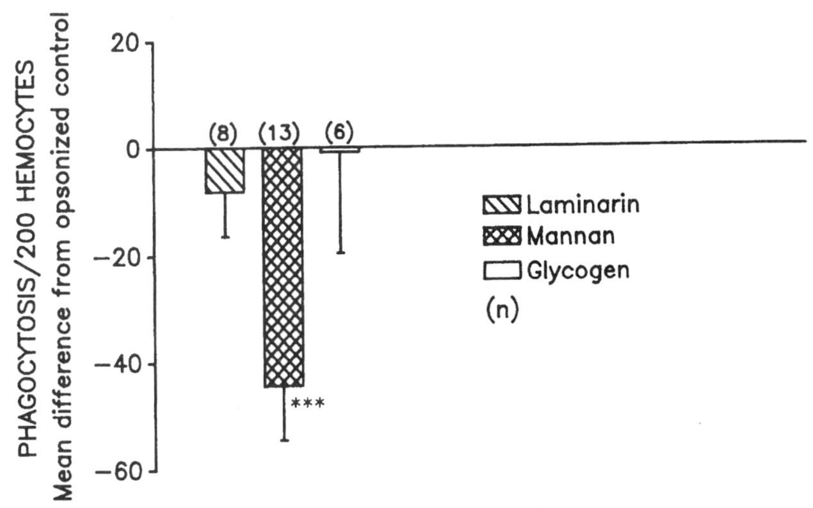

FIGURE 4. The effects of test carbohydrates on opsonization of yeast. The yeast cells were incubated for one hour in 10% resistant-strain plasma containing 1% laminarin, mannan, or glycogen; controls were incubated in 10% plasma without added carbohydrate. Levels of phagocytosis are presented relative to those obtained with control yeast cells. Paired Student's *t*-test compared to control: $^{***}p < 0.001$. (Adapted from Fryer *et al.*,[25] *Developmental and Comparative Immunology*, with permission.)

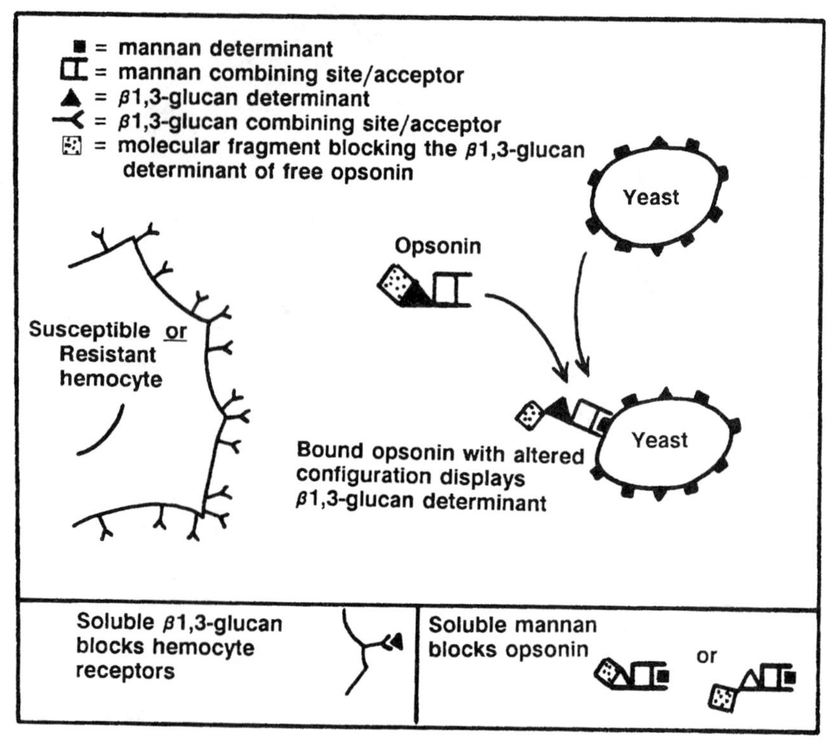

FIGURE 5. A model of the *Biomphalaria glabrata* hemocyte–opsonin–yeast system. Hemocytes from snails of different phenotypes (relative to schistosome susceptibility) express receptors (combining sites) for β,1-3-glucan molecules that are present on yeast. The soluble opsonin recognizes distinct mannan molecules on the yeast. As a result of binding to mannan, opsonin molecules reveal their own β,1-3-glucan determinants, which facilitate their recognition by the phagocytic hemocytes. As shown on *bottom left,* β,1-3-glucan in solution blocks phagocytosis. At *bottom right,* the blockage of opsonization by soluble mannan is illustrated. (Reprinted from Bayne[28] from with permission from *Bioscience.)*

native state, made the particles less "visible" to the phagocyte (a precedent for this exists in gastropod molluscs[4]), and that enzymatic processing subsequently altered the bound molecules to reveal determinants for which phagocytes had receptors. Using a fibrin matrix as substrate, we were unable to detect such putative proteolytic activity in plasma, even when assayed during and in the immediate vicinity of yeast opsonization (unpublished observations). Coincidentally, controls run in these experiments revealed the presence of an α-macroglobulin-type antiprotease activity in snail plasma, and that has been the subject of work reported elsewhere.[29] The second alternative that we considered was based on (a) our previous discovery[30] that many of the plasma proteins of this snail are "sticky" molecules, binding readily to glass, plastic, and other surfaces and (b) the knowledge that at least one plasma protein, when bound to yeast, reduces their rates of clearance in *Helix.*[4] This possibility was

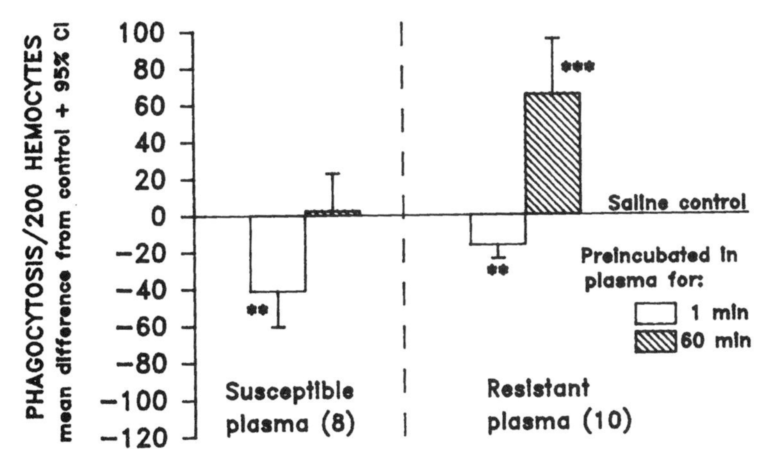

FIGURE 6. Short-term exposure of yeast cells to snail plasma reduces their subsequent recognition by phagocytes. Yeast were incubated in 10% *Biomphalaria glabrata* plasma for 1 min or for 60 min, rinsed in saline, then offered to phagocytes. Results are shown as differences from paired controls (yeast incubated in saline). Paired Student's *t*-test against control: $*p < 0.05$; $**p < 0.002$; $***p < 0.0001$. (Adapted from Fryer & Bayne[22] *Parasite Immunology,* with permission.)

that the opsonin is a minor constituent of plasma that would be underrepresented in the community of plasma molecules that bind initially to yeast, becoming enriched in a time-dependent manner as its higher binding affinity for mannan on yeast led to longer molecular retention times than those characterizing other plasma components. Repeated efforts have failed to detect enrichment of one or more molecules over the 60-minute period required for optimum opsonization. The mechanisms of this opsonic system have proven enigmatic to some degree, but we continue to pursue them and are presently exploring possible roles for carbohydrases.

Influences of Parasitism and Stress on the Opsonic and Phagocytic Activities

Evidence has been obtained for the conjecture that the yeast opsonic system is relevant to the host–parasite (i.e. schistosome) balance. It was previously known that trematode parasites often modulate the natural immune systems of their molluscan hosts (reviewed by Van der Knapp and Loker [25] and Bayne and Loker[31]). Both *in vivo* and *in vitro* activities of snail hemocytes can be altered as a consequence of infection. Decreased[32–35] and increased[33,34] phagocytic and cytotoxic activities have been reported, and microbicidal activities of hemocytes were found to be elevated in snails three weeks after infection with *S. mansoni.*[36] We asked whether or not *S. mansoni* would induce alterations in the yeast opsonization and phagocytosis systems in *B. glabrata* during the first three hours of host–parasite encounter,[37] inasmuch as this is the period in which the fate of penetrating larvae is decided.[5] The first observation (an incidental one) was that the conditions used in many laboratories to infect snails with trematode miracidia depressed opsonin titers,[38] an "immunosuppressive" effect of stress in a mollusc. Plasma from exposed, resistant snails, however, had elevated opsonic titers[37] (FIG. 7). When schistosome sporocysts and snail hemolymph (phagocytes and plasma) were coincubated for two hours *in vitro* and phagocytosis of unopsonized yeast was allowed to proceed in either the experimental plasma or physiological saline, we saw a decrease in phagocytosis by susceptible strain phagocytes; a higher proportion of resistant strain hemocytes took up yeast (FIG. 8). These changes appeared to be due to alterations in the plasma, as they were detectable only when plasma was left in for the phagocytosis phase of the experiments. The changes occurred only when sporocysts, plasma, and hemocytes were present simultaneously during the preincubation; any changes to plasma that resulted from its incubation with sporocysts for two hours *with hemocytes absent* did not subsequently influence phagocytosis (FIG. 9).

A different protocol was used to distinguish whether the effect of sporocysts on phagocytosis was mediated by the plasma or was due to influences on the phagocytes.[37] After two hours of hemolymph coincubation with or without sporocysts, the plasma was removed, the monolayers were washed (removing residual plasma and sporocysts), and the plasma was replaced, either to the wells from which it had been removed, or to alternative wells, as shown in FIGURE 10. Sporocyst effects were again seen (FIG. 11): reduced phagocytosis by susceptible cells in the presence of exposed plasma and elevated phagocytosis by resistant cells when they had been coincubated

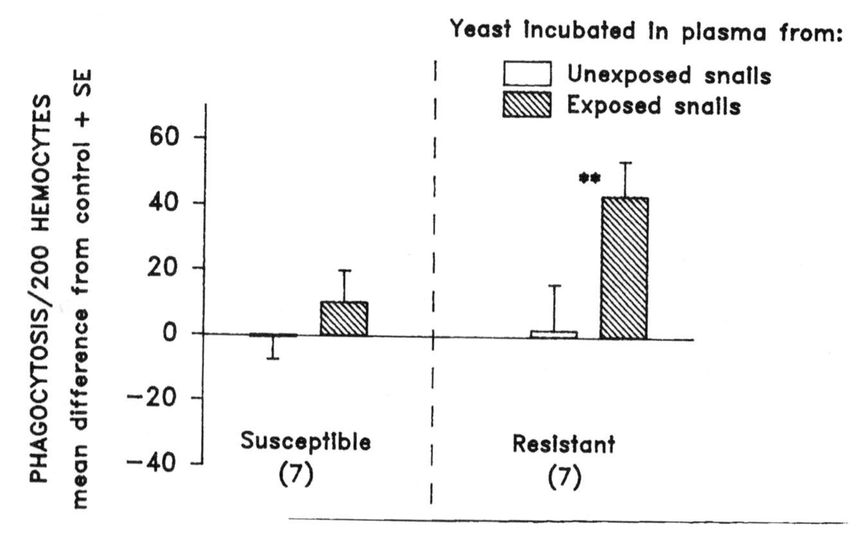

FIGURE 7. Opsonic properties of plasma from *Biomphalaria glabrata* snails that were unexposed or exposed to *Schistosoma mansoni* miracidia for 3 hr in 2 ml water before bleeding. Unexposed snails were isolated under the same conditions but without miracidia. Differences between levels of phagocytosis of yeast pretreated with unexposed or with exposed snail plasma were tested using hemocytes from naive snails. Paired Student's *t*-tests between treatment groups: $**p < 0.002$. (Adapted from Fryer & Bayne,[37] the *Journal of Parasitology,* with permission.)

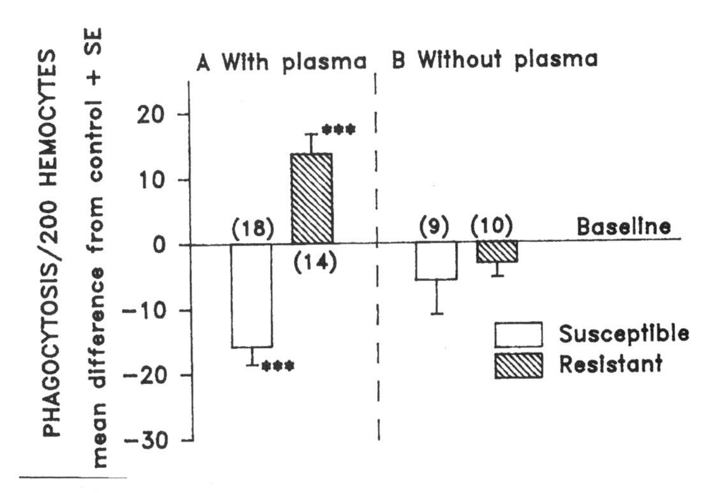

FIGURE 8. Phagocytosis by *Biomphalaria glabrata* hemocytes after incubation of whole hemolymph with *Schistosoma mansoni* sporocysts for 2 hr *in vitro*. Phagocytosis of unopsonized yeast was assessed with plasma present (**A**) and without plasma (yeast cells suspended in saline, **B**). Differences between phagocytosis values were compared with those of naive hemocytes from the same pool of hemolymph (baseline) using paired Student's t tests: $***p < 0.001$. (Adapted from Fryer & Bayne,[37] the *Journal of Parasitology*, with permission.)

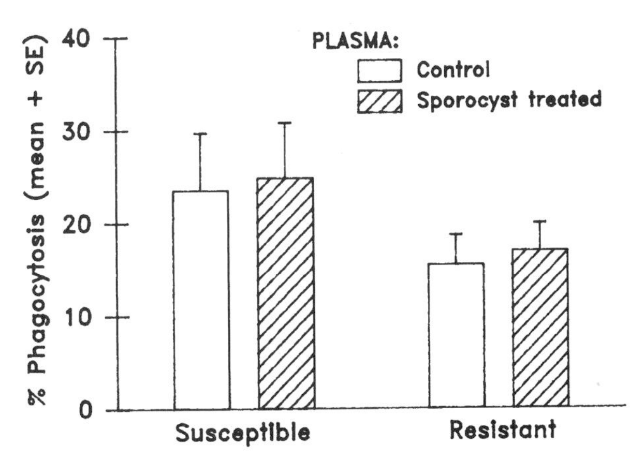

FIGURE 9. Phagocytosis by naive *Biomphalaria glabrata* hemocytes in the presence of control plasma (held for 2 hr) and plasma that has been incubated with sporocysts for 2 hr *in vitro;* $n = 11$ in each treatment group. Paired Student's t tests showed no differences in phagocytosis values between the two plasma treatments. (Adapted from Fryer & Bayne,[37] the *Journal of Parasitology,* with permission.)

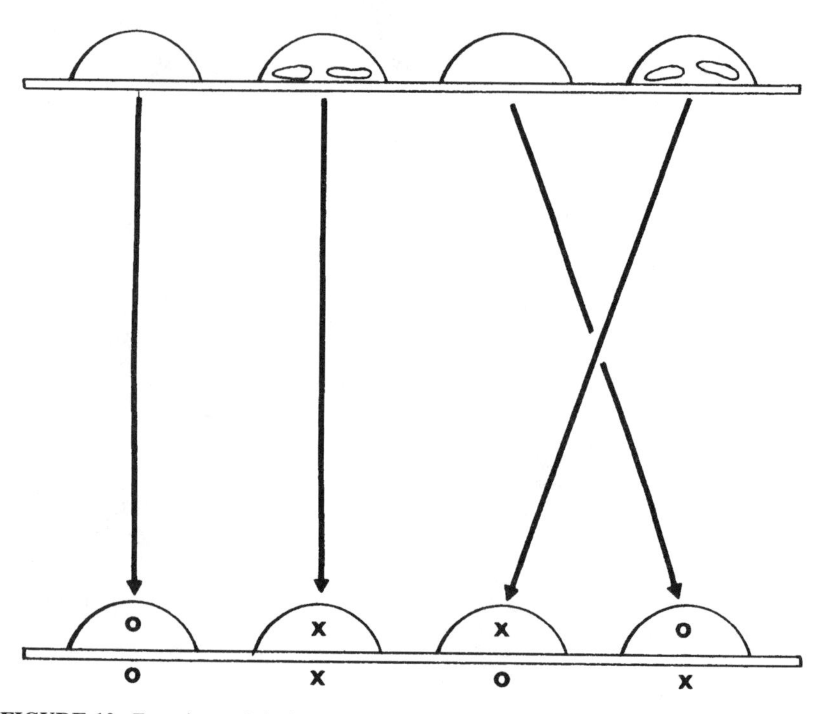

FIGURE 10. Experimental design used to determine requirements for the *in vitro* effects of sporocysts on the opsonin–phagocyte system. Hemolymph was placed on four wells of a glass slide (*upper*), and sporocysts were added to wells 2 and 4. After 2 hr, plasma and sporocysts were retrieved, the glass-adherent phagocytes were washed in physiological saline, and the plasma (without sporocysts) was returned to the wells, either to their original site or to alternatives as shown. Then yeast was added, and phagocytosis was allowed to proceed: 0 = naive; x = exposed to parasites.

with sporocysts. It therefore appears that *S. mansoni* modulated the internal defense system of its host within the first few hours after infection, and that these effects can be obtained *in vitro* only if sporocyst, hemocytes, and plasma are simultaneously present. FIGURE 12 presents an interpretation of these results.

WHAT IS GOING ON?

Abundant precedent exists for humoral lectin-type opsonins in both vertebrate and invertebrate systems. Indeed, a mannan-binding protein of human plasma is opsonic for yeast phagocytosis by human neutrophils and monocytes.[27] Among molluscs, bivalved oysters,[39] mussels,[40] and clams[41,42] have all yielded opsonic lectins from the plasma. Examples of gastropods in which plasma opsonins have been found include *Otala lactea*[43] (opsonin for yeast), *Lymnaea stagnalis*[44] (opsonin for yeast and sheep erythrocytes), and *Helix pomatia*[4] (opsonin for erythrocytes). A greater number of agglutinating lectins in molluscan plasma remain to be evaluated

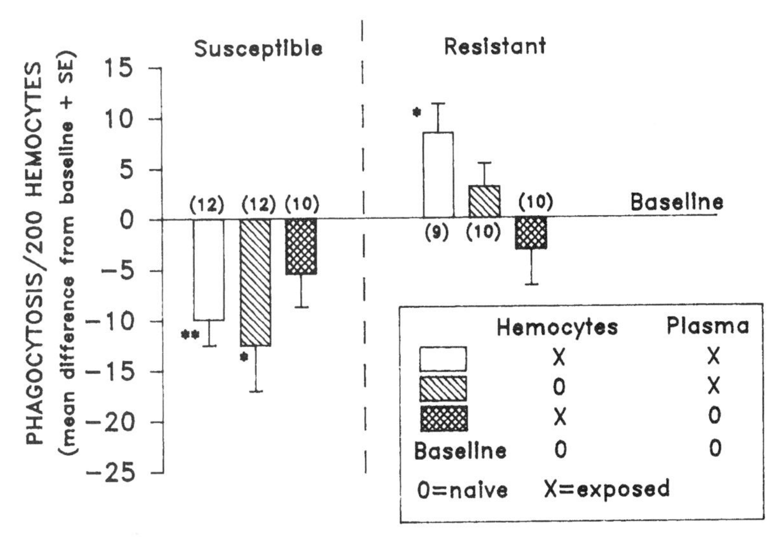

FIGURE 11. The effects of exposure of hemocytes and/or plasma to *Schistosoma mansoni* sporocysts for 2 hr *in vitro* on subsequent phagocytosis, as illustrated in FIGURE 10. Differences between phagocytosis values were compared with those of naive hemocytes and naive plasma from the same pool (baseline) using paired Student's t tests: $*p < 0.001$; $**p < 0.02$. (Adapted from Fryer & Bayne,[37] the *Journal of Parasitology,* with permission.)

for opsonic properties.[45] There is precedent also for integral membrane lectins playing significant roles in opsonin-independent phagocytosis. Abundant examples among the vertebrates (summarized by Sharon[26]) are paralleled by data from oysters,[46] in which an integral hemocyte-bound lectin appears with the same specificity as a plasma molecule, and by data from the mussel *Mytilus edulis,*[40] in which phagocytosis of yeast is inhibited by mucin. Taken together with evidence that (i) exotic lectins influence phagocytosis[47] and cytotoxicity against parasites by snail hemocytes,[48] (ii) *B. glabrata* plasma agglutinates sporocysts,[21] and (iii) knowledge that a variety of plasma proteins in this model bind rapidly to sporocysts,[30] it would seem reasonable to interpret our results as indicative of lectin-based opsonization.

Why, then, has it been impossible (unpublished observations) to even visualize the putative opsonin on a gel, much less to purify it? And why have we found it impossible to remove it by absorption from plasma? Is it likely that there is, in fact, no opsonin *per se?* Might the enhanced recognition of plasma-treated yeast be due to enzymatic activity working on the yeast surface? Host enzymes that contribute to defenses are most dramatically illustrated by lysozymes, because their effects result in total lysis of appropriate targets. But more subtle effects possibly involve modification of target surface biochemistry.[49] None of our experiments so far rules out the possibility that carbohydrases, by removing mannose residues from yeast cell surfaces, are responsible for the "opsonic effect" of resistant strain plasma. We envisage that such enzymatic activity would produce a yeast surface on which glycans would be more exposed, providing determinants for which phagocytes—as

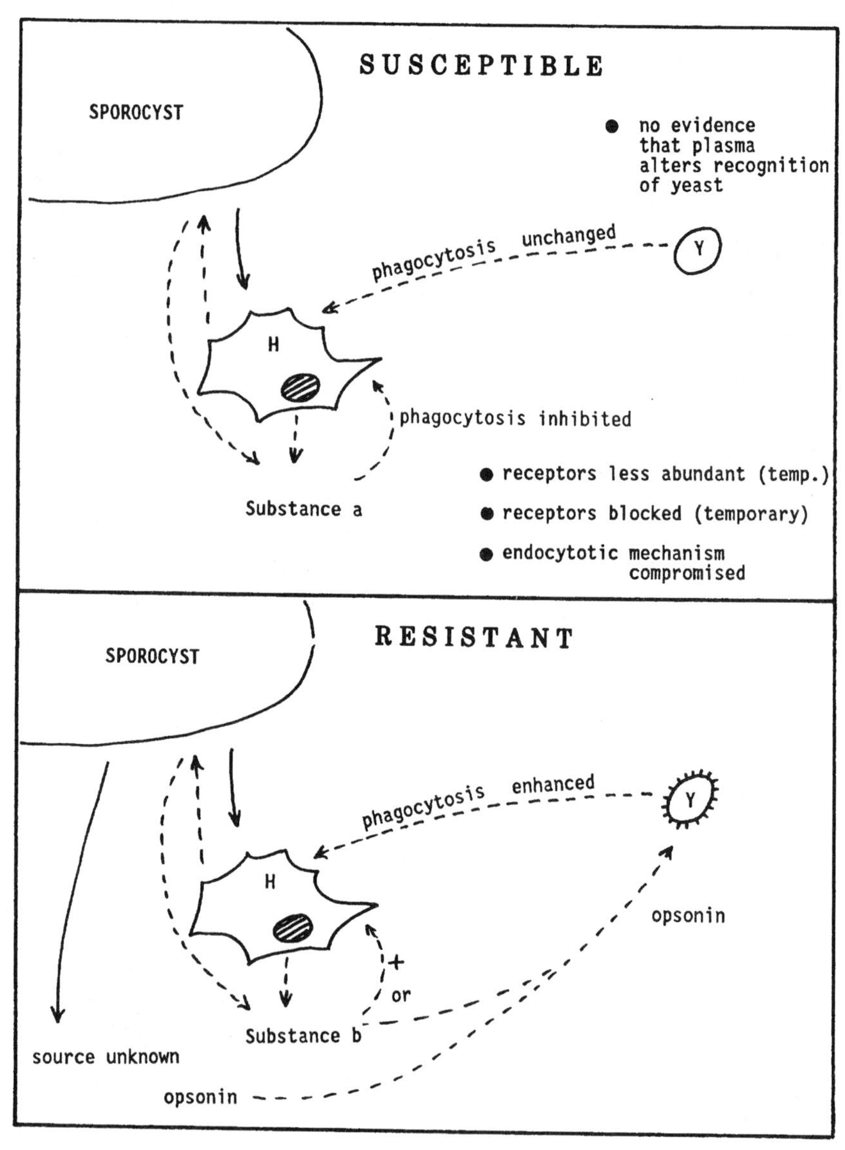

FIGURE 12. Models of the possible pathways by which sporocysts of *Schistosoma mansoni* may influence the phagocytosis of yeast by hemocytes of *Biomphalaria glabrata*. In the susceptible (M line) snails, encounters between sporocysts and hemocytes result in the release, by one or both of these partners, of a substance that depresses the phagocytic capacity of the hemocytes; this acts directly on hemocytes and not via an interaction with yeast. In the resistant (13-16-R1) snails, hemocytes, or sporocysts release a substance that enhances phagocytosis, acting either directly on hemocytes (though receptor abundance is not increased) or by means of a higher opsonin titer (source unknown). Solid lines indicate original stimuli; dotted lines indicate putative responses. (Adapted from Fryer & Bayne,[37] the *Journal of Parasitology*, with permission.)

we have already shown[25]—express receptors. As this remains speculative at present, we end by listing (unpublished) observations that are consistent with such a possibility. (1) Repeated exposure of opsonic plasma to large quantities of yeast failed to reduce opsonic activity faster than it was lost from yeast-free control aliquots. (2) Western blot analyses of plasma proteins retrieved from yeast exposed to opsonic (resistant-strain) or to nonopsonic (susceptible strain) plasmas have failed to reveal consistent, resistant-specific components. (3) When plasma proteins retrieved from yeast held for 1 minute in opsonic plasma are compared to those retrieved from yeast held 60 minutes in opsonic plasma, no evidence for enrichment of specific molecules is obtained. (4) Whereas mannan dissolved in plasma reduces opsonization strongly, it has had variable effects on the spectrum of proteins retrieved for SDS-PAGE analysis.[25] The possibility that a mannan-specific glycosidase or glycosyl-transferase is active when yeast cells are exposed to resistant-strain *B. glabrata* plasma is presently being investigated.

ACKNOWLEDGMENTS

Maggie Dykes-Hoberg, Charlie Hull, Randall Bender and David Barnes have contributed to the data and the ideas presented here. Early work by Thomas C. Cheng has inspired some of our ideas.

REFERENCES

1. Stauber, L. A. 1950. The fate of India ink injected intracardially into the oyster, *Ostrea virginica* Gmelin. Biol. Bull. **98:** 227–241.
2. Bayne, C. J. 1974. On the immediate fate of bacteria in the land snail Helix. *In* Contemporary Topics in Immunobiology, Vol. 4. Invertebrate Immunology. Edwin L. Cooper, Ed. Ch. 3: 37–45. Plenum Press. New York.
3. Crichton, R. & K. J. Lafferty. 1975. The discriminatory capacity of phagocytic cells in the chiton (*Liolophura gaimardi*). *In* Immunologic Phylogeny. W. H. Hildemann & A. A. Benedict, Eds. Adv. Exp. Med. Biol. **64:** 89–98. Plenum. New York.
4. Renwrantz, L., W. Schanke, H. Harm, H. Erl, H. Leibsch & J. Gerken. 1981. Discriminative ability and function of the immunological recognition system of the snail *Helix pomatia.* J. Comp. Physiol. **141:** 477–488.
5. Bayne, C. J. 1983. Molluscan immunobiology. *In* The Mollusca. Vol. 5. Physiology Part 2. A. S. M. Saleuddin & K. M. Wilbur, Eds. 407–486. Academic Press. San Diego.
6. McKerrow, J. H., K. H. Jeong & J. H. Beckstead. 1985. Enzyme histochemical comparison of *Biomphalaria glabrata* amoebocytes with human granuloma macrophages. J. Leuk. Biol. **37:** 341–348.
7. Cheng, T. C., A. S. Thakur & E. Rifkin. 1969. Phagocytosis as an internal defense mechanism in the Mollusca: With an experimental study of the role of leukocytes in the removal of ink particles in *Littorina scabra* Linn. *In* Symposium on Mollusca.: 547–466. Marine Biological Association of India. Bangalore, India.
8. Yoshino, T. P. & T. -L. Tuan. 1985. Soluble mediators of cytolytic activity in hemocytes of the Asian clam, *Corbicula fluminea.* Dev. Comp. Immunol. **9:** 515–522.
9. Leippe, M. & L. Renwrantz. 1988. Release of cytotoxic and agglutinating molecules by *Mytilus* hemocytes. Dev. Comp. Immunol. **12:** 297–308.
10. Dikkeboom, R., C. J. Bayne, W. P. vanderKnaap & J. M. G. H. Tijnagel. 1988. Possible role of reactive forms of oxygen in *in vitro* killing of *Schistosoma mansoni* sporocysts by hemocytes of *Lymnaea stagnalis.* Parasitol. Res. **75:** 148–154.

11. ADEMA, C. M., W. P. W. VAN DER KNAPP & T. SMINIA. 1991. Molluscan haemocyte-mediated cytotoxicity: The role of reactive oxygen intermediates. Rev. Aquat. Sci. **4:** 201–223.
12. BERTHEUSSEN, K. 1979. The cytotoxic reaction in allogeneic mixtures of echinoid phagocytes. Exp. Cell Res. **120:** 373–381.
13. PARRINELLO, N., V. ARIZZA, M. CAMMARATA & D. M. PARINELLO. 1993. Cytotoxic activity of *Ciona intestinalis* (Tunicata) hemocytes: Properties of the in vitro reaction against erythrocyte targets. Dev. Comp. Immunol. **17:** 19–28.
14. SULLIVAN, J. T., J. A. ANDREWS & R. T. CURRIE. 1992. Heterotopic heart transplants in *Biomphalaria glabrata* (Mollusca, Pulmonata): Fate of allografts. Trans. Am. Microsc. Soc. **111:** 1–15.
15 FORD, S. E., K. A. ASHTON-ALCOX & S. A. KANALEY. 1993. In vitro interactions between bivalve hemocytes and the oyster pathogen *Haplosporidium nelsoni* (MSX). J. Parasitol. **79:** 255–265.
16. FONT, W. W. 1980. Effects of hemolymph of the American oyster, *Crassostrea virginica,* on marine cercariae. J. Invertebr. Pathol. **36:** 41–47.
17. LOKER, E. S., M. E. BOSTON & C. J. BAYNE. 1989. Differential adherence of M-line *Biomphalaria glabrata* hemocytes to *Schistosoma mansoni* and *Echinostoma paraensei* larvae and experimental manipulation of hemocyte binding. J. Invertebr. Pathol. **54:** 260–268.
18. BAYNE, C. J., P. M. BUCKLEY & P. C. DEWAN. 1980. Macrophage-like hemocytes of resistant *Biomphalaria glabrata* are cytotoxic for sporocysts of *Schistosoma mansoni in vitro.* J. Parasitol. **66:** 413–419.
19. BAYNE, C. J., P. M. BUCKLEY & P. C. DEWAN. 1980b. *Schistosoma mansoni*: Cytotoxicity of hemocytes from susceptible snail hosts for sporocysts in plasma from resistant *Biomphalaria glabrata.* Exp. Parasitol. **50:** 409–416.
20. GRANATH, W. O. & T. P. YOSHINO. 1984. *Schistosoma mansoni*: Passive transfer of resistance by serum in the vector snail, *Biomphalaria glabrata.* Exp. Parasitol. **58:** 188–193.
21. LOKER, E. S., M. A. YUI & C. J. BAYNE. 1984. *Schistosoma mansoni*: Agglutination of sporocysts and formation of gels on miracidia transforming in plasma of *Biomphalaria glabrata.* Exp. Parasitol. **58:** 56–62.
22. FRYER, S. E. & C. J. BAYNE. 1989. Opsonization of yeast by the plasma of *Biomphalaria glabrata* (Gastropoda): A strain-specific, time-dependent process. Parasite Immunol. **11:** 269–278.
23. VAN DER KNAAP, W. P. W. & E. S. LOKER. 1990. Immune mechanisms in trematode–snail interactions. Parasitol. Today **6:** 175–182.
24. PHAFF, H. J. 1974. Structure and biosynthesis of the yeast cell envelope. *In* The Yeasts. A. J. Ross & J. S. Harrison, Eds. Vol. 2: 135. Academic Press. London.
25. FRYER, S. E., C. J. HULL & C. J. BAYNE. 1989. Phagocytosis of yeast by *Biomphalaria glabrata*: Carbohydrate specificity of hemocyte receptors and a plasma opsonin. Dev. Comp. Immunol. **13:** 9–16.
26. SHARON, N. 1984. Surface carbohydrates and surface lectins are recognition determinants in phagocytosis. Immunol. Today **5:** 143–146.
27. KUHLMAN, M., K. JOINER & A. B. EZEKOWITZ. 1989. The human mannose-binding protein functions as an opsonin. J. Exp. Med. **169:** 1733–1745.
28. BAYNE, C. J. 1990. Phagocytosis and non-self recognition in invertebrates. Bioscience **40:** 723–731.
29. BENDER, R. C., S. E. FRYER & C. J. BAYNE. 1992. Proteinase inhibitory activity in the plasma of a mollusc: Evidence for the presence of α-macroglobulin in *Biomphalaria glabrata.* Comp. Biochem. Physiol. **102B:** 821–824.
30. BAYNE, C. J., E. S. LOKER & M. A. YUI. 1986. Interactions between the plasma proteins of *Biomphalaria glabrata* (Gastropoda) and the sporocyst tegument of *Schistosoma mansoni* (Trematoda). Parasitology **92:** 653–664.
31. BAYNE, C. J. & E. S. LOKER. 1987. Survival within the snail host. *In* The Biology of Schistosomes: From Genes to Latrines. D. A. Rollinson & A. J. G. Simpson, Eds.: 321–346. Academic Press. New York.

32. Abdul-Salam, J. M. & E. H. Michelson. 1980. *Biomphalaria glabrata* amebocytes: Effect of *Schistosoma mansoni* infection on in vitro phagocytosis. J. Invertebr. Pathol. **35:** 241–248.
33. van der Knapp, W. P. W., E. A. Meuleman & T. Sminia. 1987. Alterations in the internal defence system of the pond snail *Lymnaea stagnalis* induced by infection with the schistosome *Trichobilharzia ocellata.* Parasitol Res. **73:** 57–65.
34. Noda, S. & E. S. Loker. 1989. Phagocytic activity of hemocytes of M-line *Biomphalaria glabrata* snails: Effect of exposure to the trematode *Echinostoma paraensei.* J. Parasitol. **75:** 261–269.
35. Bayne, C. J. 1991. Invertebrate host immune mechanisms and parasite escapes. *In* Parasite–Host Association: Co-existence or Conflict? C. A. Toft, A. Aeschlimann & L. Bolis, Eds.: 201–227. Oxford Scientific Publications. Oxford, England.
36. Douglas, J. S., M. D. Hunt & J. T. Sullivan. 1993. Effects of *Schistosoma mansoni* infection on phagocytosis and killing of *Proteus vulgaris* in *Biomphalaria glabrata* hemocytes. J. Parasitol. **79:** 280–283.
37. Fryer, E. S. & C. J. Bayne. 1990. *Schistosoma mansoni* modulation of phagocytosis in *Biomphalaria glabrata.* J. Parasitol. **76:** 45–52.
38. Fryer, S. E., M. Dykes-Hoberg & C. J. Bayne. 1989. Changes in plasma opsonization of yeast after isolation of *Biomphalaria glabrata* in small volumes of water. J. Invertebr. Pathol. **54:** 275–276.
39. Hardy, S. W., T. C. Fletcher & J. A. Olafsen. 1977. Aspects of cellular and humoral defense mechanisms in the Pacific oyster, *Crassostrea gigas. In* Developmental Immunobiology. J. B. Solomon & J. D. Horton, Eds.: 59–66. Elsevier/North Holland. Amsterdam.
40. Renwrantz, L. & A. Stahmer. 1983. Opsonizing properties of an isolated hemolymph agglutinin and demonstration of lectin-like recognition molecules at the surface of hemocytes from *Mytilus edulis.* J. Comp. Physiol **149:** 535–546.
41. Olafsen, J. A. 1988. Role of lectins in invertebrate humoral defense. *In* Disease Processes in Marine Bivalve Mollusks. W. S. Fisher, Ed. Am. Fish. Soc. Spec. Publ. **18:** 189–205.
42. Yang, R. & T. P. Yoshino. 1990. Immunorecognition in the freshwater bivalve *Corbicula fluminea* II. Isolation and characterization of a plasma opsonin with hemagglutinating activity. Dev. Comp. Immunol. **14:** 397–404.
43. Anderson, R. S. & R. A. Good. 1976. Opsonic involvement in phagocytosis by mollusc hemocytes. J. Invertebr. Pathol. **27:** 57–64.
44. Sminia, T., W. P. W. van der Knapp & P. Edelenbosch. 1979. The role of serum factors in phagocytosis of foreign particles by blood cells of the freshwater snail (*Lymnaea stagnalis*). Dev. Comp. Immunol. **3:** 37–44.
45. Renwrantz, L. 1986. Lectins in molluscs and arthropods: Their occurrence, origin and roles in immunity. *In* Immune Mechanisms in Invertebrate Vectors. A. Lackie, Ed.: 56. Zool. Soc. London Symp. Oxford Scientific Publications. Oxford, England.
46. Vasta, G. R., T. C. Cheng & J. J. Marchalonis. 1984. A lectin on the hemocyte membrane of the oyster (*Crassostrea virginica*). Cell. Immunol. **88:** 475–488.
47. Schoenberg, D. A. & T. C. Cheng. 1982. Concanavalin A-mediated phagocytosis of yeast by *Biomphalaria glabrata* hemocytes in vitro: Effects of temperature and lectin concentration. J. Invertebr. Pathol. **39:** 314–322.
48. Boswell, C. A. & C. J. Bayne. 1986. Lectin-dependent cell-mediated cytotoxicity in an invertebrate model: Con A does not act as a bridge. Immunology **57:** 261–264.
49. Cheng, T. C. 1975. Functional morphology and biochemistry of molluscan phagocytes. Ann. N.Y. Acad. Sci. **266:** 343–379.

Parasitoid-Induced Cellular Immune Deficiency in *Drosophila*[a]

TAHIR M. RIZKI AND ROSE M. RIZKI

Department of Biology
The University of Michigan
Ann Arbor, Michigan 48109-1048

INTRODUCTION

The immune response relies on the ability to distinguish non-self from self. Because endoparasites are foreign entities in their hosts, they must circumvent the hosts' capacity to recognize non-self in order to survive in the alien environment. How endoparasites evade detection and destruction by the host's immune system can reveal factors that are important to a successful defense response by the host. The variety of mechanisms used by endoparasites to bypass the challenges imposed by the insect host's cellular defense can be classified in two categories: those that avoid exciting the defense response and those that combat elements of the defense system.

Some parasites evade host hemocytes by residing within host tissues. For example, parasitic wasps (parasitoids) and parasitic nematodes develop beneath the gut epithelium or within ganglia of the central nervous system.[1–3] Other parasites avoid harmful effects of host hemocytes by utilizing developmental stages of the host in which immune responses are less likely to occur.[2,3] Parasites that develop within the host hemocoel may have surface coatings that repel, or do not arouse, a hemocyte response. That the thick fibrous layer on braconid *Cardiochiles nigriceps* eggs thwarts hemocyte attack was demonstrated by experimental removal of this surface layer to render the eggs susceptible to encapsulation by host hemocytes.[4] Likewise, surface materials on *Venturia canescens* eggs are responsible for preventing encapsulation.[5,6] In this parasitoid, 130- to 140-nm particles secreted by the calyx cells of the reproductive system adhere to long projections from the egg chorion.[7] The particles persist on the eggs up to 60 hours after oviposition in the host and are present on first instar larvae as well. The viruslike particles contain glycoproteins but no detectable nucleic acid.[8] Antibodies raised against purified *Venturia* particle proteins cross-react to a 42-kDa protein localized in the host fat body. Most of the 42-kDa protein is synthesized in host hemocytes and secreted into the hemolymph where it adheres to the surface layer of the fat body.[9] The authors concluded that the viruslike particles on the parasitoid eggs confer passive resistance to host hemocytes by virtue of the antigenic determinants they share with the host. This conclusion is compatible with the fact that foreign objects other than the parasitoid eggs are encapsulated in parasitized hosts.[10] In other parasitoid–host interactions, venom acts

[a]This research was supported by grants from the National Institutes of Health and the National Science Foundation.

synergistically with fibrous egg coats to inhibit encapsulation by host hemocytes without disrupting encapsulation of other foreign particles.[11,12]

Although encapsulation by hemocytes contains and destroys foreign objects, this defense reaction need not always be fatal for parasitoid eggs. The eggs of *Leptopilina boulardi* (strain L104), a habitual parasitoid of *Drosophila melanogaster,* adhere to the surfaces of host tissues, generally the larval gut.[13] Host hemocytes attach to the egg surfaces exposed to the hemocoel but cannot penetrate around the region of the egg attached tightly to the underlying host tissue to completely surround the eggs. The incomplete capsules are not effective in preventing escape of the parasitoid larvae, which move actively at hatching and rupture the partial capsules. This means of defense against host hemocytes is not highly effective, because survival of L104 parasitoids in *Drosophila* hosts is poor compared with parasitoids employing other defense strategies.[14,15]

In contrast to passive mechanisms for protection against host hemocytes, some parasitoids interfere with the cellular defense system of the host to ensure a suitable environment for the development of their eggs. In these insects, accessory organs of the female reproductive system synthesize materials that are coinjected with eggs into the host. The substances prevent encapsulation of not only the parasitoid eggs but also, in some cases, other foreign objects in the host hemocoel as well. That such accessory gland secretions affect hemocytes is suggested by visible effects on cell morphology in parasitized hosts.[16] More than a single factor may be required to affect host hemocytes; for example, both calyx and venom fluid of *Microplitis mediator* are necessary to inhibit filopodial elongation of the host hemocytes.[17]

Polydnaviruses that replicate in calyx cells of ichneumonid and braconid parasitoids and are injected into hosts at oviposition inhibit encapsulation of wasp eggs and play several other roles in modifying host physiology for the benefit of the developing parasitoids.[18,19] The viruses enter a variety of host tissues, including muscle, fat body, tracheal epithelium, and hemocytes.[20] Expression of polydnaviruses is detectable in hosts shortly after parasitization, but the viruses do not replicate in the parasitized hosts.[21–23] How the polydnaviruses interfere with encapsulation is not known. Impairment of hemocyte function in parasitized hosts is suggested by studies demonstrating suppression of the *in vitro* spreading behavior of hemocytes by calyx fluid and polydnaviruses.[24–26]

The most striking instance of hemocyte modification has been reported for *Drosophila melanogaster* larvae parasitized by *Leptopilina heterotoma.*[27] This parasitoid uses viruslike particles (VLPs) produced in an accessary gland (long gland), not in the calyx cells, to disrupt specifically the capsule-forming hemocytes of the host.[28,29] An evolutionary relationship between polydnaviruses in the calyx and components in venom glands of the reproductive system has been suggested.[30,31] Therefore, analysis of the *D. melanogaster–L. heterotoma* interaction is interesting and desirable for seeking relationships among the factors used by parasitoids to subjugate host cellular defenses. In addition, this parasitoid–host system offers a unique opportunity to study effects of parasitization on host hemocytes because the type of hemocytes and their responses to foreign materials in the hemocoel of the *D. melanogaster* larva have been studied extensively using mutant strains (see reviews in Rizki and Rizki[32,33]). By using these mutant strains as hosts for parasitoids, the effects of parasitization on host hemocytes can be surmised.

HEMOCYTES OF *DROSOPHILA* LARVAE

Two classes of hemocytes are found in normal *Drosophila* larvae, the plasmatocytes, which are phagocytic cells, and the crystal cells, which have prominent cytoplasmic paracrystalline inclusions containing phenol oxidases.[32,33] At pupariation, plasmatocytes undergo morphological transformation (or differentiation) from the typical spherical shape to an extremely flattened, disklike cell called the lamellocyte.[34,35] The leaflike lamellocyte is ideally suited for its function, which is to wrap around large foreign objects to form cellular capsules.

In *Drosophila* larvae with *melanotic tumor* genes, lamellocyte differentiation is induced by metabolic products from aberrant tissues and the larvae have many lamellocytes circulating in the hemolymph.[36] After the surfaces of the abnormal tissues show changes, lamellocytes begin to layer around them in a manner similar to the encapsulation of large foreign objects.[37,38] Multilayered, cellular capsules in *Drosophila* larvae are hardened and darkened due to the phenol oxidases of the crystal cells.[39–41]

Evidence for the roles of the crystal cells in the hardening and blackening processes has been obtained by genetically blocking the melanosis of tissues in melanotic tumor mutants. In double-mutant combinations using genes that either cause a deficiency or an absence of phenol oxidases, such as black cell or lozenge mutant alleles, encapsulation of the aberrant tissues created by the melanotic tumor gene proceeds. Nevertheless, the darkening and hardening processes dependent on crystal cell function are interrupted owing to the presence of the mutant genes blocking the availability of phenol oxidases.

HEMOCYTE CHANGES INDUCED BY PARASITIZATION

L. heterotoma females oviposit in the hemocoel of *Drosophila* larvae. The free-floating eggs are about 50 × 600 μm, too large to be engulfed by plasmatocytes so they should be susceptible to encapsulation by lamellocytes. When Walker[42] allowed *L. heterotoma* females to oviposit in melanotic tumor larvae, she found that the encapsulation of tissues by lamellocytes to form melanotic tumors was suppressed and that the parasitoid eggs remained free of encapsulating lamellocytes as well. This generalized blockage of encapsulation in parasitized *Drosophila* hosts was later confirmed,[13,43] and the cause of the inhibition of lamellocyte activity was identified.[27,28]

Two to four hours after melanotic tumor larvae are parasitized by *L. heterotoma*, the lamellocytes in the hemocoel lose their discoidal shape and become bipolar cells.[27] The change in cell morphology is correlated with changes in cell adhesion. Disk-shaped lamellocytes are sticky cells that adhere and clump together. The bipolar lamellocytes are nonadherent, so it is clear that they are not able to attach to foreign objects or other lamellocytes to form cellular capsules. The affected lamellocytes in parasitized *Drosophila* larvae undergo degenerative changes involving loss of cytoplasm at the elongating tips. This destruction of lamellocytes can account for the reduced frequency of lamellocytes reported for parasitized hosts rather than suppression of lamellocyte differentiation.[29]

The factor responsible for destruction of host lamellocytes is contained in the reservoir of an accessory gland of the female wasp reproductive system.[27] The source of the anti-lamellocytic factor in the female wasp was identified by injected macerated samples of various regions of the female reproductive system into *Drosophila* larvae and sampling the hemolymph of the injected larvae for the presence of bipolar lamellocytes. Only the contents of the long gland reservoir induced bipolar lamellocytes.[27] Identification of the active component in reservoir fluid was accomplished by fractionating reservoir fluid on density gradients and treating lamellocytes *in vitro* with gradient fractions.[28] Only fractions containing viruslike particles (VLPs) affected lamellocyte morphology. Therefore, in the experiments using reservoir fluid, it should be understood that the VLPs are the active factor in this fluid.

Reservoir fluid can be added to lamellocytes *in vitro* to follow the modifications in lamellocyte morphology and adhesiveness.[44] The elongating process of individual lamellocytes *in vitro* can be observed for 1 to 2 days in axenic medium as the bipolar cells continue to elongate until they become threadlike. Breakage of elongated lamellocyte remnants in undisturbed samples supports the conclusion that lamellocyte destruction in parasitized larvae occurs as the elongating tips of the bipolar cells pinch off.[27]

CYTOSKELETAL MODIFICATIONS IN LAMELLOCYTES

Cell shape is determined by the disposition of cytoskeletal elements. It was, therefore, expected that lamellocyte transition from round to bipolar involved rearrangement of microtubules. To compare microtubules of lamellocytes in unparasitized and parasitized hosts, indirect immunofluorescence studies with anti-tubulin were used.[27] The microtubules in round lamellocytes were dispersed in a radial pattern, whereas the microtubules in bipolar lamellocytes were arranged in parallel arrays along the elongating axis of the cells. When lamellocytes were treated simultaneously with wasp reservoir fluid and Vinca alkyloids which depolymerize microtubules to tubulin subunits, the cells retained the discoidal shape rather than assuming a bipolar configuration.[45] These observations were interpreted as evidence that the elongation process induced by the wasp factor requires depolymerization and repolymerization of tubulin.

A time course ultrastructural study of lamellocytes treated with wasp reservoir fluid *in vitro* was undertaken to follow events in the reorganization of the cytoskeleton. The methods for the study followed those reported previously.[28] Briefly, hemolymph from 42 *gt* w^1 tu-Sz^{ts}/y sc $gt^{XIIa}v^{81ix}$ tu-Sz^{ts} *Drosophila* larvae were collected in 1.4 ml of phosphate-buffered saline for *Drosophila.* Reservoir fluid from 35 *L. heterotoma* females was added to 900 µl of the hemocyte sample, and the remainder of the sample was retained as the control sample. A 135-µl aliquot (C_0) of hemocytes was fixed in 2.5% glutaraldehyde in cacodylate buffer before reservoir fluid was added and another 135-µl aliquot (T_0) was fixed immediately after the addition of reservoir fluid. Aliquots from the hemocyte mixtures were fixed 10, 30, 60, 120, and 180 minutes later. The cells in the samples were pelleted, postfixed in osmium, and embedded in epon. Thin sections from the epon blocks were stained with uranyl acetate–lead citrate and examined in the electron microscope.

Ultrastructural characteristics of lamellocytes from larvae with the genotype given above and processed by the same procedures used for the present study have been described previously.[28] In thin sections the lamellocytes can be recognized by their flat shape and a heavy, electron-dense deposit along the nuclear membrane. Microtubules are concentrated heavily near the nucleus of the lamellocyte. In thin sections the microtubules are dispersed in a helter-skelter fashion (FIG. 1a). Microtubules are prominent in lamellocytes incubated for 10 minutes with reservoir fluid in the medium, but few microtubules are found in the vicinity of the lamellocyte nucleus after treatment with reservoir fluid for 30 minutes (FIG. 1b). Compared with the control lamellocytes, these microtubules are very short in length.

After exposure to reservoir fluid for 60 minutes, long microtubules extend from the vicinity of the nucleus along the elongating axis of the lamellocyte (FIG. 2a). By 180 minutes posttreatment, parallel bundles of microtubules are seen in the lamellocytes (FIG. 2b). The ultrastructural observations thus confirm that wasp reservoir fluid initiates depolymerization and repolymerization of microtubules as the cells change shape from round to bipolar.

The antimitotic drug taxol has been used to study the mechanisms involved in polymerization and depolymerization of microtubules. The drug binds specifically and reversibly to polymerized tubulin and stabilizes microtubules to cold, calcium, and microtubule-disrupting drugs either by lowering the rate of dissociation or by decreasing the rate of steady-state tubulin flux.[46–48] When cells are incubated in taxol, there is a redistribution of microtubules that is characterized by the formation of microtubule bundles.[49] It has also been reported that increased tubulin synthesis in taxol-treated cells occurs as the result of an increase in tubulin mRNA.[49]

To determine whether taxol interferes with the depolymerization of lamellocyte microtubules induced by wasp reservoir fluid, *Drosophila* hemocytes *in vitro* were exposed to 2×10^{-5} M taxol in the presence of wasp material. Two series of experiments were undertaken. In one experiment taxol was mixed with *Drosophila* hemocytes, and wasp reservoir fluid was added to aliquots removed from this sample immediately 0, 15, 30, and 45 minutes later. In a second experiment, *Drosophila* hemocytes were incubated with wasp reservoir contents, and taxol was added to aliquots removed from the hemocyte sample after 0, 15, 30, and 45 minutes. Control samples containing only reservoir fluid or taxol were prepared for both experiments. The cell samples were examined by phase microscopy and photographed 4 to 4.5 hours later.

Lamellocytes incubated in taxol for 4.5 hours retained the typical circular, thin, flat shape of lamellocytes seen in freshly prepared hemocyte samples and in samples that have been incubated in phosphate-buffered saline or Schneider's culture medium for *Drosophila* cells (FIG. 3a). The lamellocytes in the samples containing wasp reservoir fluid for 4.5 hours exhibited the elongated, bipolar shape reported in previous *in vitro* studies (FIG. 3b).

When reservoir fluid was added immediately after taxol was placed in the hemocyte sample and the cells were examined four hours later, many long, thin projections extended in all directions from the lamellocyte surfaces (FIG. 4a). The typical bipolar transition induced by reservoir fluid was blocked by the presence of taxol in the incubation mixture. In the samples to which reservoir fluid was added at 15 minutes, lamellocytes with multiple surface projections and blebs at the tips were

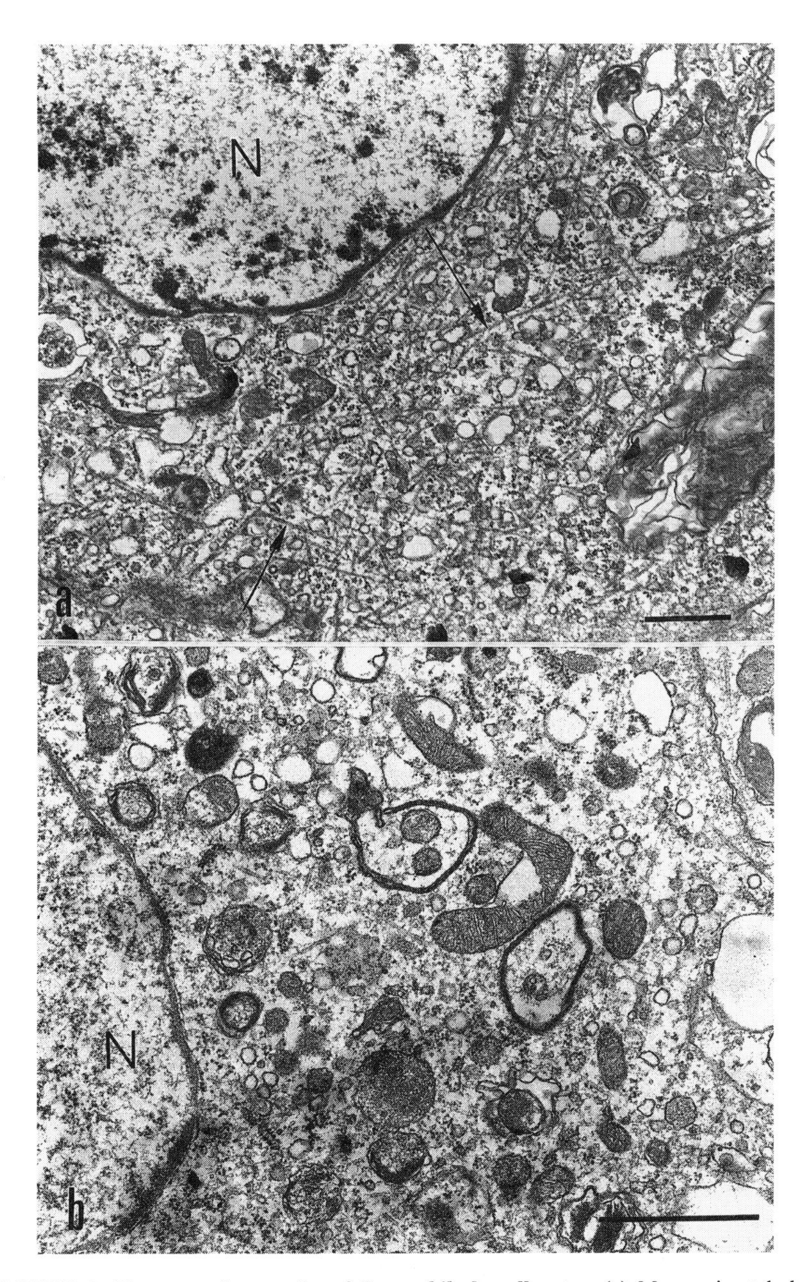

FIGURE 1. Electron micrographs of *Drosophila* lamellocytes. (a) Many microtubules (*arrows*) are apparent in the cytoplasm near the nucleus (N) in this lamellocyte from a control sample. Bar = 1 μm. **(b)** After 30 minutes' exposure to VLPs, very few short segments of microtubules are apparent in a similar region of a lamellocyte. The cell was photographed at higher magnification than (a) to resolve microtubules. Bar = 1 μm.

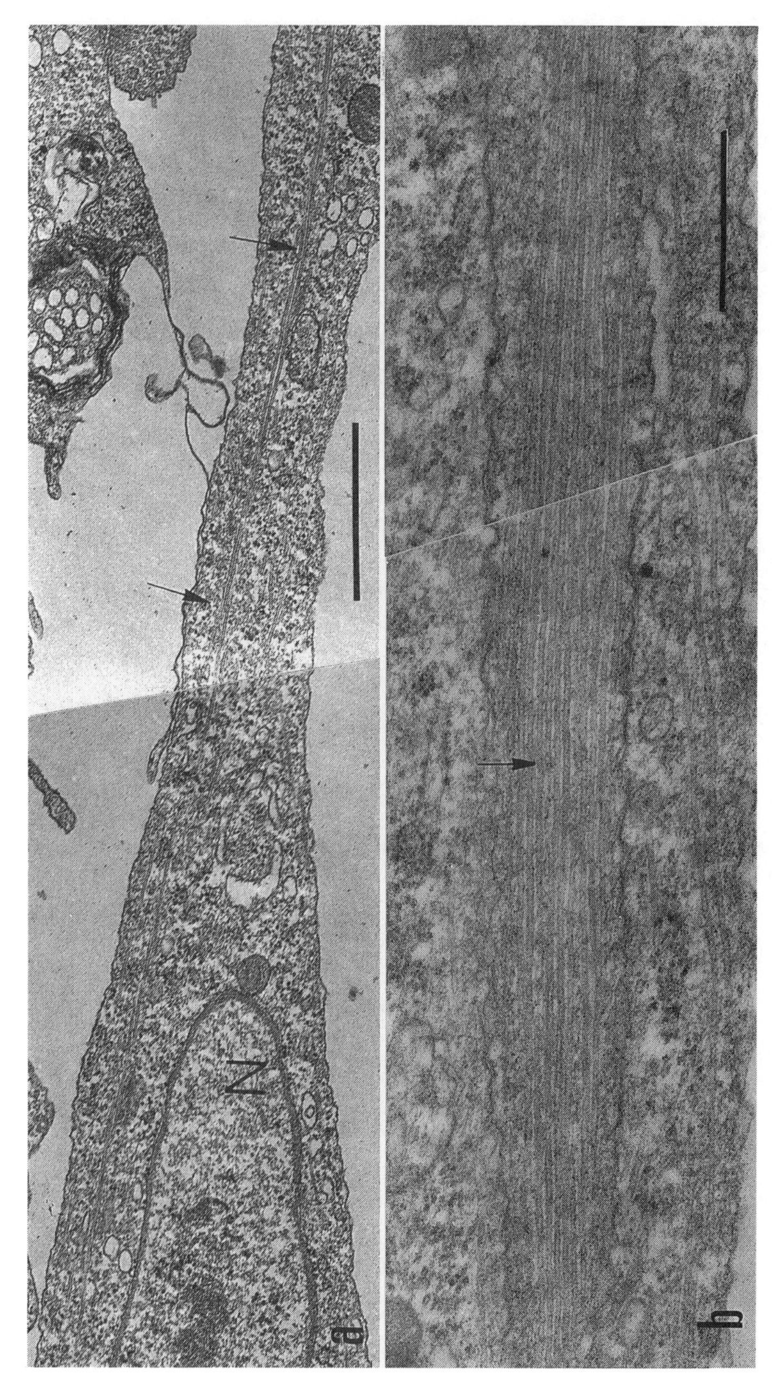

FIGURE 2. Sections of lamellocytes treated with VLPs. (a) Long stretches of microtubules (*arrows*) can be seen in the elongating axis of this lamellocyte. N = nucleus. Bar = 1 µm. **(b)** A section passing through three apposed lamellocytes in the pellet after 180 minutes' exposure to VLPs. The lamellocyte in the middle shows a parallel bundle of microtubules (*arrow*) stretching along the elongating axis. Bar = 0.5 µm.

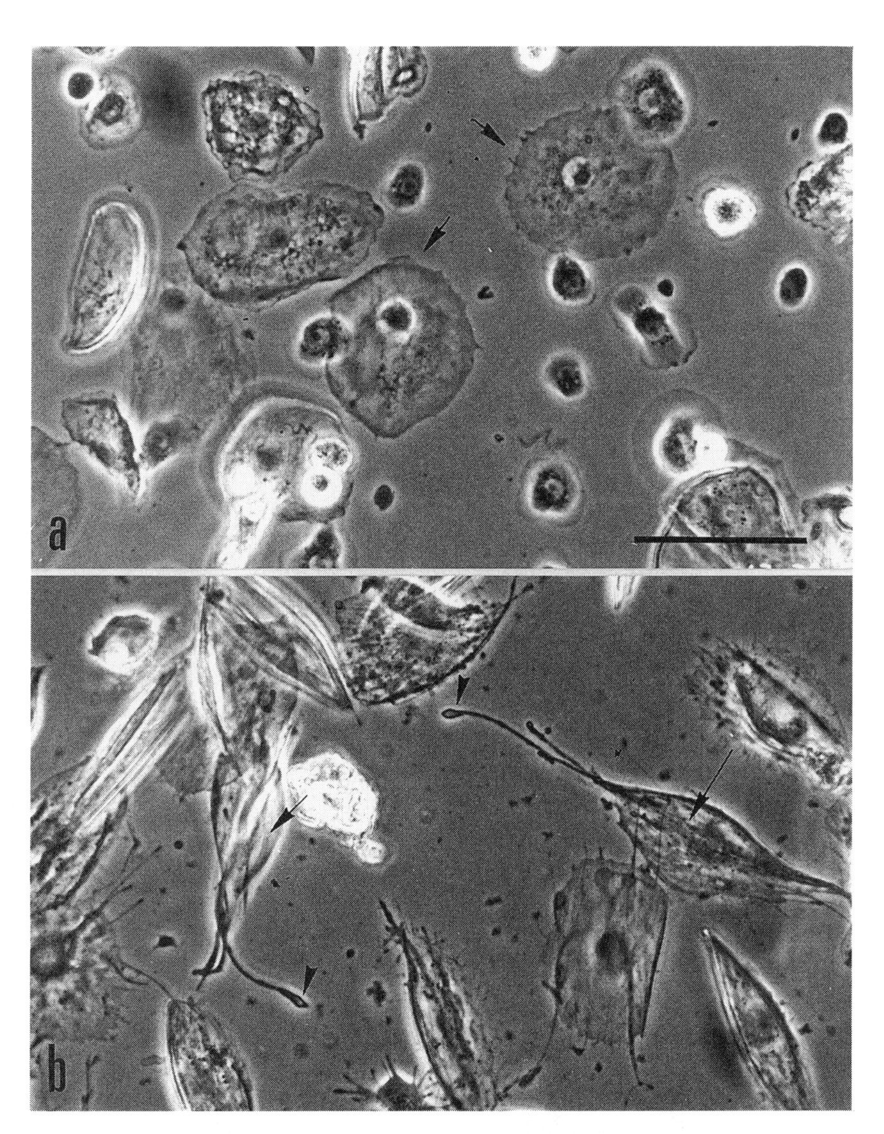

FIGURE 3. (a) Phase micrograph of a hemocyte sample treated with taxol for 4.5 hours. The lamellocytes (*arrows*) remain discoidal. Bar = 50 μm. This same magnification is used for all photographs in FIGURES 3, 4, and 5. **(b)** Hemocyte sample treated with VLPs for 4.5 hours. Note the elongated bipolar lamellocytes (*arrows*). There are cytoplasmic blebs (*arrowheads*) at the elongating tips. In electron micrographs the blebs contain numerous vesicles of various sizes.

also abundant (FIG. 4b). The surface projections from some of the lamellocytes appeared to be slightly shorter than those in the 0-minute sample. There was a noticeable decrease in the number of surface projections as well as the length of the projections when reservoir fluid was added 30 minutes after taxol (FIG. 4c). In the 45-minute sample, projections from the lamellocyte surfaces were very short, and fewer lamellocytes had the surface extensions (FIG. 4d).

In the experiment reversing the order of treatment with reservoir fluid and taxol, the sample in which reservoir fluid and taxol were added initially had lamellocytes with long surface projections, as expected from the results of the first experiment (FIG. 5a). There was a slight reduction in projection length when taxol addition was 15 minutes after reservoir fluid, and an indication of bipolar shape was apparent in some lamellocytes (FIG. 5b). Many bipolar lamellocytes were present in the sample in which taxol treatment followed reservoir fluid by 30 minutes (FIG. 5c). Some of the bipolar lamellocytes had many elongated projections, however. When taxol addition was delayed for 45 minutes, lamellocytes with the typical bipolar effect of wasp reservoir fluid were found (FIG. 5d).

The asteriated lamellocytes produced by treatment with taxol and reservoir fluid must result from microtubule elongation at numerous sites. Because taxol stabilizes microtubules, it must block or reduce the microtubule depolymerization that is initially promoted by VLPs. The elongation of microtubules that is the second function of VLPs must not be affected in the presence of taxol, so elongation occurs at the numerous randomly oriented, taxol-stabilized microtubule ends. The result is a lamellocyte with many surface projections. The length of the projections is a function of time of treatment with taxol. As microtubule stabilization proceeds, the free tubulin pool will be depleted, and tubulin will not be available for microtubule elongation. Therefore, the longer the delay in adding wasp reservoir fluid, the shorter the projections. The depolymerization and reorientation of microtubules induced by the *L. heterotoma* factor occurs within 30 minutes *in vitro* and is not reversible, because addition of taxol at 30 and 45 minutes did not inhibit the establishment of lamellocyte bipolarity. Our interpretation of the combined treatment of lamellocytes with taxol and reservoir fluid is based on the extensive studies of Horwitz[49] and her analyses of taxol effects on microtubules.

SPECIFICITY OF VLPs FOR LAMELLOCYTES

Density gradient–purified VLPs have an outer electron-dense, asymmetric coat with surface projections and a convoluted region that forms a vesicular invagination opposite the thickest portion of the electron-dense material (FIGS. 6a and 8c). When purified VLPs are added to hemolymph samples *in vitro,* the particles adhere to lamellocyte surfaces. The surface spikes of the VLP are often seen attached to the lamellocyte surface. There appears to be a disruption in the lamellocyte plasma membrane and a slight blistering at the contact site of the VLP projection (FIG. 6b). This suggests that a lytic factor at the tip of the VLP spike may dissolve the cell membrane when contact is made. Sections through VLP spikes often have a club-shaped swelling with an electron-transparent interior (FIG. 6c), which may be the source of lytic factors.

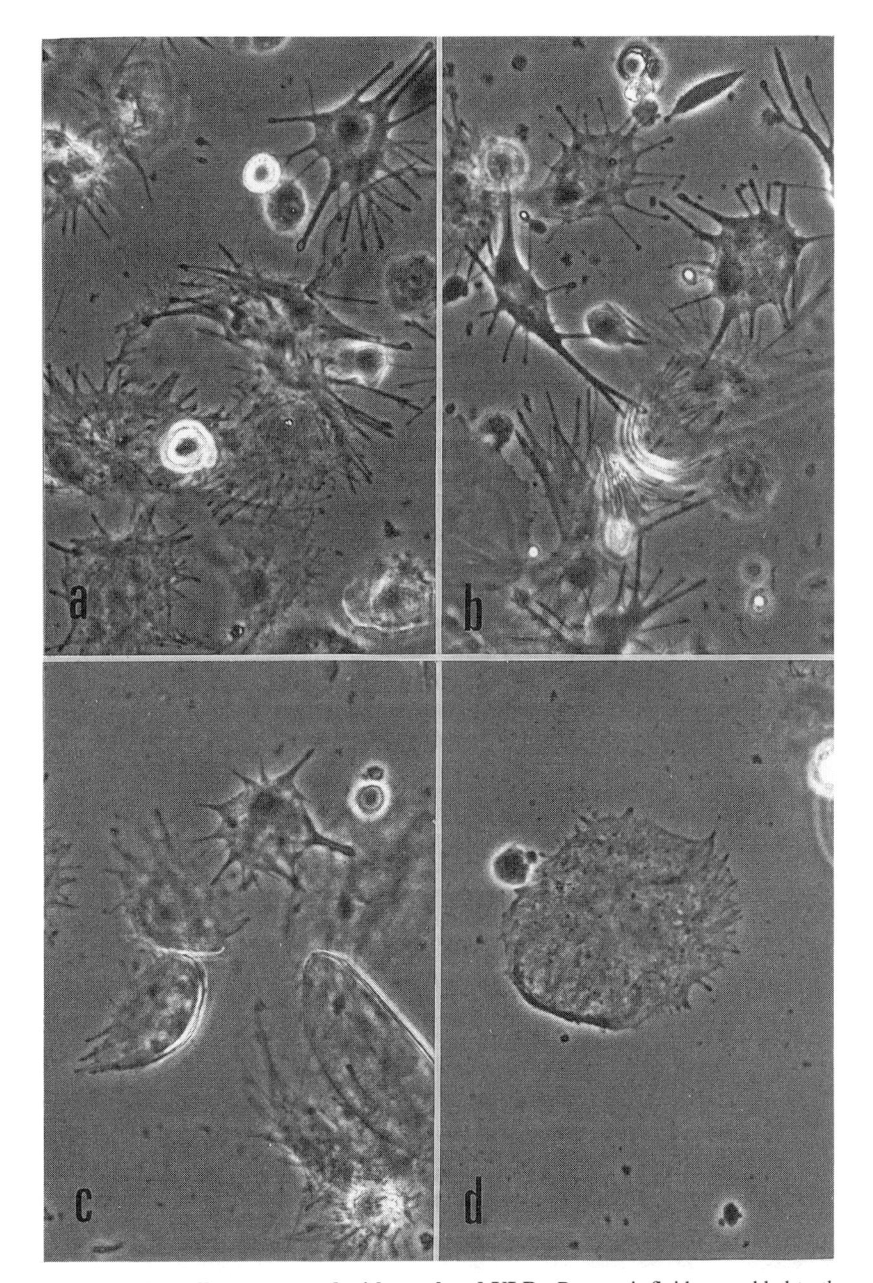

FIGURE 4. Lamellocytes treated with taxol and VLPs. Reservoir fluid was added to the sample immediately after taxol in frame **a,** after 15 minutes in frame **b,** after 30 minutes in **c,** and after 45 minutes in frame **d.** Note the decreasing length of the surface projections as the time interval between taxol treatment and VLP addition increased.

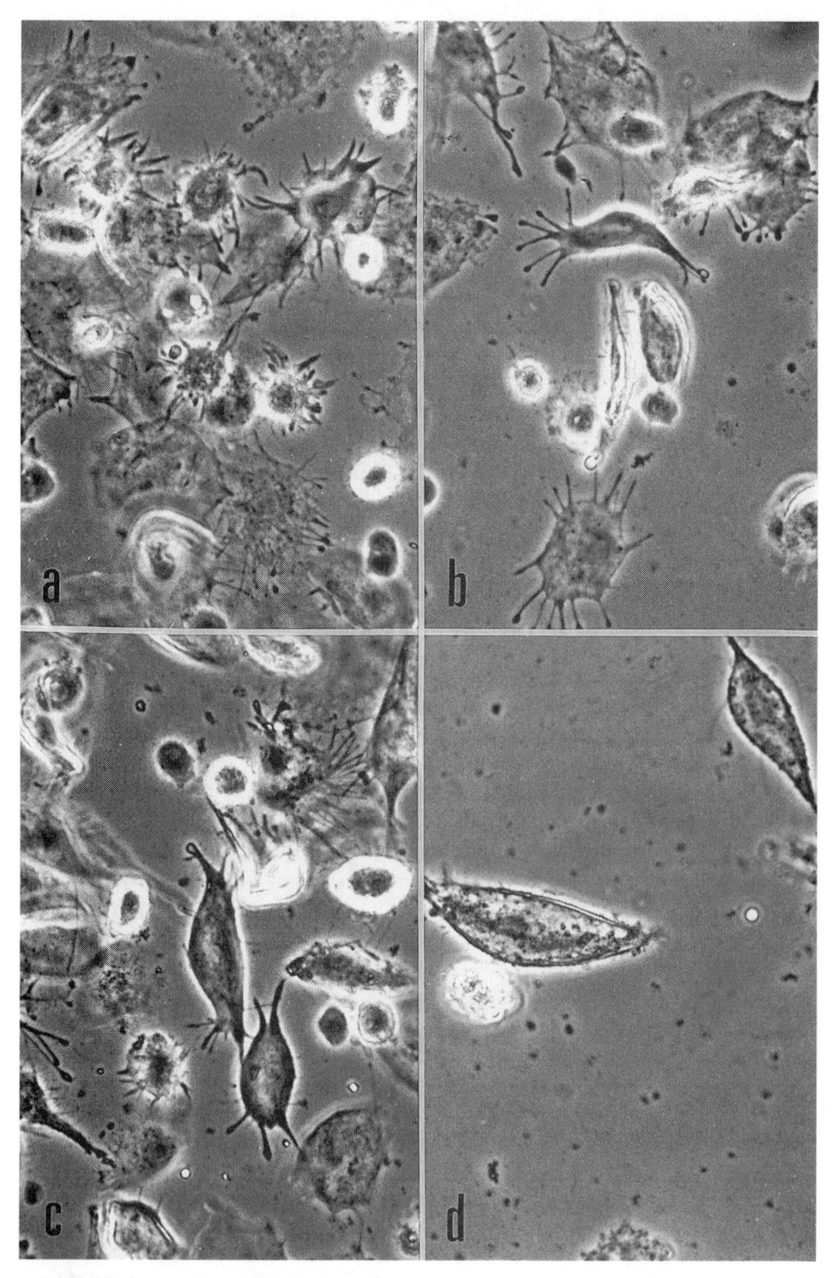

FIGURE 5. Phase micrographs of hemocytes showing the effects of VLPs and taxol. The cells were exposed to reservoir fluid, then taxol was added at 0 minutes (**frame a**), after 15 minutes (**frame b**), after 30 minutes (**frame c**), and after 45 minutes (**frame d**). Blebs can be seen at the tips of the cytoplasmic projections of lamellocytes in frames **b** and **c.** The smaller cells, showing a pronounced phase halo and lacking surface projections, are plasmatocytes.

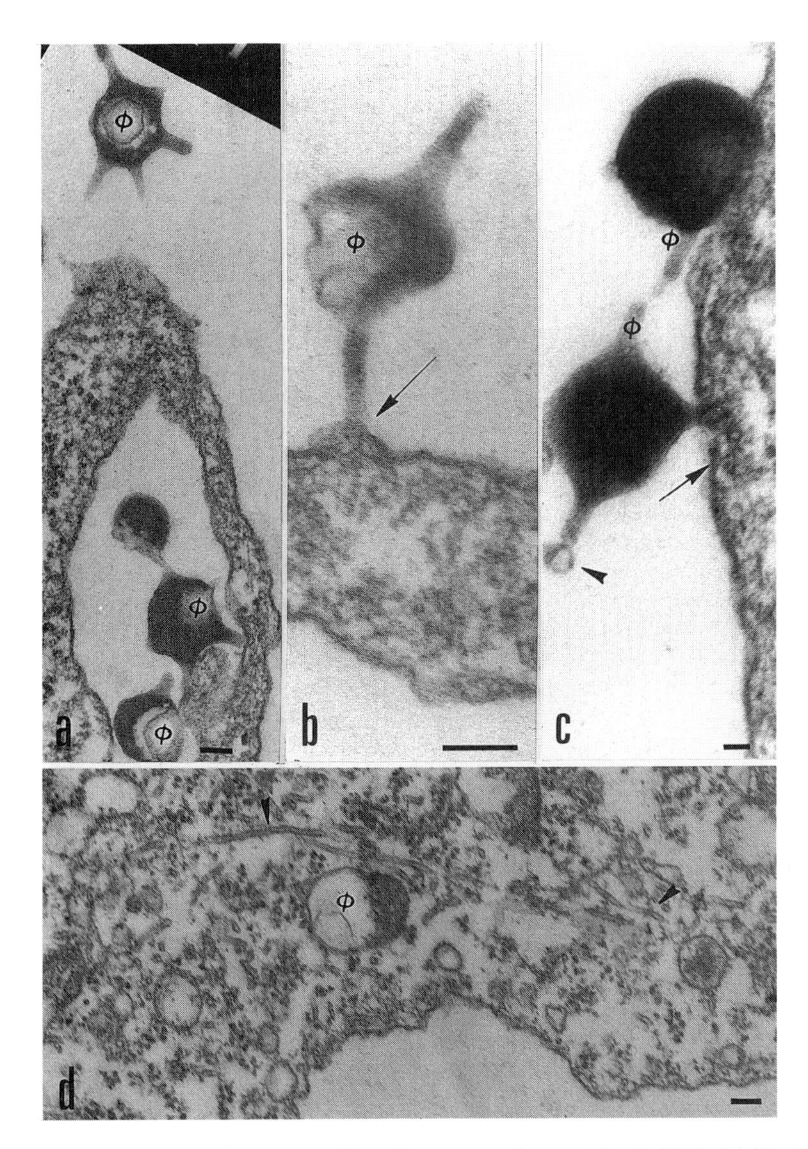

FIGURE 6. Electron micrographs of lamellocyte samples treated with VLPs (ϕ). The bar in each frame represents 0.1 μm. **(a)** Four VLPs near the surface of a lamellocyte that has folded, presumably during the pelleting or fixation processes. The upper VLP shows five surface spikes, the electron-dense outer region, and the internal electron-transparent cavity. The other VLPs in the frame illustrate the asymmetric organization of the electron-dense material of the particles. **(b)** Initial contact (*arrow*) of the VLP spike with the lamellocyte surface resulting in a blister or lysis of the plasma membrane. **(c)** The spike of the lower VLP in this frame is penetrating the lamellocyte and the continuity of the plasma membrane (*arrow*) is interrupted. The tip (*arrowhead*) of the VLP spike has a bleb. The main body of the other VLP is entering the cell. **(d)** A VLP free in the cytoplasm of a lamellocyte. Microtubules are indicated by arrowheads.

The entire VLP penetrates the lamellocyte surface without being enveloped by cell surface membranes. The particles are found free in the cytoplasm. No endocytic membrane surrounds the VLPs in the cytoplasm of the lamellocyte (FIG. 6d). Lamellocytes are not phagocytic,[50] and no evidence for VLP uptake by phagocytosis was seen for lamellocytes.

VLPs in the culture medium were phagocytosed by plasmatocytes and were found in phagocytic vesicles in these hemocytes (FIG. 7a). Destruction of VLPs in phagosomes of plasmatocytes was also seen (FIG. 7b), but no free VLPs were found in the cytoplasm of plasmatocytes. The attachment of the VLP to the lamellocyte surface and its penetration into the cell after dissolution of the plasma membrane suggest that the binding of the VLP surface spike uses a receptor site on the lamellocyte surface. Presumably, this receptor is absent in the plasmatocyte membrane and also in the inner surface of the phagosome membrane, so VLP contact with the plasmatocyte surface initiates the formation of VLP-resistant phagosomes. VLPs are confined to the phagosomes of plasmatocytes where they are inactivated or degraded. No harmful effects of VLPs on plasmatocytes have been observed. In lamellocytes, the VLPs remain free in cytoplasm. Thus, their destructive activity for the lamellocyte requires a free state in cytoplasm so that microtubule components are affected.

The difference in VLP uptake by plasmatocytes and lamellocytes agrees with earlier observations on the attachment and phagocytosis of bacteria by *Drosophila* hemocytes.[50] Bacteria injected into the hemocoel of a *Drosophila* larva adhere to surfaces of both plasmatocytes and lamellocytes, but are phagocytosed by the former

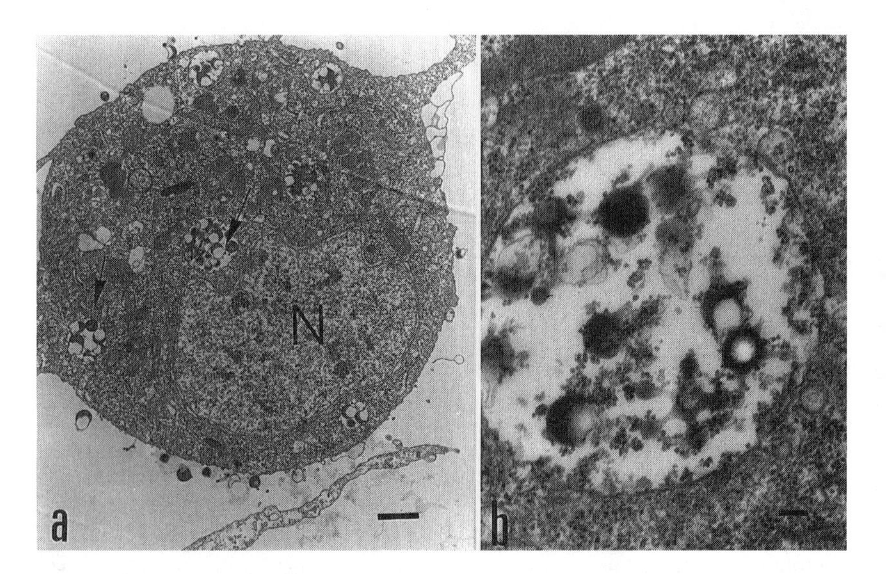

FIGURE 7. **(a)** A plasmatocyte containing several phagosomes (*arrows*) with VLPs. Also note the VLPs near the cell surface. N = nucleus. Note the lack of electron-dense material at the nuclear membrane in contrast to the nuclear membrane of the lamellocyte. Bar = 1 μm. **(b)** High magnification of a phagosome with VLPs being destroyed. Bar = 0.1 μm.

and not the latter. The differentiation of a spherical hemocyte to the discoidal lamellocyte must involve cell surface modifications[51] such that the flattened cells assume adhesive properties required for capsule formation and lose membrane fluidity needed for invagination of phagocytic cups.

The molecular nature of the *L. heterotoma* VLP has not been determined. The mature VLP isolated from the accessory gland reservoir has a complex structure. Some observations have been made on the distribution of particles in the long gland and its storage reservoir in the female wasp. The earliest recognizable stage in the formation of the VLP is found in the proximal region of the gland where large accumulations of 29-nm particles with electron-dense cores are found (FIG 8a). The small particles appear to enlarge and incorporate material that is less electron dense to give a vacuolar appearance to the particles approaching the neck of the gland connected to the duct that opens into the reservoir. Thin sections through the proximal region of the reservoir near its duct opening into the ovipositor contain VLPs with surface projections and an internal element (FIG. 8b) that resemble the gradient-purified VLPs that are active against lamellocytes *in vitro* (FIG. 8c).

The functions performed by the three types of *Drosophila* hemocytes are distinct: plasmatocytes are phagocytic, lamellocytes form capsule walls, and crystal cells are required for hardening and blackening reactions of the cellular capsule walls and wound healing.[50] Of these three cellular defense reactions, encapsulation threatens the survival of a parasite egg floating in the hemocoel. The phagocytic function of plasmatocytes and the melanization reactions of the crystal cells present no imminent danger to the parasite egg. These functions of the plasmatocytes and crystal cells remain intact in parasitized larvae and may even be beneficial in protecting the host against extraneous microbial infections and injury to the body wall.[27] Nevertheless, it is important that lamellocyte function be disrupted so that the parasite egg remains free of encapsulating hemocytes. This selective and destructive mission is accomplished by the VLPs that the female wasp injects along with its eggs into the host hemocoel.

SUMMARY

L. heterotoma females inject VLPs along with their eggs into the hemocoel of the *Drosophila* larva. The VLPs destroy lamellocytes that are potentially harmful for the parasite eggs, but the other types of host hemocytes retain their functions. The parasitized third instar *Drosophila* larvae continue to grow and pupariate, showing no outward adverse effects. Minimum disruption to the growth and development of the host is advantageous to the endoparasite egg, since survival of the endoparasite depends on the health of the host to the developmental stage at which the endoparasite begins feeding on host tissues.

The target of the VLPs must be microtubule components, because VLP entry into the lamellocyte induces depolymerization and repolymerization of microtubules. Microtubule rearrangement changes the discoidal lamellocyte to a bipolar cell. Concomitant with the modification in cell morphology, lamellocytes lose their surface adhesivity so they cannot adhere to a foreign object or to each other to form a capsule around endoparasite eggs. VLP-affected lamellocytes continue to elongate,

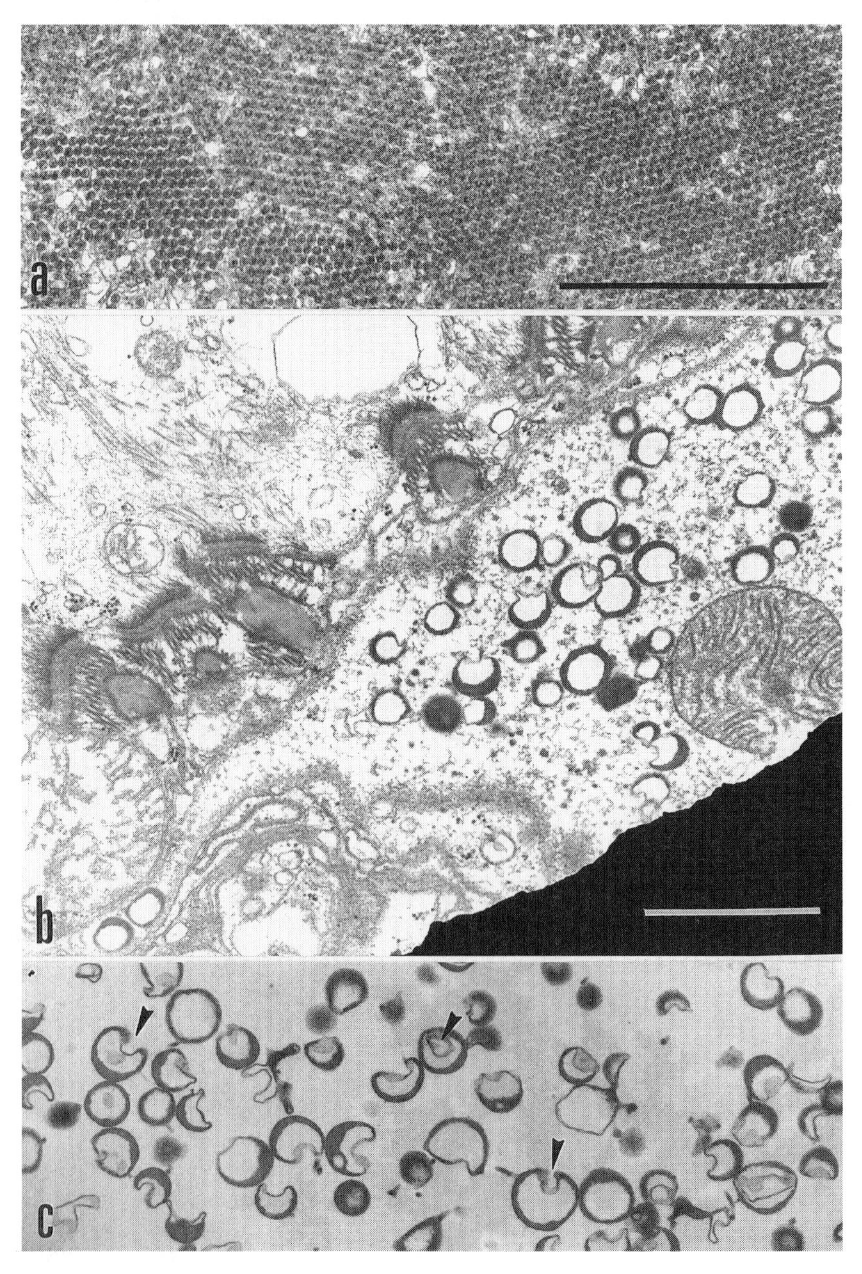

FIGURE 8. **(a)** Electron micrograph of a section passing through the proximal region of the long gland showing arrays of 29-nm particles with electron-dense centers. Bar = 1 μm. **(b)** A section passing through the proximal region of the long-gland reservoir near the duct that leads to the ovipositor. Note the profiles of the VLPs in section. Bar = 1 μm. **(c)** Section through a pellet of density gradient-purified VLPs. Internal structures can be seen in some of the VLPs, and the invagination can be seen in others (*arrowheads*). Compare these VLPs with those in frame b. Magnification is the same as b.

lose cytoplasmic contents at the poles as blebs full of microvesicles, and are eventually destroyed.

The molecular nature of the *L. heterotoma* VLP and the mechanisms underlying its specificity for lamellocyte cytoplasm are not known. An interesting consideration is the evolution of a particle whose selectivity for a host hemocyte protects the eggs of the parasitic insect. On the basis of information available at this time, it is clear that the *L. heterotoma* VLP is a useful tool for studying molecular aspects of microtubules and cytoskeleton.

REFERENCES

1. Schell, S. C. 1952. Trans. Am. Microsc. Soc. **71:** 293–302.
2. Poinar, G. O. 1968. Proc. Helminthol. Soc. Washington **35:** 161–169.
3. Klomp, H. & B. J. Teerink. 1978. Neth. J. Zool. **28:** 132–138.
4. Davies, D. H. & S. B. Vinson. 1986. J. Insect Physiol. **32:** 1003–1010.
5. Salt, G. 1973. Proc. R. Soc. London B. **183:** 337–350.
6. Bedwin, O. 1979. Proc. R. Soc. London B **205:** 267–270.
7. Rotherham, S. 1973. Proc. R. Soc. London B **183:** 179–194.
8. Feddersen, I., K. Sander & O. Schmidt. 1986. Experientia **42:** 1278–1281.
9. Berg, R., I. Schuchmann-Feddersen & O. Schmidt. 1988. J. Insect Physiol. **34:** 473–480.
10. Salt, G. 1980. Proc. R. Soc. London B **207:** 351–353.
11. Wago, H. & H. Kitano. 1985. Appl. Entomol. Zool. **20:** 103–110.
12. Kitano, H. 1986. J. Insect. Physiol. **32:** 369–375.
13. Rizki, T. M., R. M. Rizki & Y. Carton. 1990. Exp. Parasitol. **70:** 466–475.
14. Rouault, J. 1979. Compte rendu de l'Académie des Sciences, Paris. **289:** 643–646.
15. Carton, Y. & H. Kitano. 1981. Biol. J. Linn. Soc. **16:** 227–241.
16. Osman, S. E. & E. Führer. 1979. Int. J. Invert. Reprod. **1:** 323–332.
17. Tanaka, T. 1982. *In* The Ultrastructure and Functioning of Insect Cells. H. Akai, R. C. King & S. Morohoshi, Eds.: 169–172. Society of Insect Cells. Japan.
18. Edson, K. M., S. B. Vinson, D. B. Stoltz & M. D. Summers. 1981. Science **211:** 582–583.
19. Vinson, S. B., K. M. Edson & D. B. Stoltz. 1979. J. Invert. Pathol. **34:** 133–137.
20. Stoltz, D. B. & S. B. Vinson. 1979. Can. J. Microbiol. **25:** 207–216.
21. Fleming, J. G. W. & M. D. Summers. 1986. J. Virol. **57:** 552–562.
22. Blissard, G. W., J. G. W. Fleming, S. B. Vinson & M. D. Summers. 1986. J. Insect Physiol. **32:** 351–359.
23. Theilmann, D. A. & M. D. Summers. 1986. J. Gen. Virol. **67:** 1961–1969.
24. Stoltz, D. B. & D. Guzo. 1986. J. Insect Physiol. **32:** 377–388.
25. Davies, D. H., M. R. Strand & S. B. Vinson. 1987. J. Insect Physiol. **33:** 143–153.
26. Strand, M. R. & T. Noda. 1991. J. Insect Physiol. **37:** 839–850.
27. Rizki, R. M. & T. M. Rizki. 1984. Proc. Natl. Acad. Sci. USA **81:** 6154–6158.
28. Rizki, R. M. & T. M. Rizki. 1990. Proc. Natl. Acad. Sci. USA **87:** 8388–8392.
29. Rizki, T. M. & R. M. Rizki. 1992. Dev. Comp. Immunol. **16:** 103–110.
30. Tanaka, T. 1987. J. Insect Physiol. **33:** 413–420.
31. Webb, P. A. & M. D. Summers. 1990. Proc. Natl. Acad. Sci. USA **87:** 4961–4965.
32. Rizki, T. M. & R. M. Rizki. 1984. *In* Insect Ultrastructure. Vol. 2. R. C. King & H. Akai, Eds. Plenum Press. New York.
33. Rizki, T. M. & R. M. Rizki. 1986. *In* Hemocytic and Humoral Immunity in Arthropods. A. P. Gupta, Ed. John Wiley. New York.
34. Rizki, T. M. 1957. J. Morphol. **100:** 437–458.
35. Rizki, T. M. 1962. Am. Zool. **2:** 247–256.
36. Rizki, T. M. 1957. J. Morphol. **100:** 459–472.

37. Rizki, R. M. & T. M. Rizki. 1979. Differentiation **12:** 167–178.
38. Rizki, R. M. & T. M. Rizki. 1980. Wilhelm Roux' Arch. **189:** 207–213.
39. Rizki, T. M. & R. M. Rizki. 1959. J. Biophys. Biochem. Cytol. **5:** 235–240.
40. Rizki, T. M., R. M. Rizki & E. H. Grell. 1980. Wilhelm Roux' Arch. **188:** 91–99.
41. Rizki, R. M. & T. M. Rizki. 1990. J. Insect. Physiol. **36:** 523–529.
42. Walker, I. 1959. Rev. Suisse Zool. **66:** 569–632.
43. Nappi, A. J. 1975. Nature (London) **255:** 402–404.
44. Rizki, R. M. & T. M. Rizki. 1991. J. Exp. Zool. **257:** 236–244.
45. Rizki, R. M. & T. M. Rizki. 1990. Experientia **46:** 311–315.
46. Parness, J. & S. B. Horwitz. 1981. J. Cell Biol. **91:** 479–487.
47. Schiff, P. B. & S. B. Horwitz. 1980. Proc. Natl. Acad. Sci. USA **77:** 1561–1565.
48. Mole-Bajer, J. & A. S. Bajer. 1983. J. Cell Biol. **96:** 527–540.
49. Horwitz, S. B., L. Liao, L. Greenberger & L. Lothstein. 1988. *In* Resistance to Antineoplastic Drugs. D. Kessel, Ed.: 109–125. CRC Press. Boca Raton, FL.
50. Rizki, T. M. & R. M. Rizki. 1980. Experientia **36:** 1223–1226.
51. Rizki, T. M. & R. M. Rizki. 1983. Science **220:** 73–75.

Ciliate Pheromones as Early Growth Factors and Cytokines[a]

PIERANGELO LUPORINI,[b] ADRIANA VALLESI,[b]
CRISTINA MICELI,[b] AND RALPH A. BRADSHAW[c]

[b]*Department of Molecular, Cellular and Animal Biology*
University of Camerino
62032 Camerino (MC), Italy

[c]*Department of Biological Chemistry*
California College of Medicine
University of California
Irvine, CA 92717

INTRODUCTION

In the past decade, we have experienced a veritable explosion of knowledge concerning the molecules that regulate self/non-self recognition in more advanced animals, in particular the peptide hormones and cytokines responsible for cell proliferation and differentiation in mammals. Nevertheless, it is apparent that the evolutionary origins of these molecules, as well as the nature of many of the elementary processes in which they are involved, are still poorly understood. It is widely held, however, that not only are they universally synthesized throughout the animal kingdom, because the organization and integrity of the animal body depend on their activity, but also, quoting Ohno,[1] "Unicellular eukaryotes have no choice but to depend upon autocrine growth factors."

The molecular basis for the control and expression of ciliate "mating types"—morphologically identical, yet genetically distinct intraspecies classes of somatic cells—is now providing substantial experimental evidence for the hypothesis[2] that this genetic mechanism has evolved primarily to provide cells with self-recognition devices, implying an acquired cell capability to produce and, at the same time, respond to signals in an autocrine fashion. The evidence here summarized, including recent findings, suggests intriguing structure–function relationships of some of these ciliate cell type-specific signals, or pheromones, with some mammalian peptide hormones and cytokines that first appeared in animal evolution.

CILIATE PHEROMONE IDENTIFICATION

Ciliates live as individual cells that proliferate mitotically for most of their clonal life cycle. They only temporarily interrupt this proliferation in coincidence with a sexual event, which most commonly involves the formation of bicellular complexes

[a]This research was supported by the Italian Ministero dell' Università e Ricerca Scientifica e Tecnologica and CNR (P. L.), and the U.S. Public Health Service (R. A. B.)

of mating cells leading to the exchange of genetic material, before separating again to reinitiate the life cycle.

Because of their association with mating induction activity, ciliate pheromones have been identified in the extracellular environment, and for years referred to as "gamones" because it was assumed that they behave as sex factors.[3–4] Mating pairs can in fact be quickly formed by a clonal cell culture after suspension with a pheromone-containing supernatant of other cells, and using this biological assay, pheromones have been isolated from several ciliates, first *Blepharisma*[3] and then *Euplotes*[5–7] and *Dileptus.*[8] Their chemical nature has been shown to be proteinaceous in all but one striking case, that represented by gamone-2 of *B. japoncum,* which was reported to be a tryptophan-related molecule.[3]

E. RAIKOVI PHEROMONES AND THEIR GENETIC CONTROL

The system of our choice for the study of the structure and biology of ciliate pheromones is the ubiquitous, bottom-dwelling marine species *Euplotes raikovi.* Like other hypotrichs (that constitute an advanced group of ciliates), *E. raikovi* appears to have a virtually unlimited number of different mating types, and hence a capacity for producing as many different pheromones.[2] Each one is constitutively secreted from the very beginning of the clonal life cycle[2,9] and is synthesized under the control of one of a polyallelic series of genes codominantly expressed at the *mat* (mating-type) locus.[2,10]

For the degree of polymorphism, this locus does not have any counterparts in other eukaryotic single-cell organisms; instead, at least superficially, a comparison can be found with the gene complexes regulating self/non-self recognition phenomena in mushrooms,[11] plants,[12] and animals.[13] The mechanism that generates the *mat* locus polymorphism in *E. raikovi* remains to be elucidated, possibly by analysis of its structure in the genome that is unexpressed in the cell germ-line nucleus (or micronucleus).

So far, only the linear pheromone DNA sequences contained in the transcriptionally active cell somatic nucleus (or macronucleus) have been characterized.[14,15] They consist of 1100 bp and show an open reading frame encoding a 75-amino-acid precursor sequence, or prepropheromone. The pre, or signal, peptide is virtually identical in all precursors and is formed by 19 residues; the pro segment of 16–18 residues is more variable; and the mature, secreted sequence of 37–40 residues is quite diversified.

E. RAIKOVI PHEROMONE STRUCTURE

The complete amino acid sequences of seven *E. raikovi* pheromones have been determined to date and, as shown in FIGURE 1, five of them (E*r*-1, E*r*-2, E*r*-10, E*r*-11, and E*r*-20) are unique.[16] Their molecular masses (as calculated from the sequence and measured by mass spectrometry) vary from 3995.6 (E*r*-20) to 4410.9 (E*r*-1). In an optimized alignment, only seven residues of the 37 to 40 forming the mature sequence are common: the amino-terminal aspartic acid, which may reflect structural

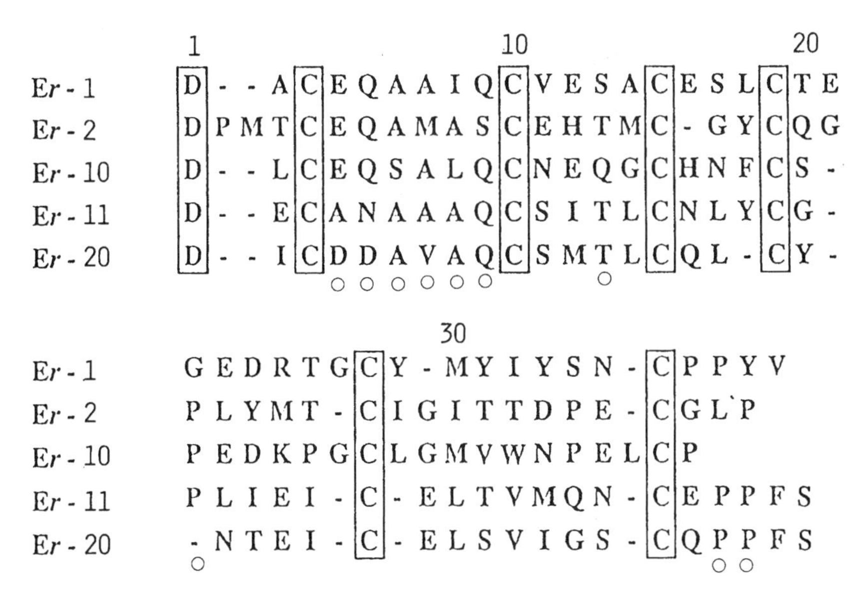

FIGURE 1. Comparison of the amino acid sequences of five pheromones from *E. raikovi.* Alignment has been maximized through insertion of gaps. The seven identical residues are enclosed in boxes, and amino acids shared by at least three sequences are marked by circles. (From Raffioni *et al.*[16] Used with permission from the *Proceedings of the National Academy of Sciences.)*

requirements for the enzyme processing the propheromone to the secreted form, and six half-cystines. There are 10 other sites in which at least three pheromones share a common residue; interestingly, seven of these are clustered in the amino-terminal region, which may thus represent a region of the molecule that is more extensively conserved and thus probably involved in a common function. In contrast, the carboxyl-terminal region appears to be quite variable and specific for each pheromone. Starting from Cys III (residue 15 in the alignment given in FIG. 1), there are only three positions with a common residue in at least three sequences (discounting the Cys residues themselves), and one to five residues may form the COOH-terminal segment extending beyond Cys VI. These sequence variations may reflect a function for the carboxy-terminal half that is more closely related to unique properties of each pheromone, such as receptor interactions. In general, the percentages of sequence identities range from only 23% to 56%, with the majority in a 25 to 30% range. Inasmuch as the amino-terminal aspartic acid and the six half-cystines contribute nearly half the identical residues, these levels of similarity are obviously minimal for a family of homologous polypeptides controlled by allelic genes.

Although markedly varied in primary structure, the main chain conformation is most likely shared in the different pheromones. The assignment of the disulfide bond pairs in two of the most sequentially divergent pheromones, E*r*-1 and E*r*-2, has in fact revealed an identical pairing pattern.[17] As illustrated in FIGURE 2, this pattern, which is likely to occur throughout the *E. raikovi* pheromone family, implies a

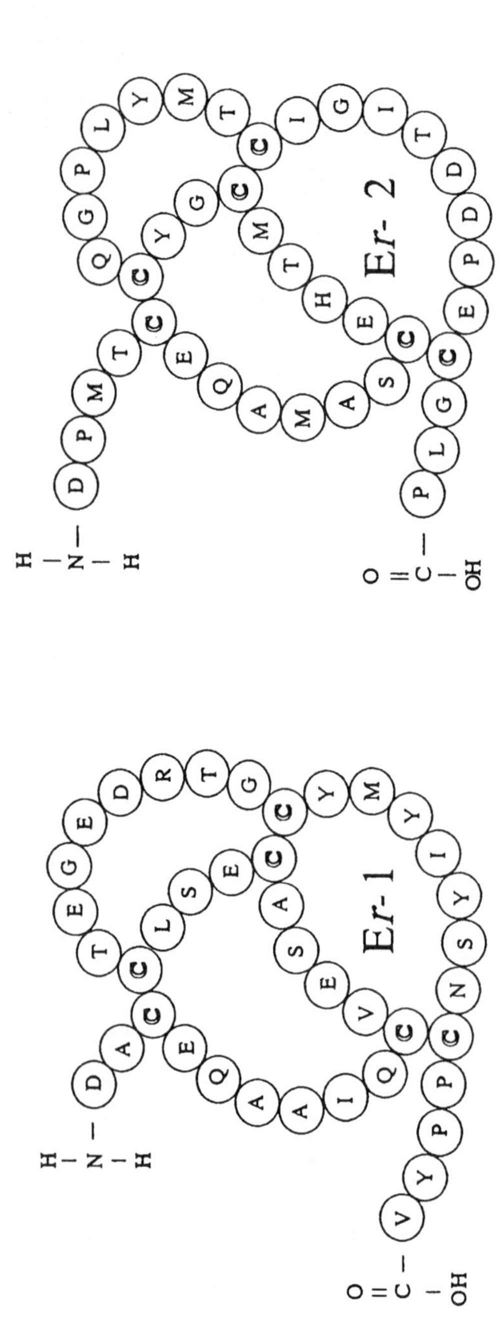

FIGURE 2. Diagrammatic representation of the amino acid sequences of the pheromones E*r*-1 and E*r*-2 that emphasizes the basic three-loop structure of *E. raikovi* pheromones. (From Stewart *et al.*[17] Used with permission from *Protein Science.*)

three-dimensional organization essentially defined by three loops, resembling that typically formed in other stable, small disulfide-rich proteins[18] such as EGF and TGF-α. The first loop is generated by the Cys(I)-Cys(IV) link and involves most of the less variable amino-terminal region; the second and third loops are closed by the Cys(III)-Cys(V) and Cys(II)-Cys(VI) links, respectively, and contain the main variations in size and distribution of hydrophobic residues.

NMR analyses of the pheromone three-dimensional structure in solution,[19] so far completed for pheromone E*r*-10 and in progress for E*r*-1 and E*r*-2, have now shown that these disulfide bridges are strategically placed to hold together four main elements of the pheromone structure: three short helices (6–9 residues long) reciprocally oriented with nearly parallel axes and a turn in the carboxy-terminal region. Additional relevant information on the pheromone conformation is expected from the completion of X-ray analyses of crystals produced for pheromone E*r*-1.[20] On the basis of an initial characterization, these crystals diffract to at least a 2.8-Å resolution and have unit cell sizes consistent with pheromone tetramers in the asymmetric unit.

E. RAIKOVI PHEROMONE ACTIVITY AND RECEPTORS

As already noted, ciliate pheromones have for years been thought of as diffusible sex factors, and, as such, capable of eliciting a biological response *only* in cells other than those of origin through binding to complementary cell receptors.[3,4] In contradiction to this widely held view, it was proposed that ciliate pheromones are, instead, self-recognition signals and primarily elicit cell responses through binding to autocrine cell receptors.[2] Because they have similar structural features, they were assumed to be capable of competing with each other in cross-binding reactions responsible for mating pair formation. To substantiate this model, we initiated studies to determine (i) if biological effects are promoted by the pheromone interactions with the cells of origin and, if so, which ones and (ii) the nature of the putative pheromone receptors involved in these interactions.

Experimental evidence has now been obtained that *E. raikovi* pheromones promote DNA synthesis and multiplication with their cells of origin; that is, these molecules behave like autocrine growth factors (A. Vallesi *et al.*, manuscript in preparation). These findings are based on observations that cells deprived of the secreted pheromone for one to two days (by repeated washing and suspension with fresh medium) and then fed, incubated with nanomolar concentrations of self pheromone, and exposed to pulses of [^{3}H]thymidine, start incorporation of the radionuclide earlier and at rates higher than cells incubated without self pheromone or with a foreign (non-self) pheromone. These effects are followed by increased cell densities and are, to a large extent, pheromone-concentration dependent.

To identify the pheromone receptors, binding and cross-linking experiments have been carried out with radioiodinated samples of pheromones E*r*-1 and E*r*-2 and cells of type I, II, and X (secreting pheromone E*r*-1, E*r*-2, and E*r*-10, respectively).[9,21] In every pheromone/cell combination (either homologous, such as between E*r*-1 and type-I cells, or heterologous, such as between E*r*-1 and type-II cells), the binding reactions were found to be saturable, specific, and reversible, involving a single class of binding sites with a K_D of 10^{-9} M. Initial molecular characterization of these sites was derived from cross-linking experiments in which interactions between radio-labeled pheromones and solubilized cell membranes were covalently stabilized and the molecular complexes were resolved by SDS polyacrylamide gel electrophoresis and autoradiography. One major band at about 38 kDa was always evident. The quaternary structure of the soluble pheromones has not been fully resolved; however, gel filtration data suggest that they may occur in solution as homodimers. In view of the fact that the pheromones are likely to participate in a 1:1 ratio with the binding sites in the formation of the complexes responsible for the 38-kDa band, the binding protein may be estimated to be about 28 kDa.

Analysis of pheromone receptor preparations eluted from an affinity column (prepared with pheromone E*r*-1 and loaded with solubilized membrane of type-I cells) has now revealed (C. Ortenzi *et al.*, manuscript in preparation) that at least one other component of about 14 kDa coexists with the 28-kDa species. In addition, both components have been shown capable of recognizing and binding the pheromone in ligand blotting experiments.

We have so far been unable to determine the sequence of this affinity-purified material by direct chemical analysis and thus learn if one component represents the dimeric form of the other, as seems probable. Nevertheless, insight into the possible structure of the 14-kDa component has come from a repeated screening of restricted cDNA libraries constructed from RNA enriched in pheromone E*r*-1 mRNA.[22] These libraries have been found to contain clones, one of which (p5/6) has been extensively characterized. It is apparently identical with the sequence encoding E*r*-1 except for two regions: one characterized by an insertion and the other by a deletion. As shown in FIGURE 3, the insertion consists of a sequence of 89 nucleotides and is located in the 5′ half of the molecule; the deletion is defined by a sequence of 78 nucleotides which, in the E*r*-1 cDNA clones, is inserted close to the polyadenylation signal of the 3′ noncoding region. Because of the insertion, a longer reading frame appears in clone p5/6 with respect to that of E*r*-1 cDNA clones. The result is the translation of a new polypeptide of 130 amino acids, with a calculated M_r of 13,730. Of the 130 residues, the 75 at the carboxy terminus are identical with those forming the E*r*-1 precursor (prepro-E*r*-1); the 55 residues at the amino terminus constitute a new sequence.

As shown in other cell systems,[23] a structurally important effect of incorporating the full sequence of the pheromone precursor into the larger product of 130 residues is likely to be the conversion of the pheromone signal peptide, normally expected to be removed during the translocation process, into a trans-membrane domain. Thus the alternately spliced sequence could encode a membrane-bound protein of 13.7 kDa, which is the estimated mass of the second receptor component purified from affinity chromatography. Evidence for this and, in particular, for a type-II orientation of the membrane-bound pheromone isoform so as to ensure that the sequence corresponding to the soluble pheromone is on the extracellular side of the cell membrane has been principally derived from immunoblot analyses carried out on solubilized cell membranes fractionated by SDS-electrophoresis and exposed to antibodies generated to soluble E*r*-1. One major band of 14 kDa was constantly apparent.

With regard to the origin of the transcripts for the secreted and membrane-bound pheromone forms, the extensive sequence identified strongly argues for a control of

```
5′ TAGCCTATAATGTCATAGA ATG GGA TGT ACA TCA GAT CTT TGT CAC ATG GCA TCA GTG
                       Met Gly Cys Thr Ser Asp Leu Cys His Met Ala Ser Val
                       1                                   10

TTA AGT AAA TAT CAG AGC CTA TTT CTA TAT CTG TGC CGT TCA AAT AAC TGT GTT GGT CCG TTA
Leu Ser Lys Tyr Gln Ser Leu Phe Leu Tyr Leu Cys Arg Ser Asn Asn Cys Val Gly Pro Leu
                        20                                  30

AAC TCT ATC AAT AGG TTC TTC TCT GGC ACA GTA CAT GTC ATA AGT TTT AGT ATC AAT TTT AGA
Asn Ser Ile Asn Arg Phe Phe Ser Gly Thr Val His Val Ile Ser Phe Ser Ile Asn Phe Arg
                    40                                      50

ATG AAC AAA CTA GCA ATT CTC GCT ATC ATC GCT ATG GTA CTC TTC AGC GCC AAC GCC▽TTC AGA
Met Asn Lys Leu Ala Ile Leu Ala Ile Ile Ala Met Val Leu Phe Ser Ala Asn Ala Phe Arg
                60                                      70

TTC CAA AGC AGA TTG AGA TCA AAT GTA GAA GCT AAG ACA GGA▼GAT GCT TGT GAG CAA GCT GCA
Phe Gln Ser Arg Leu Arg Ser Asn Val Glu Ala Lys Thr Gly Asp Ala Cys Glu Gln Ala Ala
            80                                      90

ATC CAG TGT GTT GAG TCA GCA TGT GAA AGT CTT TGT ACA GAA GGT GAA GAT AGA ACT GGC TGC
Ile Gln Cys Val Glu Ser Ala Cys Glu Ser Leu Cys Thr Glu Gly Glu Asp Arg Thr Gly Cys
        100                                     110

TAT ATG TAC ATC TAT TCT AAC TGC CCA CCT TAT GTC TAA ATCTAACTACTTTAGAATCAAACCAAATTCG
Tyr Met Tyr Ile Tyr Ser Asn Cys Pro Pro Tyr Val Stop
    120                                     130

GACAAGTTGTAGTTAACACTTACTCTGGAGTTACTTGACAAGAACTAGGATAACTTTATTGAACAAAGTTCCATCT   3′
```

FIGURE 3. Complete coding sequence and part of the flanking noncoding (regulatory?) regions of the macronuclear pheromone E*r*-1 gene of *E. raikovi.* The two framed sequences, one close to the 5′ end and the other close to the 3′ end, are alternately removed to generate two mRNAs. One of these specifies the 75-amino acid sequence of the prepropheromone; the other specifies the 130-amino acid sequence of the membrane-bound pheromone form. The two initiation translation ATG codons are boxed. The sites for the co- and post-translational processing of the pre and pro segments are indicated by an open and a solid arrow, respectively. Based on results from Miceli *et al.*[22]

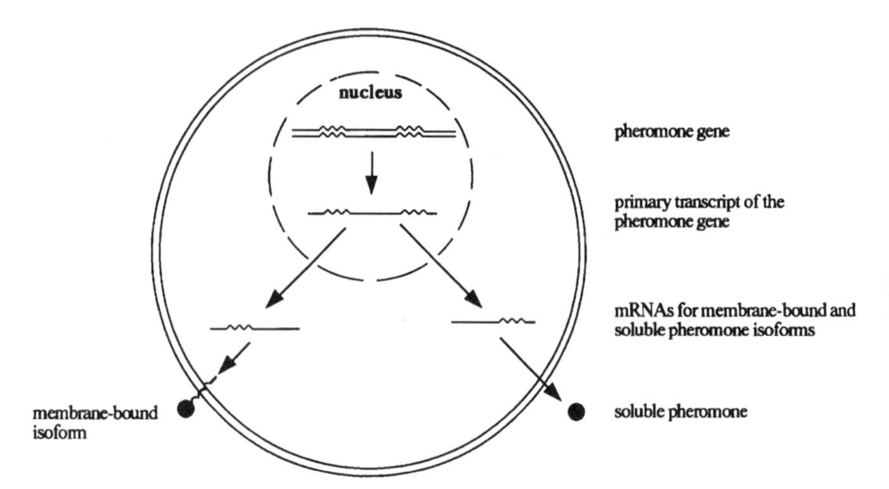

FIGURE 4. Diagrammatic representation of the control and expression of soluble and membrane-bound pheromone isoforms produced by a single cell of *E. raikovi.* The latter form is supposed to function as a binding site of the former. Based on results from Miceli *et al.*[22]

both by the same gene, through a mechanism of alternative elimination of intron-like sequences. As diagrammed in FIGURE 4, the experimental evidence accumulated to date is consistent with a process of post-transcriptional RNA splicing, like those described in a number of mammalian cells, for the production of membrane-bound and secreted forms of growth factors[24] and immunoglobulins.[25] A single gene, which in type-I cells has been estimated to be amplified to about 20,000 copies in the cell macronucleus (A. La Terza *et al.,* manuscript in preparation), would thus contain the complete information for two mRNAs, one for the secreted pheromone and the other for the membrane-bound pheromone form.

Conclusive evidence that this latter form is the same as that identified in cross-linking experiments (presumably as a dimeric form of 28 kDa) and purified by affinity chromatography has not yet been achieved (by direct chemical analysis of purified protein). However, a model, in which homologous binding reactions between membrane-bound and soluble pheromone isoforms regulate the autocrine mechanism of self recognition and activation to growth of *E. raikovi,* receives further credibility from the knowledge that a number of membrane proteins oriented with the carboxy terminus on the outside and the amino terminus on the inside of the cell membrane serve to bind specific proteins on their exterior domains, like the transferrin or the asialoglycoprotein receptors.[26]

RELATION TO MAMMALIAN CYTOKINES

Among the proteins that were assayed for their capability to inhibit autocrine pheromone-binding reactions in *E. raikovi,* human IL-2 and, to a lesser extent, other prototypic growth factors such as EGF were found to be very effective inhibitors.[21]

A 50-fold excess of IL-2 inhibited at least 90% of autocrine E*r*-1 binding reactions. In fact, a sequence comparison between IL-2 and pheromone E*r*-1 revealed intriguing similarities in two sequence segments, each surrounding one of the two IL-2 half-cystine residues required for the biological activity of this molecule.[27] One segment, presumably more important (as it also appears to be similarly maintained in different pheromones), is formed by Ile(8)-Gln-Cys-Val-Glu in E*r*-1 and Leu(56)-Gln-Cys-Leu-Glu in IL-2.

The exciting implication of a structural relatedness between *E. raikovi* pheromones and mammalian cytokines raised by this finding has recently been further strengthened by the results (A. Vallesi *et al.*, manuscript in preparation) of binding experiments of these two molecules to both *E. raikovi* cells and lymphocytes of the CTLL-2 line (which are known to be totally dependent on IL-2 for survival and consistently express a large number of IL-2 receptors). Radiolabeled pheromone E*r*-1 is capable of binding specifically to CTLL-2 cell proteins apparently equivalent in M_r with the low- and high-affinity IL-2 receptors and, reciprocally, radiolabeled IL-2 binds specifically to the *E. raikovi* 14-kDa, membrane-bound isoform. Detailed comparisons of the three-dimensional structures of these molecules, in combination with analyses of the domains of biological activity of *E. raikovi* pheromones, should help to define the sequences involved in receptor interactions.

PERSPECTIVES

The identification of growth factor/cytokine-like activity for the *E. raikovi* pheromones suggests a heretofore unsuspected link in the evolutionary development of extracellular signal transduction and emphasizes the growing importance being attached to autocrine regulation mechanisms. It should be stressed, however, that there is, as yet, no obvious, significant structural connection between these pheromones and any known growth factor. The very limited sequence relatedness noted with the IL-2 sequence only serves to pinpoint a probable common structure that is apparently recognized by both receptor systems. Because this similarity does not extend beyond two regions, it is unlikely that it represents a homologous relationship developed through divergent evolution and may be entirely fortuitous. Careful examination of the three-dimensional structures should be done before arriving at a final conclusion, as very distant relationships no longer detectable at the primary structure level can sometimes be found at the structural level, as seen, for example, in cytokines[28] and the NGF/TGFβ/PDGF superfamily.[29] Even if no evolutionary relationship is detected, it would not be surprising if a true homologue(s) for the ciliate pheromones is eventually found among the polypeptide growth factors of higher eukaryotes.

The identification of an apparent membrane-bound form of the pheromone E*r*-1 (and presumably the other *E. raikovi* pheromones as well) also provides a strong link to the growth factors. A similar form has been found for EGF (predominantly expressed in the kidney)[30] as well as TGFα and other members of the family.[31] Similar observations have been made for colony-stimulating factor[32] and tumor necrosis factor α.[33] These are also synthesized in such a way as to place the growth factor sequence on the outside of the cell, and it has been suggested that they represent examples of juxtacrine mechanisms,[34] in which both the ligand and the receptor (presumably on separate cells) are membrane bound. It should be emphasized that the alternately spliced forms of the pheromones may also participate in such interactions. However, the present evidence favors the view that they act as receptors for the soluble pheromones, which if true represents, to our knowledge, the first example of an extracellular ligand and its receptor being derived from the same gene. It has been suggested previously[35] that present-day growth factors and their receptors may have also evolved in such a fashion, and this would provide strong evidence in support of that concept.

REFERENCES

1. OHNO, S. 1987. Immortal Genes. *In* Oncogenes and Growth Factors. R. Bradshaw & S. Prentis, Eds.: 106–114. Elsevier. Amsterdam.
2. LUPORINI, P. & C. MICELI. 1986. Mating pheromones. *In* The Molecular Biology of Ciliated Protozoa. J. G. Gall, Ed.: 263–299. Academic Press. New York.
3. MIYAKE, A. 1981. Cell interaction by gamones in *Blepharisma. In* Sexual Interaction in Eukaryotic Microbes. D. H. O'Day & P. A. Horgen, Eds.: 95–129. Academic Press. New York.
4. KUHLMANN, H. W. & K. HECKMANN. 1991. J. Exp. Zool. **251:** 316–328.
5. MICELI, C., A. CONCETTI & P. LUPORINI. 1983. Exp. Cell Res. **149:** 593–598.
6. WEISCHER, A., M. FREIBURG & K. HECKMANN. 1985. FEBS Lett. **191:** 176–180.
7. AKADA, R. 1986. J. Exp. Zool. **237:** 287–290.
8. PARFENOVA, E. V., S. AFONKIN, A. L. YUDIN & R. N. ETINGOF. 1989. Acta Protozool. **28:** 11–21.
9. LUPORINI, P., C. MICELI, C. ORTENZI & A. VALLESI. 1992. Dev. Genet. **13:** 9–15.
10. LUPORINI, P., S. RAFFIONI, A. CONCETTI & C. MICELI. 1986. Proc. Natl. Acad. Sci. USA **83:** 2889–2893.
11. KUES, U. & L. A. KASSELTON. 1993. J. Cell Sci. **104:** 227–230.
12. LEWIS, D. 1976. Incompatibility in flowering plants. *In* Receptors and Recognition. P. Cuatrecasas & M. F. Greaves, Eds.: 164–198. Chapman and Hall. London.
13. TROWSDALE, J. 1993. Trends Genet. **9:** 117–122.
14. MICELI, C., A. LA TERZA & M. MELLI. 1989. Proc. Natl. Acad. Sci. USA **86:** 3016–3020.
15. MICELI, C., A. LA TERZA, R. A. BRADSHAW & P. LUPORINI. 1991. Eur. J. Biochem. **202:** 759–764.
16. RAFFIONI, S., C. MICELI, A. VALLESI, S. K. CHOWDHURY, B. T. CHAIT, P. LUPORINI & R. A. BRADSHAW. 1992. Proc. Natl. Acad. Sci. USA **89:** 2071–2075.
17. STEWART, A. E., S. RAFFIONI, T. CHAUDHARY, B. T. CHAIT, P. LUPORINI & R. A. BRADSHAW. 1992. Prot. Sci. **1:** 777–785.
18. RICHARDSON, J. 1981. Adv. Prot. Chem. **34:** 167–339.
19. BROWN L., R. S. MRONGA, R. A. BRADSHAW, C. ORTENZI, P. LUPORINI & K. WUTHRICH. 1993. J. Mol. Biol. **231:** 800–816.
20. ANDERSON, D., S. RAFFIONI, P. LUPORINI, R. A. BRADSHAW & D. EISENBERG. 1990. J. Mol. Biol. **216:** 1–2.
21. ORTENZI, C., C. MICELI, R. A. BRADSHAW & P. LUPORINI. 1990. J. Cell Biol. **111:** 607–614.
22. MICELI, C., A. LA TERZA, R. A. BRADSHAW & P. LUPORINI. 1992. Proc. Natl. Acad. Sci. USA **89:** 1988–1992.
23. WICKNER, W. T. & H. F. LODISH. 1985. Science **230:** 400–407.
24. EDWARDS, R. H., M. J. SELBY & W. J. RUTTER. 1986. Nature **319:** 784–787.
25. WALL, R. & M. KUEHL. 1983. Ann. Rev. Immunol. **1:** 393–422.
26. SINGER, S. J. 1990. Annu. Rev. Cell Biol. **6:** 247–296.
27. JU, G., L. COLLINS, K. L. KAFFKA, W. -H. TSIEN, R. CHIZZONITE, R. CROWL, R. BHATT & P. L. KILIAN. 1987. J. Biol. Chem. **262:** 5723–5731.

28. MILBURN, M. V., A. M. HESSEL, M. H. LAMBERT, S. R. JORDAN, A. E. I. PROUDFOOT, P. G. GRABER & T. N. C. WELLS. 1993. Nature **363:** 172–176.
29. MURRAY-RUST, J., N. Q. MCDONALD, T. L. BLUNDELL, M. HOSANG, C. DEFNER, F. WINKLER & R. A. BRADSHAW. 1993. Stucture **1:** 153–159.
30. RALL, L. B., J. SCOTT, G. I. BELL, R. J. CRAWFORD, J. D. PENSCHOW, D. H. NIALL & J. P. COGHLAN. 1985. Nature **313:** 228–231.
31. PANDIELLA, A. & J. MASSAGUÉ. 1993. J. Biol. Chem. **266:** 5764–5773.
32. RETTENMIER, C. W., M. F. ROUSSEL, R. A. ASHMUN, P. RALPH, K. PRICE & C. J. SCHERR. 1987. Mol. Cell. Biol. **7:** 2378–2387.
33. KRIEGLER, M., C. PEREZ, K. DEFAY, I. ALBERT & S. D. LU. 1988. Cell **53:** 45–53.
34. MASSAGUÉ, J. 1990. J. Biol. Chem. **265:** 21393–21396.
35. PFEFFER, S. & A. ULLRICH. 1985. Nature **313:** 184.

Invertebrate Cytokines[a]

GREGORY BECK AND GAIL S. HABICHT

Department of Pathology
State University of New York at Stony Brook
Health Sciences Center
Stony Brook, New York 11794-8691

INTRODUCTION

The ability to defend oneself against invaders is key to survival. Invertebrate animals have evolved myriad ways of protecting themselves from external insults.[1] Phagocytosis is a dominant mechanism by which invaders are eliminated, while antisomes, and antibacterial mediators are examples of other mechanisms in the diverse world of invertebrate inflammatory responses.[1]

In the course of evolution, vertebrates have come to rely on immunoregulatory molecules to direct both humoral and cellular immune responses. These "cytokines" are polypeptide mediators released by a variety of activated immune and nonimmune cells. These molecules have critical effects on cells of the immune system, and they exhibit hormone-like properties that affect numerous organ systems involved in host defense.[2] Interleukin (IL)-1, IL-6, and tumor necrosis factor (TNF) are major immunoregulatory cytokines with many host defense–related properties.[3] We have isolated and characterized IL-1 and TNF-like molecules from several invertebrate phyla.[2,4]

Here we report our preliminary investigations on the isolation and characterization of an invertebrate IL-6-like molecule. By studying defense mechanisms of invertebrates and their regulatory cytokines, we hope to unravel the complex web of interactions among cells and factors of vertebrate immune responses.

MATERIALS AND METHODS

Materials

Pyrogen-free water and pyrogen-free saline were obtained from Travenol (Deerfield, IL). Sterile, pyrogen-free syringes and needles were obtained from Becton-Dickinson (Rutherford, NJ). All plasticware was obtained from Falcon (Oxnard, CA). All tissue culture media were obtained from Flow Laboratories (McLean, VA). All other reagents were of analytical grade or better and were obtained from Sigma (St. Louis, MO) or Fisher (Springfield, NJ).

[a]This work was supported by a grant from the National Science Foundation (DCB 8810448).

Biochemical Characterization

(1) Collection of starfish. The starfish *Asterias forbesi* were obtained from the waters of Long Island sound at Port Jefferson, New York. All animals were used within 24 hr of collection.

(2) Isolation of coelomic fluid. Coelomic fluid was obtained from the coelom of the starfish by cutting off the tip of the arm with scissors and draining the fluid into sterile beakers as previously described.[4]

(3) Coelomic fluid preparation. The coelomic fluid was thawed and clarified by centrifugation at 2,000 × *g* for 20 min at 4°C. The fluid was then concentrated approximately 20-fold with an ultrafiltration stirred cell equipped with a diaflo ultrafilter having a nominal M_r exclusion limit of 10,000 (PM 10, Amicon, Lexington, MA). The crude concentrate was dialyzed against phosphate-buffered saline (pH 7.4) (PBS), which contained 10 μg/ml of gentamicin.

(4) Size-exclusion chromatography. Samples were applied to a G50 column (2.6 × 85 cm) in a volume of 5 ml and eluted with PBS at a flow rate of 1.2 ml/min at 4° as described in detail elsewhere.[4]

(5) Preparative isoelectric focusing. Concentrated G50 column eluates containing peaks of B9 cell proliferation activity (see below) were separated on a Rotofor cell (BioRad, Rockville Center, NY) according to the manufacturer's instructions.

Assays for IL-6 Activity

We used the B9 assay[5] for the characterization of IL-6-like activity from coelomic fluids. The B9 cell line is an IL-6-dependent murine B-cell hybridoma.[5] It is maintained in RPMI 1640 containing 10% heat-inactivated FCS, 5×10^{-5} M 2-mercaptoethanol, and ~2–5 pg/ml IL-6 (Genzyme, Boston, MA). The IL-6 assay is performed by washing the cells in RPMI 1640 three times to remove IL-6, and resuspending the cells at a concentration of 2×10^4 cells/ml in RPMI 1640 containing 20% heat-inactivated FCS, 10^{-4}M 2-mercaptoethanol, and antibiotics. The sample to be assayed (or a standard curve of human IL-6) is added to each well of a 96-well tissue culture plate in a volume of 100 μl. Control wells receive medium alone. One-hundred microliters of the cell suspension is added, and the cells are incubated at 37°C in a 5% CO_2 in air incubator for 72 hr. In an effort to eliminate the cost and hazards of radioisotopes, we have recently used a modification of the MTT assay[6] (CellTiter 96™, Promega, Madison, WI) to measure cellular proliferation. The preparation and harvesting of the plates were done exactly as described by the manufacturer.

RESULTS AND DISCUSSION

Our laboratory has been investigating the cellular host response and its control by cytokines. Cytokines include the interferons (IFNα, β, γ), the interleukins (IL; to date 13 have been described), tumor necrosis factor and lymphotoxin (TNFα and β), the colony-stimulating factors (G-CSF, M-CSF, and GM-CSF), and transforming growth factor (TGF) β.[2,3] The wide-ranging ability of cytokines to initiate and regulate host defense responses, and the critical need of the vertebrate body for these molecules, suggested to us that these molecules have been highly conserved through evolution. We reasoned that IL-1 or a similar ancestral cytokine is likely to be present in the invertebrates for several reasons. First, the molecular conservation of IL-1 and its host defense functions have been preserved in the vertebrates. Second, the cells that produce IL-1, that is, macrophages, are present throughout the animal kingdom. Echinoderms and urochordates belong to the deuterostomes as do the vertebrates, while all the other invertebrates are members of the protostomes. Our studies have demonstrated the presence of cytokines in both major groups of invertebrates.

In the early 1970s, several investigators described an activity that was directly mitogenic for murine thymocytes but not for peripheral T or B lymphocytes.[7] The factor was found in crude culture supernatants of human or murine mononuclear cells that had been stimulated with lipopolysaccharide (LPS) or the plant-derived mitogen phytohemagglutinin (PHA). They called this activity lymphocyte-activating factor (LAF).[7] It was found that LAF synergized with a submitogenic dose of PHA, or con A, when used to stimulate thymocyte proliferation. They also found that it was the adherent cells (macrophages) that were the source of LAF rather than the nonadherent lymphocytes.[7] Later investigations reported that LAF had a molecular weight (M_r) of approximately 15,000.[7] It became clear that many activities derived from stimulated macrophages overlapped in terms of their biological and biochemical properties. Mitogenic protein (MP), helper peak-1 (HP-1), T-cell-replacing factor (TRF-III), T-cell-replacing factor Mo (TRFM), B-cell-activating factor (BAF), and B-cell differentiation factor (BDF), all isolated from stimulated macrophages, had M_rs in the range of 12,000 to 18,000 and had LAF activity when assayed with murine thymocytes.[8] In 1979, after many joint studies, it was decided that these activities were all attributable to one protein, and it was called interleukin 1.[8]

The main sources of IL-1 are the mononuclear phagocytes, which include peripheral blood monocytes, peritoneal macrophages, and Kupffer cells.[3] The invertebrate coelomocyte is the correlate of the vertebrate mononuclear phagocyte.[1,3] Other cell sources in the vertebrates include keratinocytes, Langerhans cells, astrocytes and glial cells, B and T lymphocytes, large granular lymphocytes, fibroblasts, and endothelial cells to name only a few.[3,9] IL-1 can be induced by a plethora of biological, chemical, or environmental insults.[3] Some cells, like chronic lymphocytic leukemia cells, Epstein-Barr virus transformed B lymphocytes, or Raji cells constitutively produce IL-1.[10]

The immunoregulatory activities of IL-1 were explored most actively during the early studies of IL-1. The activities of IL-1 on the immune system include promotion of lymphocyte differentiation and expression of receptors, augmentation of B-lymphocyte proliferation and antibody synthesis, enhancing helper cell actions, autocrine growth factor activities, and thymocyte stimulation.[3,7,11]

Of equal importance to the immunoregulatory effects of IL-1 are its properties as a mediator of nonspecific host defense mechanisms. Activities of IL-1 include lowering of plasma iron and zinc levels, elevation of plasma copper levels, increased synthesis of several acute-phase proteins, and release of bone marrow neutrophils. Other properties of IL-1 that are crucial during the host response to attack include induction of fever, chemotaxis of neutrophils, and production of prostaglandin.[3,9]

Invertebrate IL-1 shares many biological activities with vertebrate IL-1 when assayed in several vertebrate systems. Besides thymocyte-stimulating ability, its activities include stimulation of fibroblast proliferation, protein synthesis and PGE_2 release,[4] cytotoxicity for the human melanoma cell line A375,[4] and *in vivo* induction of increases in vascular permeability.[12] Invertebrate IL-1 activity is also inhibited by polyclonal antisera to vertebrate IL-1.[13]

Biochemical characterization of IL-1 from a number of species reveals basic similarities in the structure and properties of this cytokine. Human IL-1 has a M_r of 17,500 and two major charged forms.[11] One has a pI of 7.0 (termed β), while the other has a pI of 5.0 (termed α).[3] Murile IL-1 has a M_r of 14,500 and two charged forms, pI 7.2 (β) and pI 5.5 (α).[11,14] In humans the predominant form is β, while in mice it is α.[3] In other animal species (rabbit, pigs, rat, cow), the M_r is usually in the 12,000 to 18,000 range, although large M_r forms of between 35,000 and 70,000 have been found in all species studied.[3] Isoelectric focusing studies show charge heterogeneity in all species studies with pI values in the 5.0 to 5.6 and 6.8 to 7.5 ranges.[3] Invertebrate IL-1 (isolated from Echinodermata, Annelida, and Urochordata) has a M_r of 18,000 to 22,000 and at least two major charged forms, pI 7.4 and 4.8.[2,4]

IL-1 was the best characterized cytokine in relation to an acute-phase response. It soon became apparent that IL-1 and TNFα could not account for all the activity that was observed. A distinct monokine, termed hepatocyte-stimulating factor,[15,16] was isolated that had the capacity to induce a number of other acute-phase proteins. This molecule is now known as IL-6. Like IL-1 and TNFα, IL-6 has been shown to be a major immunoregulatory molecule and inducer of the acute-phase response.[17]

IL-6 is a macrophage-derived cytokine that is also responsible for mediation of nonspecific host defense mechanisms. The pleiotropic nature of the molecule is demonstrated by the history of its isolation. It was originally described by Hirano and coworkers[18] as a T-cell-derived lymphokine that caused the terminal differentiation of activated B cells to antibody-producing cells. This activity was separable from the other B-cell growth and differentiation factors (IL-4 and 5). Another molecule was isolated from Weissenbach and coworkers[19] with antiviral activity separate from IFNβ activity and called $IFN\beta_2$. A murine T-cell-derived factor (HP1) was identified by Van Snick and coworkers[20] that was required for the survival and growth of murine B-cell hybridomas and plasmacytomas. Several groups simultaneously reported on the cloning of these molecules and the identification of them as single protein products.[21]

The calculated M_r of the mature IL-6 polypeptide is 20,781.[21] Comparison of the cDNA sequence of human and murine IL-6 shows a homology of 65% at the DNA level and 42% at the protein level.[21,22] The positions of four cysteine residues are completely conserved, suggesting that the cysteine-rich region of the protein plays a critical role in IL-6 activity.[21]

IL-6 exhibits multiple biologic activities on different target cells. Activation of T cells and thymocytes, induction of acute-phase proteins, stimulation of hematopoietic precursor cell growth and differentiation, and pyrogenic activity are only a few of its host defense–related activities.[23,24]

We have used the B9 assay[5] in an attempt to isolate an IL-6-like protein from the coelomic fluid of *Asterias forbesi.* Coelomic fluid was concentrated using ultrafiltration membranes that had a nominal molecular weight limit (nmwl) of 10,000. The retentate was applied to a Sephadex G50 column, and the fractions that had a M_r of <50,000 were collected and concentrated by ultrafiltration (10,000 nmwl). The sample was fractionated by preparative isoelectric focusing using a Rotofor cell. These experiments resulted in the isolation of a protein that had a M_r of ≈22,000 and a pI of 5.0 that was active in the B9 assay (TABLE 1). These values are similar to values that have been reported for vertebrate IL-6.[21]

Because we have found the three major host defense–related cytokines in the invertebrates, it is logical to assume that other cytokines will be found as well. TABLE 2 contains some cytokines that we feel may be found in the invertebrates and speculates on their possible functions based on their activities in the vertebrates. Indeed, certain vertebrate cytokines may lack homologues in the invertebrates, notably cytokines involved in communication of specific signals between cells participating in the more advanced humoral immune response (ILs 4, 5, and 7) or in cellular defense against viral infections (IFNα and β).[25] Whether these molecules can be found subserving other functions in invertebrates remains to be determined.

In conclusion, as in the vertebrates, cytokines seem to control many aspects of the host defense response in the invertebrates. By studying the control and actions of cytokines in the invertebrates, we hope to gain insight into the complex cellular and molecular interactions in the vertebrate immune response.

TABLE 1. Preparative Isoelectric Focusing of Pooled B9 Cell Proliferation Activity from the G50 Column Fractionation of *Asterias forbesi* Coelomic Fluid

Fraction Number	pH	IL-6 Activity (B9 Cell Proliferation MTT Assay)
Control	—	0.243 ± 0.037[a]
1	3.07	0.296 ± 0.030
2	3.55	0.272 ± 0.013
3	3.85	0.303 ± 0.019
4	4.20	0.269 ± 0.018
5	4.50	0.297 ± 0.052
6	4.78	0.305 ± 0.032
7	5.10	0.321 ± 0.058
8	5.42	0.390 ± 0.043
9	5.65	0.343 ± 0.023
10	5.94	0.312 ± 0.032
11	6.47	0.324 ± 0.054
12	6.83	0.311 ± 0.030
13	7.05	0.329 ± 0.020
14	7.27	0.293 ± 0.028
15	7.43	0.213 ± 0.008
16	7.65	0.220 ± 0.035
17	7.95	0.201 ± 0.036
18	8.40	0.102 ± 0.015
19	9.05	0.129 ± 0.006
20	9.64	0.081 ± 0.022

[a]O.D. 570; mean ± SD.

TABLE 2. Host Defense Functions of Cytokines in Vertebrates and Invertebrates

Cytokine	Major Functions[a] in the Vertebrates	Proposed Functions in the Invertebrates
IL-1α and β[b]	Induction of fever Stimulation of T-cell proliferation Major orchestrator of the inflammatory response	Stimulation of coelomocyte proliferation Stimulation of phagocytosis and chemotaxis Major mediator of host defenses
TNFα and β[b]	Activation of endothelial cells Cytostatic for various cells Major orchestrator of the inflammatory response	Stimulation of coelomocyte proliferation Inhibition of coelomocyte proliferation Major mediator of host defenses
IL-2[b]	T cell mitogen Stimulation of the release of TNF, IL-1, and TGFβ Activation of T cells and macrophages	Stimulation of coelomocyte proliferation Stimulation of release of other cytokines
IL-3	Stimulation of hematopoietic precursor cells Mitogenic for various cells Support of the proliferation of blast cells	Stimulation and differentiation of coelomocyte precursor cells
IL-6	Induction of acute-phase proteins Cytostatic for various cells Major orchestrator of the inflammatory response	Stimulation of coelomocyte proliferation Stimulation of coelomocyte migration Major mediator of host defenses
IFNγ	Activation of macrophages Induction of acute-phase proteins Induction of fever	Stimulation of coelomocyte proliferation
TGFβ	Mitogenic for various cells Inhibition of T-cell proliferation Inhibition of lymphocyte-mediated responses	Inhibition of coelomocyte proliferation
CSFs	Mitogenic for various cells Activation of macrophages Stimulation of *in vivo* hematopoiesis	Stimulation of hemocyte proliferation of higher invertebrates

[a]Reviewed in references 2 and 3.
[b]Has been described in invertebrates.[2,12,13,26]

SUMMARY

Cytokines are polypeptides released by cells involved in vertebrate host defenses that are used to communicate with similar or different cells. Interleukin (IL)-1, IL-6, and tumor necrosis factor (TNF) are major immunoregulatory cytokines. The crucial need for cytokines in the vertebrates suggests to us that they are proteins that have evolved with the host defense systems of all animals. We have isolated and characterized IL-1 and TNF-like molecules from several invertebrate phyla. These invertebrate cytokines share many biological activities with their vertebrate counterparts. An IL-6-like molecule from invertebrates has been identified and characterized by using the B9 cell stimulation assay that is routinely employed for detecting vertebrate IL-6. The advantage of using a vertebrate detection system is that it emphasizes the evolutionary continuity of the immune/inflammatory response. By studying defense mechanisms of invertebrates and their regulatory cytokines, we may be able to unravel the complex web of interactions among cells and factors of vertebrate immune responses.

REFERENCES

1. Cooper, E. 1976. Comparative Immunology. Prentice-Hall, Inc. Englewood Cliffs, NJ.
2. Beck, G. & G. S. Habicht. 1991. Immunol. Today **181:** 180–1183.
3. Oppenheim, J. J., E. Kovacs, K. Matsushima & S. Durum. 1986. Immunol. Today **7:** 45–56.
4. Beck, G. & G. S. Habicht. 1991. Mol. Immunol. **28:** 577–584.
5. Nordan, R. P. & M. Potter. 1986. Science **233:** 566–569.
6. Mosmann, T. 1983. J. Immunol. Methods **65:** 55–62.
7. Gery, I. & R. Handschumacher. 1974. Cell. Immunol. **11:** 162–170.
8. Dinarello, C. A. 1984. Rev. Infect. Dis. **6:** 51–95.
9. Oppenheim, J. J., K. Matsushima, K. Onozaki, K. Procopio & G. Scala. 1985. *In* Immune Regulation. M. Feldman & N. Mitchison, Eds.: 111–120. The Humana Press. Clifton, NJ.
10. Habicht, G. S., G. Beck & B. Ghebrehiwet. 1987. J. Immunol. **138:** 2593–2597.
11. Mizel, S. 1982. Immunol. Rev. **63:** 51–72.
12. Beck, G., R. Vasta, J. J. Marchalonis & G. S. Habicht. 1989. Comp. Biochem. Physiol. **92B:** 93–98.
13. Beck, G. & G. S. Habicht. 1986. Proc. Natl. Acad. Sci. USA **83:** 7429–7433.
14. Habicht, G. S., G. Beck, J. Benach, J. Coleman & K. Leicthling. 1985. J. Immunol. **134:** 3147–3154.
15. Gauldie, J., C. Richards, D. Harnish, P. Landsdorp & H. Bauman. 1987. Proc. Natl. Acad. Sci. USA **84:** 7251–7255.
16. Arai, K., F. Lee, A. Miyajima, S. Miyatake, N. Arai & T. Yokota. 1990. Ann. Rev. Biochem. **59:** 783–836.
17. Akia, S., T. Hirano, T. Taga & T. Kishimoto. 1990. FASEB J. **4:** 2860–2867.
18. Hirano, T., T. Taga, N. Nakano, K. Yasukawa, S. Kashiwamura, K. Shimizu, K. Nakajima, K. Pyun & T. Kishimoto. 1985. Proc. Natl. Acad. Sci. USA **82:** 5490–5494.
19. Weissenbach, J., Y. Chernajovsky, M. Zeevi, L. Shulman, H. Soreq, U. Nir, D. Wallach, M. Perricaudet, P. Tiollais & M. Revel. 1980. Proc. Natl. Acad. Sci. USA **77:** 7152–7156.
20. VanSnick, J., A. Vink, S. Cayphas & C. Uyttenhove. 1987. J. Exp. Med. **165:** 641–652.
21. Kishimoto, T. 1989. Blood **74:** 1–10.
22. Hirano, T. & T. Kishimoto. 1989. *In* Human Monocytes. M. Zembala & G. Asherson, Eds.: 217–226. Academic Press. London.
23. Wong, G. & S. Clark. 1988. Immunol. Today **9:** 137–139.
24. Le, J. & J. Vilcek. 1989. Lab. Invest. **61:** 588–602.
25. Balkwill, F. R. & F. Burke. 1989. Immunol. Today **10:** 299–304.
26. Beck, G., R. F. O'B rien & G. S. Habicht. 1989. BioEssays **11:** 62–67.

The Echinoderm Immune System

Characters Shared with Vertebrate Immune Systems and Characters Arising Later in Deuterostome Phylogeny[a]

L. COURTNEY SMITH AND ERIC H. DAVIDSON

Division of Biology
California Institute of Technology
Pasadena, California 91125

DEUTEROSTOME IMMUNE SYSTEMS VIEWED CLADISTICALLY

As is now widely appreciated, the deuterostomes, of which echinoderms, hemichordates, tunicates, and all higher chordates are the major extant groups, constitute a separate branch of the animal kingdom (see, e.g., Refs. 1–4). On structural, embryological, biochemical, and molecular bases, deuterostomes are regarded as a monophyletic group of common ancestry. They are the sister group of the coelomate protostomes, the major taxa of which are the annelids, molluscs, and arthropods. Thus, in considering the origins of the higher vertebrate immune system, it is essential to keep in mind that it is among the extant lower deuterostomes that the most relevant homologous immune system elements are to be sought.

Immune system evolution should be considered in the same way as the evolution of any other complex functional and morphological system of higher vertebrates. That is, it should be possible to analyze vertebrate immune systems in terms of sets of molecular and functional characters that have arisen at different stages of evolution. Such sets of characters will be shared among the descendants of the common ancestors from which each taxonomic subgroup, or clade, of organisms derives (unless these characters have been deleted or modified beyond recognition). In this article we briefly consider the major character sets of the mammalian immune system from a cladistic point of view, in so far as current evidence of lower vertebrate and invertebrate deuterostome immune systems permits. We then focus on cellular and molecular aspects of the sea urchin immune system, since this provides the most primitive lower deuterostome character set available for comparison. It will be evident that in fact the basal set of nonadaptive cellular immune system functions in higher vertebrates shares major features with the nonadaptive cellular immune system of echinoderms.

In TABLE 1 we have constructed an abbreviated cladistic diagram, which for reference purposes combines immune system character sets (above the horizontal axis of the table) with some selected character sets relating to other higher vertebrate organ systems and processes (below the horizontal axis). The import of this diagram, as we discussed at more length previously,[5] is that the immune system appears in

[a]Research from this laboratory was supported by National Science Foundation Grant MCB-9219330.

TABLE 1. A Cladistic Treatment of Deuterostome Immune System Characters

<table>
<thead>
<tr><th>Echinoderms</th><th>Tunicates</th><th>Jawless Fishes</th><th>Cartilaginous Fishes</th><th>Bony Fishes and Tetrapods</th></tr>
</thead>
<tbody>
<tr><td></td><td></td><td></td><td></td><td>• Rearranging Ig gene family: Ig and TcR
• Clonal amplification of B cells and Tk mediated by Th and MHC
• Affinity maturation of Ig
• Thymic T-cell selection mediated by peptide/MHC
• T cells
• MHC Class I</td></tr>
<tr><td></td><td></td><td></td><td>• MHC class II
• B cells
• Lower vertebrate Ig gene organization
• Cytotoxic cells in which specificity depends on Fc-bound IgM</td><td>• MHC Class II
• B cells
• Higher vertebrate Ig gene organization
• A variety of cell types bind Ig by several Fc receptors</td></tr>
<tr><td></td><td></td><td></td><td colspan="2">• Complement functions in conjunction with Ig to insert holes into membranes to lyse cells</td></tr>
<tr><td colspan="2" rowspan="2">• Complement-like factor(s) function as an opsonin?</td><td colspan="3">• Complement functions as an opsonin</td></tr>
<tr><td>• One component</td><td>• Six components in the classical pathway</td><td>• Nine components in the classical pathway</td></tr>
<tr><td colspan="5">• Lectins, opsonins, hemagglutinins, hemolysins
• Basic phagocyte-mediated, nonspecific defense responses: motility and chemotaxis, phagocytosis, phagosomal enzymes and reactive oxygen intermediates, encapsulation, degranulation, cytotoxic cells with preset specificities, cytokine secretion and response, receptor-mediated signal transduction systems, preset responses to pathogens
• Effective graft rejection responses (perhaps based on positive detection of simple self markers such as cell adhesion molecules in lower deuterostomes)</td></tr>
</tbody>
</table>

TABLE 1. *Continued.*

Echinoderms	Tunicates	Jawless Fishes	Cartilaginous Fishes	Bony Fishes and Tetrapods
• Embryogenesis larval body plan,[1] gut structure,[2] molecular shared characters[3]				
	• Dorsal mesoderm,[4] endocrine system,[5] embryogenesis and larval body plan,[6] CNS,[7] molecular shared characters[8]			
		Cephalochordates • Embryogenesis,[9] dorsal mesoderm,[10] CNS,[11] circulatory system,[12] liver, molecular shared character[13] **Jawless Fishes** • Embryogenesis,[14] dorsal mesoderm,[15] CNS,[16] head structure,[17] endocrine system,[18] circulatory system,[19] molecular shared character[20]		
			• CNS,[21] dorsal mesoderm,[22] endocrine system,[23] appendages,[24] motor system,[25] excretory system,[26] reproductive system,[27] head structure,[28] molecular shared characters[29]	
				• Skeletal system[30] Respiratory system[31] Dental system[32] Detailed homologies in muscular, nervous systems, and post-embryonic morphogenesis

NOTE: Immune system character sets are shown on the facing page (table modified from Smith and Davidson[51]). At each phylogenetic level, all taxa shown to the right share the listed characters or convincingly homologous versions of these characters. On this page, cladistic relations of shared character sets relating to other systems and developmental properties are indicated, for reference, in a similar format. The following provides more detail on these characters, keyed to the superscripts in the table. [1]Same basic embryonic body plan; radial cleavage; enterocoelous coelom formation; right/left coelomic asymmetry. [2]Tripartite gut. [3]Exact intron positions in actin genes. [4]Notochord; paraxial dorsal muscle bands. [5]Hatscheck's pit (probably homologous with pituitary); endostyle (probably homologous with thyroid). [6]Bilateral-tailed larva; ciliated pharyngeal gill slits; inductive neural plate specification. [7]Neural plate and dorsal nerve tube. [8]Vertebrate-type muscle actin proteins. [9]Lateral plate mesoderm surrounds coelomic cativities; embryonic induction of notochord. [10]Metameric somites. [11]Neural tube consisting of gray and white matter; dorsal and ventral nerve roots separate; motor end plates, on dorsal (though not ventral) nerves. [12]Major aorta (dorsal and ventral), veins (cardinal and subintestinal) homologous with vertebrates. [13]Creatine phosphate predominate. [14]Larval body plan resembles adult body plan; neural crest. [15]Metameric paraxial mesoderm; somites with dermatome, myotome, sclerotome. [16]Forebrain, midbrain, and hindbrain; cranial nerves differentiated from axial nerves; dorsal hollow nerve cord innervated by myotomal nerve roots with sensory and motor fibers; paired olfactory capsules; paired optic capsules. [17]Construction of head skeleton largely from neural crest derivatives. [18]Pituitary, thyroid, adrenal glands. [19]Chambered heart. [20]Noncollagenous enamel proteins in teeth; hemoglobin. [21]Enlarged forebrain. [22]Vertebrae. [23]Encapsulated supraarenal bodies; pancreas. [24]Paired pectoral and pelvic appendages. [25]Smooth muscle and striated muscle actins; myelinated nerve fibers. [26]Renal portal system. [27]Wolffian and Muellerian ducts. [28]Inclusion of anterior somites in occipital complex. [29]Elastin, histamine, oxytocin, calcitonin, α and β hemoglobins. [30]Similar internal bone deposition mechanism. [31]Lungs (or swim bladder) derived from gut outpockets. [32]Similar arrangement of teeth on jaw bones.

This partial list is collated from: Maisey[76]; Schaeffer[3]; Jefferies[2]; Gans and Northcutt[77]; Couly *et al.*[78]

higher deuterostome evolution stepwise, just as does virtually every other organ system and developmental process. As displayed in the lower half of TABLE 1, it can be seen that every major system and process of higher vertebrates was similarly built up stepwise. For example, CNS, endocrine, skeletomuscular systems, and the morphogenetic processes of development can all be regarded as the layered sum of system subelements that appeared at different stages in evolution and are of differing evolutionary antiquity. The inventions of each stage were then retained in all evolutionary descendants, and subsequently added on to. Our argument, then, is that it is most unlikely that the most sophisticated aspects of the higher vertebrate immune system, for example, immunological memory, clonal amplification, and T-cell functions, "sprang forth fully blown" at lower vertebrate, subvertebrate, or even subchordate levels of deuterostome evolution; TABLE 1 shows that no other vertebrate system has had that kind of evolutionary history.

The baseline type of cellular defense system for the deuterostomes is clearly exhibited in the echinoderms and consists of phagocyte-mediated, nonspecific cellular activation responses to challenge (see upper half of TABLE 1). Effector functions include phagocytosis, encapsulation, chemotaxis, cytotoxicity, degranulation of bactericidal compounds, lectins, hemalysins, hemaglutinins, and effective clotting reactions. The echinoderms also share with the rest of the deuterostomes preset responses to bacterial pathogens and the phenomenology of effective graft rejection (see below).

The first clearly demonstrated augmentation to this basic, nonspecific phagocytic type of immune system is the appearance of a complement component in the jawless fishes or cyclostomes.[6,7] Although Varner *et al.*[8] had reported short sequence similarities between a hagfish protein believed to be an immunoglobulin (Ig) molecule and shark and mammalian Ig sequence elements, when the gene was cloned and sequenced, it became clear that it was actually a complement protein.[6,7] The circulating complement component in hagfish functions as an opsonin for efficient removal of invading microorganisms by phagocytosis.[9] The elasmobranchs or cartilaginous fishes have an abbreviated classical cascade consisting of six components rather than nine, and it functions in a manner similar to that in the higher vertebrates, that is, in conjunction with Ig to insert holes into membranes to destroy foreign cells.[10,11] Phagocytosis experiments using sea urchin coelomocytes have suggested that there may be a complement-like opsonin system in these animals.[12–15] Thus, when sheep red blood cells were coated with IgM and complement, specifically C3b or iC3b, they were taken up by the phagocytic coelomocytes at a significantly faster rate than were red cells coated with IgM but lacking any C3 products. These phagocytosis experiments were verified by using purified complement components rather than serum as the complement source, and by showing that inhibitors for mammalian complement (such as heat, yeast cell wall components, cobra venom factor, hydrazine, and ammonium) destroy or inhibit the complement-like opsonin activity in the sea urchin. It is an interesting thought that echinoderms and cyclostomes might have a similar opsonization mechanism that is related to a family of higher vertebrate molecules. There is no direct evidence whatsoever that complement components are present in sea urchins, and such a demonstration would require convincing sequence data. Nor are there currently data available on any complement-like tunicate components.

The immune systems of sharks and other elasmobranchs possess many important components of higher vertebrate immune systems, as shown in TABLE 1. These include the major histocompatibility complex (MHC) class II,[16] B cells with high concentrations of circulating Ig,[17,18] and an effective cellular system that binds circulating antibodies by Fc receptors.[19] The regulation of Ig production and secretion in these animals is an interesting question, because affinity maturation of antibodies and clonal amplification do not occur.[20–22] Neither T cells, nor functions that could be assigned to T cells, have been identified in these fish. Future studies on the elasmobranch immune system will clarify this question, and the mechanisms revealed will undoubtedly expand our understanding of the regulatory mechanisms employed by the mammalian Ig system.

The final step in the immunophylogenetic cladogram of TABLE 1 brings us to the bony fishes and tetrapods, which exhibit a higher vertebrate type of immune system that shows adaptive specificity, as exemplified by the mammalian system. Attributes of this system include MHC class I and II, T cells, B cells, rearranging Ig family genes, thymic selection, and clonal amplification. However, even in the higher vertebrates the baseline characters identified in the echinoderms can be identified. They are represented by macrophage- and granulocyte-type cells that have nonspecific effector functions. The higher vertebrate phagocyte system has been extensively augmented by the addition of, and integration with, the lymphocyte system that functions by non-self recognition and that displays both adaptive specificity and memory.

Two points become obvious from an examination of the upper half of TABLE 1. First, the adaptive immune system is a vertebrate invention. This system, as it functions in the higher vertebrates, is clearly not present in the subvertebrate deuterostomes. Second, there is a large gap in the appearance of major modifications and additions to the basic phagocyte system between the echinoderms and the cartilaginous fishes, which have B cells, Ig, and MHC class II. It might be supposed that the tunicates or protochordates will be found to have some chordate-like addition or modification to the nonspecific cellular defense system that functions in the echinoderms. However, at present there is no evidence that these animals have immune system potentialities other than those equivalent to the echinoderm systems. It is interesting that the jawless fishes, which are true vertebrates, have not been shown to possess any of the attributes of the higher vertebrate immune system except for complement. Though more information is obviously required, it appears so far that the cyclostomes function essentially on an invertebrate type of defense system using lectins, opsonins, and phagocytosis. To obtain illumination on the origins of higher vertebrate immune functions, it is clear that research on the immune systems of echinoderm, protochordate, and cyclostome groups will be enormously rewarding.

BASELINE IMMUNE FUNCTIONS EXHIBITED IN ECHINODERMS

Allograft Rejection

Let us now turn to a description of the nonspecific, phagocyte-mediated immune system in the echinoderms that is the baseline defense system of the deuterostome clade. Hildemann and his colleagues first demonstrated a functioning immune system in echinoderms by showing that sea cucumbers and sea stars would reject allografts and accept autografts.[23,24] Later, Coffaro and Hinegardner[25] and Coffaro[26,27] performed large-scale analyses of allograft rejections using *Lytechinus pictus*, a small sea urchin (first set, $n = 213$; second set, $n = 83$; third party, $n = 79$). In a reanalysis of the data from Coffaro's dissertation (Fig. 1), the rejection kinetics show that the immune system in these animals operates by a nonspecific, activation type of mechanism. This type of system is defined by the rejection kinetics of the third-party allografts, which were superimposable on those for the second set, even though both second-set and third-party allografts showed increased rejection rates. These results show that the host animals could not discern differences between second-set tissues to which they had been presensitized, and third-party tissues that they had not previously contacted. An activated state was maintained in these organisms for about six months, at which point the median rejection time for second-set allografts began to increase.[26] These experiments prove that sea urchins activate their cellular defense systems in response to contact with the first-set allografts and then nonspecifically respond in an augmented manner to any foreign tissue. These are characteristics of an immunological defense system that does *not* possess specific immunological memory, does *not* function by non-self recognition,[5] and may be responding to the absence of self.

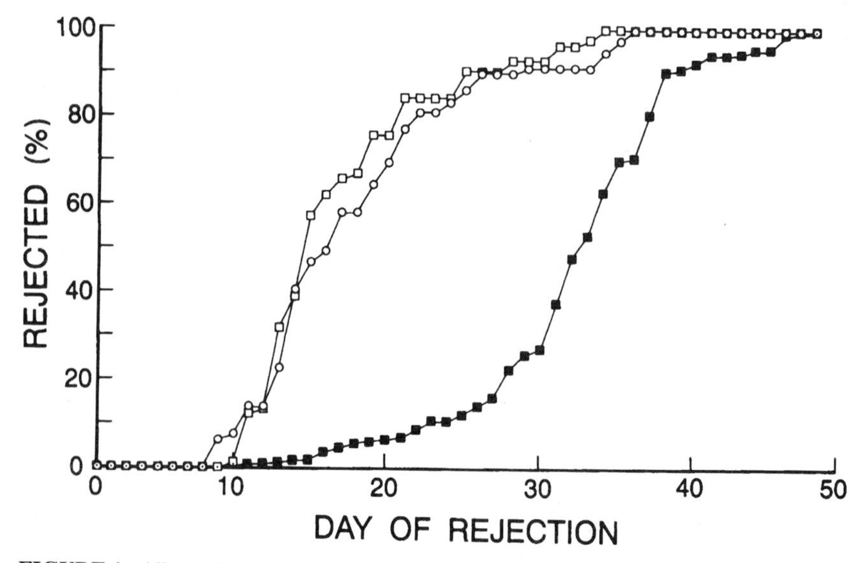

FIGURE 1. Allograft rejection kinetics in *Lytechinus pictus*. The data for this figure came from Coffaro.[26] First-set allografts, *filled squares*; second-set allografts, *open squares*; third-party allografts, *circles*. (Taken from Smith & Davidson,[5] originally published by Immunology*today*, Elsevier Trends Journals. Used with permission.)

Clearance of Injected Substances

The coelomocytes, found in the fluid that fills the coelomic cavity of echinoderms, mediate the immune functions in these animals. In sea urchins, the coelomocyte population is composed of phagocytes, vibratile cells, and colorless and red spherule cells.[28–30] These cells have many functions that are similar to those of the phagocytic type of cell found in higher vertebrates such as macrophages and granulocytes. Immune memory and recognition specificity have been investigated in echinoderm coelomocytes by studying the phagocytic activities of these cells. Clearance rates of various substances repeatedly injected into the coelom were followed over time to identify the capabilities of the coelomocytes to remove the injected substances from the coelomic cavity. A variety of substances have been tested, which include foreign echinoderm cells, variously treated red blood cells, yeast, bacteria, bacteriophage, foreign proteins, carbon, carmine, Sephadex beads, and latex beads.[12,29,31–37] In all of these studies, it was clear that, although echinoderm coelomocytes were very good at removing injected substances, they did not function faster nor with specificity upon repeated injections. These results agree with the allorejection data of Coffaro discussed above; echinoderm immunity is mediated by nonspecific coelomocyte activation.

Clot Formation

In addition to phagocytosis, clot formation is a very important defense response to injuries in sea urchins. This is because these animals have an inflexible test and cannot employ muscular contraction to aid in closing wounds as in sea stars and sea cucumbers.[29,38] Two types of clotting methods are mediated by different coelomocyte types. The vibratile cells, which have a single flagellum and actively swim through the coelomic fluid, are full of granules. When these cells degranulate or lyse in response to injury, the normally liquid coelomic fluid is turned into a gel, which effectively stops "bleeding."[29,38] Cellular clots are formed by the phagocytes. In response to injury, these amoeboid cells rapidly alter their petaloid lamellipodia into filopodia that become intertwined to create a net. This net catches other cells and forms plugs to seal wounds.[39–42] Phagocyte shape transformation that leads to clotting is sensitive to Ca^{2+},[43] and it may be regulated by Ca^{2+} influxes that are induced by transduced signals in response to injury or infection.

Other Coelomocyte Immune Functions

The phagocytes and red spherulous cells are directly involved in several defense functions in the sea urchin. *In vivo,* the phagocytes and red spherulous cells have been shown to infiltrate into allografted,[24,26] injured,[26,44,45] and infected tissues.[46,47] In addition to the phagocytic activity of the coelomocytes, the phagocytes encapsulate or form syncytia around foreign particles that are too large to phagocytose[36,40] or form a wall to isolate an infected region from the rest of the animal.[29,46] Lysosomal enzymes,[48,49] hydrogen peroxide,[50] and prophenyloxidase[51] have been shown to operate in echinoderm coelomocytes, all of which are important in post-phagocytic degradation. In addition, a variety of undefined opsonins, hemagglutinins, hemolysins, and cytolysins[14,52–57] (for review on invertebrate lectins, see Vasta[58]) function in echinoderm coelomic fluid and on coelomocyte surfaces including the complement-like opsonins discussed above. The red spherulous cells contain echinochrome A in their granules, which are degranulated in the presence of bacteria.[29] Echinochrome A is a naphthaquinone that has antibacterial[59,60] and antialgal[47] activities. In mixed allogeneic cultures of sea urchin phagocytes, cytotoxic reactions have been demonstrated,[61] although this result has not been repeated.[62]

Coelomocytes may act through a cytokine system. Diffusible cytokines, secreted by activated coelomocytes, may function to amplify the activation response to other cells located throughout the animal, or to attract other cells to a site of infection or injury. Sea star factor is a component in coelomocyte lysates that is very effective in activating coelomocytes both *in vitro* and *in vivo.*[63,64] Cell proliferation activity has been demonstrated in fractions of sea star coelomic fluid or coelomocyte lysates when added to cultures of mouse thymocytes, fibroblasts,[65] or sea star coelomocytes.[66] Receptors with signal-transducing activities on echinoderm coelomocytes may function in activation and chemotaxis and have been implied from the analysis of profilin expression in activated coelomocytes in sea urchins (see below).

A MOLECULAR PARAMETER OF COELOMOCYTE ACTIVATION

Profilin Expression in Sea Urchin Coelomocytes

We have recently cloned, sequenced, and characterized the sea urchin profilin gene from an activated coelomocyte cDNA library.[67] Profilin is one of many cytoplasmic proteins that bind to and interact with actin (see reviews by Aderem,[68] and Stossel[69]). Monomer(G-)actin is released from a depolymerizing filament in the form of ADP-G-actin, which is 5 to 10 times less likely to be reincorporated into a polymerizing filament than ATP-G-actin.[70] Recent work has shown that profilin catalyzes the adenosine nucleotide exchange on G-actin,[71] which essentially promotes filament polymerization. Profilin is also involved in regulating the inositol triphosphate (IP_3) second-messenger system. It binds to phosphoinositol-(4,5)-bisphosphate (PIP_2) and blocks the activity of nonactivated (nonphosphorylated) phospholipase Cγ1 (PLCγ1) to hydrolyze PIP_2 into IP_3 and diacyl glycerol (DAG).[72,73] When the cell is not receiving and transducing signals, profilin is bound to PIP_2; IP_3 and DAG are not formed; the actin cytoskeleton is quiescent; and the cell does not change shape or move.[68] When a signal is transduced across the membrane by a receptor or receptor complex, a variety of proteins are phosphorylated and activated including PLCγ1. Activated PLCγ1 displaces profilin from PIP_2, and hydrolyzes PIP_2 to IP_3 and DAG, which have a variety of downstream activities on the activating cell. The displacement of profilin from PIP_2 changes its concentration in the nearby cytoplasm,[74] which promotes filament polymerization by catalyzing the nucleotide exchange on ADP-G-actin. Essentially, profilin functions to couple and coordinate signal transduction with actin filament polymerization and cytoskeletal modifications that appear as cell activation with accompanying changes in behavior.

Profilin Transcripts Are Elevated in Activated Coelomocytes

We have recently found that the numbers of transcripts from the sea urchin profilin gene in coelomocytes increase during activation and peak in concentration at about one day after systemic challenge or injury.[67] Probe excess transcript titrations (see Lee *et al.*[75] for detailed methods description) were performed on coelomocyte RNA isolated from variously treated sea urchins (FIG. 2). These treatments were administered

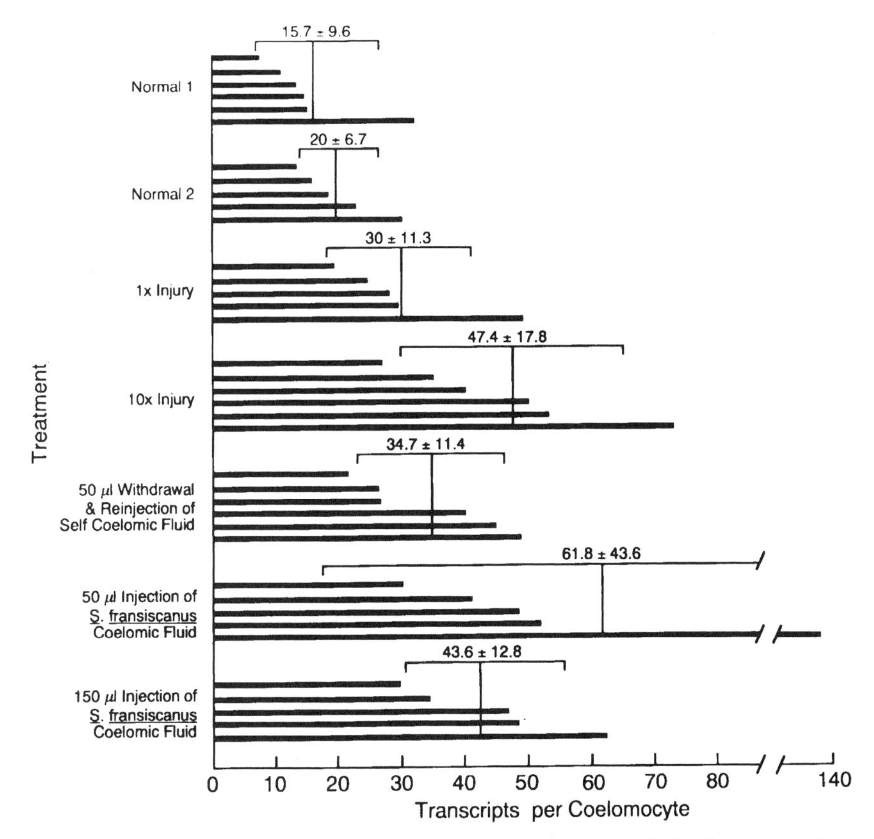

FIGURE 2. Profilin transcript prevalences in coelomocytes from individual sea urchins after various treatments. The sea urchins were treated (treatments are listed on the left) on days 1, 2, 4, and 5; and the cells were collected 24 hours after the last treatment. Probe excess transcript titrations were performed on total RNA isolated from the coelomocytes (see Lee *et al.*[75] or Smith *et al.*[67] for methods). One or ten times injury is 1 or 10 needle-hole injuries in the peristomial membrane per day on each day of treatment. Means and standard deviations are shown. p values from Student's t test analysis for each experimental group compared to the combined normal groups are as follows: $p \leq 0.0005$, animals receiving 150 ml of foreign coelomic fluid; $p \leq 0.005$, animals receiving 10 needle-hole injuries per day; $p \leq 0.025$, animals receiving one needle-hole injury per day or those having 50 ml of their own coelomic fluid withdrawn and reinjected or those receiving 50 ml of foreign coelomic fluid. (Taken from Smith *et al.*[67] *Molecular Biology of the Cell* © 1992. The American Society for Cell Biology. Used with permission.)

on days 1, 2, 4, and 5, and included needle-hole injuries through the peristomial membrane, self-coelomic fluid that was withdrawn and reinjected, or injections of coelomic fluid from a congeneric species. Results of the transcript titrations from individual sea urchins show significant increases in the numbers of profilin transcripts per coelomocyte from those animals receiving treatments compared to the unmanipulated normal controls. Statistical analyses of these data suggest that there is a hierarchy in the cellular responses that depends on the degree of injury, and that there is no specifically augmented response to injections of xenogeneic cells (see legend to FIG. 2). Sea urchin coelomocytes are very sensitive to perturbations of the coelomic cavity and show increases in the transcript prevalence of the profilin gene in response to minor challenges. The increase in profilin transcript prevalence in these cells correlates with activation, which includes increases in motility, phagocytosis, and secretion. It is implied from other studies of profilin that sea urchin profilin functions to couple the transduction of activation signals to changes in the cytoskeleton.[67]

SUMMARY

In summary, the characters of the echinoderm immune system that we review here can be considered to illuminate the baseline nonadaptive immune systems that were our original deuterostome heritage. We still retain—and greatly rely upon—similarly functioning, nonadaptive cellular defense systems. It is worth stressing that sea urchins are long lived, normally healthy animals that display remarkable abilities to heal wounds and combat major infections. From an external point of view, their immune systems obviously work very well. Thus, their cellular defense systems are extremely sensitive, and they respond rapidly to minor perturbations, all without any specific adaptive capabilities. These systems probably function through the transduction of signals conveying information on injury and infection, just as do the equivalent systems that underlie and back up our own adaptive immune systems, and that provide the initial series of defenses against pathogenic invasions.

Many extremely interesting questions remain regarding the evolution of the deuterostome immune response. Are the echinoderm and tunicate systems the same, or have the protochordates augmented the basic phagocyte system with an as yet unidentified chordate-like character? Do the jawless fishes produce Igs that would make them similar to the sharks, or are they vertebrates without an Ig system that essentially rely on an invertebrate-like, nonspecific, activated phagocyte type of immune system? How do sharks regulate their immune system without T cells and MHC class I? How do they avoid producing autoantibodies? Future research will not only answer these questions, but those answers will also be enlightening with regard to the origins of the mammalian immune system in which ancient functions and subsystems remain.

REFERENCES

1. HYMAN, L. H. 1959. The Invertebrates: Smaller Coelomate Groups. McGraw-Hill. New York.

2. Jefferies, R. P. S. 1986. The Ancestry of the Vertebrates. British Museum (Natural History). London.
3. Schaeffer, B. 1987. Deuterostome monophyly and phylogeny. *In* Evolutionary Biology, vol. 21. M. K. Hecht, B. Wallace & G. T. Grance, Eds.: 179–235. Plenum Press. New York.
4. Davidson, E. H. 1991. Spatial mechanisms of gene regulation in metazoan embryos. Development **113:** 1–26.
5. Smith, L. C. & E. H. Davidson. 1992. The echinoid immune system and the phylogenetic occurrence of immune mechanisms in deuterostomes. Immunol. Today **13:** 356–362.
6. Ishiguro, H., K. Kobayashi, M. Suzuki, K. Titani, S. Tomonaga & Y. Kurosawa. 1992. Isolation of a hagfish gene that encodes a complement component. EMBO J. **11:** 829–837.
7. Hanley, P. J., J. W. Hook, D. A. Raftos, A. A. Gooley, R. Trent & R. L. Raison. 1992. Hagfish humoral defense protein exhibits structural and functional homology with mammalian complement components. Proc. Natl. Acad. Sci. USA **89:** 7910–7914.
8. Varner, J., P. Neame & G. W. Litman. 1991. A serum heterodimer from hagfish *(Eptatretua stoutii)* exhibits structural similarity and partial sequence identify with immunoglobulin. Proc. Natl. Acad. Sci. USA **88:** 1746–1750.
9. Raftos, D. A., R. W. Hook & R. L. Raison. 1992. Complement-like protein from the phylogenetically primitive vertebrate, *Eptatretus stoutii,* is a humoral opsonin. Comp. Biochem. Physiol. **103B:** 379–384.
10. Jensen, J. A., E. Festa, D. S. Smith & M. Cayer. 1981. The complement system of the nurse shark: Hemolytic and comparative characteristics. Science **214:** 566–569.
11. Jensen, J. A. & E. Festa. 1981. The 6-component complement system of the nurse shark. *In* Aspects of Developmental and Comparative Immunology. J. B. Solomon, Ed.: 485–486. Pergamon Press. Elmsford, NY.
12. Bertheussen, K. 1981. Endocytosis by echinoid phagocytes in vitro. I. Recognition of foreign matter. Dev. Comp. Immunol. **5:** 241–250.
13. Bertheussen, K. 1982. Receptors for complement on echinoid phagocytes. II. Purified human complement mediates echinoid phagocytosis. Dev. Comp. Immunol. **6:** 635–642.
14. Bertheussen, K. 1983. Complement-like activity in sea urchin coelomic fluid. Dev. Comp. Immunol. **7:** 21–31.
15. Bertheussen, K. & R. Seljelid. 1982. Receptors for complement on echinoid phagocytes. I. The opsonic effect of vertebrate sera on echinoid phagocytosis. Dev. Comp. Immunol. **6:** 423–431.
16. Kasahara, M., M. Vasquez, K. Sato, E. C. McKinney & M. Flajnik. 1992. Evolution of the major histocompatibility complex: Isolation of class II A cDNA clones from cartilaginous fish. Proc. Natl. Acad. Sci. USA **89:** 6688–6692.
17. Litman, G. W., J. P. Rast, M. J. Shamblott & R. N. Haire. 1993. Phylogenetic diversification of immunoglobulin genes and the antibody response. Mol. Biol. Evol. **10:** 60–72.
18. Marchalonis, J. J. & S. F. Schluter. 1990. Origins of immunoglobulins and immune recognition molecules. Bioscience **40:** 758–768.
19. McKinney, E. C. 1992. Shark lymphocytes: Primitive antigen reactive cells. Ann. Rev. Fish Dis. **2:** 43–51.
20. Mäkelä, O. & G. W. Litman. 1980. Lack of heterogeneity in anti-hapten antibodies of a phylogenetically primitive shark. Nature **287:** 639–640.
21. Shankey, T. V. & S. W. Clem. 1980. Phylogeny of immunoglobulin structure and function. VIII. Intermolecular heterogeneity of shark 19S IgM antibodies to pneumococcal polysaccharide. Mol. Immunol. **17:** 365–375.
22. Hinds, K. R. & G. W. Litman. 1986. Major reorganization of immunoglobulin V_H segmental elements during vertebrate evolution. Nature **320:** 546–549.
23. Hildemann, W. H. & T. G. Dix. 1972. Transplantation reactions of tropical Australian echinoderms. Transplantation **15:** 624–633.
24. Karp, R. D. & W. H. Hildemann. 1976. Specific allograft reactivity in the sea star *Dermasterias imbricata.* Transplantation **22:** 434–439.

25. COFFARO, K. A. & R. T. HINEGARDNER. 1977. Immune response in the sea urchin *Lytechinus pictus*. Science **197:** 1389–1390.
26. COFFARO, K. A. 1979. Transplantation immunity in the sea urchin. Ph.D. thesis, University of California. Santa Cruz, CA.
27. COFFARO, K. A. 1980. Memory and specificity in the sea urchin *Lytichinus pictus*. *In* Phylogeny of Immunological Memory. M. J. Manning, Ed.: 77–80. Elsevier/North-Holland. Amsterdam.
28. BOOLOOTIAN, R. A. & A. C. GIESE. 1958. Coelomic corpuscles of echinoderms. Biol. Bull. **115:** 53–63.
29. JOHNSON, P. T. 1969. The coelomic elements of sea urchins *(Strongylocentrotus)* I. The normal coelomocytes: Their morphology and dynamics in hanging drops. J. Invert. Pathol. **13:** 25–41.
30. KARP, R. D. & K. A. COFFARO. 1980. Cellular defense systems in the echinodermata. *In* Phylogeny of Immunological Memory. M. J. Manning, Ed.: 257–282. Elsevier/North-Holland. Amsterdam.
31. REINISCH, C. L. & F. B. BANG. 1971. Cell recognition: Reactions of the sea star *(Asterias vulgaris)* to the injection of amebocytes of sea urchin *(Arbacia punctulata)*. Cell. Immunol. **2:** 496–503.
32. WARDLAW, A. C. & S. E. UNKLES. 1978. Bactericidal activity of coelomic fluid from the sea urchin *Echinus esculentus*. J. Invert. Pathol. **32:** 25–34.
33. COFFARO, K. A. 1978. Clearance of bacteriophage T4 in the sea urchin *Lytechinus pictus*. J. Invert. Pathol. **32:** 384–385.
34. BERTHEUSSEN, K. 1981. Endocytosis by echinoid phagocytes in vitro. II. Mechanisms of endocytosis. Dev. Comp. Immunol. **5:** 557–564.
35. YUI, M. A. & C. J. BAYNE. 1983. Echinoderm immunology: Bacterial clearance by the sea urchin *Strongylocentrotus purpuratus*. Biol. Bull. **165:** 473–486.
36. DYBAS, L. & P. V. FRANKBONER. 1986. Holothurian survival strategies: Mechanisms for the maintenance of a bacteriostatic environment in the coelomic cavity of the sea cucumber, *Parastichopus californicus*. Dev. Comp. Immunol. **10:** 311–330.
37. PLYTYCZ, B. & R. SELJELID. 1993. Bacterial clearance by the sea urchin *Strongylocentrotus droebachiensis*. Dev. Comp. Immunol. **17:** 283–289.
38. BINYON, J. 1972. Excretion and the role of amoebocytes. *In* Physiology of Echinoderms.: 24–45. Pergamon Press. New York.
39. ENDEAN, R. 1966. The coelomocytes and coelomic fluids. *In* Physiology of Echinodermata. R. A. Boolootian, Ed.: 301–328. John Wiley & Sons. New York.
40. JOHNSON, P. T. 1969. The coelomic elements of sea urchins *(Strongylocentrotus)* III. In vitro reaction to bacteria. J. Invert. Pathol. **13:** 42–62.
41. EDDS, K. T. 1977. Dynamic aspects of filopodial formation by reorganization of microfilaments. J. Cell Biol. **73:** 479–491.
42. EDDS, K. T. 1993. Cell biology of echinoid coelomocytes. I. Diversity and characterization of cell types. J. Invert. Pathol. **61:** 173–178.
43. HENSON, J. H. & G. SHATTEN. 1983. Calcium regulation of the actin-mediated cytoskeletal transformation of sea urchin coelomocytes. Cell Motil. **3:** 525–534.
44. HEATFIELD, B. M. & D. F. TRAVIS. 1975. Ultrasturctural studies of regenerating spines of the sea urchin *Strongylocentrotus purpuratus*. I. Cells without spherules. J. Morphol. **145:** 13–50.
45. HEATFIELD, B. M. & D. F. TRAVIS. 1975. Ultrasturctural studies of regenerating spines of the sea urchin *Strongylocentrotus purpuratus*. II. Cells with spherules. J. Morphol. **145:** 51–72.
46. HÖBAUS, E. 1979. Coelomocytes in normal and pathologically altered body walls of sea urchins. *In* Proceedings of the European Colloquium on Echinoderms. M. Jangoux, Ed.: 247–249. A. A. Balkema. Rotterdam, the Netherlands.
47. JOHNSON, P. T. & F. A. CHAPMAN. 1970. Abnormal epithelial growth in sea urchin spines *(Strongylocentrotus franciscanus)*. J. Invert. Pathol. **16:** 116–122.
48. CANICATTI, C. 1990. Lysosomal enzyme pattern in *Holothuria polii* coelomocytes. J. Invert. Pathol. **56:** 70–74.

49. Canicatti, C. & J. Tschopp. 1990. Holoenzyme-A, one of the serine proteases of *Holothuria polii* coelomocytes. Comp. Biochem. Physiol. **96B:** 739–742.
50. Ito, T., T. Matsutani, K. Mori & T. Nomura. 1992. Phagocytosis and hydrogen peroxide production by phagocytes of the sea urchin *Strongylocentrotus nudus*. Dev. Comp. Immunol. **16:** 287–294.
51. Roch, P., C. Canicatti & S. Sammarco. 1992. Tetrameric structure of the active phenoloxidase evidenced in the coelomocytes of the echinoderm *Holothuria tubulosa*. Comp. Biochem. Physiol. **102B:** 349–355.
52. Ryoyama, K. 1973. Studies on the biological properties of coelomic fluid of sea urchin. I. Naturally occurring hemolysin in sea urchin. Biochem. Biophys. Acta **320:** 157–165.
53. Ryoyama, K. 1974. Studies on the biological properties of coelomic fluid of sea urchin. II. Naturally occurring hemagglutinin in sea urchin. Biol. Bull. **146:** 404–414.
54. Canicatti, C. 1990. Hemolysins: Pore-forming proteins in invertebrates. Experientia **46:** 239–243.
55. Canicatti, C. 1991. Binding properties of *Paracentrotus lividus* (Echinoidea) hemolysin. Comp. Biochem. Physiol. **98A:** 463–468.
56. Stabili, L., L. Pagliara, M. Metrangolo & C. Canicatti. 1992. Comparative aspects of echinoidea cytolysins—the cytolytic activity of *Spherichinus granularis* (Echinoidea) coelomic fluid. Comp. Biochem. Physiol. **101A:** 553–556.
57. Canicatti, C., P. Pagliari & L. Stabili. 1992. Sea urchin coelomic fluid agglutinin mediates coelomocyte adhesion. Eur. J. Cell **58:** 291–295.
58. Vasta, G. R. 1992. Invertebrate lectins: Distribution, synthesis, molecular biology, and function. *In* Glycoconjugates: Composition, Structure, and Function. H. J. Allen & E. C. Kisallus, Eds.: 593–634. Marcel Dekker. New York.
59. Messer, L. I. & A. C. Wardlaw. 1980. Separation of coelomocytes of *Echinus esculentus* by density gradient centrifugation. *In* Proceedings of the European Colloquium on Echinoderms. M. Jangoux, Ed.: 319–323. A. A. Balkema. Rotterdam, the Netherlands.
60. Service, M. & A. C. Wardlaw. 1984. Echinochrome-A as a bactericidal substance in the coelomic fluid of *Echinus esculentus*. Comp. Biochem. Physiol. **79B:** 161–165.
61. Bertheussen, K. 1979. The cytotoxic reaction in allogeneic mixtures of echinoid phagocytes. Exp. Cell Res. **120:** 373–381.
62. Dales, R. P. 1992. Phagocyte interactions in echinoid and asteroid echinoderms. J. Mar. Biol. Assoc. U.K. **72:** 473–482.
63. Bang, F. B. & A. Lemma. 1962. Bacterial infection and reaction to injury in some echinoderms. J. Insect. Pathol. **4:** 401–414.
64. Prendergast, R. A. & M. Suzuki. 1970. Invertebrate protein simulating mediators of delayed hypersensitivity. Nature **227:** 277–279.
65. Beck, G., R. F. O'Brien & G. S. Habicht. 1986. Isolation and characterization of a primitive interleukin-1-like protein from an invertebrate, *Asterias forbesi.* Proc. Natl. Acad. Sci. **83:** 7427–7433.
66. Beck, G. & G. S. Habicht. 1990. Primitive cytokines: Harbingers of vertebrate defense. Immunol. Today **12:** 180–183.
67. Smith, L. C., R. J. Britten & E. H. Davidson. 1992. SpCoel1: A sea urchin profilin gene expressed specifically in coelomocytes in response to injury. Mol. Biol. Cell **3:** 403–414.
68. Aderem, A. 1992. Signal transduction and the actin cytoskeleton: The roles of MARCKS and profilin. TIBS **17:** 438–443.
69. Stossel, T. P. 1993. On the crawling of animal cells. Science **260:** 1086–1094.
70. Pollard, T. D. 1986. Rate constants for the reaction of ATP- and ADP-actin with the ends of actin filaments. J. Cell Biol. **103:** 2747–2754.
71. Goldschmidt-Clermont, P. J., M. I. Furman, D. Wachsstock, D. Safer, V. T. Nachmias & T. D. Pollard. 1992. The control of actin nucleotide exchange by thymosine and profilin. A potential regulatory mechanism for actin polymerization in cells. Mol. Biol. Cell **3:** 1015–1024.
72. Machesky, L., P. J. Goldschmidt-Clermont & T. O. Pollard. 1990. The affinities of human platelet and *Acanthamoeba* profilin isoforms for polyphosphoinositides account

for their relative abilities to inhibit phospholipase C. Cell Regul. **1:** 937–950.

73. GOLDSCHMIDT-CLERMONT, P. J., J. W. KIM, L. M. MACHESKY, S. G. RHEE & T. D. POLLARD. 1991. Regulation of phospholipase Cγ1 by profilin and tyrosine phosphorylation. Science **251:** 1231–1233.
74. HARTWIG, J. H., K. A. CHAMBERS, K. L. HOPCIA & D. J. KWIATOWSKI. 1989. Association of profilin with filament-free regions of human leukocyte and platelet membranes and reversible membrane binding during platelet activation. J. Cell Biol. **109:** 1571–1579.
75. LEE, J. J., R. J. CALZONE, R. J. BRITTEN, R. C. ANGERER & E. H. DAVIDSON. 1986. Activation of sea urchin actin genes during embryogenesis. Measurement of transcript accumulation from five different genes in *Strongylocentrotus purpuratus.* J. Mol. Biol. **188:** 173–183.
76. MAISEY, J. G. 1986. Heads and tails: A chordate phylogeny. Cladistics **2:** 201–256.
77. GANS, C. & R. NORTHCUTT. 1983. Neural crest and the origin of vertebrates: A new head. Science **220:** 268–274.
78. COULY, G. F., P. M. COLTEY & N. M. LE DOUARIN. 1993. The triple origin of skull in higher vertebrates: A study in quail-chick chimeras. Development **117:** 409–429.

Allorecognition and Humoral Immunity in Tunicates[a]

DAVID A. RAFTOS[b]

Immunobiology Unit
School of Biological and Biomedical Sciences
University of Technology
Sydney, N.S.W., 2007, Australia

INTRODUCTION: PHYLOGENY AND NEO-DARWINISM

Reconciliation of Molecular and Traditional Phylogenies

Recent molecular analyses have brought into question the validity of traditional animal phylogenies.[1] Classical evolutionary trees, which are based on morphological and embryological criteria, represent the phylogenesis of the animal kingdom as a gradual, branching process that incorporated two major animal lineages, the protostomes and deuterostomes (FIG. 1A).[1] However, sequence comparisons of highly conserved molecules such as 18S ribosomal RNA (rRNA) now suggest that the majority of modern coelomate and pseudocoelomate phyla arose extremely rapidly from common triploblastic ancestors during the early Cambrian period (FIG. 1B).[1,2] Such molecular phylogenies indicate that the phylum Chordata is not the apex of a gradually evolving deuterostomate lineage, but arose suddenly at approximately the same time as the other major coelomate groups.[1]

This modern view of animal phylogeny is consistent with contemporary reinterpretations of Darwinian evolutionary theory. Neo-Darwinism, particularly the theory of punctuated equilibrium, supports the concept of rapid, profound change.[3] The sudden radiation of metazoans by such rapid evolutionary diversification might imply that physiological mechanisms such as the integrated immune system of vertebrates are not the result of a gradual stepwise evolution, which would have resulted in the retention of readily identifiable intermediate forms among other extant phyla.

The Chordate Lineage

Molecular phylogenies bring into question the use of morphological and embryological criteria in discerning relationships over large phylogenetic distances.

[a]This work has been sponsored by grants from the Australian Research Council, the U.S. National Science Foundation, the Australian-American Educational Foundation (Fulbright Postdoctoral Fellowship), the American Association of Immunologists (Frederick B. Bang Fellowship in Marine Invertebrate Immunology), the University of Technology, Sydney, and Macquarie University.

[b]Address for correspondence: Immunobiology Unit, School of Biological and Biomedical Sciences, University of Technology, Sydney, P.O. Box 123, Broadway, N.S.W., 2007, Australia.

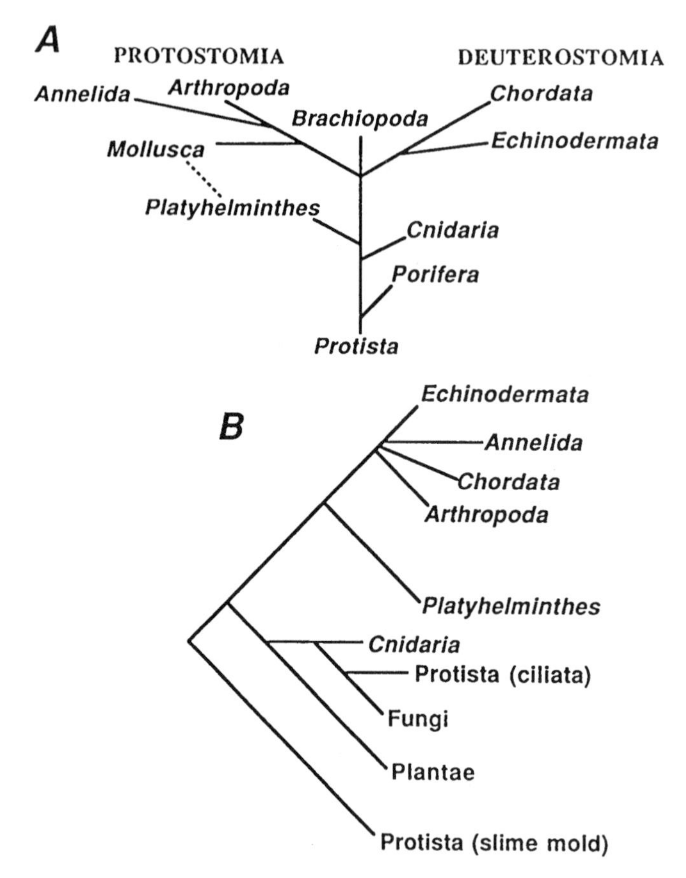

FIGURE 1. Phylogenetic trees for the metazoa. (**A**) Classical phylogeny based on traditional zoological comparisons of morphology and embryology. (**B**) an evolutionary tree based on partial sequences of 18S rRNAs (modified from Field *et al.*[1]).

Nevertheless, these zoological parameters appear to be robust when comparing animals within phyla. Both classical and contemporary phylogenies place tunicates within the phylum Chordata.[1,4] A number of estimates suggest that the ancestors of these filter-feeding invertebrates diverged from the common chordate lineage shortly after the major Cambrian radiations.[1] As such, tunicates are the most closely related invertebrate group to the true vertebrates.[4] It is this relationship that has spurred considerable interest in the immunological functions of tunicates.

Their evolutionary relatedness to vertebrates suggests that tunicates are the most likely invertebrates to express immunological reactions that are precursors of the vertebrate integrated immune response. Nevertheless, there is still no evidence to suggest that tunicates, or any other invertebrates, use immunoglobulin antibodies as humoral defense molecules within an integrated immune system.[5] Given that observation, two fundamental questions must be addressed. First, do invertebrates possess any other mechanisms that are obvious progenitors of the vertebrate integrated immune system? Second, what form of humoral defense is employed by tunicates in the absence of an antibody-mediated system? This review begins to address both of those questions. It describes two types of defense reaction—histocompatibility and adaptive inflammation—that have been characterized in solitary tunicates of the genus *Styela.*

HISTOCOMPATIBILITY RESPONSES

Kinetics of Graft Rejection in **Styela plicata**

The extraepithelial tunic tissue of the solitary tunicate *Styela plicata* has proven to be a useful medium for tissue transplantation studies.[6] FIGURE 2 shows that the majority of tunic allografts exchanged between individuals from within the same population are rejected, while autografts usually survive indefinitely. The median rejection time for first-set allografts is 38 days.[6]

Morphology of Allograft Rejection

Both allografts and autografts in *S. plicata* initially heal into their graft beds after transplantation (FIG. 3a).[7] Fusion is facilitated by the deposition of new tunicin fibers

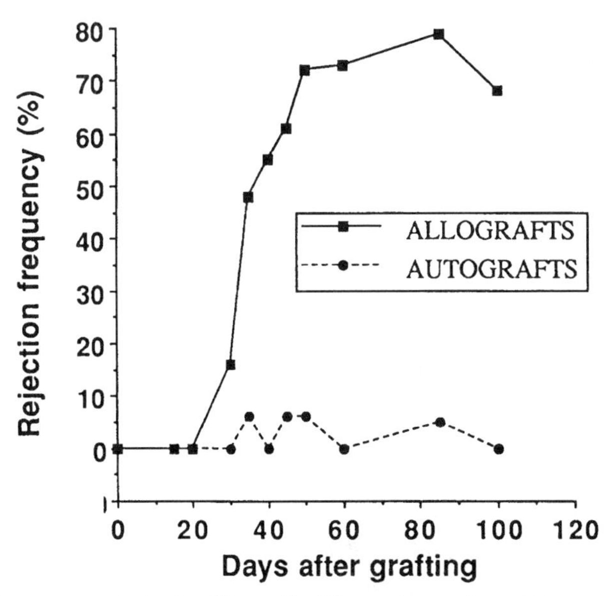

FIGURE 2. Frequency of rejection (%, $n > 9$) of first-set autografts and first-set allografts in *S. plicata.*

at the interface between graft and host. This extracellular, fibrous material comprises the majority of the tunic. Amoeboid cells and blood vessels represent only minor components of tunic tissue. Tunicin fiber deposition during the fusion of grafts to their hosts appear to be undertaken by highly vesiculated morula cells. These host cells aggregate at the graft/host interface soon after transplantation and have previously been implicated in tunic deposition.[8]

The fusion of autografts to their graft beds occurs to the degree that they rapidly become indistinguishable from the surrounding tissue (FIG. 3a). In contrast, the majority of allografts are subjected to a series of rejection reactions after initially fusing to the host.[7] Those rejection reactions culminate in the degeneration and elimination of allografts. The first stage of allograft rejection is characterized by the

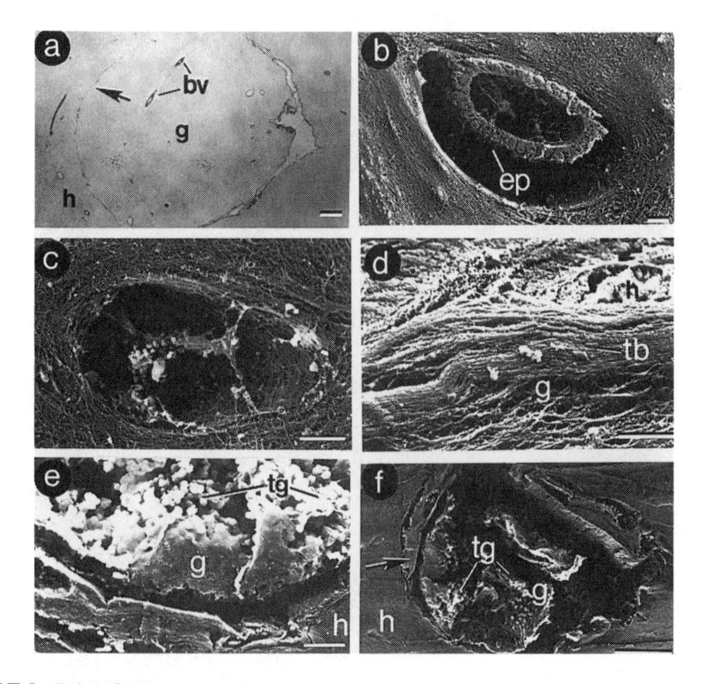

FIGURE 3. Brightfield and scanning electron micrographs of the allograft rejection response in *S. plicata.* **(a)** Cross section of an autograft 30 days after grafting. The graft is fused to the surrounding host tissue. The fusion interface between graft and host is shown by the arrow. Bar = 0.5 mm. **(b)** A blood vessel with intact cellular epithelium in an allograft 10 days after grafting. Bar = 10 μm. **(c)** A degenerative blood vessel within an allograft 35 days after grafting. The cellular epithelium has been destroyed. Bar = 10 μm. **(d)** Dense tunicin boundary deposited between an allograft and its host 30 days after grafting (immediately before graft separation and degeneration). Bar = 10 μm. **(e)** Formation of granules of extracellular tunicin material during the degeneration of an allograft 35 days after grafting. Bar = 100 μm. **(f)** Low-power micrograft of a degenerative allograft showing necrosis and detachment of the graft from its host *(arrow).* Bar = 500 μm. Abbreviations: bv, blood vessel; ep, vascular epithelium; g, graft; h, host; tb, tunicin boundary; tg, tunicin granules.

destruction of blood vessels within the transplanted tissue (FIG. 3b,c).[7] Dense boundaries of extracellular material are then deposited at the interface between graft and host, effectively amputating the allografts (FIG. 3d). Finally, the fibrous extracellular matrix of detached transplants undergoes a gradual necrosis that results in the elimination of the graft (FIGS. 3e,f).[7]

Cellular Restriction of Allorecognition

The discriminative process that allows the distinction of allografts from autografts appears to involve the recognition of allogeneic cellular determinants followed by the cytotoxic destruction of allogeneic cells within grafts.[9] Injecting allogeneic cells into naive hosts specifically preimmunizes them to subsequent tissue transplants (FIG. 4). This indicates that cellular determinants are the targets of recognition. A capacity for cytotoxic activity in response to the recognition of allogeneic cellular determinants is implied by the demonstration of allogeneic cellular cytotoxicity *in vitro.*[10]

These data suggest that the necrosis of extracellular tunic tissue that follows the cytotoxic destruction of allogeneic graft cells is essentially nonspecific and results directly from cytotoxic activity. That lack of specificity during the final degenerative

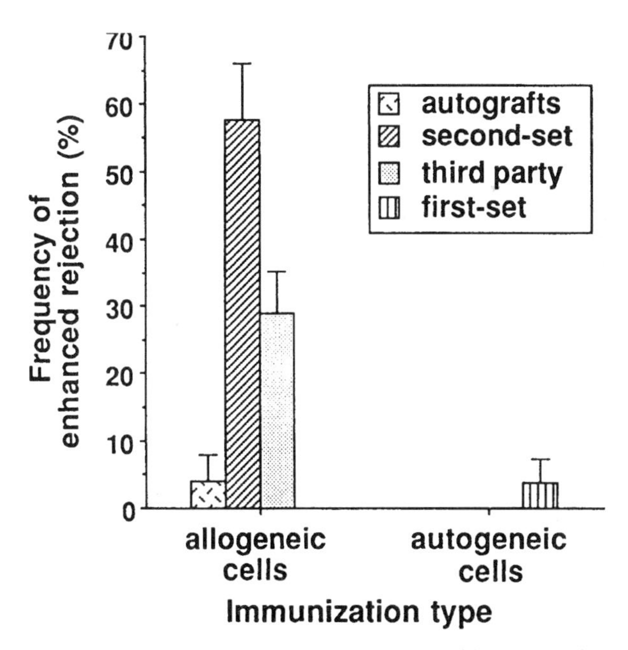

FIGURE 4. Effect of immunizing *S. plicata* with injections of either allogeneic or autogeneic hemocytes on the frequency of enhanced rejection (%, $n > 24$) among autografts and various types of allograft. Grafts were deemed to have been rejected by an enhanced response if they were eliminated within 30 days of grafting. This time point is known to accurately discriminate between rapid (memory-derived) and normal (first-set) responses.

phase of rejection is also indicated by experiments in which autografts were artificially depleted of their cellular components by irradiation. Those cell-depleted autografts underwent rejection in a manner analogous to that of allografts.[9] Again, this suggests that the recognition and destruction of cellular determinants is the discriminatory phase of allograft rejection.

Genetics of Histocompatibility

The molecular nature of the cellular determinants that stimulate allograft rejection has not been determined. Genetic analyses, however, have shown that alloantigenicity is determined by a single histocompatibility locus that incorporates multiple alleles.[11] The genetic "rules" that govern the recognition of allogeneic tissue in *Styelids* are identical to those of vertebrates. Reciprocal first-set allografting experiments have shown that the frequency of unilateral rejection, where one animal rejects while the other retains reciprocal allografts, is consistent with the specific identification of alloantigens.[12] Hence, a single disparate haplotype is sufficient to stimulate rejection. Only tunicates of identical tissue type are mutually compatible.

On the basis of these "rules" for allorecognition, comparative tissue typing has been used to demonstrate the monofactorial basis of alloantigenicity in *S. plicata.*[11] Analysis of rejection patterns in reciprocal allografting matrixes allowed the identification of a number of individuals that were homozygous for histocompatibility tissue type. Those homozygotes were used as donors to define the histocompatibility genotypes of large numbers of tunicates. This comparative tissue typing indicated that there are five distinct haplotypes within the test population. That limited number of alleles is inconsistent with polyfactorial inheritance. Moreover, Hardy-Weinberg analysis of genotype frequencies implies that the identified tissue haplotypes are encoded by a single gene locus.[11]

Implication of Lymphocyte-like Cells in Allograft Rejection

The identification of a causative role for cellular recognition in graft rejection has allowed a specific host cell type to be implicated in the allograft rejection process.[8] Large numbers of lymphocyte-like cells (LLCs) congregate around necrotic graft blood vessels during the cellular depletion phase of rejection (Figs. 5a,b). LLCs can be detected crossing the graft bed, and thus appear to be of host origin. Differential counts of cells at the graft/host interface have shown that LLCs are the only host cell type that accumulates around and within allografts but not autografts.[8] Such accumulation indicates that LLCs are responsible for the specific recognition and destruction of allogeneic cells that initiates graft rejection.[8]

Specific Alloimmune Memory

The involvement of LLCs in rejection has also helped to support the contention that histocompatibility in *Styelids* is an adaptive process.[8] LLCs have been shown to

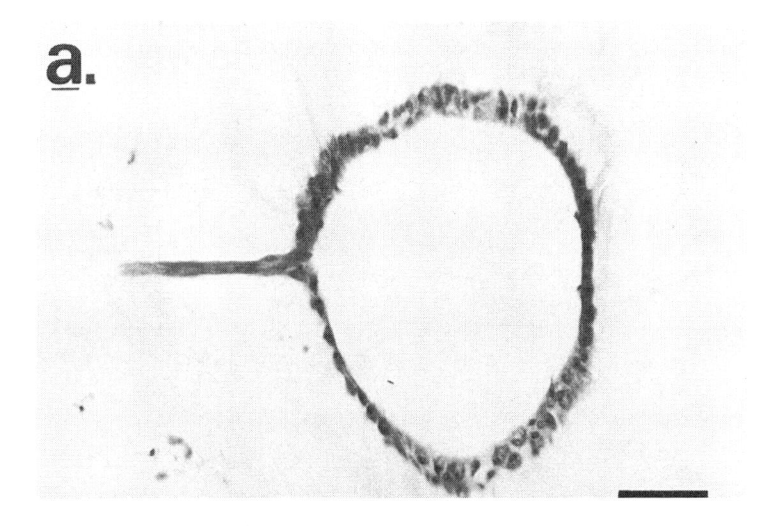

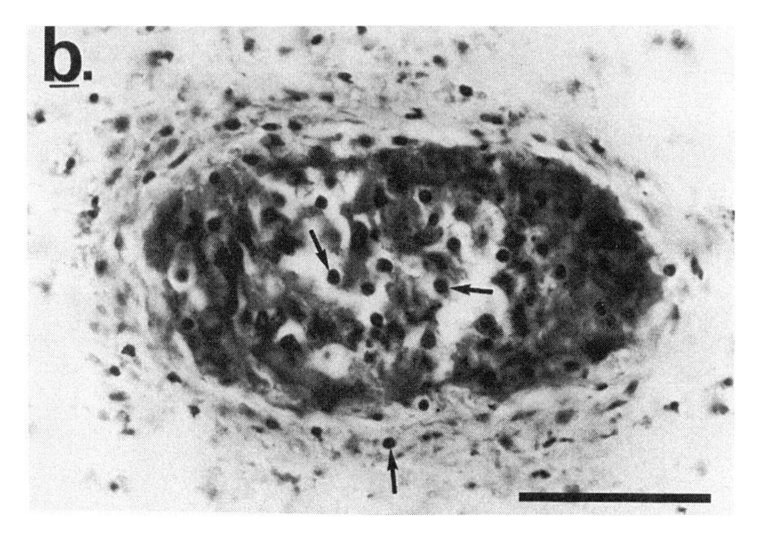

FIGURE 5. Brightfield micrographs of blood vessels within *S. plicata* allografts. (**a**) Blood vessel 15 days after grafting showing intact cellular epithelium and no evidence of host cell cytotoxic activity. (**b**) Lymphocyte-like cells from the host (*arrows* show examples) accumulating around a degenerating blood vessel in an allograft 30 days after grafting. Bars = 100 μm.

proliferate in response to allogeneic challenge.[13] Immunohistochemical analysis of bromodeoxyuridine-labeled cells within hematopoietic crypts of the body wall revealed a significantly greater number of proliferative LLCs in allogeneically immunized tunicates when compared to autogeneic controls (FIG. 6).[13] This enhanced proliferation after allogeneic immunization may account for the specifically increased accumulation of LLCs that can be detected within second-set allografts.[13] It may also be responsible for the increased rate of allograft rejection that is stimulated by immunizing tunicates with either hemocytes or tunic transplants.[6,9]

Although such responses suggest that *S. plicata* has the capacity for alloimmune memory, the data indicating immunological specificity are complicated by cross-reactivity to some third-party allografts. At all of its levels of expression, alloimmune memory extends to approximately 25% of third-party grafts.[6,8,9,13] To explain such third-party responses in the context of specific memory, it should be noted that the observed cross-reactivity closely approximates the level that would be expected to result from chance similarity between second-set and third-party donors given the limited allogeneic polymorphism in the test population.[11] This explanation is sup-

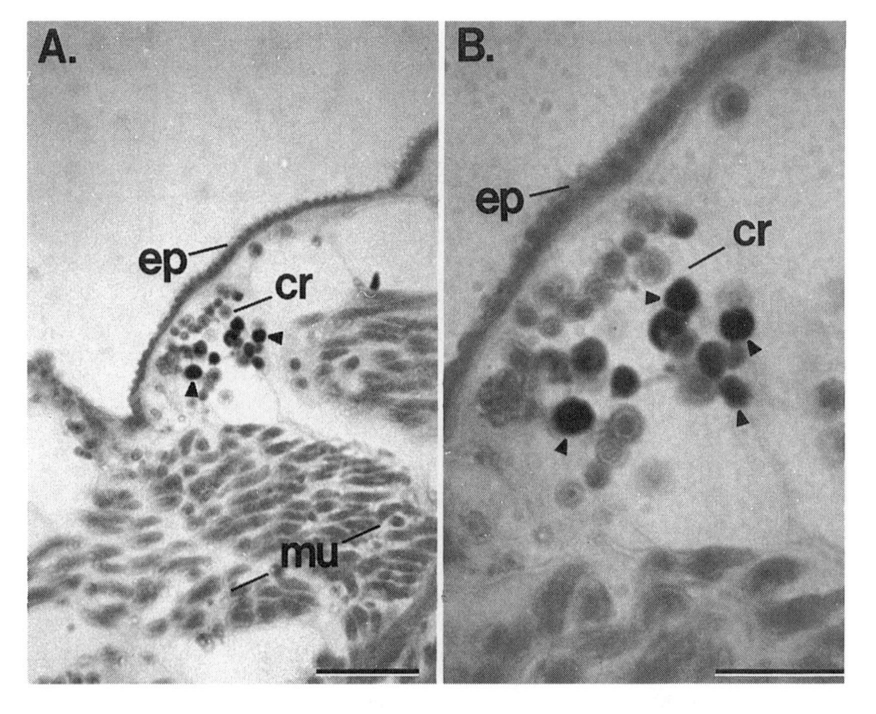

FIGURE 6. Immunohistochemical analysis of cell proliferation in bromodeoxyuridine (BrDu)-pulsed body wall tissue from allogeneically immunized *S. plicata.* Both micrographs show darkly stained, BrDu-positive lymphocyte-like cells within hematopoietic crypts of the body wall. Scale bars: **A** = 100 μm; **B** = 20 μm. Abbreviations: cr, crypt of proliferating cells; ep, body wall epithelium; mu, muscle.

ported by allografting experiments that have shown that cross-reactive third-party tissue usually shares haplotypes with preimmunizing (second-set) donors.[6] Those experiments tend to confirm the existence of alloimmune memory rather than confound it.

Comparison of Tunicate and Vertebrate Histocompatibility Systems

The histocompatibility system of *S. plicata* bears at least superficial similarity to its vertebrate counterpart. In both systems, the rejection of allogeneic tissue follows the specific recognition of cellular determinants that are encoded by discrete, polymorphic histocompatibility loci. As in vertebrates, lymphocyte-like cells in tunicates appear to be responsible for the recognition of allogeneic differences, to which they respond with proliferation and cytotoxicity. That cellular activity correlates with the expression of alloimmune memory.

Such similarities may reflect homology between the histocompatibility mechanisms of tunicates and vertebrates. Indeed, it is difficult to identify any other physiological system in invertebrates from which the integrated immune system of vertebrates could have arisen. That difficulty in identifying alternative precursors may, however, simply reflect a persistent underestimation of the power and speed of evolutionary diversification. Neo-Darwinist theories suggests that profound change can be extremely rapid.[3] It may be that the integrated immune system of vertebrates evolved from a functionally unrelated, and hence well-disguised system with such rapidity that detectable intermediates will never be conclusively identified. It must also be stressed that the mechanistic similarities between tunicate and vertebrate histocompatibility reactions are superficial. There is no molecular evidence to support a close relationship between the two systems.

Functions of Histocompatibility Systems

Although they have superficially similar mechanisms, allograft rejection in tunicates and vertebrates may subserve different physiological functions.[14] There is little doubt that the major role of histocompatibility antigens in vertebrates is antigen presentation. Vertebrate transplantation rejection reactions are an inevitable but biologically irrelevant consequence of antigen presentation and recognition.

In tunicates there is no data regarding antigen presentation. Nevertheless, it is clear that, unlike vertebrates, tunicate histocompatibility reactions subserve a direct and essential physiological function. Allogeneic recognition prevents fusion between conspecifics in the close aggregations that typify natural populations of tunicates.[14] That direct role for allorecognition has been confirmed by the identification of extremely rare, fused pairs of *S. plicata in situ.* Transplantation experiments have shown that those fused animals always share histocompatibility tissue types. Moreover, allozyme electrophoresis has confirmed that the shared histocompatibility tissue types of fused animals do not arise from sibship or chimerism. This suggests that fused pairs represent extremely rare, chance associations between mutually compatible individuals and, by corollary, that the histocompatibility system of *Styel-*

ids otherwise prevents the fusion of conspecifics. The involvement of histocompatibility reactions may help to preserve genetic identity or prevent the transmission of pathogens between fused animals.[14] Such selection pressures could represent fundamental forces that drove the evolution of allorecognition in many phyla. Alternatively, the histocompatibility system of *S. plicata* could be the result of direct adaptation to the specific life history of tunicates.

HUMORAL DEFENSE MOLECULES IN TUNICATES

Absence of Immunoglobulin

Although the precursors of vertebrate integrated immunity have not yet been identified among invertebrates, it is unlikely that tunicates and other invertebrates use rearranging immunoglobulins as humoral defense molecules. However, tunicates possess many of the fundamental cellular effector mechanisms that are used by antibody in vertebrates. This suggests that the recent advent of clonally derived immunoglobulins exploited pre-existing cellular mechanisms of defense that employed recognition molecules other than antibody. One such recognitive mechanism is described below.

Phagocytosis and Opsonization

The most readily identified effector mechanism on which invertebrates rely is phagocytosis. Tunicate hemocytes are avidly phagocytic. That phagocytic activity depends on the capacity of humoral factors to opsonize foreign antigens. FIGURE 7 shows that the ingestion of Congo red–stained yeast by *Styelid* phagocytes is significantly enhanced by preincubating target cells with hemolymph.[15,16] Kinetic studies suggest that opsonization enhances both the rate and total capacity for phagocytosis.

Opsonic factors in tunicate hemolymph act by recognizing and binding target cells. That ability to bind antigen has been confirmed by experiments in which the opsonic activity of hemolymph was depleted by absorption with target cells. The interaction implied by this depletion is relatively specific. Absorbing hemolymph with sheep red blood cells cannot eliminate opsonic activity directed against yeast (FIG. 7).[15]

Once bound to their antigens, opsonins must have the capacity to stimulate the ingestion of target–opsonin complexes. Phagocytosis may be elicited by conformational changes to opsonins that result from their binding to antigen. Those conformational changes might facilitate interaction with specific opsonic receptors on tunicate hemocytes. The existence of such receptors is implied by the species specificity that is evident for opsonization in *Styelids.*[15]

It is also clear that recognition of foreign targets by opsonic factors in *Styelid* hemolymph can be inhibited by selected carbohydrates and EDTA. Of the eight monosaccharides tested so far, *N*-acetyl galactosamine and α-D-galactose effectively inhibit opsonization (unpublished data).[15] This indicates that the opsonic factors in

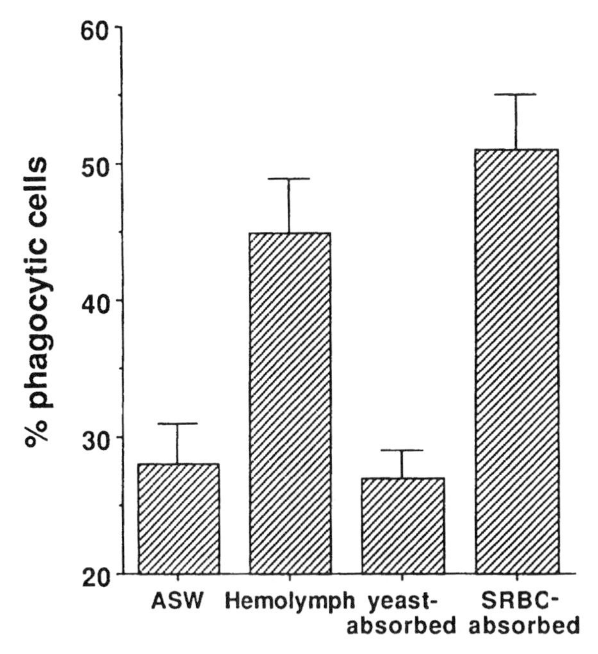

FIGURE 7. Opsonic activity of *S. clava* hemolymph. Percentage (n = 6) of *S. clava* hemocytes ingesting (within 30 min) Congo red–stained yeast cells that had been incubated with either control medium (ASW), unfractionated hemolymph, hemolymph that had been absorbed twice with yeast cells, or hemolymph that had been absorbed with glutaraldehyde-fixed sheep red blood cells (SRBC).

Styelid hemolymph are lectins that bind to target cells via carbohydrate recognition units that are specific for galactosyl configurations. The strong inhibitory effect of EDTA on opsonization suggests that this lectin-like recognitive activity is divalent cation dependent.

Isolation of Tunicate Opsonins

Gel filtration and chromatofocusing have been used to isolate the opsonic factors from the hemolymph of two *Styelid* species, *S. clava* and *S. plicata.* A single peak of opsonic activity is evident among FPLC and conventional gel filtration fractions.[17,18] That peak can comprise multiple proteins with molecular weights of 14 to 18 kDa. Further characterization by isoelectric focusing or chromatofocusing has identified two major opsonic proteins that have been designated tunIL1-α (pl 5.5) and tunIL-β (pl 7.0).[18] Preliminary data suggests that those two molecules are of similar structure, so that one may represent a truncated form or allelic variant. TunIL1-β is the predominant form and can often be isolated to unity by gel filtration alone.[17] This protein exhibits similar EDTA sensitivity and carbohydrate specificity to that of unfractionated hemolymph, and so probably accounts for most of the lectin-mediated opsonic activity that is evident in hemolymph.[16]

TunIL-β from *S. clava* has recently been subjected to NH_2-terminal and internal amino acid sequencing. Oligonucleotide primers based on amino acid sequence data have been used in polymerase chain reactions (PCR) to amplify cDNA from a library constructed using *S. clava* hemocyte mRNA. Amplified products within the appropriate size range to encode 14- to 18-kDa proteins have been identified and cloned. Those products are currently being sequenced to determine the primary structure of tunIL1-β.

Additional Biological Activities of tunIL1

Lectins such as tunIL1 have been identified in many invertebrate species including tunicates.[19,20] They have also frequently been associated with defensive reactions.[19] Although the requirement for purified material has often prevented the strict correlation of lectins with opsonization, there are a number of reports that confirm that invertebrate lectins can act as powerful humoral opsonins.[19,21] Hence, it is not surprising that tunIL1 from *Styelids* uses lectin-like activity to opsonize target cells, even though this represents the first correlation of opsonization with lectin-mediated recognition in tunicates. TunIL1, however, is set apart from many other invertebrate lectins by the additional biological activities that it has been shown to subserve in its native species. Those additional functions, which are described in detail below, may reflect an adaptive divergence of lectin activity in tunicates. Alternatively, lectins in other invertebrates may exert diverse biological activities that have not yet been identified.

TunIL1 Stimulates Tunicate Cell Proliferation

TunIL1-β, but not tunIL1-α, can stimulate the proliferation of *Styelid* pharyngeal explants.[22,23] The pharynx is a major hematopoietic organ in tunicates. Proliferative responses to tunIL1 in pharyngeal tissue are dose dependent and can be eliminated by removing tunIL1 from culture medium (FIG. 8). TunIL1-β is an extremely potent mitogen, being active at concentrations as low as 150 ng/ml.[22]

TunIL1 Is a Chemoattractant

Both tunIL1-α and -β can stimulate the migration of tunicate hemocytes.[23,24] Again, enhanced migration is dose dependent, with as little as 150 ng/ml eliciting a migratory response (FIG. 9). Checkerboard analysis has confirmed that the enhanced migration elicited by tunIL1 reflects a directed movement of hemocytes along concentration gradients and so represents true chemotaxis.[24] It does not simply result from an increase in the undirected amoeboid activity of hemocytes resulting from metabolic stimulation (chemokinesis).

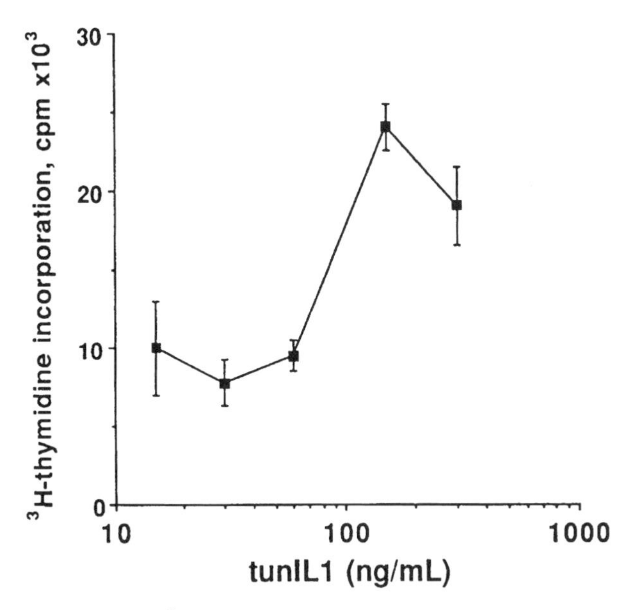

FIGURE 8. Incorporation of [^{3}H]thymidine by pharyngeal explants of *S. clava* cultured for three days in various concentrations of tunIL1-β. Bars = SEM; *n* > 8.

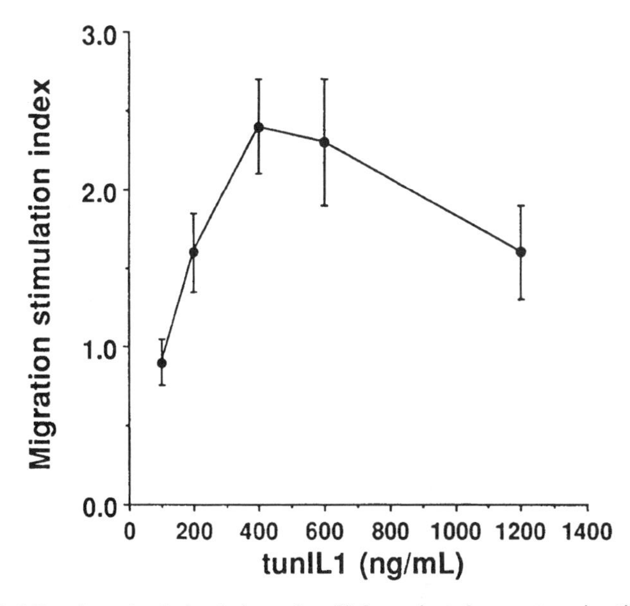

FIGURE 9. Migration stimulation indexes (*n* > 3) for tunicate hemocytes migrating through chemotaxis membranes toward various concentrations of tunIL1 placed in lower wells of chemotaxis chambers.

Metabolic Activation by tunIL1

Incubation of tunicate hemocytes with tunIL1-β significantly enhances their capacity to incorporate labeled amino acids and glucose (unpublished data). This suggests that tunIL1 can stimulate metabolic pathways of hemocytes. That enhanced metabolic activity may also be responsible for the increase in phagocytosis that is evident when tunIL1 is used to activate hemocytes directly rather than to opsonize antigenic targets.[16,23] Hemocytes that have been incubated with tunIL1 and then extensively washed have a far greater propensity to ingest nonopsonized target cells than untreated controls. This phagocytic activation can be explained either by the integration of tunIL1 into phagocyte membranes where it could act as a cell surface opsonin or by the ability of tunIL1 to directly activate hemocytes in the absence of antigen. The latter alternative is supported by experiments in which the ingestion of latex beads, which do not contain carbohydrate determinants and cannot be recognized or bound by tunIL1, is enhanced by activating hemocytes directly with tunIL1.[16,23]

Adaptive Secretion of tunIL1

The secretion of tunIL1 can be significantly increased by stimulating tunicate hemocytes with antigen.[23,25] Culture supernatants conditioned by tunicate hemocytes

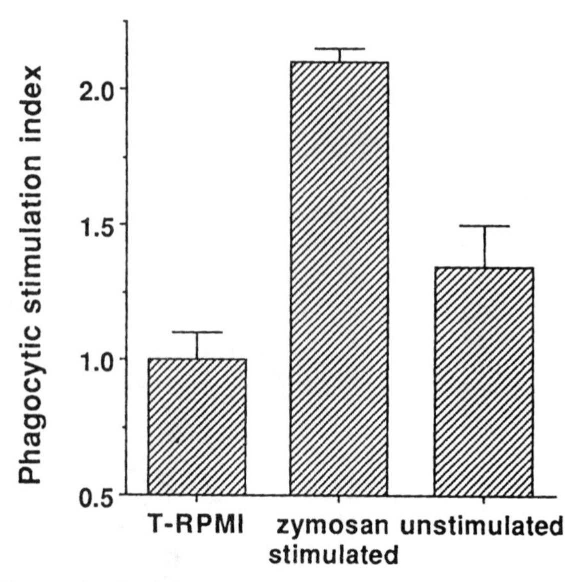

FIGURE 10. Phagocytic stimulation indexes for tunicate hemocytes exposed to yeast cells that had been incubated with either culture medium alone (T-RPMI), culture supernatants from hemocytes stimulated with zymosan, or culture supernatants from unstimulated hemocytes. Bars = SEM; $n > 7$.

that had been stimulated with zymosan are capable of enhancing proliferation and phagocytosis (FIG. 10).[25] Electrophoretic analysis indicates that this stimulatory activity is associated with the adaptive release of tunIL1-β by antigen-activated hemocytes.[23,25]

TunIL1 as a Central Regulator of Defensive Responses

The ability to enhance phagocytosis, cell proliferation, chemotaxis, and metabolic activity, as well as its capacity for adaptive release upon antigenic stimulation, suggests that tunIL1 may be central to a nonclonal, but inducible defense system in tunicates (FIG. 11).[26] A mechanism can be envisaged where tunIL1 recognizes carbohydrate moieties on the surface of pathogenic invaders. That recognitive activity would result in the ingestion of antigen and, in turn, could stimulate the release of additional tunIL1 from phagocytic hemocytes. Adaptive secretion might further enhance phagocytic clearance of pathogen via opsonization and by directly activating phagocytes. Increased localized tunIL1 concentrations may also act to recruit additional hemocytes to the site of infection via chemotaxis and enhanced cell proliferation. Recruited hemocytes would contribute to antigen clearance and to additional tunIL 1

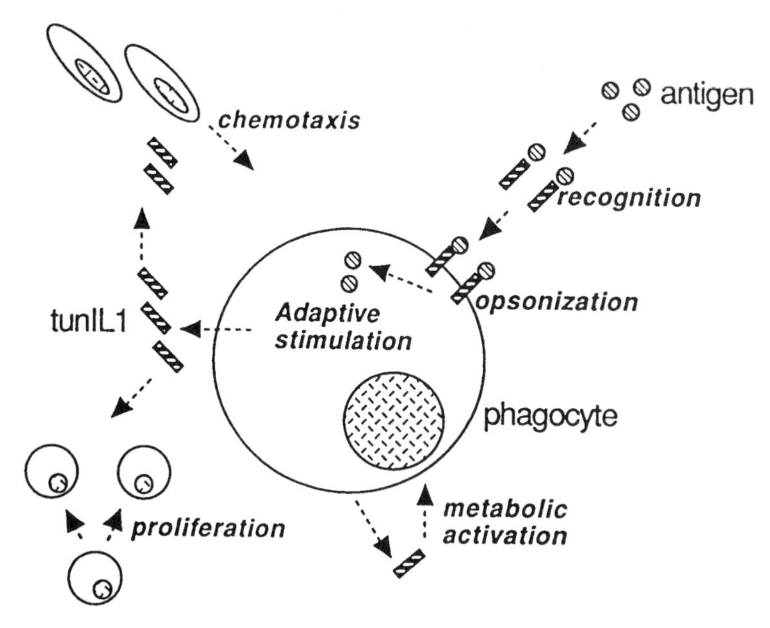

FIGURE 11. This schematic model depicts the interaction of cellular functions that may be mediated by tunIL1 molecules during defensive reactions in tunicates. tunIL1 acts as a recognition molecule that facilitates the phagocytosis of foreign cells by tunicate blood cells. In turn, recognition may increase the secretion of tunIL1, resulting in enhanced metabolic activity and the recruitment of additional blood cells to the site of infection via chemotaxis and cell proliferation.

secretion. In the absence of antibody-mediated immune responses, this system of adaptive inflammation is likely to be a critical defense mechanism in tunicates.

Possible Homologies of tunIL1 to Vertebrate Immune-Related Molecules

The evolutionary affinities of tunIL1 remain unclear. They will only be thoroughly defined when complete nucleotide and deduced amino acid sequence data becomes available. Hence, it should be emphasized that the following discussion is conjectural.

As its name suggests, tunIL1 shares a number of physicochemical and functional characteristics with the mammalian inflammatory cytokine, interleukin 1 (IL1).[18] tunIL1 was first identified by Beck *et al.* on the basis of its similarity to IL1.[18] They showed that both tunIL1 and IL1 have monomeric molecular weights of 14 to 18 kDa. tunIL1 and IL1 also segregate into a number of allotypes with predominant forms that have similar isoelectric points (IL1-β and tunIL1-β, pl 7.0–7.2; IL1-α and tunIL1-α, pl 5.5–6.0). As with tunIL1, IL1 can stimulate cell proliferation; its production and secretion is induced by zymosan; and it has been implicated, albeit arguably, in chemotaxis and phagocytosis.[27] When tested in mammalian systems, tunIL1 can stimulate the proliferation of murine thymocytes and L929 fibroblasts, as well as increasing the vascular permeability of rabbit skin (unpublished data).[18] These functions are also traditionally associated with mammalian IL1. It even appears that tunIL1 and IL1 share lectin-like activity. A number of reports suggests that mammalian IL1 can act as a lectin by binding to specific carbohydrates.[28,29]

tunIL1 also bears similarities to members of the C-type lectin family. C-type lectins are typified by divalent cation dependence and galactosyl specificities.[30] They share sequence homology in their carbohydrate-binding domains, particularly at 18 invariant residues that apparently confer a tertiary structure amenable to carbohydrate binding.[30] C-type lectins have been identified in many invertebrates and vertebrates. Among the mammalian representatives of this molecular family are the mannose-binding proteins (MBPs). MBPs are acute-phase reactants, the production and secretion of which are dramatically enhanced in the early, acute stages of infection.[31] The physiological functions of MBPs and other acute-phase reactants are largely unknown. However, the recent implication of MBPs in antibody-independent opsonization suggests that they may subserve a similar role in the regulation of defensive responses to that of tunIL1 in tunicates.[31] There is no structural data to suggest that tunIL1 is homologous to MBPs or other C-type lectins. However, a member of the C-type lectin family that exhibits similar molecular weight, divalent cation-dependence, and carbohydrate specificity to tunIL1 has recently been identified in another tunicate species.[32]

FUTURE DIRECTIONS

The research described here is being used as a basis for studies in our laboratory that will further characterize invertebrate defense systems *per se.* This applied research has two specific aims. First, we will test the effect of environmental pollution on the assays of tunicate cell function that have already been developed. The estuarine habitat and filter-feeding life history of *Styelid* tunicates may make them sensitive biomonitors of environmental contamination. Second, we hope to identify regulatory proteins that are comparable to tunIL1 in commercially exploited invertebrate species. This research may lead to the development of methods to artificially modulate the defense systems of farmed invertebrates in order to enhance their resistance to disease.

ACKNOWLEDGMENTS

The studies of *Styelid* tunicates described here were undertaken in collaboration with David Briscoe, Noel Tait, Elizabeth Kingsley (School of Biological Sciences, Macquarie University, N.S.W., Australia), Edwin Cooper, Karen Kelly, Dan Stillman (Department of Anatomy and Cell Biology, UCLA School of Medicine, University of California at Los Angeles, California), Greg Beck, Gail Habicht (Department of Pathology, State University of New York, Stony Brook, NY), Robert Raison, and Monika Burandt (Immunobiology Unit, University of Technology, Sydney, N.S.W., Australia).

REFERENCES

1. Field, K. G., G. J. Olsen, D. J. Lane, S. J. Giovannoni, M. T. Ghiselin, E. C. Raff, N. R. Pace & R. A. Raff. 1988. Science **239:** 748–753.
2. Hillis, D. M. & M. T. Dixon. 1991. Q. Rev. Biol. **66:** 11–446.
3. Gould, S. J. & N. Eldredge. 1977. Paleobiology **3:** 115–151.
4. Berril, N. J. 1955. The Origin of Vertebrates. Oxford University Press. New York.
5. Smith, L. C. & E. H. Davidson. 1992. Immunol. Today **13:** 356–361.
6. Raftos, D. A., N. N. Tait & D. A. Briscoe. 1987. Dev. Comp. Immunol. **11:** 343–351.
7. Raftos, D. A. 1990. Cell Tiss. Res. **261:** 389–296.
8. Raftos, D. A., N. N. Tait & D. A. Briscoe. 1987. Dev. Comp. Immunol. **11:** 713–726.
9. Raftos, D. A. 1990. Dev. Comp. Immunol. **15:** 241–249.
10. Kelly, K., E. L. Cooper & D. A. Raftos. 1992. J. Exp. Zool. **262:** 202–208.
11. Raftos, D. A. & D. A. Briscoe. 1990. J. Hered. **81:** 160–166.
12. Raftos, D. A., N. N. Tait & D. A. Briscoe. 1988. Transplantation **45:** 1123–1126.
13. Raftos, D. A. & E. L. Cooper. 1991. J. Exp. Zool. **260:** 391–400.
14. Kingsley, E., D. A. Briscoe & D. A. Raftos. 1989. Biol. Bull. **76:** 282–289.
15. Kelly, K., E. L. Cooper & D. A. Raftos. 1993. Dev. Comp. Immunol. **17:** 29–39.
16. Beck, G., R. F. O'Brien, G. S. Habicht, D. L. Stillman, E. L. Cooper & D. A. Raftos. Cell. Immunol. **146:** 284–299.
17. Kelly, K., E. L. Cooper & D. A. Raftos. 1992. Comp. Biochem. Physiol. **103B:** 749–753.
18. Beck, G., G. R. Vasta, J. J. Marchalonis & G. S. Habicht. 1989. Comp. Biochem. Physiol. **92B:** 93–97.
19. Vasta, G. R. 1991. *In* Phylogenesis of Immune Functions. G. W. Warr & N. Cohen, Eds.: 73–102. CRC Press. Boca Raton, FL.
20. Vasta, G. R., G. W. Warr & J. J. Marchalonis. 1982. Comp. Biochem. Physiol. **73:** 887–900.
21. Renwrantz, L. & A. Stahmer. 1983. J. Comp. Physiol. **149:** 535–546.
22. Raftos, D. A., E. L. Cooper, G. S. Habicht & G. Beck. 1991. Proc. Natl. Acad. Sci. USA **88:** 9518–9522.

23. Kelly, K., E. L. Cooper & D. A. Raftos. 1993. Zool. Sci. **10:** 57–64.
24. Raftos, D. A., D. L. Stillman, G. S. Habicht, E. L. Cooper & G. Beck. In preparation.
25. Raftos, D. A., E. L. Cooper, D. L. Stillman, G. S. Habicht & G. Beck. 1992. Lymphokine Cytokine Res. **11:** 235–240.
26. Raftos, D. A. & R. L. Raison. 1992. Today's Life Sci. **14:** 16–20.
27. Dinarello, C. A. & J. G. Cannon. 1986. Prog. Immunol. **6:** 449–457.
28. Muchmore, A. V. & J. M. Decker. 1987. J. Immunol. **138:** 2541–2546.
29. Brody, D. T. & S. K. Durum. 1989. J. Immunol. **143:** 1183–1187.
30. Drickamer, K. 1988. J. Biol. Chem. **263:** 9557–9560.
31. Ezekowitz, R. A. B. 1991. Curr. Top. Biol. **1:** 60–62.
32. Suzuki, T., T. Takagi, T. Furukohri, K. Kawamura & M. Nakauchi. 1990. J. Biol. Chem. **265:** 1274–1281.

Specificity and Memory in Invertebrates[a]

EDWIN L. COOPER

Laboratory of Comparative Immunology
Department of Anatomy and Cell Biology
School of Medicine
University of California
Los Angeles, California 90024

INTRODUCTION

What is Memory?

According to a popular dictionary, memory refers specifically to the mind's ability for retention and revival of past thoughts, images, ideas, and so forth. Memory, however, can include not only the nervous system but also processes employed by the immunological network. Gray has recently asserted that during the last five years there has been a resurgence of interest in immunological memory.[1] This active or passive memory is a specific function of specialized cells formed during the adaptive immune response to an antigenic stimulus. Memory cells, demonstrable by adoptive transfer, are small lymphocytes that behave similarly during cell-mediated responses. In B-cell humoral responses, one can quantitate memory by measuring antibody levels from about the fifth day after the initial stimulus, during the primary response, when it can first be detected (precipitation, agglutination, or complement fixation) and again during the peak response. If a booster injection of antigen is given, the body, already primed and ready with its larger clone of specific memory cells, responds more rapidly, more strongly, and for a longer period.

The Challenge of Invertebrate Memory

This topic was deemed so important nearly 15 years ago that a major symposium was devoted to the evolution of memory.[2] It is difficult to treat all information on memory as it applies to invertebrate immunity[3]; thus, this paper will be restricted to cellular responses among invertebrates, where a memory component has been demonstrable, and will focus on cellular and humoral aspects in earthworms. The concern with memory here and ten years ago is in marked contrast to a recent assertion that to search for immunological memory (or perhaps to be preoccupied with its presence or absence *vis à vis* other crucial and relevant immune mechanisms) is a "damp exercise."[4] Crucial experiments must be performed in greater depth to strengthen the existing evidence for memory. First, the challenge times for second-

[a]This work was partially supported by the National Science Foundation and the Environmental Protection Agency.

set grafts should be varied so we can gain an appreciation for the persistence and strength of memory. Second, because of the cell diversity of invertebrate defense systems, experiments should be performed to show the kinds of cells responsible for transferring the response. Third, it should be our aim to test the specificity of first sensitization by observing the rejection time of second grafts following various kinds of sensitization. Finally, we should then begin to identify subpopulations of invertebrate immune or defense cells.

Two major groups of invertebrates have been classified according to how the mouth is formed. The groups include the Protostomia (e.g., Annelida, Mollusca, Arthopoda) and the Deuterostomia (e.g., Echinodermata and Tunicata) (FIG. 1). I have presented this definition of Protostomes and Deuterostomes elsewhere, but am reminded of its necessity and utility here.[5] First, we often forget that the invertebrates are classified into two major groups. Second, on the basis of this classification rests the structure of our facts and ideas related to the origin of memory. For purposes of

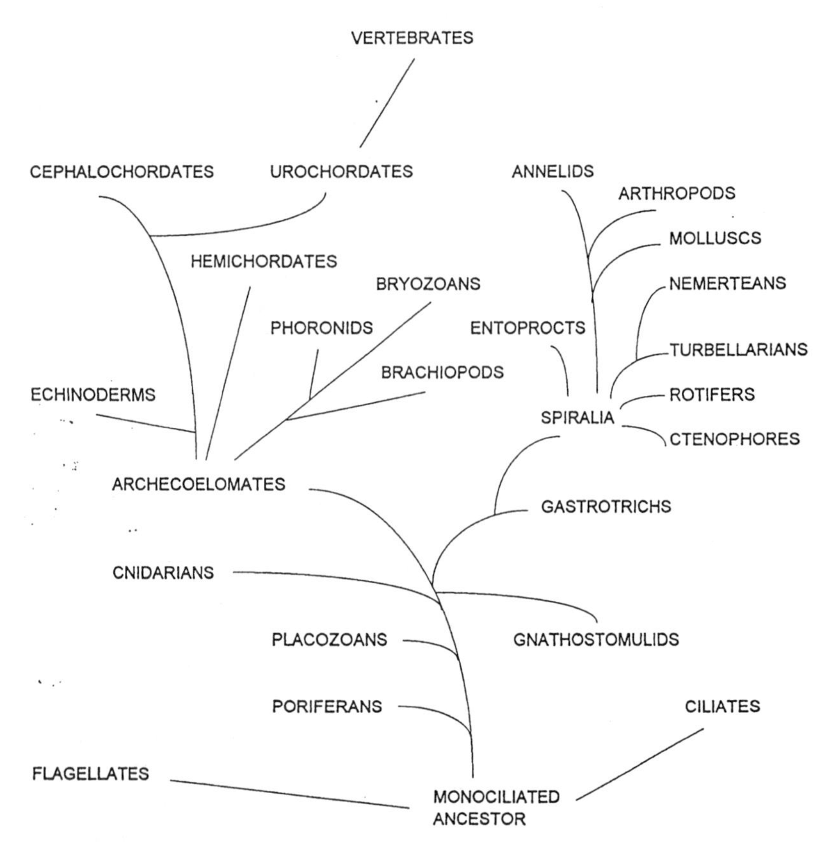

FIGURE 1. Phylogenetic tree showing the invertebrates and their relationships as well as the position of humans (vertebrates) in this scheme. The echinoderms and chordates represent the deutoerostomes, and the annelids, molluscs, and arthropods represent the protostomes.

order, these classifications are necessary; however, such orderliness is often transient. Thus, we rest with the present system, assuming that vertebrate anamnestic responses evolved from the deuterostome side.

MEMORY IN METAZOANS: FROM SPONGES TO ECHINODERMS

The existence of a memory-type response has now been traced to the earliest of the metazoa, the sponges, which possess potential immunocytes, notably amoebocytes. Three components as minimal criteria for immunological competence have been identified.[6] Since then, with respect to invertebrates, these criteria have changed little despite sporadic investigations. These include (1) cytotoxic or antagonistic reactions following sensitization; (2) selective or specific reactivity; and (3) inducible memory or selective altered reactivity on secondary contact. In this presentation because of lack of space, I will focus on the earthworm (Protostomia, Annelida). However, readers should not ignore the wealth of literature that deals with simple and complex metazoa and where memory has been demonstrated: (a) Porifera,[6] (b) Cnidaria,[7,8] (c) Nemertea,[9] and (d) Echinodermata.[10–13]

OLIGOCHAETE ANNELIDS (EARTHWORMS)

Introduction to the Intrafamilial Xenograft Model **Lumbricus < > Eisenia**

Since the early 1960s the earthworm has emerged as the prototypic invertebrate for studying the early characteristics of cellular immunity. It is well known that three laboratories have been active in deciphering the phenomenon of graft rejection. Clearly, first-set graft rejection does occur in allogeneic and xenogeneic situations. Moreover, there is an accelerated rejection response when second sets are performed. At least three approaches have been used to show the activity of secondary host effector cells as memory cells. These include (1) gross changes in the response to a second transplant, (2) adoptive transfer of the response by coelomocytes, and (3) quantitation of the coelomocyte response to secondary grafts. One of the most recent reviews was by Cooper (1986),[14] and further advances are found in a full book devoted to immunity in annelids by Vetvicka *et al.* (1993).[15]

Gross Manifestations of Memory to Xenografts: The "2/3" Response Represents Positive/Negative or Low/High Responders

In the first major work on earthworms published in *Transplantation,* I observed the destruction of first-set xenografts at times ranging from 6 to 147 days. Most second-set xenografts were destroyed by accelerated reactions.[16] Nevertheless, a significant number showed the opposite effect, that is, prolonged survival in relation to first-set grafts, here referred to as the "2/3" reaction. Two important features emerged: (1) acute rejectors (11 days or less) of first sets always showed prolonged survival of second sets, whereas chronic rejectors generally showed accelerated destruction of repeat grafts; (2) short intervals (0–4 days) between first- and second-set grafts produced accelerated rejections, while extended intervals (33–71 days) led to prolonged survival of repeat grafts. Early onset of rejection of second-set grafts usually was followed by accelerated breakdown, whereas late onsets of rejection were followed by prolonged survival. There was no apparent donor specificity in the rejection times of repeat grafts, which probably reflects the predominance of many common species-specific xenogeneic antigens coupled with an apparent lack of precise host receptors for their detection.

Quantitation of Cellular Anamnesis in the Xenograft System

The quantitative response of coelomic cells associated with first- and second-set *Eisenia* xenografts transplanted to *Lumbricus* hosts at 20°C has been compared with autografts and nonspecific wounds.[17] Coelomocyte numbers were significantly lower in response to first- and to second-set xenografts. Coelomocytes also increased in association with autografts and nonspecific wounds, but the reaction is short-lived and essential for early wound healing and repair. Such nonspecific increases are different from subsequent, specific immunologic, longer-lasting coelomocyte responses. First-set xenografts induced a relatively slow increase in coelomocytes, which declined after three to four days post grafting. By contrast, second-set xenografts caused an accelerated rise in coelomocytes, usually 20 to 30% greater than the maximum coelomocyte response induced by first-set xenografts. The mean survival time for first-set xenografts (non-self) was 17±1 days, but repeat second sets were rejected in an accelerated time of 6±1 days. Autografts (self) are never destroyed. After priming with a first-set xenograft, this heightened coelomocyte reaction to a second-set xenograft was interpreted as an anamnestic response. The memory response is measurable in two ways: grossly as accelerated rejection of repeat xenografts and, at the cellular level, heightened coelomocyte numbers. Specific cellular immunity is demonstrable phylogenetically at the level of annelid worms.

At 20°C, the temperature at which the study was conducted, xenograft rejection was completed in a mean time of 17 days, approximately 10 days beyond the decline in coelomocyte numbers under the graft. If a second graft was added at this time (17 days post-first xenograft), the second xenograft was rejected in an accelerated mean time of 6 days, indicating a memory component to the rejection response.[17] Counts of coelomocytes taken from under such second grafts or second wounds varied considerably from those associated with first grafts or wounds. After second xenografts, coelomocyte numbers were 20 to 30% higher than after first grafts, peaking one to two days after grafting. Second transplants (including autografts) and wounds were also given five days after the initial procedure, and results were similar to those found in the 17-day group. Such accelerated and heightened responses to second grafts suggest the presence of a memory component reflected in the coelomocyte response.

Adoptive Transfer of the Memory Response to Xenografts

The involvement of coelomocytes in graft rejection is based on additional experimental evidence demonstrating adoptive transfer.[18–20] In these experiments, host

L. terrestris (A) was first xenografted with *E. foetida,* then coelomocytes were harvested at five days post-transplantation and injected into ungrafted *L. terrestris* (B). This second host was then xenografted with the same *E. foetida* donor used to induce immunity in A. Because *L. terrestris* shows only negligible allograft responses late after transplantation,[21] no early coelomocyte allo-incompatibility was expected before the action of primed coelomocytes against the *E. foetida* graft on *L. terrestris* (B). The second host (B) showed accelerated rejection of its first transplant, thus demonstrating short-term memory and confirming the adoptive transfer of graft rejection capacity by primed coelomocytes. The response is cell mediated, since transfer of coelomic fluid alone, free of coelomocytes, is ineffective. Coelomocytes from unprimed *L. terrestris* or from *L. terrestris* primed with saline are also unable to transfer the response. How specific this adoptive transfer response is remains to be determined. For example, if we graft a *L. terrestris* host with *E. foetida,* then transfer the coelomocytes and graft the new host with a third species, what is the response, how long does it last, and what is the role of temperature in regulation? Is memory longer lived at lower temperatures?

What is the Duration of the Memory Response to Allografts?

For a long time after the demonstration of memory in the 2/3 response to xenografts, there was controversy concerning whether memory existed, because in this instance, there were both short- and long-term or low and high responders. To reconcile these opposing expressions of memory (negative/positive; low responder/high responder?), Cooper and Roch did, in effect, lay to rest the controversy that had ensued.[22] There was indeed memory in earthworms, demonstrable as shortened responses to second-set grafts, but the memory appears to be short-lived. On *Lumbricus terrestris,* they performed second-set allografts from the same donor to the same recipient, showing that they did undergo accelerated rejection if transplanted less than 10 days after the first-set allografts. In contrast, second-set allografts, transplanted at longer intervals (20 to 90 days) show no accelerated rejection. Accelerated rejection of allotransplants under these kinetic conditions agrees with earlier intrafamilial xenotransplant results obtained in earthworms affirming an essential experimental variable: the time of second-set transplantation. Accelerated rejection, weak specificity, and short-term "memory" are three characteristics of the earthworm's allogeneic cellular defense/immune system. These reactions probably result from intense responses (although of short duration) of coelomocytes activated by first transplants that are still present at the time of second-set grafting but absent or inactive at later periods.

Is the Incorporation of [^{3}H]Thymidine into Coelomocytes an Indicator of Memory?

Coelomocytes are stimulated by mitogens and by putative transplantation antigens. Lemmi demonstrated the *in vivo* synthesis of DNA, measured by [^{3}H]thymidine incorporation into coelomocytes of xenografted *L. terrestris* at various times post grafting, with a peak incorporation at 4 days.[23] Roch *et al.,* investigated *in vitro*

DNA synthesis of *E. fetida* coelomocytes after wounding, allografting, and xenografting.[24] Peak incorporation of [^{3}H]thymidine occurred four days later, but showed decreasing stimulation indices (in brackets): allografts [15] → xenografts [5] → wounds. [4] Responses following second grafts or wounds were more complex, with a peak at day 2 for xenografts [4], at days 2 and 6 for allografts [4], and at day 6 for wounds [3]. Xenografting (particularly second sets) stimulated DNA synthesis in greater numbers of small coelomocytes (<10 μm) than in larger cells (>10 μm).

Stimulation of coelomocytes by both wounding and grafting is greater than that induced by mitogens, as pointed out by Valembois and Roch.[25] Four days after xenografting, coelomocytes (13%) incorporated [^{3}H]thymidine *in vitro;* after second xenografts, fewer (6%) were labeled. After first wounds, 6.5% were labeled, and after second wounds only 4%. Small cells appeared to be stimulated more by grafting than by wounding. Larger coelomocytes were stimulated neither by grafts nor wounds and were characterized by abundant cytoplasm, a few free ribosomes, polyribosomes, and moderate amounts of endoplasmic reticulum. Vesicular elements incorporated [^{3}H]thymidine more frequently than did small coelomocytes characterized by a high nucleocytoplasmic ratio and many free ribosomes. After second xenografts, however, the situation is reversed, with more smaller cells labeled than large cells. Coelomocytes do respond to mitogens, wounds, and graft responses that show some degree of specificity and perhaps proliferation. According to Lemmi and Cooper,[26] the proliferative event probably lasts three to four days. Using thymidine, of high specific activity to derange the DNA in a "thymidine suicide" experiment, revealed no derangement, suggesting that increased uptake of label was due to induced coelomocyte proliferation. Coelomocytes respond *in vitro* by directional migration to both bacterial and tissue antigens,[27] and the responding coelomocytes (92–94%) were neutrophils (type-I granulocytes).[28]

Is There Evidence That Coelomocytes Can Cooperate in Effecting a Response?

With respect to graft rejection, strong suggestive evidence supports the interaction of cells at an early crucial period.[29] The recognition of foreign tissue antigens, the earliest phase of graft rejection in earthworms, has been subjected to analysis by confronting host leukocytes with foreign erythrocytes. Only rabbit and rat erythrocytes significantly prevented healing of allografts when grafts were transplanted and erythrocytes injected simultaneously. In contrast, autografts and allografts transplanted on worms injected one or two days before grafting were never affected. Because earthworms readily produce higher titers of erythrocyte agglutinins at 24-hr post-injection than at later times, we propose a hypothetical scheme of earthworm leukocyte interactions that may occur during the early phases of graft healing and of agglutinin synthesis. Injected worms that succeed in healing allografts did not show any modification of the rejection intensity, confirming the fact that the steps of recognition and blocking of a special leukocyte population occurs rapidly after introduction of antigen. As evidenced by mitogenic stimulation, cooperative phenomena exist between earthworm leukocytes, suggesting similar interactions during graft rejection responses.

The Hapten Carrier Effect: An Indicator of Earthworm Immunity

When the phenomenon of memory was first demonstrated, it was obtained from studies of cell cooperation in the immune response to hapten–carrier complexes. Although the hapten–carrier conjugates used in these studies may be considered unnatural immunogens, they are, in fact, true analogues of antigens that occur in nature. The latter are composed of multiple antigenic determinants, each of which can be considered a hapten. Usually when an animal is given a primary immunization with a hapten–protein conjugate, for example, dinitrophenyl coupled to bovine serum albumin (DNP-BSA), and then is given a second injection of the same immunogen at a later time, antibodies against the hapten (DNP) are formed at a rate typical of the secondary response. If the second injection is given with DNP coupled to a non-cross-reacting protein carrier; for example, ovalbumin (OA), a secondary anti-DNP response is not usually manifested. However, if the animal that receives the second injection of DNP-OA has been previously primed with OA itself, then a substantial anti-DNP response will be elicited. This is the carrier effect that suggests that recognition of both hapten and carrier is required for the secondary response.

Laulan *et al.*[30] asked the question of *Lumbricus terrestris,* Is it capable of a new response by synthesizing specific substances against a synthetic small molecule (hapten)? In order to show the inducible character of these substances, it is necessary to use an antigen unknown by the earthworms and with no chance of being encountered in the natural environment. They immunized earthworms with two synthetic haptens. The results show that the substances elaborated by *L. terrestris* in response to an immunization have all the characteristics that can define them (inducibility, specificity, and memory) as antibody-like substances. They immunized earthworms with conjugates made of one of two different synthetic haptens (400 MW) and a carrier protein: bovine serum albumin (BSA) of keyhole limpet hemocyanin (KLH). The presence of anti-hapten substances in coelomic fluid was tested against each iodinated hapten derivative (^{125}I-hapten). The ^{125}I-hapten–substance complexes were separated from the free derivatives by polyethylene glycol (PEG) 6000 precipitation or by gel filtration. They found that earthworms synthesized specific substances against the immunizing hapten. The importance of the response depended on the carrier protein and on the amount of introduced immunogen. A kinetic study of first and second immunization showed that these substances, elaborated in response to the immunization with the synthetic hapten, were synthesized by cells that kept the immunological memory.

How Does Worm Cell Behavior Indicate a Memory Response?

It is essential that we recognize the very important and crucial role played by cells in generating an immune response in earthworms. The Czech earthworm school has done much to identify and characterize the nature of the responsible cells.[31] First, coelomocytes are capable of phagocytosis *in vitro* as detected by flow cytometry using FITC-labeled synthetic 2-hydroxy-ethyl-methacrylate particles (FITC-HEMA). They found increased phagocytic activity after *in vitro* stimulation by a protein antigen (arsanylated human serum albumin) and after preincubation of particles in cell-free coelomic fluid. Because antigen is essential for establishing a response, Bilej *et al.*[31] went further to characterize the antigen-binding properties of coelomocytes by means of quantitative autoradiography and direct measurement of radioactivity. They found that the antigen-binding capacity was significantly increased after stimulation by injecting 10 μm of arsanilic acid coupled to human serum albumin (ARS-HSA) into the coelomic cavity. One clever innovation was to embed the antigen in 3% agar gel to prevent it from being expelled from the dorsal pores. Moreover, they also found that preincubation of coelomocytes with nonlabeled proteins reduced the binding of radiolabeled antigen. The highest level of inhibition was found when the same protein was used for preincubation. These results indicate that antigen-binding properties are to some extent specific. More recently this group has developed monoclonal antibodies to the specific antigen-binding protein in earthworms.[32]

To follow the fate of antigens, two approaches have been used.[33,34] First, as a follow-up of the previous study, antigen-binding properties of coelomocytes revealed that direct measurement of radioactivity gives a good quantitative estimation of antigen recognition, but does not identify the specific cells that are involved.[35] Thus in the first study, they attempted to define the antigen-reactive cell lineage. A hapten–carrier system labeled by colloidal gold has been used with electron microscopy. The labeled antigen was bound mainly (though not exclusively) by agranular cells, probably neutrophilic coelomocytes.[36] The particles were disseminated throughout the cytoplasm without apparent condensation at or close to the cell surface. Enhanced antigen binding and internalization by cells after the *in vivo* presensitization were quantitatively supported, as viewed by electron microscopy, by flow cytometric analysis using FITC-labeled antigen. Both observations are higher after the stimulation in comparison with controls. Moreover, antigen binding can be inhibited by preincubation of coelomocytes with various nonlabeled similar proteins, although the most potent inhibitor was the immunizing antigen. In a companion work, they sought to relate the function of phagocytosis to the antigen.

Phagocytosis, one of the oldest defense mechanisms in the animal kingdom, is usually linked with induction of an oxidative burst. Production of highly reactive oxygen radicals represents an effective way of destroying engulfed microorganisms. An extremely sensitive method for monitoring phagocytic activity, especially the process of antigen degradation, is provided by chemiluminescent measurements. The process of phagocytosis and the subsequent destruction of ingested materials is linked with the production and release of highly reactive oxygen metabolites such as superoxide anion (O_2^-), hydrogen peroxide (H_2O_2), singlet oxygen ($^1O_2$), and hydroxyl radicals (OH). These metabolic pathways, termed the respiratory burst, represent an efficient mechanism of antimicrobial and antitumor activity of phagocytes. The degeneration of the labile oxygen components results in light emission, that is, chemiluminescence, that can be amplified through addition of exogenous luminescent substrates such as luminol and detected by a modified sensitive spectrophotometer (luminometer). The results were interesting in that, despite the high phagocytic activity, the chemiluminescent assay did not reveal increased production of oxygen radicals. Ultrastructural changes of engulfed particles observed by electron microscopy may be caused by intracellular lytic enzymes.[34]

Now that we knew that antigen was degraded and which cells were the most likely to house them, the obvious experiment was to test for memory. It is generally accepted that the main source of coelomocytes is the lining of the coelomic cavity. Nevertheless, there is disagreement as to whether free coelomocytes proliferate. According to some workers, free coelomocytes are a terminal population that cannot divide. In contrast, several papers indicate mitotic activity of coelomocytes under stress conditions or in response to transplantation antigens and to mitogens with an increased level of DNA synthesis. For the first time, Bilej *et al.* stimulated earthworms by injecting them with 10 μm of arsanilic acid coupled to human serum albumin in 3% agar gel (ARS-HSA).[37] Cells were collected thereafter at several intervals. The proliferative activity was observed and was measured by incorporation of [^{3}H]TdR. They found that the cells of the lining of the coelomic cavity respond immediately to stimulation as measured by [^{3}H]TdR uptake. What is proposed is that the mesenchymal lining acts as a source of precursor cells that after antigenic stimulation proliferate, differentiate, and then enter the coelomic cavity to fulfill their defensive functions. Only after second contact with antigen can this predetermined subpopulation undergo further mitotic cycles. The continued antigenic pressure could explain, in part, the necessity of a certain specificity of the proliferative response.

In Vitro *Agglutinin Production by Earthworm Leukocytes*

Using another approach, Stein and Cooper demonstrated a kind of secondary and specific response to various erythrocytes.[38] Leukocytes of the earthworm *Lumbricus terrestris* secrete agglutinins *in vitro,* as shown by measuring agglutinin titers of the culture medium and by observing secretory rosette formation by leukocytes with erythrocytes. Leukocytes form the highest percentages of secretory rosettes with rabbit erythrocytes (RBC) with other RBC species in the following order: rat, guinea pig, mouse, calf, sheep, horse, goat. Leukocytes displayed allotypic specificity by forming rosettes selectively with erythrocytes from different individual rabbits. Eight sugars inhibited rosette formation, along with the polysaccharide mannan and the glycoproteins thyroglobulin and bovine submaxillary mucin. Cyclohexamide did not affect rosette formation, suggesting that agglutinins may be preformed and stored in leukocytes before secretion. Leukocytes also formed E-type rosettes with erythrocytes, but apparently utilized different receptors from those of secretory rosettes because they were not inhibited by the same sugars.

SPECIFICITY AND MEMORY IN RELATION TO IMMUNOLOGIC THEORY

An attempt to resolve this question was proposed in two reviews,[39,40] both in response to a paper challenging the presence or absence of specificity and memory in invertebrates. In the terminology of Klein, invertebrate responses should be deemed either anticipatory or nonanticipatory.[41] Actually with respect to theory, the invertebrates are crucial to two points: (1) the origin of the immune response and (2) once evolved, how is it manifested?

ARE THERE ALTERNATIVES TO CLONAL SELECTION, AN ESSENTIAL COMPONENT OF SPECIFICITY AND MEMORY?

The clonal selection paradigm has organized immunological thinking for more than three decades.[42–44] According to clonal selection, antigens are responsible for organizing the immune system and subsequent responses, since only those lymphocytes that bear appropriate receptors matching the antigens will be contacted by them. This, of course, leads to their production of clones like those originally stimulated (the survival of the fittest)—clonal selection is a Darwinian corollary.[45] Cohen asserts that in contrast to "clonal selection, the germ-line effectively encodes a primitive internal image of bacteria, viruses, and the context of inflammation. These images do not depend on the antigen receptors of lymphocytes. For example, components of complement can recognize some microorganisms directly, targeting them for phagocytosis or lysis. Natural killer (NK) cells can respond to bacteria; macrophages and other white blood cells have invariant germ-line receptors for lipopolysaccharides, muramyl dipeptide cell wall elements, and other distinctly bacterial molecules. Many different cells recognize viral nucleic acids and the interferons that are elaborated as a consequence of viral infection. This primitive information arms cells with the capacity to recognize and respond to invaders: to secrete cytokines; to migrate [toward], adhere, and penetrate tissues; to engulf bacteria and viruses; and to activate enzyme systems and generate toxic molecules and free radicals that can kill invaders. The germ-line picture of infection and infectious agents developed over evolutionary time as a result of the fact that parasites are not only packages of antigenic variation, but also they are constrained by invariant structures and programs dictated by their ecology, that is, their need to exist in defined anatomical and biochemical niches, to reinfect new hosts, and so forth. These obligatory manifestations of parasitic life were exploited by the germ line to evolve an internal picture of infection. Charles Janeway reasons that the immune system evolved to discriminate infectious *non-self* from noninfectious *self*.[46] FIGURES 2 and 3 illustrate the main features of the clonal selection and cognitive paradigms. The diagram of the clonal selection paradigm (FIG. 2) shows its appealing simplicity, whereas the cognitive paradigm is more complicated.[11]

The contrasts in the two paradigms are of interest in the context of invertebrate immunology as we struggle with the problem of specificity and memory. Why? Cohen's proposition offers evidence to support much of the work being done on invertebrates today, not ignoring the consequences of clonal selection but not emphasizing it either.[40,47] The cognitive paradigm would seem to encompass the so-called nonanticipatory responses of invertebrates as proposed by Klein.[41] Moreover, it conforms with, but may be a variant of, the nonclonal/clonal receptors as these appeared during evolution.[47] From the work on invertebrates, there is a fair amount of evidence to support all of the characteristics that contribute to the cognitive paradigm.[40] The list is fairly extensive, but two recent examples in the context of mammals as mentioned by Cohen are of interest: natural killer cells[48–60] and the work on cytokines.[61–64]

It must be kept in mind that the above arguments pertain to invertebrates and how their responses may function similarly to those of mammals. (The question of homology/analogy is not raised in this context.[40]) This completely removes any

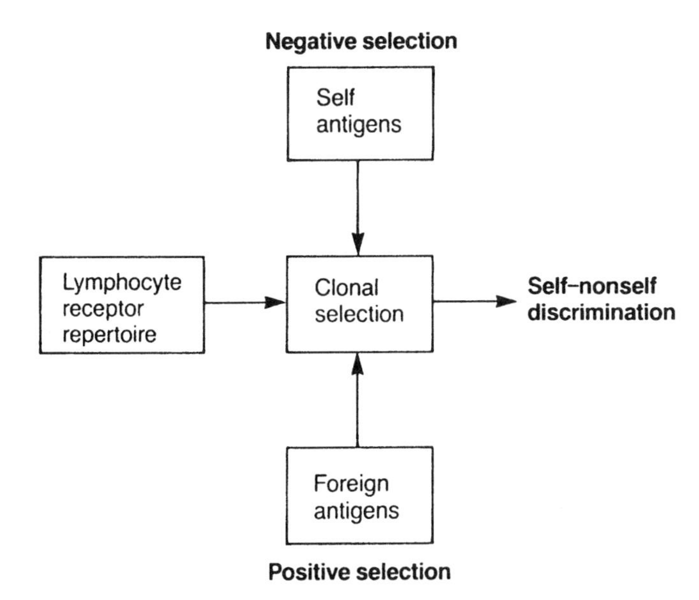

FIGURE 2. The clonal selection paradigm. This paradigm proposes that the receptor repertoires of B and T cells are organized by two antigenic forces: the self and the foreign antigens, which positively select the remaining clones that have complementary receptors. The output of the system is the discrimination between the self antigens, which are ignored, and the foreign antigens impinging on the system, which are rejected.

relationship with B lymphocytes and T lymphocytes and their variable regions of Igs and T-cell receptors (TCR), the so-called variable region molecules or VRMs. According to Stewart[65] the VRMs are unique to vertebrates, and their immune system is thought to have arisen from cell-adhesion molecules (CAMS). This seems to negate a rather popular view among immunologists that the immune system evolved to discriminate noninfectious *self* from infectious *non-self.*[46] It fails to recognize, however, the earliest of molecules, Thy-1, as well as the CAMS or their near kin, which guide invertebrate cells during development where there is a fair amount of evidence that it exists in invertebrates.[66] This has been criticized.[67,68]

FINAL COMMENT

The immune and the nervous systems share one characteristic, the phenomenon of memory. Whereas the nervous system monitors by means of nonmigratory cells and their products, monitoring by the immune system is by means of mobile cells and circulating substances, a characteristic that makes the problem of understanding immunologic memory more difficult. Both the immune and nervous systems learn and become more effective after repeated stimuli, a comparison that may be meaningless or that may prod immunologists to greater insights by basing theory on the apparently more simple functioning of the nervous system.[69,70]

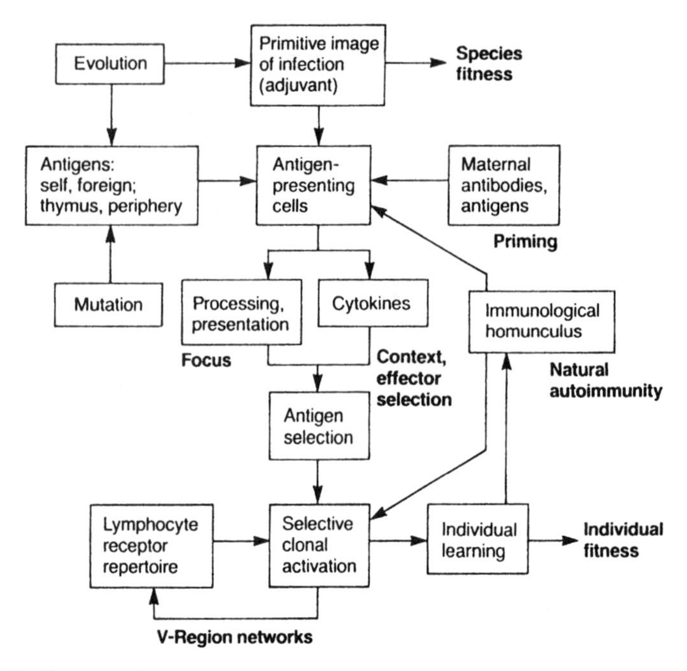

FIGURE 3. The cognitive paradigm. This paradigm proposes that clonal activation, which is tailored by the antigenic experience of each individual, is subject to the intervention of antigen selection, a process which expresses the evolutionary adaptations of the species. Antigens, self and foreign, are determined by evolution and by mutations. Evolution has also encoded, within the germ line, mechanisms of innate immunity that constitute a primitive image of infection. The agents of this primitive image include the APCs: macrophages, dendritic cells, endothelial cells, tissue cells, and even B and T cells themselves. These cells ignore, destroy, or process and present certain epitopes and elaborate various cytokines. These activities provide focus, sense context, and activate particular effector mechanisms. The outcome is the selction of certain antigens and molecular motifs as signals that function to activate selected clones. This form of antigen selection, in addition, is primed by maternal antibodies (an image of the infectious environment) and by maternal antigens. The immunological homunculus (an image of the key self antigens) is formed by contact with self antigens in the thymus and in the periphery. It includes T and B cells that recognize a limited number of dominant self antigens and the regulatory cells that interact with the autoimmune T and B cells. The homunculus is the expression of natural regulated autoimmunity; it influences the functioning of the APC, the specificity of clonal activation, and the behavior of the regulatory networks, some of which are connected by V-region idiotypes. The ouput of the system is fitness, which profits from both foreign and self recognition.

REFERENCES

1. Gray, D. 1993. Immunological memory. Ann. Rev. Immunol. **11:** 49–77.
2. Manning, M. J., Ed. 1980. Phylogeny of Immunological memory. Elsevier/North-Holland Biomedical Press. Amsterdam.
3. Cooper, E. L. 1982. Invertebrate Defense Systems: An Overview. *In* The Reticuloendothelial System. A Comprehensive Treatise. Vol. 3. N. Cohen & M. M. Sigel, Eds. Plenum Press. New York.

4. COOPER, E. L. 1992. Reply to: The echinoid immune system revisited. Immunol. Today. **14:** 92.
5. COOPER, E. L. 1992. Overview of immunoevolution. Boll. Zool. **59:** 119–128.
6. HILDEMANN, W. H., I. S. JOHNSON & P. L. JOKIEL. 1979. Immunocompetence in the lower metazoan phylum: Transplantation immunity in sponges. Science **204:** 420–422.
7. HILDEMANN, W. H., R. L. RAISON, G. CHEUNG, C. J. HULL, L. AKAKA & J. OKAMOTO. 1977. Immunological specificity and memory in a scleratinian coral. Nature **270:** 219–223.
8. SALTER-CID, L. & C. H. BIGGER. 1991. Alloimmunity in the gorgonian coral *Swiftia exserta.* Biol. Bull. **181:** 127–134.
9. LANGLET, C. & J. BIERNE. 1984. Immunocompetent cells requisite for graft rejection in Lineus (Invertebrata, Nemertea). Dev. Comp. Immunol. **8:** 547–557.
10. HILDEMANN, W. H. & T. G. DIX. 1972. Transplantation reactions of tropical Australian echinoderms. Transplantation **15:** 624–633.
11. KARP, R. D. & W. H. HILDEMANN. 1976. Specific allograft reactivity in the sea star *Dermasterias imbricata.* Transplantation **22:** 434–439.
12. SMITH, L. C. & E. H. DAVIDSON. 1992. The echinoid immune system and the phylogenetic occurrence of immune mechanisms in deuterostomes. Immunol. Today **13:** 356–361.
13. SMITH, L. C., R. J. BRITTEN & E. H. DAVIDSON. 1992. SpCoell: A sea urchin profilin gene expressed specifically in coelomocytes in response to injury. Mol. Biol. Cell. **3:** 403–414.
14. COOPER, E. L. 1986. Evolution of histocompatibility. *In* Immunity in Invertebrates. M. Brehelin, Ed. Springer Verlag. Heidelberg.
15. VETVICKA, V., P. SIMA, E. L. COOPER, M. BILEJ & P. ROCH. 1993. The Immunology of Annelids. CRC Press. Boca Raton, FL.
16. COOPER, E. L. 1968. Transplantation immunity in annelids. I. Rejection of xenografts exchanged between *Lumbricus terrestris* and *Eisenia foetida.* Transplantation **6:** 322–337.
17. HOSTETTER, R. K. & E. L. COOPER. 1973. Cellular anamnesis in earthworms. Cell. Immunol. **9:** 384–392.
18. DUPRAT, P. 1967. Etude de la prise et du maintien d'un greffon de paroi du corps chez le lombricien *Eisenia fetida.* Ann. Inst. Past. **113:** 867–881.
19. BAILEY, S., B. J. MILLER & E. L. COOPER. 1971. Transplantation immunity in annelids. II. Adoptive transfer of the xenograft reaction. Immunology **21:** 81–86.
20. VALEMBOIS, P. 1971. Role des leucocytes dans l'acquisition d'une immunite antigreffe specifique chez les lombriciens. Arch. Zool. Exp. Gen. **112:** 97–104.
21. COOPER, E. L. 1969. Chronic allograft rejection in *Lumbricus terrestris.* J. Exp. Zool. **171:** 69–73.
22. COOPER, E. L. & P. ROCH. 1986. Second-set allograft responses in the earthworm *Lumbricus terrestris.* Transplantation **41:** 514–520.
23. LEMMI, C. A. E. 1982. Characteristics of primitive leukocytes equipped with receptors for xenogeneic grafts. *In* Developmental Immunology: Clinical Problems and Aging. E. L. Cooper & M. A. B. Brazier, Eds. Academic Press. New York.
24. ROCH, P., P. VALEMBOIS & L. DUPASQUIER. 1975. Responses of earthworm leukocytes to concanavalin A and transplantation antigens. Adv. Exp. Biol. Med. **64:** 45–54.
25. VALEMBOIS, P. & P. ROCH. 1977. Identification par autoradiographie des leucocytes stimules a la suite de plaies ou de greffes chez un ver de terre. Biol. Cell **28:** 81–82.
26. LEMMI, C. A. E. & E. L. COOPER. 1981. Induction of coelomocyte proliferation by xenografts in the earthworm *Lumbricus terrestris.* Dev. Comp. Immunol. **1:** 73–80.
27. MARKS, D. H., E. A. STEIN & E. L. COOPER. 1979. Chemotactic attraction of *Lumbricus terrestris* coelomocytes to foreign tissue. Dev. Comp. Immunol. **3:** 277–285.
28. LINTHICUM, D. S., D. H. MARKS, E. A. STEIN & E. L. COOPER. 1977. Graft rejection in earthworms: An electronmicroscopic study. Eur. J. Immunol. **7:** 871–876.
29. COOPER, E. L. & P. ROCH. 1984. Earthworm leukocyte interactions during early stages of graft rejection. J. Exp. Zool. **232:** 67–72.
30. LAULAN, A., A. MOREL, J. LESTAGE, M. DELAAGE & P. CHATEAUREYNAUD. 1985. Evidence

of synthesis by *Lumbricus terrestris* of specific substances in response to an immunization with a synthetic hapten. Immunology **56:** 751–758.
31. Bilej, M., J-P. Scheerlinck, T. VandenDriessche, P. De Baetselier & V. Vetvicka. 1990. The flow cytometric analysis of in vitro phagocytic activity of earthworm coelomocytes (*Eisenia foetida;* Annelida). Cell Biol. Int. Rep. **14:** 831–837.
32. Tuckova, L., J. Rejnek, M. Bilej, H. Hajkova & A. Romanovsky. 1991. Monoclonal antibodies to antigen-binding protein of annelids *Lumbricus terrestris.* Comp. Biochem. Physiol. **100B:** 19–23.
33. Bilej, M., P. Rossmann, T. VandenDriessche, J.-P. Scheerlinck, P. De Baetselier, L. Tuckova, V. Vetvicka & J. Rejnek. 1991. Detection of antigen in the coelomocytes of the earthworm, *Eisenia foetida* (Annelida). Immunol. Lett. **29:** 241–246.
34. Bilej, M., P. De Baetselier, I. Trebichavsky & V. Vetvicka. 1991. Phagocytosis of synthetic particles in earthworms: Absence of oxidative burst and possible role of lytic enzymes. Folia Biol. **37:** 227–233.
35. Bilej, M., L. Tuckova, J. Rejnek & V. Vetvicka. 1990. In vitro antigen-binding properties of coelomocytes of *Eisenia foetida* (Annelida). Immunol. Lett. **26:** 183–188.
36. Linthicum, D. S., E. A. Stein, D. H. Marks & E. L. Cooper. 1977. Electron microscopic observations of normal coelomocytes from the earthworm *Lumbricus terrestris.* Cell Tissue Res. **185:** 315–330.
37. Bilej, M., P. Sima & J. Slipka. 1992. Repeated antigenic challenge induces earthworm coelomocyte proliferation. Immunol. Lett. **32:** 181–184.
38. Stein, E. A. & E. L. Cooper. 1988. *In vitro* agglutinin production by earthworm leukocytes. Dev. Comp. Immunol. **12:** 531–547.
39. Marchalonis, J. J. & S. F. Schluter. 1990. On the relevance of invertebrate recognition and defense mechanisms to the emergence of the immune response of vertebrates. Scand. J. Immunol. **32:** 13–20.
40. Cooper, E. L., B. Rinkevich, G. Uhlenbruck & P. Valembois. 1992. Invertebrate immunity: Another viewpoint. Scand. J. Immunol. **35:** 247–266.
41. Klein, J. 1989. Are invertebrates capable of anticipatory immune response? Scand. J. Immunol. **29:** 499–505.
42. Cohen, I. R. 1992. The cognitive principle challenges clonal selection. Immunol. Today **13:** 441–444.
43. Cohen, I. R. 1992. The cognitive paradigm and the immunological homunculus. Immunol. Today **13:** 490–494.
44. Burnet, F. M. 1959. The Clonal Selection Theory of Acquired Immunity. Cambridge University Press. Cambridge, England.
45. Cooper, E. L. 1982. Did Darwinism help comparative immunology? Am. Zool. **22:** 890.
46. Janeway, C. A., Jr. 1992. The immune system evolved to discriminate infectious nonself from non infectious self. Immunol. Today **13:** 11–16.
47. Janeway, C. A. 1989. Approaching the asymptote? Evolution and revolution in immunology. Cold Spring Harbor Symp. Quant. Biol. **54:** 1–13.
48. Tyson, C. J. & C. R. Jenkin. 1974. The cytotoxic effect of hemocytes from the crayfish *(Parachaeraps bicarinatus)* on tumor cells of vertebrates. Aust. J. Exp. Biol. Med. Sci. **52:** 915–923.
49. Parrinello, N. & D. Rindone. 1980. Studies on the natural hemolytic system of the annelid worm *Spirographis spallanzanii* Viviani (polychaeta). Dev. Comp. Immunol. **5:** 33–42.
50. Valembois, P., P. Roch & D. Boiledieu. 1980. Natural and induced cytotoxicities in sipunculids and annelids. *In* Phylogeny of Immunological Memory. M. J. Manning, Ed.: 47–55. Elsevier/North-Holland. Amsterdam.
51. Decker, J. M., A. Elmholt & A. V. Muchmore. 1981. Spontaneous cytotoxicity mediated by invertebrate mononuclear cells toward normal and malignant vertebrate targets: Inhibition by defined mono- and disaccharides. Cell. Immunol. **59:** 161–170.
52. Valembois, P., P. Roch & D. Boiledieu. 1982. Cellular defense systems of the Platyhelminthes, Nemertea, Sipunculida and Annelida. *In* The Reticuloendothelial System: A Comprehensive Treatise. Vol 3. N. Cohen & M. M. Sigel, Eds. Plenum. New York.

53. Luquet, G. & M. Leclerc. 1983. Spontaneous and induced cytotoxicity of axial organ cells from *Asterias rubens* (Asterid-echinoderm). Immunol. Lett. **6:** 339–342.
54. Söderhäll, K., A. Wingren, M. W. Johansson & K. Bertheussen. 1985. The cytotoxic reaction of hemocytes from the freshwater crayfish, *Astacus astacus.* Cell. Immunol. **94:** 326–332.
55. Leippe, M. & L. Renwrantz. 1988. Release of cytotoxic and agglutinating molecules by *Mytilus* hemocytes. Dev. Comp. Immunol. **12:** 297–308.
56. Franceschi, C., A. Cossarizza, D. Monti & E. Ottaviani. 1991. Cytotoxicity and immunocyte markers in cells from the freshwater snail *Planorbarius corneus* L. (Gastropoda pulmonata): Implication for the evolution of natural killer cells. Eur. J. Immunol. **21:** 489–493.
57. Porchet-Henneré, E., T. Dugimont & A. Fischer. 1992. Natural killer cells in a lower invertebrate, *Nereis diversicolor.* Eur. J. Cell Biol. **58:** 99–107.
58. Suzuki, M. M. & E. L. Cooper. 1994. Allogeneic killing by earthworm effector cells. Submitted to Natural Immunity.
59. Suzuki, M. M. & E. L. Cooper. 1994. Killing of intrafamilial and xenogeneic targets by earthworm effector cells. Submitted to Immunology Letters.
60. Suzuki, M. M., E. L. Cooper, G. S. Eyambe, A. J. Goven, L. C. Fitzpatrick & B. J. Venables. 1994. Effects of exposure to polychlorinated biphenyls (PCBs) on natural cytotoxicity of earthworm coelomocytes. Submitted.
61. Raftos, D. A., D. L. Stillmann & E. L. Cooper. 1991. Interleukin-2 and phytohemagglutinin stimulate the proliferation of tunicate cells. Immunol. Cell Biol. **69:** 225–234.
62. Raftos, D. A., E. L. Cooper, G. S. Habicht & G. Beck. 1991. Invertebrate cytokines: Tunicate cell proliferation stimulated by an interleukin 1-like molecule. Proc. Natl. Acad. Sci. USA **88:** 9518–9522.
63. Raftos, D. A., E. L. Cooper, D. L. Stillman, G. S. Habicht & G. Beck. 1992. Invertebrate cytokines: Release of interleukin-1-like molecules form tunicate hemocytes stimulated with zymosan. Lymphokine Cytokine Res. **11:** 235–240.
64. Beck, G., R. F. O'Brien, G. S. Habicht, D. L. Stillman, E. L. Cooper & D. A. Raftos. 1993. Invertebrate cytokines III: Invertebrate interleukin-1-like molecules stimulate phagocytosis by tunicate and echinoderm cells. Cell. Immunol. **146:** 284–299.
65. Stewart, J. 1992. Immunoglobulins did not arise in evolution to fight infection. Immunol. Today **13:** 396–395.
66. Cooper, E. L. & M. H. Mansour. 1989. Distribution of Thy-1 in invertebrates and ectothermic vertebrates. *In* A. E. Reif & M. Schlesinger, Eds. Cell Surface Antigen Thy-1 Immunology. Neurol. Ther. Appl. Marcel Dekker. New York.
67. Langman, R. E. 1989. The Immune System. Evolutionary principles guide our understanding of this complex biological defense system. Academic Press. New York.
68. Cohn, M. & R. Langman. 1990. The protection: The unit of humoral immunity selected by evolution. Immunol. Rev. **115:** 1–131.
69. Cooper, E. L. 1991. Evolutionary development of neuroendocrineimmune system. Adv. Neuroimmunol. **1:** 83–96.
70. Cooper, E. L. 1992. Perspectives in neuroimmunomodulation: Lessons from the comparative approach. Anim. Biol. **1:** 169–180.

Tumor-Suppressor Genes, Hematopoietic Malignancies and Other Hematopoietic Disorders of *Drosophila melanogaster*[a]

ELISABETH GATEFF[b]

Genetisches Institut
Johannes Gutenberg-Universität
55099 Mainz, Germany

The main components of the *Drosophila* circulatory system are the dorsal vessel with the attached hematopoietic organs (lymph glands), the hemolymph, and the hemocytes.[1] Two blood cell types originate in the hematopoietic organs, the plasmatocytes and the crystal cells.[2,3] At the end of larval life and in particular during the larval–pupal transition, plasmatocytes differentiate into podo- and lamellocytes. This differentiation is characterized by a steady increase of mitochondria, rER, and primary and secondary lysosomes.[3] The maturation of the plasmatocytes is further characterized by increased phagocytosis. Thus, plasmatocytes and their descendants are involved in phagocytosis, encapsulation, and melanization of foreign bodies. *Drosophila* plasmatocytes can, therefore, be compared with vertebrate macrophages in their capacity to recognize "self" from "non-self," to phagocytose, and to encapsulate. The nonphagocytic crystal cells seem to be able to coagulate and melanize hemolymph.[3] Blood cells also play an important role during metamorphosis during which most cells are of the lamellocyte type and are involved in sequestering disintegrating larval tissues.

Hereditary hematopoietic disorders were reported in the earliest days of *Drosophila* genetics. The first lethal mutation, affecting hematopoiesis and designated as *1(1)7* was discovered in 1916 by C. B. Bridges. The studies by Stark[4–7] of this mutant revealed a plasmatocyte tumor that she classified as a lymphorsarcoma. The loss of the mutant strain posed the questions whether the mutant plasmatocyte growth was truly malignant and whether *Drosophila* can develop true malignancies in general.

In mutagenesis experiments Gateff[8–14] identified four genes, which in the homozygously mutated state cause the malignant transformation of the larval plasmatocytes (TABLE 1). Two further blood tumor mutants are the *Tum*[1,25] and the *air*[16] or *hen*[17] mutations. With the exception of the *Tum*[1] mutation, which displays a dominant conditional lethality, the remaining five mutations show a recessive mode of inheritance and, thus, by definition can be classified as tumor-suppressor genes. On the other extreme is a recently found gene mutation that causes blood cell deficiency (E. Gateff, unpublished). In addition, about 40 gene mutations are known to induce

[a]The generous support of the Deutsche Forschungsgemeinschaft SFB 304 is gratefully acknowledged.

[b]Address for correspondence: Genetisches Institut, Johannes Gutenberg-Universität, Saarstr. 21, 55099 Mainz, Germany.

TABLE 1. Tumor Suppressor Genes of *Drosophila melanogaster* Causing Malignant Transformation of Plasmatocytes

Designation of Gene and Reference	Locus	Number of Alleles
Lethal (1) malignant blood neoplasm[11] (*1(1)mbn*)	8D1-4	2
Lethal (2) malignant blood neoplasm[12] (*1(3)mbn*)[a]	n.d.	1
Lethal (3) malignant blood neoplasm-1[13] (*1(3)mbn-1*)	65A1,2	26
Lethal (3) malignant blood neoplasm-2[14] (*1(3)mbn-2*)	n.d.	1
Tumorous lethal[1] (*Tum*[1])[15] allele of hopscotch	1-34,5 cM[b]	1
Aberrant immune response 8 (*air8*)[16] or *hen*[17]	7C4-9	2

[a]Mutant stock lost, but blood cell line, designated *mbn-2,* available.
[b]cM: centi Morgan.

abnormal encapsulation and melanization of larval tissues by lamellocytes, a process that normally takes place only in defence reactions.[18]

Thus, the powerful *Drosophila* genetics provides the unique opportunity (i) to dissect genetically the development and differentiation of the primordial immune system, (ii) to identify genes instrumental in the recognition of "self" and "non-self," and (iii) to study genes coding for antibacterial peptides.

This review paper deals with the following two tumor-suppressor genes, lethal (1) malignant blood neoplasm (*1(1)mbn*)[11] and lethal (3) malignant blood neoplasm-1 (*1(3)mbn-1*)[13] (TABLE 1) and the mutant lethal (3) hematopoietic organs missing (*1(3)hem*), which lacks blood cells (E. Gateff, unpublished). Furthermore, the humoral immune response of the above three mutants and the tumorous blood cell line *in vitro mbn-2*[19] will be presented.

THE *1(1)mbn* MALIGNANT BLOOD CELL TUMOR PHENOTYPE AND DEVELOPMENTAL CAPACITIES OF THE TUMOROUS HEMOZYTES

The hemizygous *1(1)mbn* male mutant larvae exhibit extremely enlarged hematopoietic organs (FIG. 1a) below their transparent dorsal cuticle. Histological preparations confirm this observation (compare FIG. 1d and e) and show further that most larval tissues are partially or fully destroyed (compare FIG. 1b and c). Within the wild-type hematopoietic organs, the majority of the cells are proplasmatocytes[3] (FIG. 1d). Procrystal cells constitute less than 5% of the primordial blood cell population and, thus, cannot be recognized at the low magnification in FIG. 1d. In the mutant hematopoietic organs, in addition to the proplasmatocytes and procrystal cells, numerous plasmato-, podo-, and lamellocytes, normally found only in the hemolymph, are encountered[20] (FIG. 1e and FIG. 2d).

In wild-type hematopoiesis the proliferation rate of the progenitor blood cells in

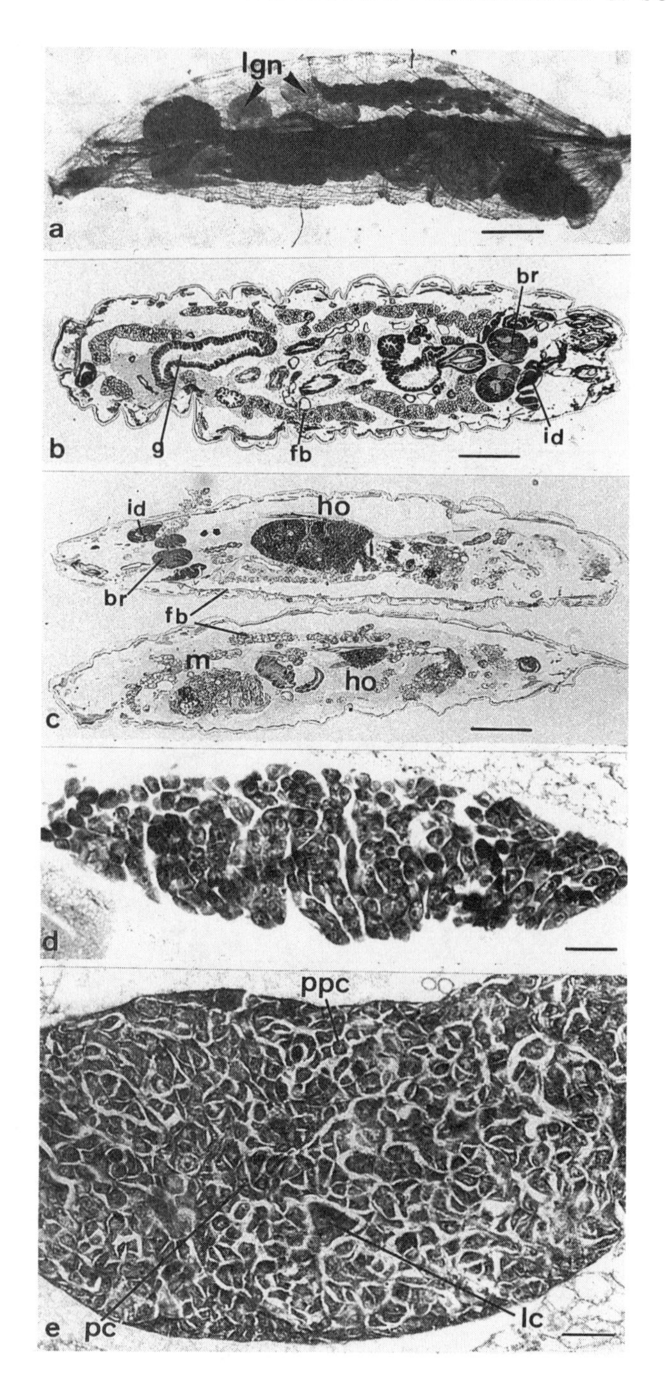
lgn
a
br
id
b
g
fb
id
ho
br
fb
m
ho
c
d
ppc
e
pc
lc

FIGURE 1. (a) Whole montat of *1(1)mbn* mutant larva. Note the enlarged hematopoietic organs (ho) below the transparent integument. **(b,c)** Longitudinal histological sections of ready to pupate, third-instar, wild-type and mutant larvae. Most organs are destroyed by the tumorous blood cells in the mutant in c. **(d,e)** Longitudinal histological sections of wild-type and mutant hematopoietic organs from mature larvae. In the wild-type section, small, basophilic proplasmatocytes prevail, while in the mutant, in addition to the small proplasmatocytes (ppc), larger plasmato-, podo-, and lamellocytes are precent (pc, lc). Abbreviations: br, brain; fb, fat body; g, gut; ho, hematopoiectic organs; id, imaginal disc; lc, lamellocyte; m, melanotic mass, pc, plasmatocyte; ppc, proplasmatocyte. Histological preparations: hematoxylin and eosin. Bars: 400 μm (a,b,c); 20 μm (d,e).

the hematopoietic organs and the blood cells in the hemolymph is strictly regulated.[3] In the mutant hematopoietic organs and hemolymph, in contrast, the blood cells proliferate uncontrolled.[20] And indeed, comparing the numbers of [^{3}H]thymidine-labeled cells in the autoradiograms of a wild-type and a mutant hematopoietic lobe shows many more labeled cells in the mutant than in the wild-type counterpart (FIG. 2a, b). Consequently, more blood cells are released from the mutant than from the wild-type hematopoietic organs into the hemolymph, where they continue to divide and, thus cause a three- to sevenfold increase of the hemocyte count as compared to that in the wild type.

In conclusion, the malignant nature of the mutant plasmatocytes *in situ* is revealed (i) by their autonomous, lethal growth and (ii) by their inability to recognize "self" from "non-self," leading to invasion and destruction of the larval tissue by phagocytosis (FIG. 1c).

In order to show that the malignant growth, observed *in situ,* is truly autonomous and is not induced by extrinsic factors present in the mutant larvae, such as, for instance, abnormal basement membranes, we implanted mutant hematopoietic tissue pieces into wild-type host.[8,9] In general the autonomy of *Drosophila* malignant cells is expressed by their ability to grow in a similarly malignant fashion after implantation *in vivo* into wild-type hosts as *in situ.* Small tissue pieces of *1(1)mbn* hematopoietic organs, implanted into the abdomens of wild-type female flies, grew autonomously in an invasive, lethal fashion, just as *in situ,* and caused the death of the hosts in 8 to 10 days. Such hosts showed after dissection as well as in histological preparations (FIG. 3b, e) large, loose cell masses within the body cavity, consisting of plasmato-, podo-, lamellocytes. The cells invaded the ovaries (FIG. 3b), the single egg chambers (FIG. 3c), the thoracic muscles (FIG. 3d), and the head capsule (not shown). The mutant plasmatocytes can be subcultured in wild-type female hosts for numerous transfer generations, where they show a high division rate, as demonstrated by the [^{3}H]thymidine incorporation into a great number of nuclei (FIG. 3e). Thus, the mutant hematopoietic organs represent a lethal, malignant, and transplantable neoplasm. In contrast, wild-type hematopoietic organs implanted into female flies do not grow excessively nor do they kill their hosts.

The molting hormone ecdysone plays a crucial role in all normal developmental processes.[21,22] The exposure of dividing cells, such as the imaginal disc cells, to ecdysone at the end of the third larval instar causes cessation of cell division and the beginning of morphogenesis. Upon ecdysone exposure, all cells in the larval hematopoietic organs cease to divide, enter the hemolymph, and attain the capacity to phagocytose only the degrading larval cells, but not the developing adult structures.

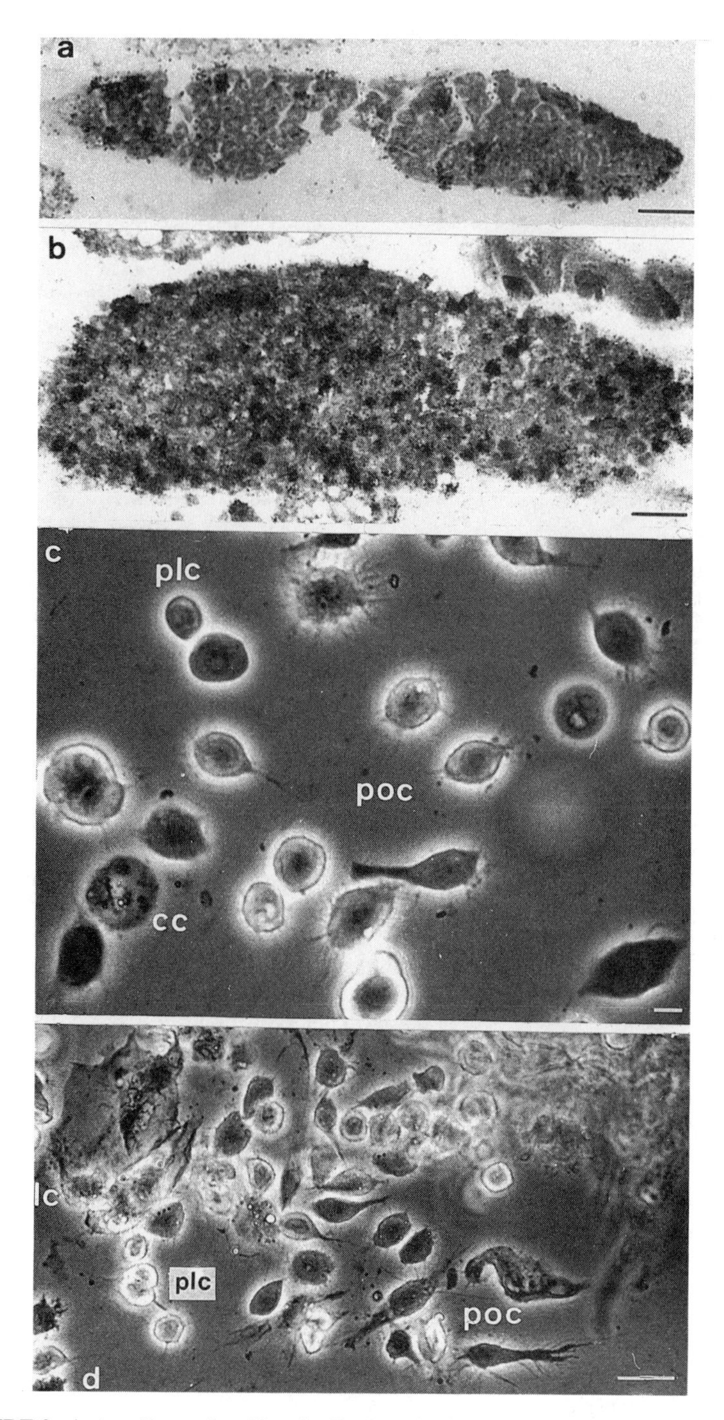

FIGURE 2. Autoradiographs of longitudinal sections through the first hematopoietic lobe of **(a)** third-instar, wild-type larva and **(b)** third-instar *1(1)mbn* larva. Compare the number of cells showing [^{3}H]thymidine incorporation in a and b. Phase-contrast photomicrographs of **(c)** wild-type blood cells from the hemolymph and **(d)** mutant blood cells derived from the hematopoietic organs, showing plasmatocytes (plc), podocytes (poc), and lamellocates (lc). Bars: 40 μm (a,b); 5 μm (c); 20 μm (d).

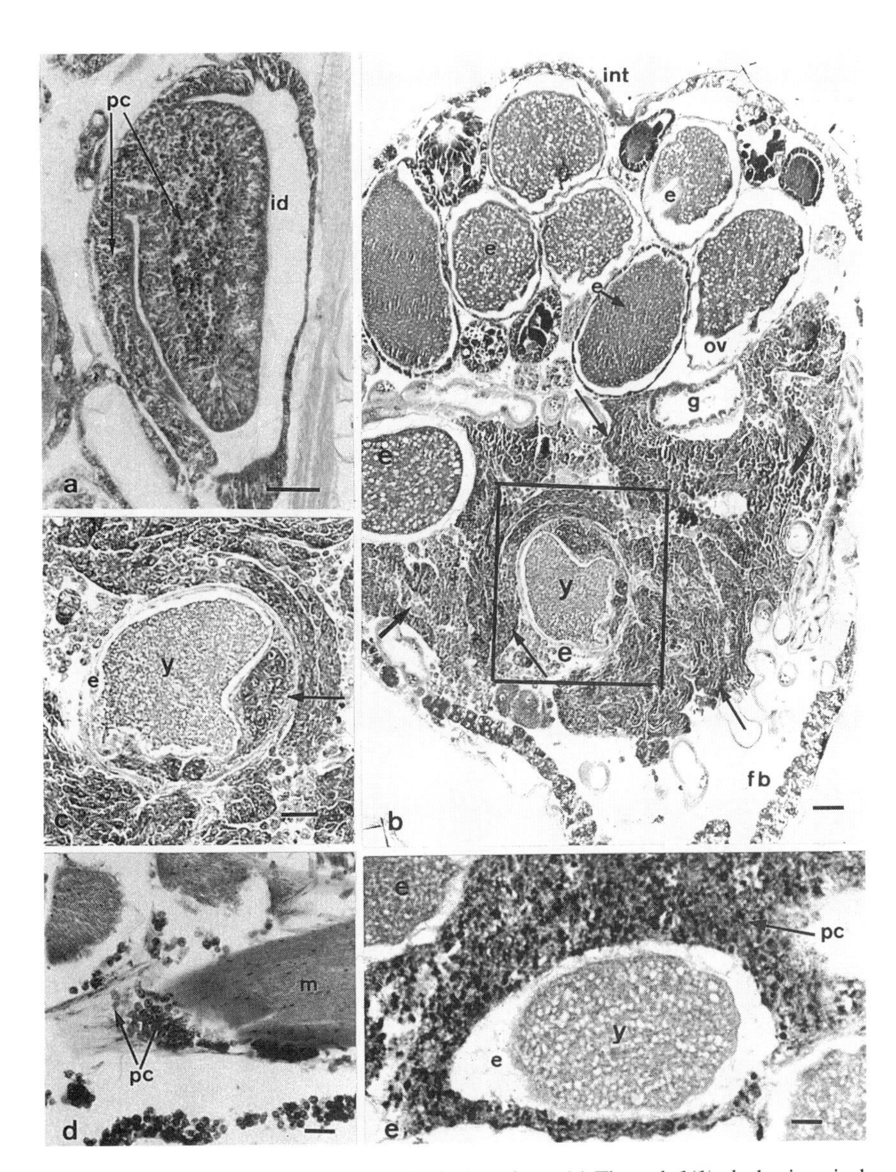

FIGURE 3. Photomicrographs of histological sections: (**a**) Through *1(1)mbn* leg imaginal disc (id) *in situ,* showing invasion of mutant plasmatocytes (pc). (**b**) Through abdomen of wild-type female fly. The malignant plasmatocytes (*arrows*) have destroyed almost all egg chambers (e) of the lower ovary, while the upper ovary (ov) is intact. (**c**) Egg chamber (boxed detail from **b**), showing areas of invasion into the yolk (y). (**d**) Thoraric muscle (m) invaded by mutant plasmatocytes (pc). (**e**) Autoradiograph of malignant plasmatocytes (pc), surrounding an egg chamber (e) and exhibiting heavy [^{3}H]thymidine incorporation. Abbreviations: e, egg chamber; fb, fat body; g, gut; id, imaginal disc; int, integment; m, muscle; ov, ovary; pc, plasmatocytes; y, yolk. Histological preparations: hematoxylin and eosin. Bars: 100 μm (a,b); 60 μm (c,d,e).

Malignant *Drosophila* cells behave towards ecdysone at metamorphosis in two ways.[8–10] They either respond to the hormone by cessation of cell division or they continue to grow throughout metamorphosis and adult development and cause the death of the developing adult before or shortly after eclosion. The response of the *1(1)mbn* plasmatocytes to ecdysone *in situ* and after culture *in vivo* can be tested by implantation of the tissues or cells into mid and late third-instar, wild-type larvae (96-hr, 124-hr). In three experimental series (TABLE 2) the following tissues were implanted: (i) tissue pieces of *1(1)mbn* hematopoietic organs, (ii) *1(1)mbn* plasmatocytes cultured *in vivo* in wild-type female adults for 27 transfer generations, and (iii) tissue pieces of wild-type hematopoietic organs.

Implantation of tissue pieces of wild-type hematopoietic organs into 96-hr- or 124-hr-old wild-type larvae had no effect on the development of the wild-type hosts (TABLE 2; experimental series III). In contrast, tissue pieces of *1(1)mbn* hematopoietic organs (experimental series I) and *1(1)mbn* plasmatocytes subcultured *in vivo* in adult female abdomens for 27 transfer generations (experimental series II) induced various developmental disorders in the hosts. It was generally observed that mutant implants caused more severe developmental defects in the 96-hr-old than in the 124-hr-old, wild-type hosts. For instance, in experimental series I, from the 19 young wild-type larvae, injected with pieces of mutant hematopoietic organs, 18 pupariated. Of these, seven completed adult development, while in the remaining 11 animals development was interrupted at various stages of adult development ranging from death shortly after puparium formation and up to pharate adults. Of the injected older larvae, on the other hand, 13 completed adult development while only seven animals were arrested in adult development.

In experimental series II, 96-hr- and 124-hr-old, wild-type larvae were injected with mutant plasmatocytes that had been subcultured *in vivo* in female adults for 27 transfer generations. None of the injected 96-hr-old larvae pupariated. In comparison, of the 13 older larvae, 11 pupariated, but only three animals completed adult development (TABLE 2). Apparently, the prolonged *in vivo* subculture of the mutant plasmatocytes had increased their degree of malignancy.

Histological preparations of injected larvae, which failed to pupariate (TABLE 2; experimental series I and II), revealed invasion of the malignant plasmatocytes into the imaginal discs, and thus, their destruction (compare FIG. 4a, b). These experiments prove, once more, the autonomy of the malignant mode of growth of the mutant plasmatocytes also in the wild-type larvae. The brain hemispheres (FIG. 4b) and the ventral ganglion (not shown) exhibit extensive neuronal degeneration when compared with the brain hemispheres of control larvae (FIG. 4a). The neuronal degeneration is, however, not caused by the invasion of mutant plasmatocytes. A possible explanation for the degeneration may be the destruction of the imaginal discs and the muscles, both of which send during normal development numerous axonal processes into the central nervous system. In the injected animals, due to the extensive damage, no axons form. The degeneration of the nervous system and in particular of the neurosecretory cells, which normally provide the neurohormonal stimulus for ecdysone secretion, is most probably responsible for the failure of the larvae to pupariate, since no defects could be detected in the ring gland, the ecdysone-secreting organ.

A wide range of developmental abnormalities were encountered among the

TABLE 2. Developmental Capacity of Wild-Type and *1(1)mbn* Hematopoietic Organs and *in Vivo* Subculted Plasmatocytes in 96-hr- and 124-hr-old, Wild-Type Larvae

		96-hr-old, Wild-Type Host Larvae				124-hr-old, Wild-Type Host Larvae			
Experimental Series	Type of Implanted Tissue/Cells	No. of Implanted Larvae	No. of Pupariated Larvae	No. of Eclosed Adults	Affected Tissue/Organ[a]	No. of Implanted Larvae	No. of Pupariated Larvae	No. of Eclosed Adults	Affected Tissue/Organ[a]
I	*1(1)mbn* hematopoietic organs	19	18	7	ID,M NS,A	20	20	13	M,A
II	*1(1)mbn* blood cells from 27th transfer generation *in vivo*	17	0	0	ID,M NS	13	11	3	ID,M NS,A
III	Wild-type hematopoietic organs	10	10	10	—	12	12	12	—

[a] A, abdomen; ID, imaginal disc; M, muscle; NS, nervous system.

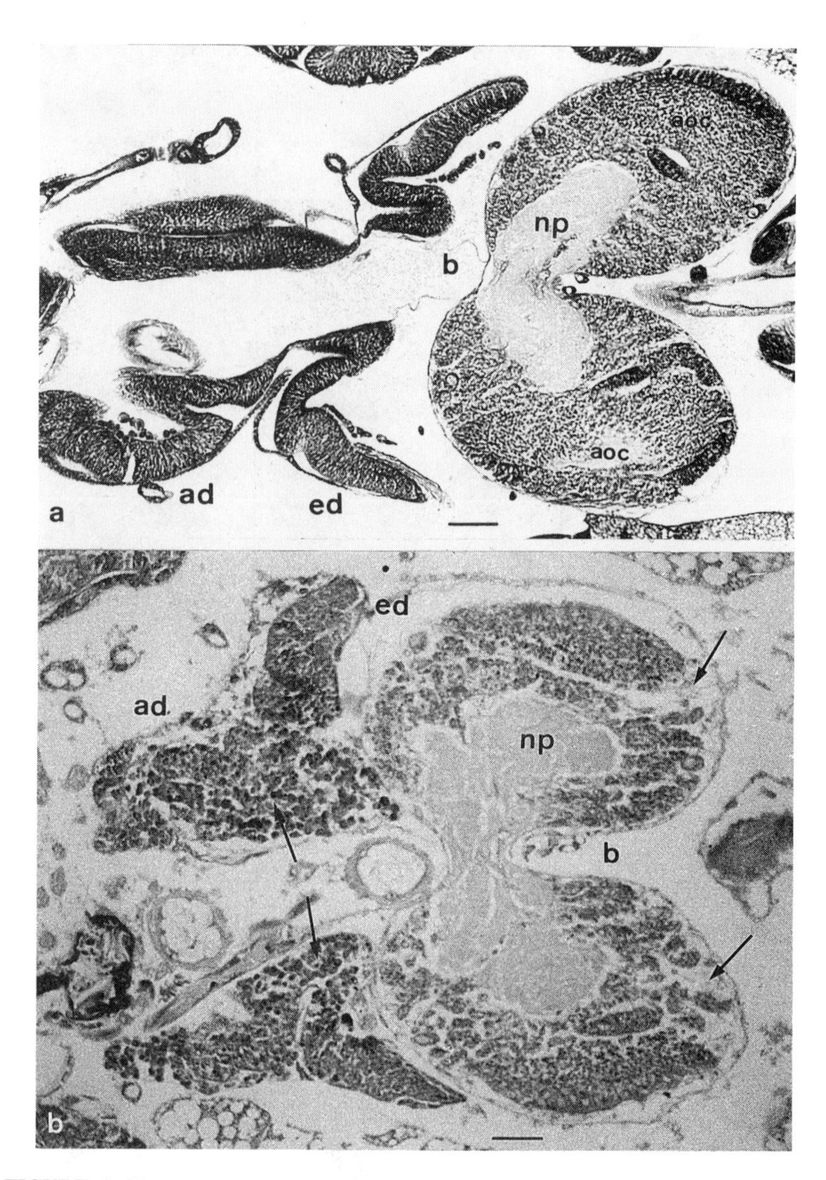

FIGURE 4. Photomicrographs of sagital histological sections through the brain hemispheres and the eye-antennal imaginal discs of **(a)** Mature wild-type larva injected at the mid-third instar (96-hr) with hematopoietic organs from a ready to pupate wild-type larva. No developmental defects can be observed in the brain **(b)** and the imaginal disks (ad, ed). Wild-type larva of an age comparable to a, injected at the age of 96-hr with a *1(1)mbn* hematopoietic organ tissue piece. Note the invasion of mutant plasmatocytes into the eye (ed) and antennal imaginal disc (ad) and the complete destruction of the imaginal disc epithelium (compare with a). Within the cellular cortex of the brain hemispheres, extensive neuronal degeneration (*arrow*) can be observed, but no blood cell invasion. Histological preparations: hematoxylin and eosin. Abbreviations: ad, antennal imaginal disc; aoc, adult optic center anlage; b, brain hemisphere; ed, eye imaginal disc; np, neuropile. Bars: 20 μm.

larvae, which had pupariated but did not eclose as adults. Some of the larvae died shortly after pupariation, apparently as a result of the extensive damage to the imaginal discs, the nervous system, and the muscles. Larvae with less severely affected imaginal discs and central nervous systems reached different adult developmental stages. In all pharate adults the abdominal cuticle was partially or completely absent (not shown). Histology of such animals showed further severe defects of the thoracic musculature. In contrast to the wild-type, in which the fibers of the longitudinal indirect and dorsoventral flight muscles are straight (FIG. 5a), the experimental animals show highly distorted, folded muscle fiber bundles (FIG. 5b, c). These results show that the mutant plasmatocytes must interfere in a stage- and tissue-specific manner with imaginal disc, muscle, and abdominal histoblast development. These results suggest further that blood cells not only play a role in defense but also are involved in the normal growth and differentiation of larval and adult tissues and organs.

TIME OF DETERMINATION OF THE NEOPLASTIC MODE OF GROWTH

The question of when it is during development that the *1(1)mbn* plasmatocytes become determined for malignant growth was studied by transplanting wild-type and mutant hematopoietic organ primordia from embryos, first-, and second-instar larvae into the abdomen of wild-type female flies. If the *1(1)mbn* primordial blood cells are already determined for neoplastic growth in the embryo or early larva, then malignant growth should ensue in the abdomens of female flies. Since the embryonic and first larval instar hematopoietic organ primordia cannot be handled individually because of their small size, we implanted wild-type and mutant anterior halves from 18-hr-old embryos and first-instar larvae into the abdomens of wild-type flies. After two to three weeks of *in vivo* culture, the implants were dissected and examined. Within the wild-type implants, salivary glands, muscles, nervous systems, and imaginal discs were encountered. Hematopoietic organs, however, were not found. In the mutant implants, all the above-mentioned organs were also present, and no abnormal hemocyte growth could be observed.

The obvious conclusion from these results is that the primordial *1(1)mbn* plasmatocytes in the embryonic and first-instar larval hematopoietic organs are not yet determined for malignant growth. Further transplantation experiments showed that the critical developmental period during which the wild-type *1(1)mbn*$^+$ gene function is required for tumor suppression takes place during the late second-larval instar.

IMMUNE RESPONSE OF THE *1(1)mbn* LARVAE

In wild-type larvae and adults upon bacterial inoculation, a number of antibacterial peptides are synthesized[23–25] in the fat body.[26] Hemolymph from inoculated and uninoculated *1(1)mbn* larvae exhibit strong antibacterial activity in the inhibition zone assay, indicating that either all or some of the genes coding for antibacterial peptides are constitutively expressed (J. H. Postlethwait, D. Hoffmann, and E.

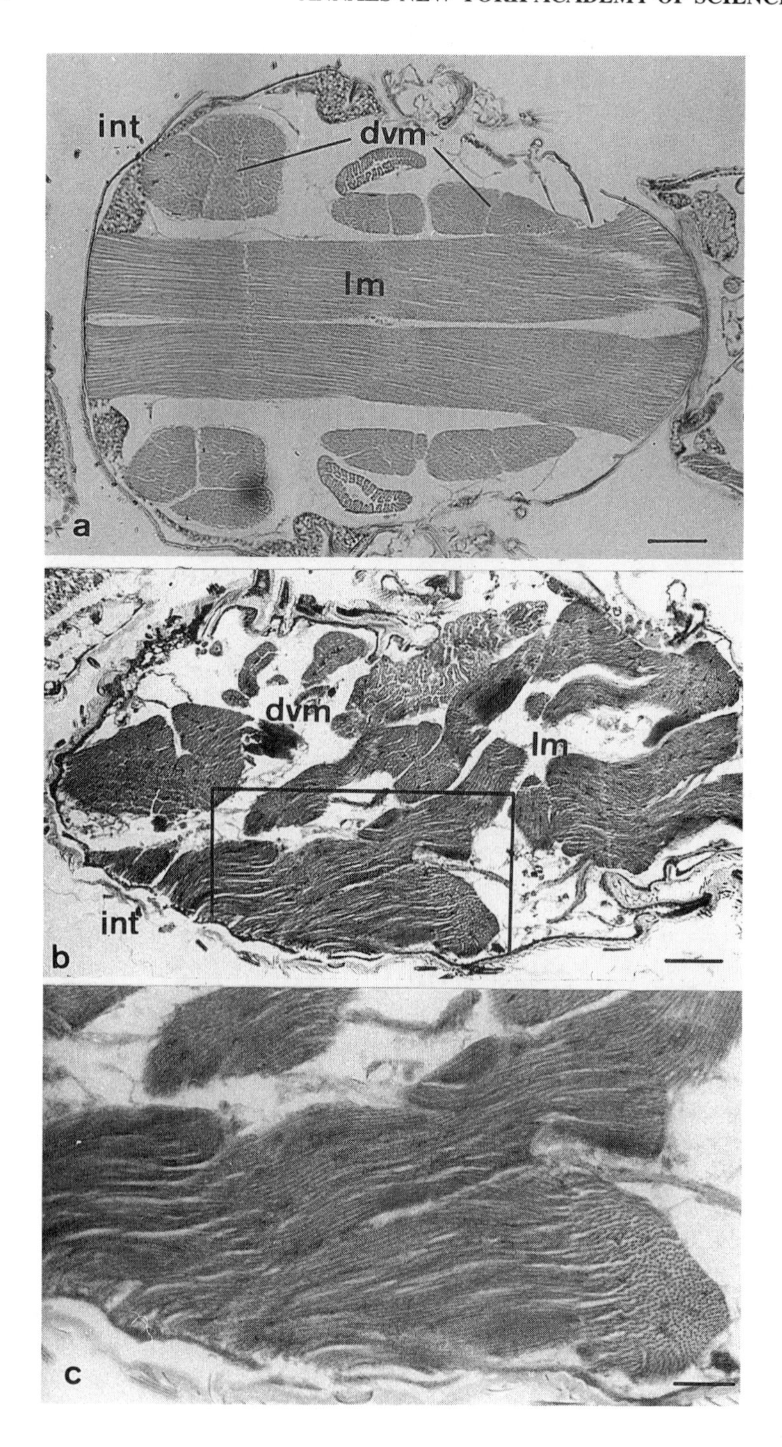
int
dvm
lm
a
dvm
lm
int
b
c

FIGURE 5. (**a**) Photomicrograph of longitudinal histological section through adult thorax derived from a mature (124-hr) wild-type larva injected with wild-type hematopoietic organs. (**b**) Longitudinal histological section through the thorax of a pharate adult, derived from a wild-type larva injected shortly before puparium formation with a tissue piece of a *1(1)mbn* hematopoietic organ. (**c**) Enlargement of boxed area in b. Compare the normal longitudinal (lm) and dorso-ventral (dvm) indirect flight muscles with the highly distorted counterparts in b and c. Abbreviations: dvm, dorso-ventral flight muscle; int, integument; lm, longitudinal flight muscle. Histological preparations: hematoxylin and eosin. Bars: 100 μm (a,b); 50 μm (c).

Gateff, unpublished results). However, Northern analysis showed that cecropin gene expression in the mutant is inducible and, thus, normally regulated (J. H. Postlethwait, personal communication). Cecropin monoclonal antibodies likewise decorate only induced *1(1)mbn* blood cells (T. Trenzcek and E. Gateff, unpublished results). Which of the known or still unknown antibacterial peptide gene(s) is (are) then constitutively expressed in this mutant? Preliminary results with diptericrin monoclonal antibody, raised in the laboratory of J. Hoffmann (Strasbourg), show all *1(1)mbn* hemocytes of inoculated as well as uninoculated larvae to recognize a diptericin-like moiety (T. Trenzcek and E. Gateff, unpublished results). These intriguing results show that in the mutant the diptericin genes, in contrast to the wild-type, seem to be uncoupled from the coordinated regulation of cecropin and other antibacterial peptide genes. Further studies should identify which of the known and still unknown antibacterial peptide genes are constitutively expressed in the *1(1)mbn* mutant. Of course the most important question remains to be solved; namely, in what way is the expression of these genes directly or indirectly regulated by the *1(1)mbn* tumor-suppressor gene? The molecular analysis of the *1(1)mbn* gene, presently under way, should show the relationship between the *1(1)mbn* mutation and the unregulated expression of the diptericrin gene and possibly of other antibacterial peptide genes.

THE *1(3)mbn-1* BLOOD CELL TUMOR PHENEOTYPE AND DEVELOPMENTAL CAPACITIES OF THE TUMOROUS PLASMATOCYTES

Like the *1(1)mbn* mutation, the *1(3)mbn-1* gene mutation[14] (TABLE 1) also causes the malignant transformation of the plasmatocytes.[28] Mature mutant larvae harbor about 10^6 hemocytes in their hemolymph as compared with 5×10^4 blood cells in wild-type larvae (compare FIGS. 1b and 6a). The hematopoietic organs are 50 times larger than their wild-type counterparts. In contrast to the *1(1)mbn* mutant hematopoietic organs, in which about 50% of the cells are plasmato-, podo-, and lamellocytes (see above), in the *1(3)mbn-1* hematopoietic organs the cell population consists almost exclusively of neoplastic proplasmatocytes whose excessive growth is responsible for the immense size increase of the hematopoietic organs. Only very few nontumorous crystal cells can be encountered among the many proplasmatocytes. Throughout the third-larval instar, proplasmatocytes become released from the

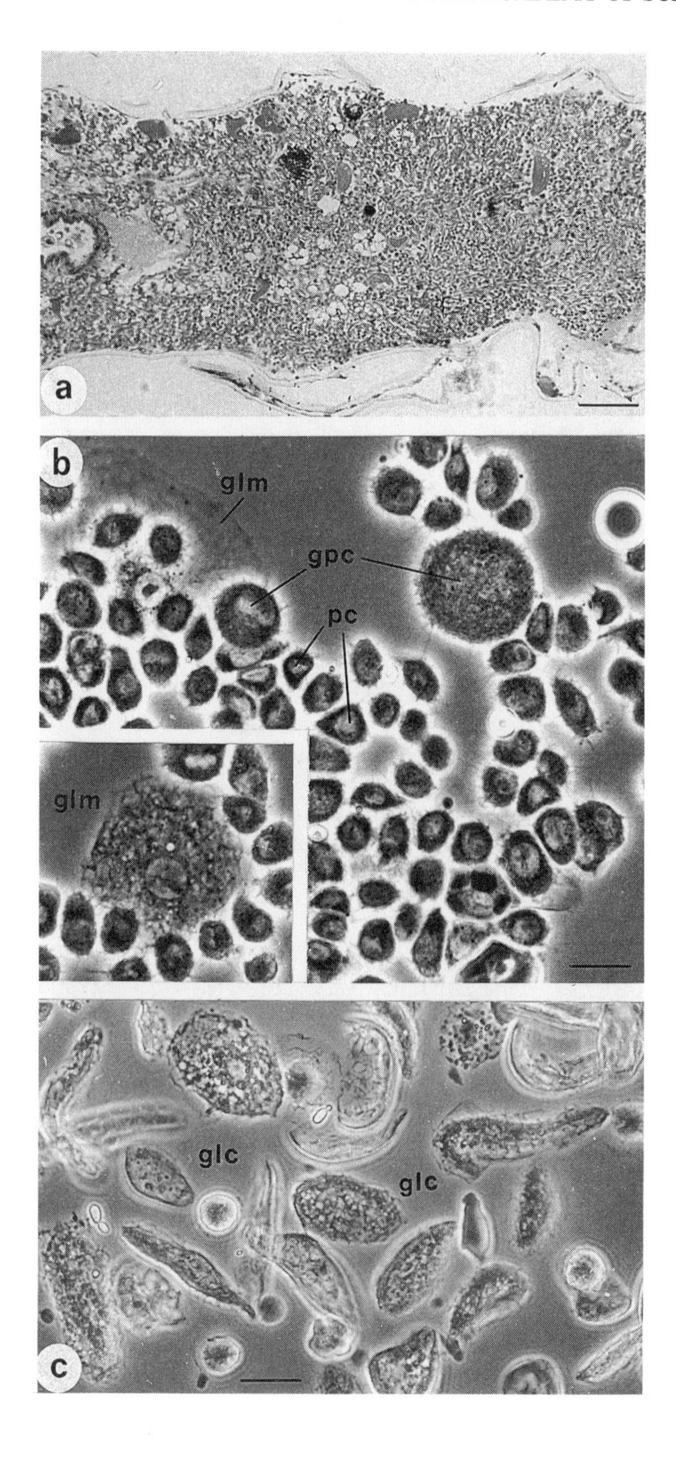
a
b
glm
gpc
pc
glm
glc
glc
c

FIGURE 6. Photomicrographs of (**a**) longidutinal histological section through a *1(3)mbn-1* larva. Compare with FIGURE 1b, c, and d. Note the tightly packed tumorous plasmatocytes and the almost complete destruction of the larval tissues. Hematoxylin and eosin. (**b**) Phase contrast of *1(3)mbn-1* blood cells from a larva as in a, showing predominantly plasmatocytes (pc), a giant plasmatocyte (gpc), and a giant lamellocyte (glc). The insert shows a giant lamellocyte. (**c**) Phase contrast photomicrograph of *1(3)mbn-1* blood cells from a tumor consisting primarily of giant lamellocytes (glc) and very few plasmatocytes (pc). Bars: 100 μm (a); 20 μm (b,c).

hematopoietic organs into the hemolymph where some differentiate into plasmatocytes while others remain undifferentiated, forming secondary nests of hematopoiesis and thus contributing to the 10- to 15-fold increase of the blood cell count in the mutant hemolymph as compared to that of the wild-type. Thus, contrary to the *1(1)mbn* blood tumor, in which about half of the cells are differentiated plasmato-, podo-, and lamellocytes, the *1(3)mbn-1* hematopoietic tumor consists primarily of plasmatocytes and immature proplasmatocytes (FIG. 6b). Mutant larvae, usually delayed by 3 to 4 days in puparium formation, show most larval tissues destroyed and the body cavity tightly filled with large masses of blood cells (FIG. 6a). As in the *1(1)mbn* mutant, the destruction of the larval tissues is due to the incapacity of the mutant plasmatocytes to recognize self from non-self. They actively invade the larval tissues via the basement membrane and bring about their destruction by phagocytosis. An unusual feature of the *1(3)mbn-1* hematopoiesis is that most plasmatocytes never differentiate into podo- and lamellocytes. Instead, 2 to 3% of the plasmatocytes engage in an abnormal cellular growth that results in giant plasmatocyte-like cells (FIG. 6b), with an enormously enlarged cytoplasm in which large amounts of primary and secondary lysosomes and a small, eccentrically located nucleus[27] are found. Giant plasmatocytes differentiate into giant lamellocytes by an extreme flattening process (FIG. 6b). Conglomerates of giant lamellocytes can yield small melanotic masses. Lamellocytes of normal size are absent.

In conclusion, the *1(3)mbn-1* blood cell tumor *in situ* is composed of large numbers of immature, dividing proplasmatocytes and plasmatocytes, which, unable to recognize self from non-self, invade and phagocytose the larval tissues. Thus, the vast majority of the *1(3)mbn-1* tumorous blood cells seem to be arrested in their differentiation at the proplasmatocyte–plasmatocyte stage. The above-described tumor phenotype applies to 9 of the 26 *1(3)mbn-1* alleles. In the remaining alleles we regularly observe a higher number of giant plasmatocytes and giant lamellocytes (FIG. 6c), which form large melanotic masses in the body cavity.[28] Taking all allelic mutant phenotypes together, we consider it safe to conclude that the *1(3)mbn-1* gene must be involved in some way in the differentiation of the cells of the plasmatocyte line. Depending on the type of mutation, differentiation into podo- and lamellocytes is either arrested at the proplasmato–plasmatocyte stage or a highly aberrant differentiation into giant plasmato- and lamellocytes takes place.

We tested further the capacity of the cells in the *1(3)mbn-1* hematopoietic organs to grow after implantation into wild-type larvae and adults, and in no case found malignant growth, pointing to nonautonomy. Apparently a factor or factors present in the wild-type larva and adult but absent in the mutant inhibits the unrestrained, malignant growth of the *1(3)mbn-1* plasmatocytes *in vivo.*

The molecular cloning and sequencing of the *1(3)mbn-1* gene are completed.[28] The deduced amino acid sequence of the putative protein shows no significant homology to any presently known protein. Investigations are under way that should shed light on the wild-type gene function in tumor suppression and tumorigenesis.

IMMUNE RESPONSE OF *1(3)mbn-1* MUTANT LARVAE

The antibacterial immune response of the mutant larvae, as measured in the inhibition zone assay, is fully inducible (J. H. Postlethwait, D. Hoffman, and E. Gateff, unpublished results). The question of whether antibacterial peptide genes can also be induced in the mutant blood cells was approached by the *in situ* hybridization of cecropin and diptericin monoclonal antibodies to induced and noninduced, mutant blood cells (T. Trenzcek and E. Gateff, unpublished results). Contrary to the situation in the fat body, in which these genes are inducible, in the mutant blood cells they seem to be expressed constitutively, most probably at a very low level.

PHENOTYPE OF THE MUTANT LETHAL (3) HEMATOPOIESIS MISSING *(1(3)hem)*

The homozygous *1(3)hem* animals die shortly after puparium formation (E. Gateff, unpublished). Anatomical and histological investigations revealed the complete absence of primordial blood cells in the hematopoietic organs (compare FIG. 7a and b) and of plasmatocytes and crystal cells in the hemolymph (compare FIG. 8a, b, and c). A further aberration is found in the mutant brain, where unlike the dividing wild-type optic neuroblasts arranged in compact sheath in the lateral portion of each brain hemisphere (FIG. 8d), the mutant optic neuroblasts never divide, and instead become extremely large (FIG. 8e), exhibiting giant chromosomes in their enormous nuclei. Imaginal discs are also abnormal. The above-described phenotype allows the conclusion that the wild-type *1(3)hem⁺* gene product must be in some way instrumental in the cell type-specific inhibition of cell division. In context with this discussion, it is interesting to note that blood cells do not seem to be required for the completion of larval development. Moreover, we did not observe an increased death rate among the mutant larvae resulting from infection, since they showed the induction of a perfectly normal immune response in the inhibition zone assay (J. H. Postlethwait, D. Hoffmann, and E. Gateff, unpublished results). One of the inducible antibacterial peptides was shown in a Northern analysis to be cecropin (J. H. Postlethwait, personal communication).

What then is the function of the larval blood cells aside from their capacity to phagocytose and encapsulate? In *Manduca sexta,* Dunn *et al.*[29] proposed that hemocytes partially degrade bacteria by phagocytosis, releasing peptidoglycans into the hemolymph, which induce the synthesis of antibacterial peptides in the fat body. However, recent *in vitro* studies by the same investigators with isolated fat body showed that blood cells are not required for the induction of antibacterial peptides. On the other hand, *in vitro* cultured *mbn-2* tumorous plasmatocytes[20] have recently

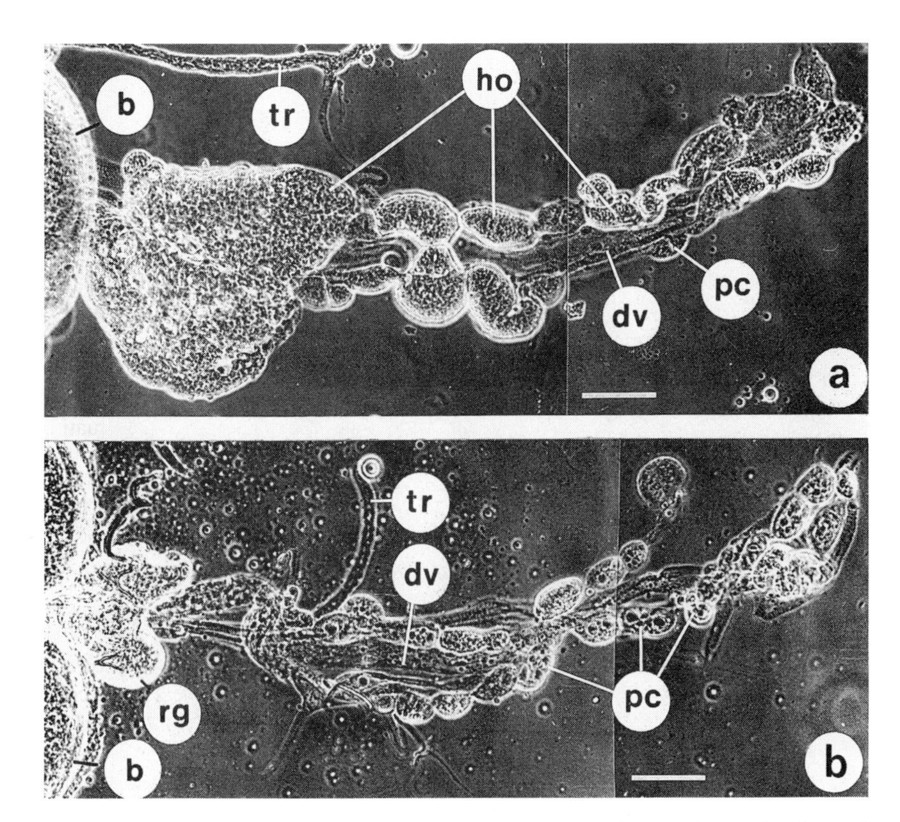

FIGURE 7. Phase-contrast photomicrograph of wholemounts of (**a**) Wild-type dorsal vessel (dv) with hematopoietic organs (ho) and pericardial cells (pc). (**b**) *l(3)hem* dorsal vessel (dv) with pericardial cell (pc). Hematopoietic organs are missing. Abbreviations: b, brain; dv, dorsal vessel; ho, hematopoietic organs; pc, pericardial cell; rg, ring gland; tr, trachea. Bars: 50 μm.

been shown to synthesize cecropin[30] and diptericin[31] upon bacterial stimulation. Another line of preliminary evidence, suggesting the constitutive expression of certain antibacterial peptides in blood cells, comes from *in situ* hybridization of the cells with monoclonal antibodies against cecropin and diptericin (see above). Blood cells from wild-type *l(1)mbn* and *l(3)mbn-1* larvae and the *mbn-2* blood cell line *in vitro* react with monoclonal antibodies against cecropin and diptericin in the induced as well as the noninduced stage. With the exception of noninduced *l(1)mbn* plasmatocytes, which do not express cecropin (see above), all other blood cells express cecropin constitutively. These observations suggest that blood cells may produce the above antibacterial peptides steadily in small amounts just like recent findings by P. E. Dunn (personal communication) have shown the synthesis of cecropin-like peptides in other than immune tissues of *Manduca sexta,* such as the gut, the salivary glands, the malpighian tubules, and the epidermis. Further studies of the regulation

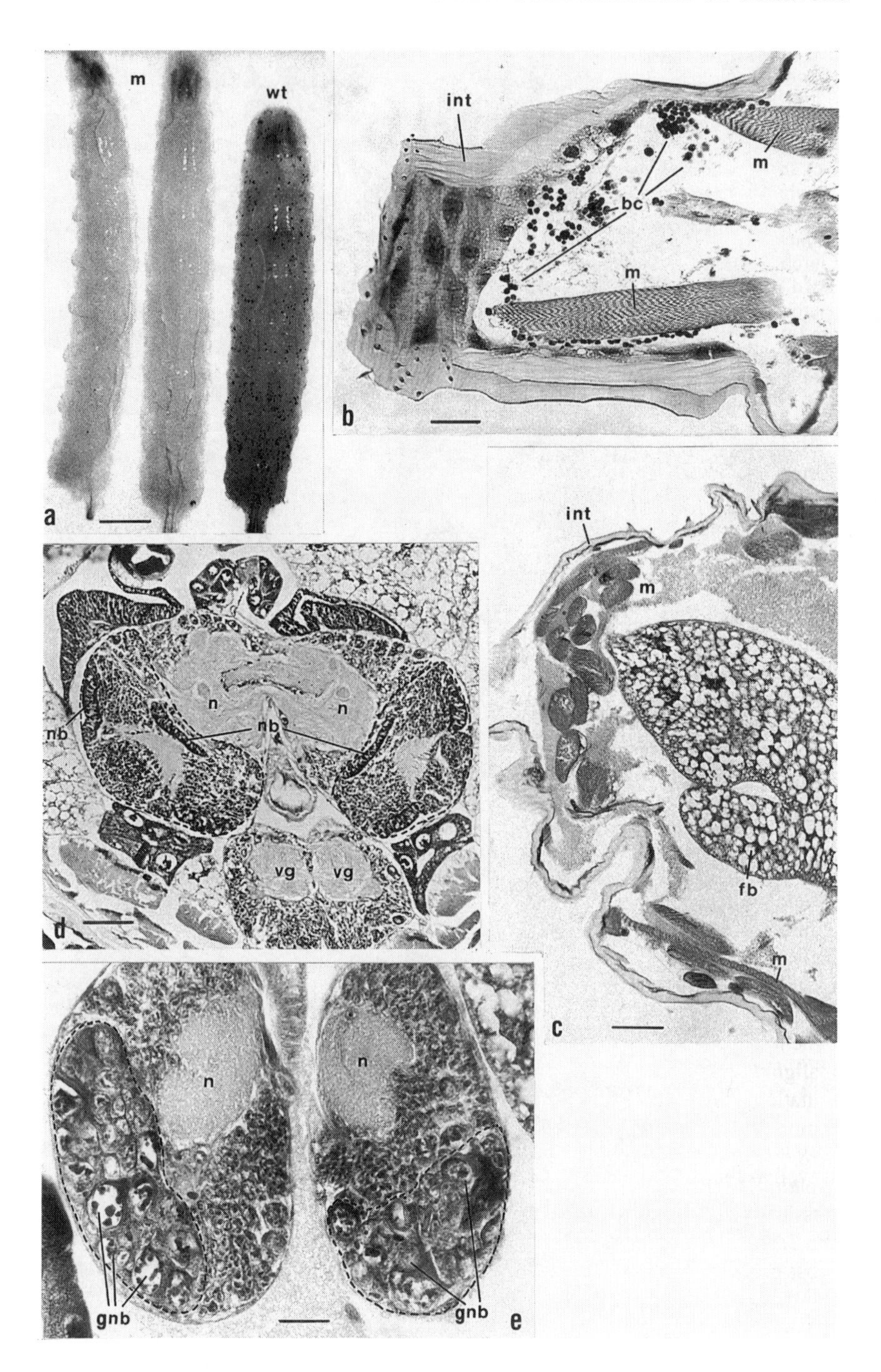
m
wt
a
int
bc
m
m
b
int
m
fb
m
c
n
n
nb
nb
vg
vg
d
n
n
gnb
gnb
e

FIGURE 8. **(a)** Wholemount of two mature *1(3)hem* larvae (m) and a wild-type larva (wt) of comparable age after treatment with 70°C water for 1 minute. Note the many black spots below the wild-type cuticle, representing melanized crystal cells, and the absence of the black spots (e.g., crystal cells) in the two mutant larvae. **(b)** Photomicrograph of longitudinal, histological section through the hind portion of a mature wild-type larva, showing blood cells (bc). **(c)** Photomicrograph of a comparable histological section as in (b) from a *1(3)hem* larva showing no blood cells. **(d)** Photomicrograph of sagital histological section through brain hemispheres of a mature, wild-type larva. **(e)** Comparable sagital histological section through *1(3)hem* brain hemispheres, showing within the adult optic anlagen small numbers of giant neuroblasts (gnb) (delimited by stippled lines). Compare with the normal neuroblasts in d. bc, blood cells; fb, fat body; gnb, giant neuroblast; int, integument; m, muscle; n, neuropile, nb, neuroblast; vg, ventral ganglion. Histological preparations: hematoxylin and iosin. Bars: 0.5 μm (a); 50 μm (b,c,d,e).

of these genes in different tissues are needed in order to understand their function in normal development and defense.

CONCLUDING REMARKS

The above three mutants, representing extreme hematopoietic disorders, namely, malignant growth of hemocytes on the one hand and complete absence of hemocytes on the other, show the wide range of possibilities that *Drosophila* offers for the identification and characterization of the hereditary factors involved in the differentiation of the components of the highly efficient primary cellular and humoral immune systems. Characteristic of normally differentiated hemocytes is their capacity to recognize self from non-self. Because in the two tumor mutants this process is obviously disturbed, it can be assumed that the two genes may directly or indirectly interfere with the gene product(s) instrumental in the process of recognition. In turn, this implies that, in the mutants, hemocyte differentiation is arrested at particular developmental stages. As described above, the plasmatocytes in the two mutants exhibit different states of morphological differentiation: *1(1)mbn* plasmatocytes appear morphologically differentiated, while depending on the allelic mutation, *1(3)mbn-1* blood cells are either arrested at the proplasmatocyte–plasmatocyte stage or engage in an abnormal differentiation, resulting in giant plasmato-, podo-, and lamellocytes (FIG. 6). The *Drosophila* tumor mutants show unequivocally that the malignant transformation of specific cell types is causally related to only one gene mutation and no additional mutations are required for the expression of the malignant tumor phenotype.[8–10] The recently cloned hematopoietic tumor-suppressor gene *air8*,[16] also called *hen*,[17] was shown to encode a homologue of the human ribosomal protein S6, which is assumed to have a regulatory function during hematopoiesis, possibly via its developmentally regulated phosphorylation. On the other extreme is the *1(3)hem* mutation which lacks blood cells altogether. This interesting gene is most probably involved in cell division of the primordial blood cells and the primordial adult optic neuroblast. The lack of blood cells does not seem to affect the development of larval tissues, since mutant larvae reach the third larval instar and pupariate, after which they die. However, the adult primordia, such as the imaginal discs, are severely affected, which suggests that blood cells may play a role in their growth and differentiation. The transplantation experiment with *1(1)mbn* hematopoietic organs suggests such a role not only in imaginal disc but also in muscle development. Thus, the molecular and cellular analysis of the above three mutants will certainly aid us in understanding the factors directing the growth and differentiation of *Drosophila* hemocytes and their role in the primary immune response.

ACKNOWLEDGMENTS

I am grateful to Thomas Schubert for his able technical assistance and to Uta Kestner for the typing of the manuscript.

REFERENCES

1. Rizki, T. M. 1978. The circulatory system and associated cells and tissues. *In* The Genetics and Biology of *Drosophila,* Vol. 2b. M. Ashburner & T. R. F. Wright, Eds.: 397–448. Academic Press. London.
2. Bairati, A. 1964. L' ultrastruttura dell' organo dell'emolinfa nella'larva di *Drosophila melanogaster.* Z. Zellforsch. **61:** 669–802.
3. Shrestha, R. & E. Gateff. 1982. Ultrastructure and cytochemistry of the cell-types in the larval hematopoietic organs and hemolymph of *Drosophila melanogaster.* Dev. Growth Diff. **24:** 65–82.
4. Stark, M. B. 1918. A hereditary tumor in the fruit fly *Drosophila.* J. Cancer Res. **3:** 297–301.
5. Stark, M. B. 1919. A hereditary tumor. J. Exp. Zool. **27:** 509–529.
6. Stark, M. B. 1935. A hereditary lymphosarcoma in *Drosophila.* Collection of papers of the New York Homeopathic Medical College and Flower Hospital. **1:** 397–400.
7. Stark, M. B. 1937. The origin of certain hereditary tumors in *Drosophila.* Am. J. Cancer **31:** 253–267.
8. Gateff, E. 1978. Malignant and benign neoplasms of *Drosophila. In* The Genetics and Biology of *Drosophila.* Vol. 2B. M. Ashburner & T. R. F. Wright, Eds.: 181–261. Academic Press. London.
9. Gateff, E. 1978. Malignant neoplasms of genetic origin in the fruit fly *Drosophila melanogaster.* Science **200:** 1446–1459.
10. Gateff, E. & B. C. Mechler. 1989. Tumor suppressor genes of *Drosophila melanogaster. In* Critical Reviews on Oncogenesis, Vol. 1. E. Pimentel, Ed.: 221–245. CRC Press. Miami.
11. Gateff, E. 1974. *1(1)mbn*: Lethal (1) malignant blood neoplasm. Dros. Inf. Serv. **51:** 21.
12. Gateff, E. 1977. *1(2)mbn*: Lethal (2) malignant blood neoplasm. Dros. Inf. Serv. **52:** 4.
13. Gateff, E. 1977. *1(3)mbn*: Lethal (3) malignant blood neoplasm. Dros. Inf. Serv. **52:** 5.
14. Gateff, E. Unpublished mutant.
15. Corvin, H. O. & W. D. Hanratty. 1976. Characterization of a unique, lethal, tumorous mutation in *Drosophila.* Mol. Gen. Genet. **144:** 345–347.
16. Watson, K. L., K. D. Konrad, D. F. Woods & P. J. Bryant. 1992. *Drosophila* homologue of the human S6 ribosomal protein is required for tumor suppression in the hematopoietic system. Proc. Natl. Acad. Sci. USA **89:** 11302–11306.
17. Stewart, M. J. & R. Denell. 1993. Mutations in the *Drosophila* gene encoding ribosomal protein S6 cause tissue overgrowth. Mol. Cell. Biol. **13:** 2524–2535.
18. Sparrow, J. C. 1978. Melanotic "tumors." *In* The Genetics and Biology of *Drosophila.* Vol. 2b. M. Ashburner & T. R. F. Wright, Eds.: 277–307. Academic Press. London.
19. Gateff, E., L. V. Gissman, R. Shresta, N. Plus, H. Pfrister, J. Schröder & H.

ZUR HAUSEN. 1980. Characterization of two tumorous blood cell lines of *Drosophila melanogaster* and the viruses they contain. *In* Invertebrate Systems *in Vitro.* E. Kurstak, K. Maramorosch & A. Dübendorfer, Eds.: 517–533. Elsevier/North-Holland. Amsterdam.
20. SHRESTA, R. & E. GATEFF. 1982. Ultrastructure and cytochemistry of the cell-types in the tumorous hematopoietic organs and the hemolymph of the mutant *lethal (1) malignant blood neoplasm* (*1(1)mbn*) of *Drosophila melanogaster.* Dev. Growth Diff. **24:** 83–98.
21. KRISHNAKUMARAN, A. & H. A. SCHNEIDERMAN. 1964. Developmental capacity of the cells of an adult moth. Exp. Zool. **157:** 293–305.
22. KRISHNAKUMARAN, A., S. J. BERRY, H. OBERLANDER & H. A. SCHNEIDERMAN. 1967. Nuclic acid synthesis during insect development. II. Control of DNA synthesis in *Ceeropia* silkworm and other saturnid moths. J. Insect. Physiol. **13:** 1–57.
23. ROBERTSON, M. & J. H. POSTLETHWAIT. 1986. The humoral antibacterial response of *Drosophila* adults. Dev. Comp. Immunol. **10:** 167–176.
24. BOMAN, H. G., I. NILSSON & B. RASMUSON. 1972. Inducible antibacterial defence system in *Drosophila.* Nature **237:** 232–234.
25. WICKER, C., J. M. REICHHART, D. HOFFMAN, D. HULTMARK, Ch. SAMAKOVLIS & J. HOFFMANN. 1990. Characterization of a *Drosophila* cDNA encoding a novel member of the diptericin family of immune peptides. J. Biol. Chem. **265:** 22493–22498.
26. FAYE, J. & G. R. WYATT. 1980. The synthesis of antibacterial proteins in isolated fat body from *Ceeropia* silkmoth pupae. Experentia **36:** 1325–1329.
27. SHRESTA, R. & E. GATEFF. 1986. Ultrastructure and cytochemistry of the tumorous blood cells in the mutant *lethal (3) malignant blood neoplasm (l(3)mbu)* of *Drosophila melanogaster.* J. Invert. Pathol. **48:** 1–12.
28. KONRAD, L., G. BECKER, A. SCHMIDT, T. KLÖCKNER, G. KAUFER-STILLGER, S. DRESCHERS, T. FRIEDBERG, J. ENGELS, J. E. EDSTRÖM, A. HOTZ-WAGENBLATT & E. GATEFF. 1993. Molecular cloning, structure and possible function of the blood cell tumor suppressor gene, *lethal (3) malignant blood neoplasm-1* (*1(3)mbn-1*) of *Drosophila melanogaster.* Dev. Biol. Submitted.
29. DUNN, P. E., W. DAI, M. R. KANOST & C. GENG. 1985. Soluble peptidoglycan fragments stimulate antibacterial protein synthesis by fat body by larvae of *Manduca sexta.* Dev. Comp. Imunol. **9:** 559–568.
30. SAMAKOVLIS, Ch., B. ÄSLING, H. G. BOMAN, E. GATEFF & D. HULTMARK. 1992. *In vitro* induction of cecropin genes—An immune response in a *Drosophila* blood cell line. Biophys. Biochem. Res. Commun. **188:** 1169–1175.
31. KAPPLER, Ch., M. MEISTER, M. LAGNEUX, E. GATEFF, J. A. HOFFMANN & J. M. REICHHART. 1993. Insect immunity. Two 17-bp repeats nesting a κ B-related sequence confer inducibility to the diptericin gene and bind a polypeptide in bacteria-challenged *Drosophila.* EMBO J. **12:** 1564–1568.

Chemical Toxicity and Host Defense in Earthworms

An Invertebrate Model[a]

A. J. GOVEN,[b] L. C. FITZPATRICK,[b], AND B. J. VENABLES[c]

[b]*Environmental Effects Research Group*
Department of Biological Sciences
University of North Texas
Denton, Texas 76203

[c]*TRAC Laboratories*
113 Cedar Street
Denton, Texas 76201

INTRODUCTION

Understanding toxic potential and mechanisms of action of environmental xenobiotics is fundamental for assessing risk to public and environmental health. Although numerous methods have been developed for screening chemicals and studying their modes of action on a variety of acute toxic endpoints and subchronic–chronic processes, only relatively recently have scientists become aware of the broad spectrum of xenobiotics that alter immune function and of the immune system's potential as a target organ system for use in assessing the toxicity of exposure to chemicals.[1] Because there is considerable information on the innate, nonspecific, specific and molecular components of the immune system, immune responses are especially well suited for comparative analyses that emphasize mechanisms of chemical toxicity. Additionally, since immune responses are important host defense mechanisms, their modulation may result in increased incidence of infections that could influence the survival of individuals and their populations. Subcellular, cellular, and organismal immunological indicators of either exposure to or effects of chemicals, especially those that can be demonstrated at exposure levels shown to be nontoxic by traditional toxicity evaluation, can be used as assessment tools. Such immunological indicators fall within the category of measurement endpoints defined as biomarkers.[2]

Although complexity of the immune system has increased during the course of animal evolution, certain aspects of immunity have been conserved phylogenetically, and immunocytes in one form or another can be found in all phyla above Porifera. Additionally, there is sufficient basic and comparative information on the immune systems of several invertebrate taxa to support using their immune function to study sublethal toxicity of chemicals. For these reasons, invertebrates are amenable for development of panspecific biomarkers for assessment of environmentally relevant

[a]Portions of this research and manuscript preparation were supported by grants from the National Institute of Environmental Health Sciences (ES-34811), the Texas Advanced Technology Program, and the U.S. Environmental Protection Agency.

chemicals on immune defense systems. Furthermore, invertebrates are fundamental components of all ecosystems, dominating animal biomass and pathways of energy flow and nutrient recycling. As such, they represent the ecological front line in pathways of exposure to environmental xenobiotics. Evaluation of chemically induced alteration in their immune function offers the potential for development of biomarkers that are not only quite sensitive to sublethal exposure but also are ecologically realistic representations of the bioavailability of toxicants in the environment. Detection of alterations in the immunocompetence of these front-line organisms may serve as valuable early warning symptoms of deleterious effects that threaten survival of individuals or populations and long-term ecosystem health and stability. Among invertebrates, earthworms possess a number of attributes that make them an especially appropriate choice for development of immune-based biomarkers for investigation of effects of chemicals on host immune defense systems.

RATIONALE FOR THE EARTHWORM MODEL

In 1988 we began to develop a model immunoassay system using earthworms, *Lumbricus terrestris,* as nonvertebrate surrogates for assessing immunotoxic potential of chemicals and understanding their modes of action. Development of an earthworm immune-based system of biomarkers was based on a need for rapid, sensitive, cost-effective, and socially uncontroversial surrogate immunoassay protocols that could be used as an adjunct or complement to existing protocols with mammals. Such an assay system would be used to screen chemicals to determine if further mammalian tests should be performed.

Earthworms were selected for several reasons: (1) Their immune functions appear to be sufficiently analogous or homologous to those in vertebrates for use in screening chemicals for immunotoxicity in higher organisms, including mammals[3–6]; (2) being virtually ubiquitous and ecologically important soil organisms, they are valuable *in situ* sentinels for use in assessing risks to public and environmental health; and (3) earthworm behavior and morphology enable their direct exposure to complex environmental mixtures and matrices of pollutants. Additionally earthworms are easy and inexpensive to maintain and conduct immunological experiments with, their basic biology is well known, and they are currently used in standardized acute toxicity protocols for laboratory and *in situ* bioassays.[7]

EARTHWORM IMMUNE SYSTEM

The earthworm immune system is housed in the coelom, which contains coelomic fluid and coelomocytes. Coelomocytes are, like mammalian leukocytes, sensitive to foreign materials. They are active in (1) Innate immune reactions such as lysozyme production,[8] (2) nonspecific immune reactions such as phagocytosis and inflammation, (3) the more complex cellular immune responses responsible for such reactions as graft rejection, and (4) humoral immune responses including synthesis and secretion of agglutinins and lytic factors.

Earthworms possess efficient nonspecific mechanisms of disposing of foreign material, equivalent to the responses found in vertebrates. Earthworm coelomocytes search out, phagocytose, and destroy non-self material.[9] Additionally, coelomocytes mount a well-defined generalized inflammation in response to tissue injury as part of tissue repair.[10,11]

Earthworm coelomocytes synthesize and secrete an array of humoral factors that participate in immune responses. The most important humoral factors are agglutinins and lysins. Agglutinins are specifically induced by and react with antigen.[12] Agglutinins function to aggregate foreign materials and may serve as opsonins providing an efficient mechanism for phagocytosis.[13] Presence and strength of agglutinins in coelomic fluid can be determined in the same manner as vertebrate antibody and is reported as a titer. Coelomocytes synthesizing and secreting agglutinins are assayed by determining the ability of cells to form secretory rosettes (SR) with erythrocytes (antigens). SR are formed by coelomocytes releasing agglutinins in response to erythrocyte stimulation that results in multiple layers of erythrocytes adhering to the coelomocyte.[14–16] Lytic factors inhibit growth of bacteria and thus are important to earthworm defense against pathogens.[17,18] Bacteriostatic and bactericidal effects can be induced by inoculating earthworms with sublethal numbers of bacteria, resulting in the immunization of animals and resistance to challenge.[14]

Cell-mediated immune response has been demonstrated in earthworms by transplantation studies. Transplantation experiments have demonstrated that earthworms are capable of recognizing and rejecting foreign tissue grafts while accepting autografts.[19,20] Xenografts are rejected more vigorously than allografts. Rejection of a second transplanted graft from the same donor is accelerated, suggesting a memory component.[21,22]

TIER APPROACH TO ASSESS CHEMICAL IMMUNOTOXICITY

The complexity of the immune system and the fact that no single immunoassay is sufficient for screening or assessing mechanisms for toxic action by xenobiotics had led to tiered approaches with multiple assays and endpoints.[2,23] A typical tier approach involves two panels of immunoassays: Tier 1, a panel of relatively simple parameters of innate immunity (II), nonspecific immunity (NSI), humoral-mediated immunity (HMI), and cell-mediated immunity (CMI), which with limited effort can be used to screen xenobiotics; and Tier 2, a comprehensive panel of additional NSI, HMI, CMI, and host-resistance challenge endpoints. The latter can be used to further test chemicals that produce immune alteration in Tier 1, usually at lower dose levels, and to provide in-depth information on modes of action and the specific nature of toxic effects.

Our approach to develop immunoassays using earthworms follows the National Toxicology Program's (NTP) two-tiered immunoassay protocol using mice.[23] The NTP's objective was to develop a panel of immunoassays having sufficient sensitivity and comprehensiveness to detect subtle immunologic effects from exposure to chemicals and relate them to alterations in host resistance to disease-producing organisms (e.g., bacteria, viruses, and parasites) or tumor cells.

In our work, coelomocytes from earthworms exposed to chemicals initially are assayed using Tier 1 tests (TABLE 1). Although the probability of detecting immunotoxicity in Tier 1 is high, only limited information on the specificity of the immune defect may be discernable. For this reason, chemicals causing measurable effects in Tier 1 are reexamined using relevant, corresponding Tier 2 assays (TABLE 2). Because Tier 1 assays parallel and are predictive of those in Tier 2, this approach provides for efficient, cost-effective collection of the most relevant information concerning immunotoxicity.

SELECTION OF IMMUNOASSAYS

We have targeted alterations of the innate, nonspecific, and specific immune responses by chemicals in earthworms for developing a suite of immunological biomarkers to determine immunomodulatory effects of chemicals on host defense systems. Criteria used to select immunoassays were that they (1) be coelomocyte based, (2) measure earthworm immune functions that are homologous or analogous to those in vertebrates, (3) predict secondary immunodeficiency effects such as reduced host resistance, and (4) have low natural variability.

Coelomocyte-based immunoassays were chosen for development because these cells are the earthworm's immunocytes and because of our development of a noninvasive extrusion technique for their collection that has proven to be superior to the traditional coelomic cavity puncture method.[24] The extrusion technique produces clean samples of viable immunocompetent cells in numbers large enough for multiple immunoassays to be performed on individual earthworms and produces little

TABLE 1. Earthworm Tier 1 Immunoassays for Use in Screening Chemicals for Immunomodulatory Effects

Parameter	Procedure
Innate Immunity	
• Enzyme function	• Lysozyme activity
Immunopathology	
• Cytology	• Complete coelomocyte count
• Body mass	• Total body weight
• Inflammation	• Wound healing
Nonspecific Immunity	
• Phagocytosis	• Ingestion of erythrocytes/fluorescent beads
• Natural killer cell activity	• Allogeneic/xenogeneic target cells
Humoral-Mediated Immunity	
• Agglutinins/lytic factors	• Agglutination/lytic titers
• Rosette formation	• Secretory rosettes
Cell-Mediated Immunity	
• Allo-antigen recognition	• Primary/secondary allogeneic tissue recognition/rejection

TABLE 2. Earthworm Tier 2 Immunoassays for Comprehensive Analysis of Immunomodulatory Effects of Chemicals

Parameter	Procedure
Immunopathology	
• Immune cell quantitation	• Flow cytometric analysis of coelomocytes
Nonspecific Immunity	
• Phagocytosis	• Coelomocyte ingestion of bacteria/yeast
• Bactericidal activity	• Coelomocyte bacterial killing
• Enzyme function	• Nitroblue tetrazolium dye reduction
Humoral-Mediated Immunity	
• Challenge response	• Bacterial agglutination
• Antimicrobial activity	• Bacteriostatic activity of coelomic fluid
Cell-Mediated Immunity	
• Mixed leukocyte response	• Allogeneic/xenogeneic coelomocyte stimulation
• Blast-transformation	• Coelomocyte mitogen/antigen stimulation
Host Resistance Challenge Models	
• Bacterial models	• Resistance to bacterial challenge

trauma, allowing for cells to be harvested by sequential extrusion of the same animal at six-week intervals without affecting normal coelomocyte cytological or immunological parameters.[24] The extrusion technique allows for monitoring earthworm immune function during long-term (subchronic/chronic) exposure to chemicals. Immunoassays that were homologous or analogous to those in vertebrates were developed because they would allow for the prediction of potential chemical effects on vertebrate immune responses. Additionally, the wealth of knowledge on vertebrate immunity could be used to elucidate chemical effects on invertebrate immunity. Immunoassays that are diagnostic or predictive of secondary immunodeficiency effects, such as reduced host resistance to disease following initial chemical-induced immunologic lesions, are valuable because of their use in public health risk assessment. Also, these immunoassays examine functional immunologic measurements that correlate with host resistance to foreign challenge. Immunological biomarkers of chemical stress should be stable.[25] They must have an appropriate mix of the competing characteristics of variability in response to pollutants and consistency in response to nonchemical influences (i.e. habitat changes) to achieve maximum signal-to-noise ratio (chemical response: natural variability). Finally, they should measure something of ecological significance, be easily and unambiguously quantifiable, biologically appropriate for expected exposure routes, and, if possible, indicative of specific classes of chemicals.[25]

Herein we report on the immunomodulatory effects of chemicals on earthworms, *L. terrestris,* as indicated by immunoassays measuring innate immunity as lysozyme activity; cytology as total coelomocyte counts (TCC), differential coelomocyte counts (DCC), and coelomocyte viability; NSI as phagocytosis and nitroblue tetrazolium (NBT) dye reduction; and HMI as SR formation.

ASSESSMENT OF IMMUNOTOXIC EFFECTS OF CHEMICALS ON EARTHWORM HOST DEFENSE

Cytological Biomarkers

Cytological parameters of TCC, DCC, and coelomocyte viability represent a logical set of easily measured, sensitive, and stable biomarkers for assessing acute and chronic immunotoxicity. Chemical effects on these basic cytological parameters indicate effects on both the overall health of an earthworm and potential problems in NSI and specific immunity as a complete blood count would in mammals.

Effects of PCB on TCC, DCC, and Coelomocyte Viability

The extrusion technique allows for multiple coelomocyte collection from individual earthworms at six-week intervals with normal cytological (TCC, DCC, viability) parameters at each extrusion (Figs. 1,2).[24] This technique is especially important in the study of chronic exposures on an individual basis by examination of chemical effects on regeneration of immunocompetent coelomocytes.

Much of our cytological work has been conducted following earthworm exposure to Aroclor 1254,[26] a commercial mixture of polychlorinated biphenyls (PCB) with known vertebrate immunotoxic effects.[27] For laboratory exposure we used a five-day filter paper contact exposure to 10 μg PCB/cm^2. This represents a sublethal exposure well below the 300 μg/cm^2 LC_{50} and results in a$_{50}$whole-body tissue concentration of approximately 100 to 200 μg/g dry mass (LD_{50} = 1140 μg/g dry mass). Measurement of body burden concentrations of PCB at 0, 6, 12, and 18 weeks postexposure resulted in 91.2, 41.0, 30.2, and 15.7 μg PCB/g dry mass, respectively.

Acute effects of PCB on cytological parameters were evident on coelomocytes collected immediately after exposure. Significant differences in viability (Fig. 1) between coelomocytes from exposed and unexposed earthworms suggest that PCB was somewhat toxic to coelomocytes. DCC for controls agree with reports that basophils, acidophils, and transitionals represent 60, 30, and 3% of the total coelomocyte population for normal earthworms.[5] Although TCC (Fig. 1) were not immediately (0 hr depuration) affected by PCB exposure, DCC (Fig. 2) showed a striking decrease in percentage of basophils, a cell responsible for HMI, and a concomitant increase in acidophils, a phagocytic cell.

Chronic PCB effects were measured by the ability of earthworms to repopulate their coelomic cavity with immunocompetent coelomocytes as determined by TCC (Fig. 1) on coelomocytes collected by secondary extrusion at 6, 12, and 18 weeks postexposure. There were significant differences in TCC between exposed and unexposed earthworms at six weeks. At 12 weeks, exposed earthworms extruded too few cells to obtain accurate cell counts. By 18 weeks, when PCB tissue concentration was lowest (15.7 μg/g dry mass), TCC returned to normal, indicating a repopulation of coelomic cavity. Viability (Fig. 1) and DCC (Fig. 2) of coelomocytes collected from exposed earthworms followed the same pattern: Abnormal cytological patterns at six weeks; too few cells to assay at 12 weeks; and return to normal parameters at 18 weeks postexposure.

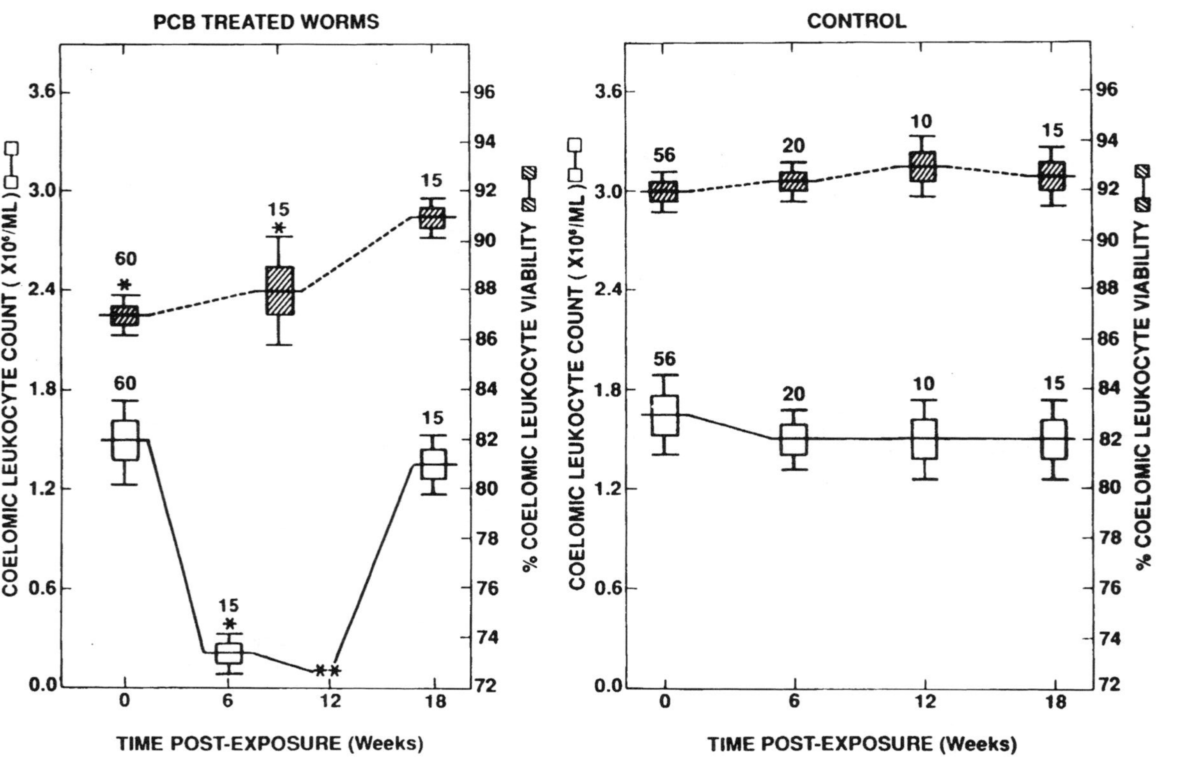

FIGURE 1. Effects for five-day PCB (Aroclor 1254) filter paper exposure (10 μg/cm^2) on total numbers (*open rectangles*) and viability (*shaded rectangles*) of coelomocytes collected from *Lumbricus terrestris.* All exposed and unexposed earthworms were extruded at 0 hours, and 6, 12, and 18 weeks postexposure. Horizontal line is the mean, rectangles are standard error, and vertical lines represent 95% confidence interval. Sample sizes are indicated by numbers above assay points. Asterisks indicate means significantly different from controls. Double asterisks at 12 weeks indicate too few cells were extruded to obtain reliable data.

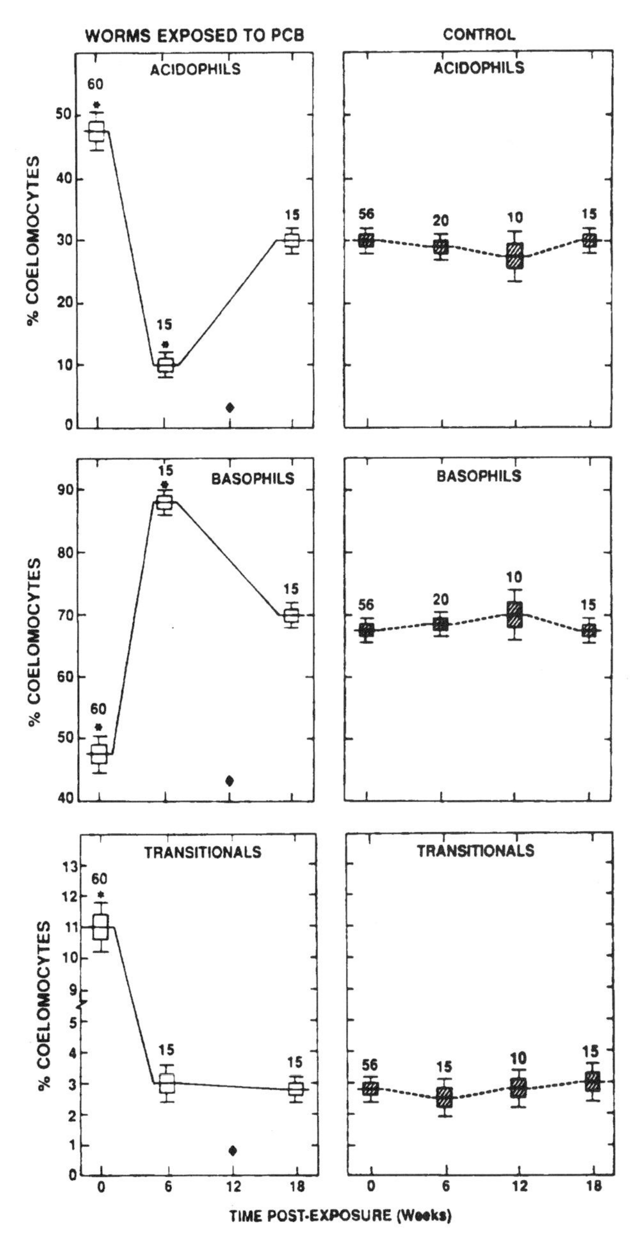

FIGURE 2. Comparison of mean acidophil, basophil, and transitional percentages between earthworms exposed for five days to 10 μg/cm^2 PCB (Aroclor 1254) on filter paper (*open rectangles*) and controls exposed to saline (*shaded rectangles*). Coelomocytes were collected by extrusion at 0 hours, and 6, 12, and 18 weeks postexposure. Asterisks indicate significant difference from controls and diamond at 12 weeks depuration indicates too few cells were extruded to obtain reliable data. Symbols same as FIGURE 1. Sample sizes are indicated by numbers above assay points.

Because PCB-exposed earthworms were unable to extrude normally until 18 weeks postexposure when the PCB tissue concentration was at its lowest, we suggest that the xenobiotic affected the ability of coelomopoietic tissues to produce normal numbers of coelomocytes. Experiments conducted using mammals demonstrated similar effects of PCB exposure characterized by a decreased number of circulating leukocytes.[28,29] Inhibition of rates by which the coelomic cavity is repopulated with immunocompetent coelomocytes reflects the ability of coelomopoietic tissue to replace killed or immunocompromised coelomocytes and portend chronic immunodeficiency in earthworms exposed to chemicals.

Innate Immunity

Effect of Cu on Lysozyme Activity

Lysozyme, a host physiological factor, is an enzyme capable of bactericidal activity via action on peptidoglycan of gram-positive bacterial cell walls and functions as a component of an organism's innate or natural antimicrobial defense mechanism.[30] We have demonstrated lysozyme activity to be present in earthworm coelomic fluid and coelomocyte extracts.[8] This lysozyme activity appears to be similar to that found in serum and leukocyte extracts of various mammals, including humans (FIG. 3). Correlation between increasing lysozyme activity in coelomocyte extracts and increasing coelomocyte numbers suggests that the enzyme is a product of coelomocytes,[8] as it is believed serum lysozyme activity results from release of the enzyme by neutrophils populating the vasculature.[31] Phylogenetic conservation indicates that lysozyme is a constituent of the primitive innate immune defense mechanism associated with the granulocyte, monocyte–macrophage system of mammals, and with coelomocytes in earthworms. Reduced lysozyme activity in earthworms due to chemical exposure could be used to predict chemical immunotoxicity in higher organisms.

We assessed the sensitivity of earthworm lysozyme activity as an assay for chemical immunotoxicity using Cu^{27} (as $CuSO_4$), which is known to inhibit vertebrate lysozyme activity.[32] Earthworms exposed to sublethal concentrations of Cu^{27} (LC_{50} = 2.58 μg Cu^2/cm^2) at 10°C for five days using the filter paper contact method showed significantly reduced lysozyme in both coelomic fluid and coelomocyte extracts (FIG. 4). Coelomic fluid lysozyme activity decreased to 40 and 50% of controls after exposure to 0.5 and 1.0 μg/cm^2 Cu^2, respectively. Lysozyme activity of coelomocyte extracts decreased to 54% of controls after exposure to 1.0 μg/cm^2 Cu^2. Although exposure to 0.5 μg/cm^2 Cu^2 did not significantly reduce lysozyme activity of cell extracts, exposure reduced enzyme activity to 72% of controls. Tissue concentration of Cu^2 was 28.5 and 73.1 μg/g dry mass in earthworms exposed to 0.5 and 1.0 μg/cm^2 of Cu^2, respectively, indicating a dose–response relationship between Cu^2 and reduction in lysozyme activity.

It is likely that Cu^2 interacts with or binds to lysozyme in a manner adversely affecting the functional conformation of the enzyme, leading to attenuation or inactivation. Metal ions, including Cu^{27}, tend to bind to basic proteins such as lyso-

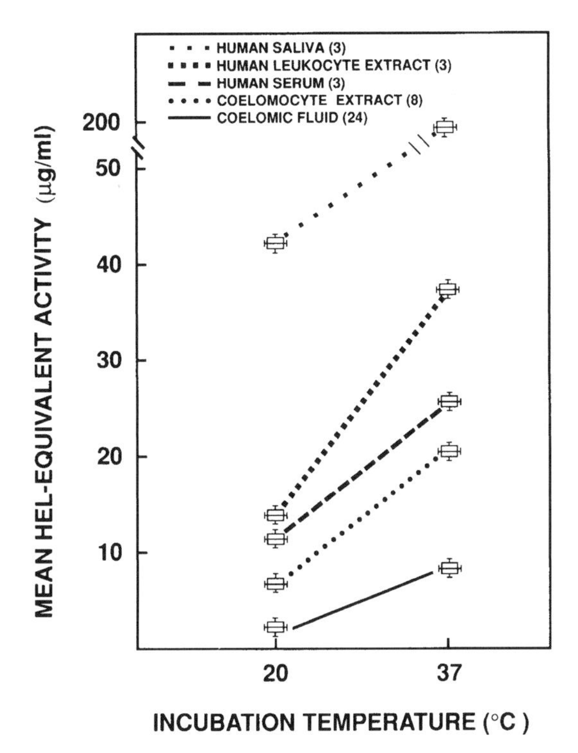

FIGURE 3. Lysozyme activity, measured as hen's egg lysozyme (HEL) equivalence (μg/ml) in human saliva, leukocyte extract and serum, and earthworm (*Lumbricus terrestris*) coelomocyte extract and coelomic fluid at 20 and 37°C incubation temperature for 24 hours. Symbols same as those in FIGURE 1. Numbers represent the sample size.

zyme,[33] leading to chemical reactions that catalyze hydrolysis of peptide bonds or disulfide linkages or breakage of hydrogen bonds,[32] causing structural changes in enzymes.

Reduced lysozyme activity in earthworms suggests an immunosuppression that could result in lowered resistance to bacterial challenge. The earthworm lysozyme activity assay appears to be sufficiently sensitive for measuring sublethal effects of chemicals on an important innate immune function common to diverse animals, including humans.

Nonspecific Immunity

Assessment of chemical effects on phylogenetically conserved immune responses such as immunocyte spreading (activation), phagocytosis, and NBT dye reduction should allow for prediction of xenobiotic effects on homologous vertebrate nonspecific responses.

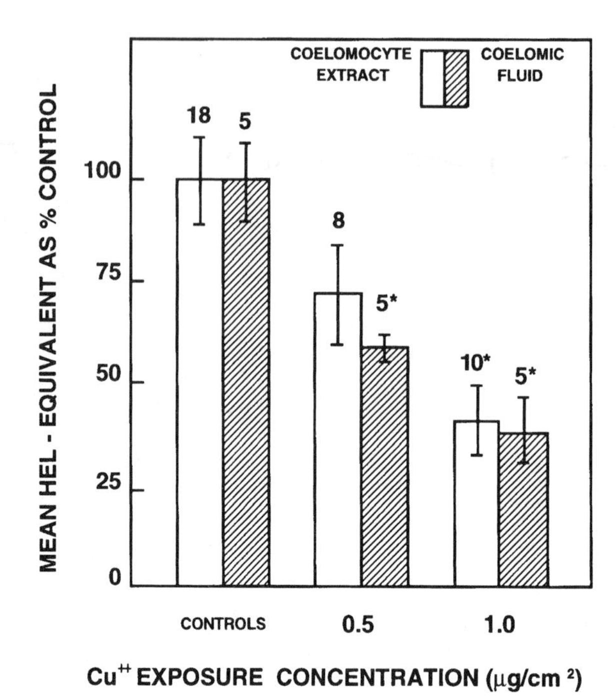

FIGURE 4. Suppression of lysozyme activity, measured as hen's egg lysozyme (HEL) equivalence (μg/ml), of earthworm (*Lumbricus terrestris*) coelomic fluid and coelomocyte extract after five-day filter paper contact exposure to sublethal ionic concentrations (0.5 and 1.0 μg/cm^2) of Cu^2 ($CuSO_4$). Expressed as percentage of control. Vertical lines represent standard errors. Numbers represent sample size. *Significant difference from controls as measured by the one-way Dunnett's test, $\alpha = 0.005$.

Effect of PCB on Coelomocyte Phagocytosis

As with cytological biomarkers, much of our work on phagocytosis is conducted using coelomocytes sequentially extruded from earthworms after a five-day filter paper exposure to a sublethal concentration (10 μg/cm^2) of PCB.[26] This method permits determination of acute effects of PCB immediately after exposure and chronic effects measured during postexposure PCB depuration.

PCB exposure resulted in both acute and chronic inhibition of coelomocyte phagocytic competence (FIG. 5). Coelomocytes harvested immediately after exposure demonstrated significantly reduced ability to ingest RRBC compared to cells collected from unexposed animals. Chronic PCB effects were assessed on cells collected by secondary extrusion at six, 12, and 18 weeks postexposure. Phagocytic competence was significantly inhibited at six weeks of depuration. Assays were not possible at 12 weeks owing to the influence of PCB exposure on the ability of earthworms to extrude cells. At 18 weeks, coelomocytes showed normal phagocytic

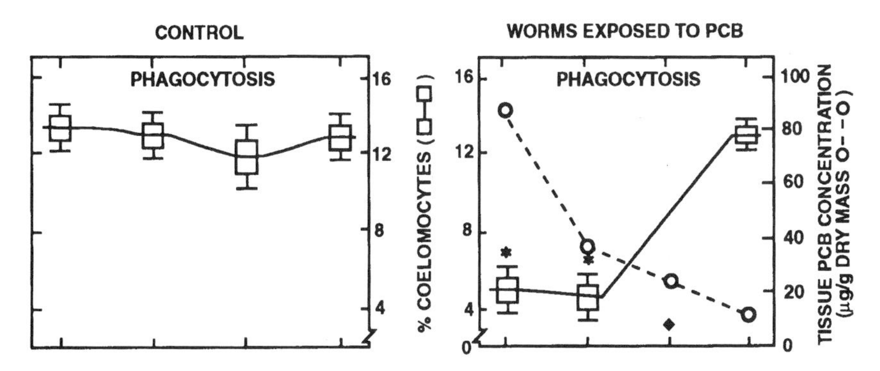

FIGURE 5. Effect of five-day PCB (Aroclor 1254) filter paper exposure (10 μg/cm^2) on phagocytic activity of coelomocytes collected from *Lumbricus terrestris.* All exposed and unexposed earthworms were extruded at 0 hours, and 6, 12, and 18 weeks postexposure. Asterisks indicate significant difference from controls. Diamond at 12 weeks indicates too few cells were extruded to obtain reliable data. Symbols same as FIGURE 1.

immune function. These results illustrate that PCB can mediate a loss of coelomocyte phagocytic activity with recovery only after PCB depuration.

Phagocytosis, a well-documented nonspecific immune function of coelomocytes,[9] is the most primitive of the protective responses in animals. Suppression of the phagocytosis by coelomocytes after exposure to PCB may be homologous to suppression of mammalian cell phagocytosis. Mammals exposed to PCB have decreased immunoglobulin (Ig) levels.[27] Reduction in opsonin synthesis (Ig) could decrease phagocytosis. We have shown (discussed below) that after PCB exposure, coelomocytes secrete lower levels of agglutinins as measured by SR formation. A reduction in agglutinins, which may function as opsonins in earthworms, may account for the suppressed phagocytosis by coelomocytes, as reduced Ig production does in mammalian leukocytes.

Effect of Refuse-Derived Fuel-Fly Ash (RDFF) on NBT Dye Reduction

The major roles of phagocytes are ingestion and killing of microorganisms, principally by the oxygen-dependent "respiratory burst" involving superoxide anion (O_2^-) and hydrogen peroxide (H_2O_2) production.[34] Activation of oxygen metabolism of phagocytosis is a useful marker of phagocytic ingestion and an important process related to killing of ingested microbes.[35]

The NBT dye reduction assay has been used to evaluate the ability of phagocytes to catabolize and kill phagocytosed bacteria by the "respiratory burst."[35] This assay, which indirectly measures O_2^- production by colorimetry, was developed to detect metabolic defects associated with chronic granulomatous disease in humans. The NBT assay with earthworms would complement our use of coelomocyte phagocytic activity and identify xenobiotics that interfere with intracellular oxidative bactericidal activity of phagocytic cells without affecting their stimulation and/or actual phagocytosis.

We have demonstrated that earthworm coelomocytes have the ability to reduce NBT dye in a nearly linear fashion over incubation times, as expected for leukocytes collected from mice and humans (FIG. 6).[36] Thus, it appears that the responsible cellular mechansisms are broadly conserved phylogenetically. Differences in dye reduction between coelomocytes and neutrophils, in particular, may be explained by difference in incubation temperature (10°C for earthworm vs. 37°C in mammals) and

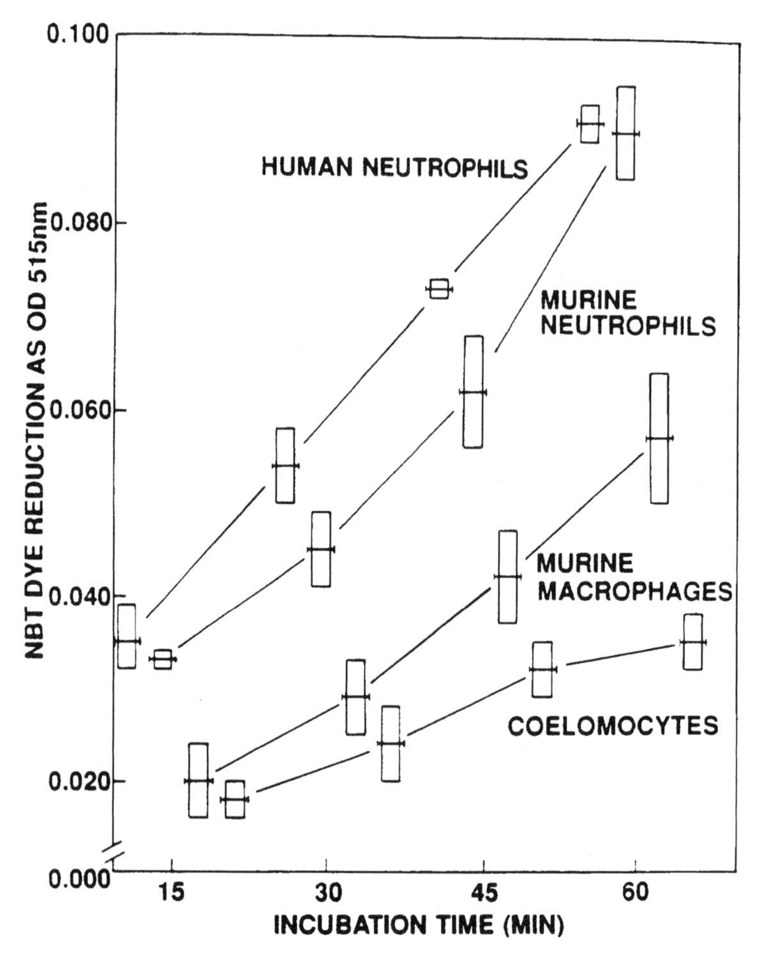

FIGURE 6. NBT dye reduction by murine and human neutrophils, murine macrophages, and earthworm (*Lumbricus terrestris*) coelomocytes measured after four incubation periods. Horizontal lines and vertical rectangles represent means and standard errors, respectively. The number of humans and individual mice used for each of the four NBT assays is given under the cell type. The number for earthworms represents pooled coelomocyte samples taken from 26 different groups of five earthworms.

enrichment of respective cell populations. Enrichment of mammalian cells produced a bias toward cell types involved in phagocytosis and catabolism of microorganisms, whereas collection of coelomocytes by the extrusion technique resulted in a mixture of cells having varied levels of involvement in phagocytosis.

Sensitivity of the earthworm NBT dye reduction assay to chemical exposure was determined using coelomocytes harvested from earthworms exposed for five days to sublethal RDFF: commercial soil mixtures of 10:90, 30:70, 50:50, and 70:30 by dry mass.[36] Earthworms were exposed in one-liter glass jars with metal caps within an environmental chamber at 10°C without light. Five earthworms were housed in each jar, which contained 150 g dry weight of the RDFF: commercial soil mixture

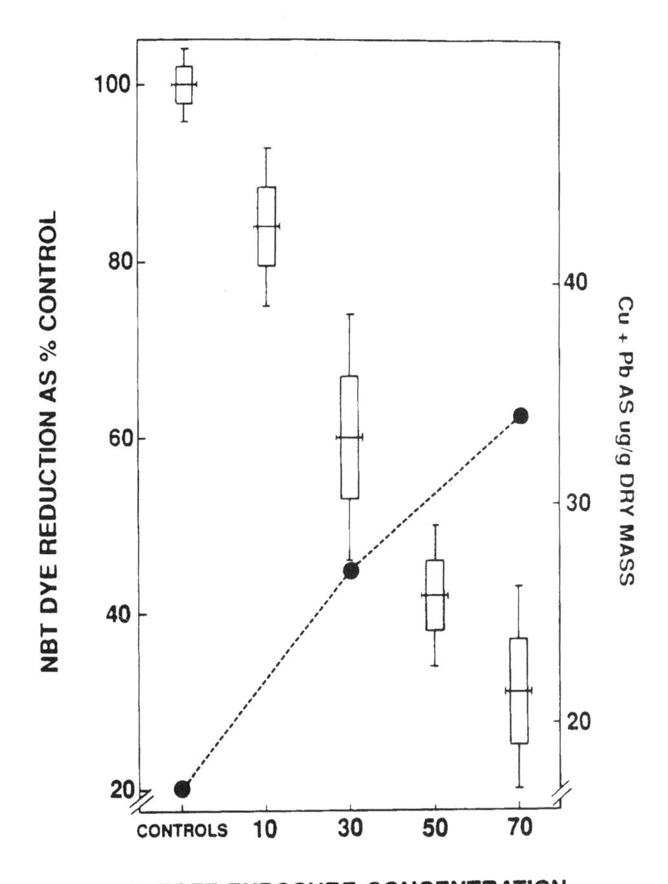

FIGURE 7. NBT dye reduction after 60 minutes incubation by coelomocytes from earthworms (*Lumbricus terrestris*) exposed for five days to commercial soil (controls) and refuse-derived fuel fly ash (RDFF) mixtures with commercial soil of 10, 30, 50, and 70%. Expressed as percent of control $OD_{515\ nm}$. Symbols same as those in FIGURE 1. Data based on pooled coelomocyte samples from six different groups of five earthworms for each exposure concentration. Dashed line represents corresponding tissue levels of metals (Cu + Pb).

hydrated with 10 ml deionized water. Controls were exposed similarly, but to 150 g of soil only. Concentrations of principal RDFF heavy metals (Cd, Cu, Cr, Zn, Ni, and Pb) were determined for parent material and acid-digested tissues in selected earthworms by atomic absorption spectrometry.

NBT reduction by coelomocytes was influenced significantly by RDFF exposure concentration.[36] Expressed as percent of controls, NBT reduction was inversely related to RDFF concentrations (FIG. 7). Dye reduction by coelomocytes from 30, 50, and 70% RDFF exposure groups was significantly lower by 40, 59, and 64%, respectively, than that of controls. All exposure groups, except 50 and 70%, were significantly different from each other. Suppression of NBT dye reduction exceeded the 25% value used in clinical medicine to define immune suppression in humans.[37]

Concentrations of Zn, Pb, Cu, Cr, Ni, and Cd were 2342, 610, 470, 104, 49, and 20 mg/kg dry mass, respectively, in undiluted RDFF. Metals were analyzed in control earthworms and in those exposed to 30 and 70% RDFF concentrations—extremes of the range, where there were significant effects on NBT reduction. Tissue concentrations of Cu and Pb were significantly higher than those of the controls. The other metals were not significantly different between controls and exposed earthworms.

Because NBT reduction in phagocytic cells occurs by a chemical reaction between the dye and O_2^-, and O_2^- is produced by the one-electron reduction of O_2 (a reaction catalyzed by NADPH oxidase), suppression of dye reduction suggests that heavy metals (known enzyme toxicants) interfered in the pathway leading to O_2^- formation. Both Cu and Pb are known to affect glucose-6-phosphate dehydrogenase (G-6-PD),[38] the enzyme in the hexose monophosphate shunt involved in conversion of $NADP^+$ to NADPH, which is the reducing agent in O_2^- production.[39] Additionally, Pb has been linked to G-6-PD deficiency in mammals[40] and suppresses resistance to bacterial infection in mice.[41] The latter accords with increased susceptibility to infection found in children poisoned by Pb[40] and our preliminary data suggesting decreased resistance to infection with *Aeromonas hydrophilia* in the manure worm, *Eisenia foetida,* after a five-day exposure to RDFF concentrations of 30, 50, and 70%.

Suppression of NBT dye reduction occurred at tissue concentrations of Cu and Pb similar to those in some natural populations (roadside, mining sites). The assay also showed toxicity of Pb and/or Cu at exposure levels below those affecting growth and reproduction and causing mortality.[42] Thus, the earthworm NBT assay appears to be sufficiently sensitive for measuring a sublethal effect of heavy metals on an important nonspecific immune function of cells (the ability to resist infection by killing microorganisms oxidatively) common to a wide diversity of animals, including important wildlife, at realistic environmental concentrations and below those reported to produce other forms of toxicity.

Application of Nonspecific Immunoassays to Hazardous Waste Site Soils

We have performed a preliminary assessment of sublethal toxicity of soils from a Superfund hazardous waste site (HWS) using nonspecific immunoassays on *L. terrestris* exposed *in situ* to soils having different contamination levels.[43] The HWS was used to mix and batch agrochemicals including herbicides and pesticides for resale. Earthworm mortality was highly correlated to tissue and soil concentrations of chlordane and DDT.[44] Earthworms were categorized as having been exposed *in situ* to high, intermediate, or low contamination levels according to *in situ* morbidity and mortality data.[45] We also examined immunologic effects in *L. terrestris* exposed in our laboratories to the highly contaminated soil diluted with artificial soil to 5%.[43]

Coelomocyte spreading, phagocytosis of antigenic rabbit red blood cells (RRBC), and reduction of NBT dye were used as assays for nonspecific immune function (TABLE 3). Spreading indicates membrane activation, leading to phagocytosis or enhanced ingestion of particulate antigens. Reduction in any of these responses would indicate a suppression in NSI. Coelomocytes from earthworms exposed *in situ* for 48 hours to HWS soils of low, intermediate, and high levels of contamination demonstrated suppressed coelomocyte activation (spreading) and ability to reduce NBT dye at all three exposure levels, while phagocytosis was reduced at intermediate and high levels. There was an apparent dose–response effect that correlated with the exposure level. All three NSI functions were suppressed in coelomocytes from earthworms exposed in the laboratory for five days to a 5% dilution of HWS soil. Results indicate a good correspondence between field and laboratory exposure.

SPECIFIC IMMUNITY

Coelomocytes synthesize and secrete agglutinins that participate in the humoral immune response. Agglutinins, although not related to antibody, exhibit analogous functional properties. Agglutinins are specifically induced by antigen,[12] function to aggregate foreign material, and may serve as opsonins.[13] Inhibition of the humoral response in earthworms due to chemical exposure can result in an immunosuppression and lead to earthworms developing similar problems as seen in vertebrates, namely reduced host resistance to infection.

TABLE 3. Effects of *in Situ* Exposure to Hazardous Waste Site Soil of Low, Intermediate, and High Levels of Contamination and Laboratory Exposure to 5% Dilution of HWS Soil on Coelomocyte Activation (Spreading), Phagocytosis, and Nitroblue Tetrazolium (NBT) Dye Reduction

Immune Parameter	*In Situ* Exposure			Laboratory Exposure 5% HWS Soil
	L	I	H	
Spreading	76	22	45	67
Phagocytosis	100	44	13	28
NBT	67	43	13	45

NOTE: HWS, hazardous waste site, L, low; I, intermediate; H, high. Expressed as percent normal unexposed controls.

Effect of PCB on HMI

PCB effects on HMI were assayed using SR formation to enumerate agglutinin producing coelomocytes.[46] SR are coelomocytes binding at least two layers of four or more erythrocytes to their surface with a secreted agglutinin. SR formation and PCB (Aroclor 1254) concentrations in coelomocytes during a five-day filter paper exposure (10 μg/cm^2) and depuration are shown in FIGURE 8. Although PCBs were below detection at 48 hours, SRs were 88% normal. By 64 hours when PCBs were first detected, SR formation was only 55%. Maximal suppression (45% normal) occurred at five days, coincident with maximum PCB concentration in coelomocytes. Recovery of SR formation paralleled PCB loss from coelomocytes. To determine relations between nominal PCB filter paper exclusive concentrations, tissue levels, and SR formation (i.e. dose–response), earthworms were exposed to 2.5, 5.0, 10.0, and 40 μg PCB/cm^2 for five days, then analyzed (FIG. 9). Corresponding whole-earthworm concentrations were about 56, 76, 185, and 221 μg/g dry mass. SR formation against erythrocytes was significantly influenced by PCB tissue levels as low as 76 μg/g dry mass (5.0 μg/cm^2) and dependent on exposure concentrations from 2.5 to 10 μg/cm^2. Immunotoxicity occurred at considerably lower levels than acute LC_{50}/LD_{50} levels, suggesting that the SR immunoassay is sensitive to PCB

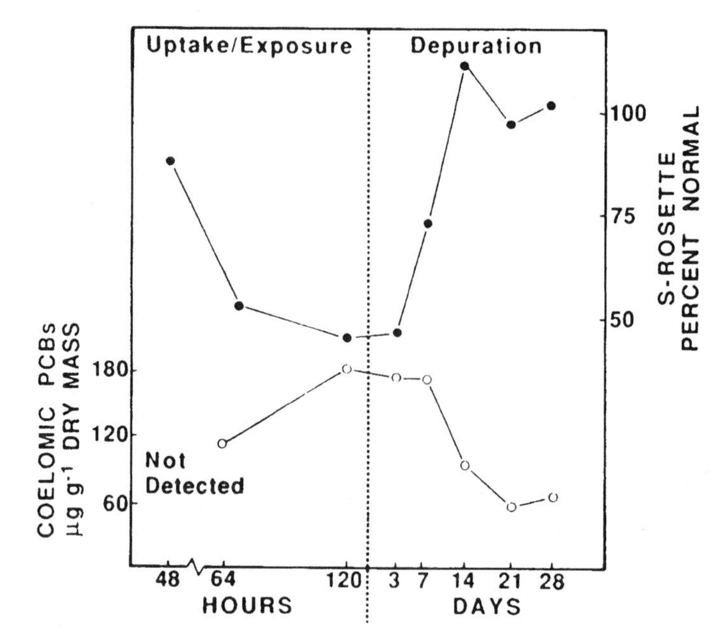

FIGURE 8. Relation between humoral immune function as demonstrated by secretory rosette (SR) formation (*shaded circles*) and levels of PCB (*open circles*) in coelomocytes collected from earthworms after five-day PCB (Aroclor 1254) filter paper exposure (10 μg/cm^2). SR formation expressed as percent normal rosette formation by unexposed control earthworm coelomocytes.

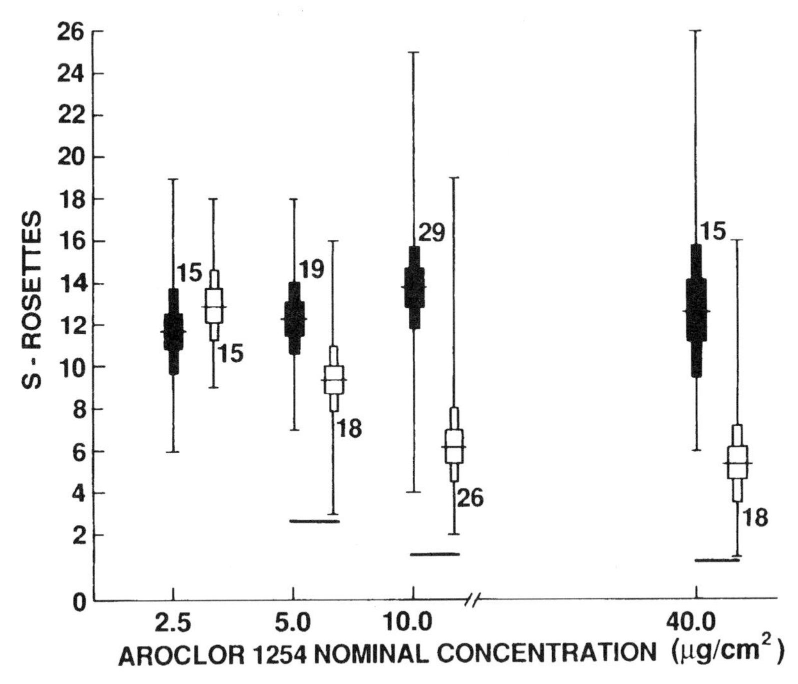

FIGURE 9. Relation between secretory rosette formation by earthworm *L. terrestris* coelomic leukocytes and nominal filter paper exposure concentrations of Aroclor 1254. Wide vertical rectangles, narrow vertical rectangles, and vertical lines are ±SE, 95% C.I., and range, respectively. Sample sizes given above controls (*solid*) and below experimentals (*open*). Underlines denote significant differences between controls and experimentals.

levels within an order of magnitude of those reported for a wide variety of wildlife from contaminated areas.

SR formation indicates the ability of coelomic leukocytes to produce agglutination factors in response to foreign challenge such as bacterial or fungal infection. These factors serve to aggregate particular antigen and act as opsonin to facilitate phagocytosis. As such, they are functionally analogous to antibodies produced in mammals. They are inducible by exposure to antigen (recognition) and respond anamnestically (memory) in sensitized (immunized) mammals and earthworms. Thus, it should be possible to predict effects of chemicals on recognition/memory, processing, and effector phases of humoral immunity in mammals by studying their influence on agglutinin production, release, and activity in earthworms.

CONCLUSIONS

On the basis of our work with earthworms, invertebrates have considerably greater potential for use in immunotoxicology than is generally realized. They are sufficiently complex with immune processes that are broadly conserved phylogenetically for use as surrogates for vertebrates in studying the immunotoxic effects of environmental pollutants. Among invertebrates, earthworms possess a number of attributes that make them an appropriate choice for investigation of effects of chemicals on host immune defense systems and development of immune-based biomarkers, especially for terrestrial toxicology. Earthworm coelomocyte-based biomarkers have been shown to be sensitive indicators of sublethal immunotoxicity of single chemicals and complex mixtures. Development of a comprehensive suite of indicator measurement endpoints with invertebrates, representing various levels of integration within a tiered framework should provide for cost-effective assessment of potential risks to both humans and wildlife from terrestrial pollution.

ACKNOWLEDGMENT

We thank Mrs. Sue Rowan for preparing the manuscript.

REFERENCES

1. Dean, J. H., M. J. Murray & E. C. Ward. 1986. Toxic responses of the immune system. *In* Casarett and Doull's Toxicology. C. D. Klaassen, M. O. Amdur & J. Doull, Eds.: 245–286. McMillan Publishing Company. New York.
2. Weeks, B. A., D. P. Anderson, A. P. Dufour, A. Fairbrother, A. J. Goven, G. P. Lewis & G. Peters. 1992. Immunological biomarkers to assess environmental stress. *In* Biomarkers. R. J. Hugget, R. A. Kimerle, P. M. Mehrle & H. L. Bergman, Eds.: 211–234. Lewis Publishers. Chelsea, MI.
3. Hostetter, R. K. & E. L. Cooper. 1974. Earthworm coelomocyte immunity. *In* Contemporary Topics in Immunobiology. E. L. Cooper, Ed. Vol. **4:** 91–107. Plenum Press. New York.
4. Cooper, E. L. 1974. Phylogeny of leukocytes: Earthworm coelomocytes in *in vitro* and *in vivo*. *In* Lymphocyte Recognition and Effector Mechanisms. K. Lindahl-Kiessling & D. Osoba, Eds.: 155–162. Academic Press. New York.
5. Cooper, E. L. & E. A. Stein. 1981. Oligochaetes. *In* Invertebrate Blood Cells. N. A. Ratcliffe & A. F. Rowley, Eds.: 75–140. Academic Press. New York.
6. Cooper, E. L. 1976. The earthworm coelomocyte: A mediator of cellular immunity. *In* Phylogeny of Thymus and Bone Marrow-Bursa Cells. R. K. Wright & E. L. Cooper, Eds.: 9–18. Elsevier, North Holland. Amsterdam, the Netherlands.
7. Reinecke, A. J. 1992. A review of ecotoxicological test methods using earthworms. *In* Ecotoxicology of Earthworms. P. W. Greig-Smith, M. B. Becker, P. J. Edwards & F. Heimbach, Eds.: 7–19. Intercept Ltd. Andover, United Kingdom.
8. Goven, A. J., S. C. Chen, L. C. Fitzpatrick & B. J. Venables. 1993. Environ. Toxicol. Chem. In press.
9. Stein, E. A., R. R. Avtaliion & E. L. Cooper. 1977. J. Morphol. **153:** 467–477.
10. Valembois, P. 1974. Cellular aspects of graft rejection in earthworms and some other metazoa. *In* Contemporary Topics in Immunology. E. L. Cooper, Ed.: 75–90. Plenum Press. New York.
11. Keng, L. B. 1985. J. Pathol. **71:** 221–248.
12. Wojdani, A., E. A. Stein, C. A. Lemmi & E. L. Cooper. 1982. Dev. Comp. Immunol. **6:** 407–410.
13. Stein, E. A. & E. L. Cooper. 1982. Dev. Comp. Immunol. **5:** 415–425.
14. Stein, E. A. & E. L. Cooper. 1982. Agglutinins as receptor molecules: A phylogenetic

approach. *In* Developmental Immunology: Clinical Problems and Aging. E. L. Cooper & M. A. B. Brazier, Eds.: 85–98. Academic Press. New York.
15. STEIN, E. A. & E. L. COOPER. 1988. Dev. Comp. Immunol. **12:** 531–547.
16. MOHRIG, W., E. KANSCHKE & M. EHLERS. 1984. Dev. Comp. Immunol. **8:** 471–476.
17. VALEMBOIS, P., P. ROCH, M. LASSEGUES & P. CASSAND. 1982. J. Invert. Pathol. **40:** 67–69.
18. LASSEGUES, M., P. ROCH & P. VALEMBOIS. 1989. J. Invert. Pathol. **53:** 1–6.
19. COOPER, E. L. 1969. Science **166:** 1414–1415.
20. COOPER, E. L. 1971. Transplant. Proc. **3:** 214–216.
21. COOPER, E. L. & P. ROCH. 1986. Transplantation **41:** 514–520.
22. BAILY, S., B. J. MILLER & E. L. COOPER. 1971. Immunology **21:** 81–86.
23. LUSTER, M. I., A. E. MUNSON, P. T. THOMAS, M. P. HOLSAPPLE, J. D. FENTERS, K. L. WHITE, L. D. LAUER, D. R. GEMOLEC, G. L. ROSENTHAL & J. H. DEAN. 1988. Fund. Appl. Toxicol. **10:** 2–19.
24. EYAMBE, G. S., A. J. GOVEN, L. C. FITZPATRICK, B. J. VENABLES & E. L. COOPER. 1991. Lab. Anim. **25:** 61–67.
25. SUTER II, G. W. 1989. Ecological endpoints. *In* Ecological Assessment of Hazardous Waste Sites: A Field and Laboratory Reference. W. Warren-Hicks, B. R. Parkhurst & S. S. Baker, Jr., Eds.: 2.1–2.26. Environmental Protection Agency. Washington, D.C.
26. GOVEN, A. J., G. S. EYAMBE, L. C. FITZPATRICK, B. J. VENABLES & E. L. COOPER. 1993. Environ. Toxicol. Chem. **12:** 863–870.
27. LEE, T. P. & K. J. CHANG. 1985. Health effects of polychlorinated biphenyls. *In* Immunotoxicology and Immunopharmacology. J. H. Dean, M. I. Luster, A. E. Munson & M. Amos, Eds.: 415–422. Raven Press. New York.
28. CARTER, J. W. & J. CLANCY. 1980. Immunopharmacology **2:** 341–347.
29. FISHBEIN, L. 1974. Ann. Rev. Pharmacol. **14:** 139–156.
30. SALTON, M. R. J. 1975. Bacteriol. Rev. **21:** 82–99.
31. ZUCKER, S., D. J. HANES, W. R. VOGLER & R. Z. EANES. 1970. J. Lab. Clin. Med. **75:** 83–92.
32. FEENEY, R. E., R. M. LEONARD & D. D. EUFERNIO. 1956. Arch. Biochem. Biophys. **61:** 72–83.
33. JOLLES, P. & J. JOLLES. 1984. J. Mol. Cell. Biochem. **63:** 165–189.
34. DRUTZ, D. J. & J. MILLS. 1984. Immunity and infection. *In* Basic and Clinical Immunology. D. P. Sites, J. D. Stobo, H. H. Fudenberg & J. V. Wells, Eds.: 197–222. Lange Medical Publications. Los Altos, CA.
35. BRAUNDE, A. I. 1981. Mechanisms of natural resistance to infection. *In* Medical Microbiology and Infectious Disease. A. I. Braunde, C. E. Davis & J. Fierer, Eds.: 739–756. W. B. Saunders. Philadelphia, PA.
36. CHEN, S. C., L. C. FITZPATRICK, A. J. GOVEN, B. J. VENABLES & E. L. COOPER. 1991. Environ. Toxicol. Chem. **10:** 1037–1043.
37. MADERAZO, E. G. & P. A. WARD. 1980. Leukocyte function test. *In* Methods in Immunodiagnosis. N. R. Rose & P. E. Bigazzi, Eds.: 53–63. John Wiley and Sons. New York.
38. VALLEE, B. J. & D. D. ULMER. 1972. Biochemical effects of mercury, cadmium and lead. *In* Annual Review of Biochemistry. E. S. Snell, P. D. Boyer, A. Meister & R. L. Rinsheimer, Eds. Vol. **41:** 91–128. Annual Reviews. Palo Alto, CA.
39. ABSOLOM, D. R. 1986. Basic methods for the study of phagocytosis. *In* Methods in Enzymology. G. Disabato & J. E. Verse, Eds. Vol. **132:** 95–180. Academic Press. New York.
40. STOKINGER, H. E. 1981. The metals. *In* Patty's Industrial Hygiene and Toxicology. G. D. Clayton & F. E. Clayton, Eds. Vol. **2A:** 1493–2060. John Wiley and Sons. New York.
41. HEMPHILL, F. E. & M. L. KAEBERLE. 1971. Science **172:** 1031–1032.
42. MA, W. C. 1982. Pedobiologia **24:** 109–119.
43. VENABLES, B. J., L. C. FITZPATRICK & A. J. GOVEN. 1992. Earthworms as indicators of ecotoxicity. *In* Ecotoxicology of Earthworms. P. W. Greig-Smith, H. Becker, P. J. Edwards & F. Heimbach, Eds.: 197–206. Intercept Ltd. Andover, United Kingdom.
44. CALLAHAN, C. A. & G. LINDER. 1992. Assessment of contaminated soils using earthworm

test procedures. *In* Ecotoxicology of Earthworms. P. W. Greig-Smith, H. Becker, P. J. Edwards & F. Heimbach, Eds.: 187–196. Intercept Ltd. Andover, United Kingdom.

45. GREENE, J. C., C. L. BARTELS, W. J. WARREN-HICKS, B. R. PARKHURST, G. L. LINDER, S. A. PETERSON & W. E. MILLER. 1989. Protocols for short-term toxicity screening of hazardous waste sites. U.S. Environmental Protection Agency. Corvallis, OR.
46. RODRIGUEZ-GRAU, J., B. J. VENABLES, L. C. FITZPATRICK & A. J. GOVEN. 1989. Environ. Toxicol. Chem. **8:** 1201–1207.

Immunomodulation by Didemnins

Invertebrate Marine Natural Products

DAVID W. MONTGOMERY,[a,b,c] GARY K. SHEN,[a]
ELLEN D. ULRICH,[c] AND CHARLES F. ZUKOSKI[a,c]

[a]*Department of Surgery*
[b]*Department of Pharmacology*
University of Arizona College of Medicine
Tucson, Arizona 85742
and
[c]*Department of Veterans Affairs Medical Center*
Tucson, Arizona 85723

The didemnins are a family of seven amino acid, cyclic depsipeptides isolated from invertebrate marine tunicates of the family Didemnidae. They were initially identified as potent antiproliferative and antiviral activities,[1-3] present in crude extracts of these organisms, and subsequently purified.[1] The prototypic member of this family, didemnin B (DB), has a potency in the nanomolar range to inhibit *in vitro* growth of a variety of human and animal tumor cell lines. DB also demonstrated strong *in vivo* anti-tumor activity in mouse models[3] as well as against several primary human tumors in the stem cell assay.[4] These promising early results led to the currently ongoing phase II clinical trials of this drug as an antineoplastic agent.

DB also exerts potent immunologic activity. Lymphocyte proliferation *in vitro* stimulated by mitogens,[5,6] alloantigens,[5] the lymphokines IL-2 and IL-4,[6,7] and prolactin[8] were inhibited by nanomolar and subnanomolar concentrations of this compound. Evaluation of DB for immunosuppressive effects indicated strong activity in a variety of animal models. Drug doses in the μg/kg range inhibited the graft-versus-host reaction[5] and prolonged survival of skin grafts in mice[9] as well as heart grafts in mice and rats.[10,11] Another effect of DB, marked enhancement of antibody production *in vitro* and *in vivo,* has also been reported.[12] A variety of biochemical effects of DB on cultured cells has been documented. For example, it inhibits protein and DNA synthesis at concentrations higher than those required to inhibit lymphocyte proliferation.[5,13] IL-2 production is not inhibited by DB,[7] and immunoglobulin synthesis by hybridoma cells lines[12] is enhanced. Therefore, it is unlikely that inhibition of protein synthesis explains the antiproliferative effects of this compound, as suggested by others.[7,13] DB blocks cell cycle G1-S phase transition at nanomolar concentrations, but cell cycle analysis indicated that this is not a cycle-specific effect.[13] We have previously suggested[5] that DB acts early in the lymphocyte activation process to prevent proliferation.[5] However, signal transduction events leading to IL-2 synthesis, or IL-2 and IL-4 signal coupling in dependent cell lines, appears to be unaffected. Therefore, some other pathway must be involved in DB action. One possibility may be effects of DB on protein phosphorylation, since it has been reported to inhibit specific protein phosphorylation induced by phorbol esters in mouse skin.[14]

Although no clear-cut mechanism has been defined that explains the diverse actions of DB, our work in the Nb2 node lymphoma tumor cell line indicated that some, but not all, of the effects of DB are mediated by a high-abundance, low-affinity cytosolic receptor for this drug.[15] This is similar to findings reported for cyclosporine, FK-506, and rapamycin.[16] which bind to cytosolic proteins recently named the "immunophilins." Drug–immunophilin complexes are proposed to mediate, at least in part, the immunosuppressive actions of these drugs by the inhibition of the phosphatase calcineurin, consequently interfering with activation of key nuclear transcription factors required for IL-2 gene expression. To examine the possibility that DB may also interact with members of the immunophilin family, we have assessed normal human thymocytes and murine splenocytes for similar, DB-binding proteins.

MATERIALS AND METHODS

Animals and Reagents

Male Balb/c mice were obtained from Harlan Sprague-Dawley (Indianapolis, Indiana) and maintained in an AALAS-approved animal facility. Tritiated DB (^{3}H-DB, 2.7 mCi/mg), didemnin A, DB, and acyclo-DB were the generous gifts of Kenneth L. Rinehart, Jr. FK-506 was generously provided by Drs. Thomas E. Starzl and Charles W. Putnam. Cyclosporin A (CSA) in cremaphore vehicle (Sandimmune IV) was obtained from Sandoz (East Hanover, NJ). All other reagents were from commercial sources. Stock solutions of didemnins were prepared in DMSO to 1 mg/ml, and FK-506 was dissolved to 1 mg/ml in absolute ethanol.

Lymphocyte Cell Suspensions

Mouse splenocytes were prepared aseptically and suspended to 10^7 ml in RPMI 1640 with 1% fetal calf serum (FCS) as previously described.[5] Viability was assessed by trypan blue dye exclusion. Human thymus tissue was obtained from pediatric cardiac surgery patients at the Arizona Health Sciences Center in Tucson by Drs. G. Sethi and J. Copeland. Single-cell suspensions of thymocytes were prepared as previously described[17] and cultured identically to the splenocyte suspensions.

Measurement of [^{3}H]DB Binding

Cell suspensions were prewarmed to 37°C in 5% CO_2 in air immediately after preparation. The binding assay was performed in 5-ml sterile polypropylene tubes (Falcon 2029, Becton-Dickenson, Lincoln Park, NJ). Each tube received 20 μl [^{3}H]DB diluted in RPMI 1640 with 1% FCS and 5% DMSO (assay medium) to the desired concentration. An additional 20 μl volume of either medium alone ([^{3}H]DB-only tubes) or medium with added cold DB (hot + cold tubes). Then 500 μl of cell suspension containing 5×10^6 cells was added, and tubes were mixed gently. The final DMSO concentration was 0.02%. Tubes were covered, then incubated in 5% CO_2 in air at 37°C. Cells were harvested by dilution with 4 ml of ice-cold phosphate-buffered saline (PBS), collected by vacuum filtration onto 2.5-cm glass-fiber filters (Whatman GF/C), and washed three times with 4 ml each of cold PBS. Filters were air dried, and amount of [^{3}H]DB bound was measured by liquid scintillation spectrometry. Specific binding was calculated by subtracting binding in the presence of 100-fold excess unlabeled DB from total binding in the absence of competitor.

Time Dependence of [^{3}H]DB Binding

Murine splenocytes were incubated at 37°C with either [^{3}H]DB (50,000 cpm/tube) or [^{3}H]DB plus 1 μg/tube unlabeled DB. At each time point (15 min–8 hours) thereafter, three tubes of cells under each condition were harvested. Specific [^{3}H]DB binding was computed as above. To characterize the time dependence of dissociation of bound [^{3}H]DB, replicate samples were incubated for 4 hours as described above to obtain binding equilibrium. Cells were washed three times with culture medium, resuspended in 500 μl of medium with the addition of 1 μg unlabeled DB to prevent reassociation of label. Cells were again incubated at 37°C, and samples were harvested over the course of 16 hours.

Saturation Isotherm and Scatchard Analysis

Triplicate samples of murine splenocytes of human thymocytes were incubated with increasing concentrations of [^{3}H]DB in the presence or absence of 1 μg/tube of unlabeled DB (1.66×10^{-3} M). Maximum amounts of added [^{3}H]DB varied between experiments (from 5×10^5 cpm to 1.4×10^6 cpm/tube), and all concentrations were measured directly by scintillation counting. After 3 hours incubation at 37°C, cells were harvested, and [^{3}H]DB binding determined. The binding constants K_d and β_{max} were calculated from Scatchard plots of transformed data.[18]

Specificity of [^{3}H]DB Binding

Competitive binding experiments were run with another active didemnin, didemnin A (DA); the inactive analogue acyclo-DB; and the immunosuppressives CSA, FK-506, and dexamethasone. Each ligand was tested at concentrations of 1, 10, 100, and 1000 ng/tube in the presence of 20,000 cpm (5 ng) [^{3}H]DB/tube. Unlabeled DB was used as a positive control for competition in each experiment. Murine splenocytes were incubated with ligands for 3 hours at 37°C, and cells were harvested as before.

Subcellular Fractionation

Murine splenocytes (1×10^8 cells) were incubated for 3 hours at a concentration of 10^7 ml in the presence of 20,000 cpm/ml [^{3}H]DB or the same amount of [^{3}H]DB in the presence of added, unlabeled DB (2 μg/ml). Cells were washed three times in cold RPMI 1640 with 0.3% BSA to remove unbound drug, then resuspended in 1 ml ice-cold 25 mM Tris-HCl, pH 7.4, 0.32 M sucrose, 25 μg/ml leupeptin, 2 mM EDTA, and 5 mM *n*-ethyl maleimide. After sonicating on ice (five 15-sec pulses) at a power setting of 12 (Virsonic 50, Virtis Co., Gardiner, NY), the lysate was diluted to 5 ml with cold Tris-sucrose (above) and centrifuged (4°C) 10 min at 400 × *g* to remove unbroken cells and nuclei. The supernatant was decanted and centrifuged (4°C) at 20,000 × *g* for 30 min to sediment mitochondria, and the remaining supernatant was centrifuged (4°C) for 60 min at 100,000 × *g* to produce the cytosolic fraction. The pellets from the 20,000 × *g* and 100,000 × *g* spins were resuspended in 200 μl PBS plus 1% NP-40, 0.1% SDS and counted in their entirety. Aliquots of the homogenate and 100,000 × *g* supernatants were also counted to determine the amounts of [^{3}H]DB present. Specifically bound [^{3}H]DB was computed by subtracting nonspecific counts from total cpm bound in the parallel fractions.

RESULTS

Kinetics of [^{3}H]DB Binding and Dissociation

The time dependence of [^{3}H]DB association and dissociation in mouse splenocytes is shown in FIGURE 1. The amount of [^{3}H]DB specifically bound increased logarithmically, attaining a maximum by 4 hours of incubation (top panel). The half-time for association derived from linear regression of ln [^{3}H]DB bound specifically versus time was 1.1 hours. In contrast, nonspecific binding was extremely rapid, achieving a maximum at 15 minutes and remaining constant thereafter at approximately 15% of total binding. For dissociation-time measurements, cells were incubated until maximum specific binding was achieved (3 hours), washed to remove free drug, then sampled over the course of 16 hours. The initial rate of [^{3}H]DB dissociation was logarithmic for the first 4 hours, and the dissociation half-life was 5.4 hours. Dissociation slowed thereafter, and by 16 hours 25% of the initial [^{3}H]DB bound continued to be associated with the cells. Nonspecifically bound [^{3}H]DB dissociated very rapidly over the first 15 minutes, with very little further decrease over the next 16 hours. Therefore, [^{3}H]DB binding is time dependent, with the rate of association being much greater than that of dissociation. No attempts were made to determine affinity constants from these data, as subsequent experiments indicated that DB must cross the cellular plasma membrane to interact with its cytosolic binding site, and this transport rate has not been determined.

Saturation Isotherms

FIGURE 2 shows curves of total, specific, and nonspecific [^{3}H]DB binding to murine splenocytes over a wide range of drug concentrations (0 to 350 ng [^{3}H]DB/tube). Specific [^{3}H]DB binding was saturable, and specific binding was 70% of total binding at saturation. Nonspecific binding increased linearly with [^{3}H]DB concentration and was not saturable. The Scatchard plot of these data was linear, indicating a single class of binding site was present, and the computed dissociation

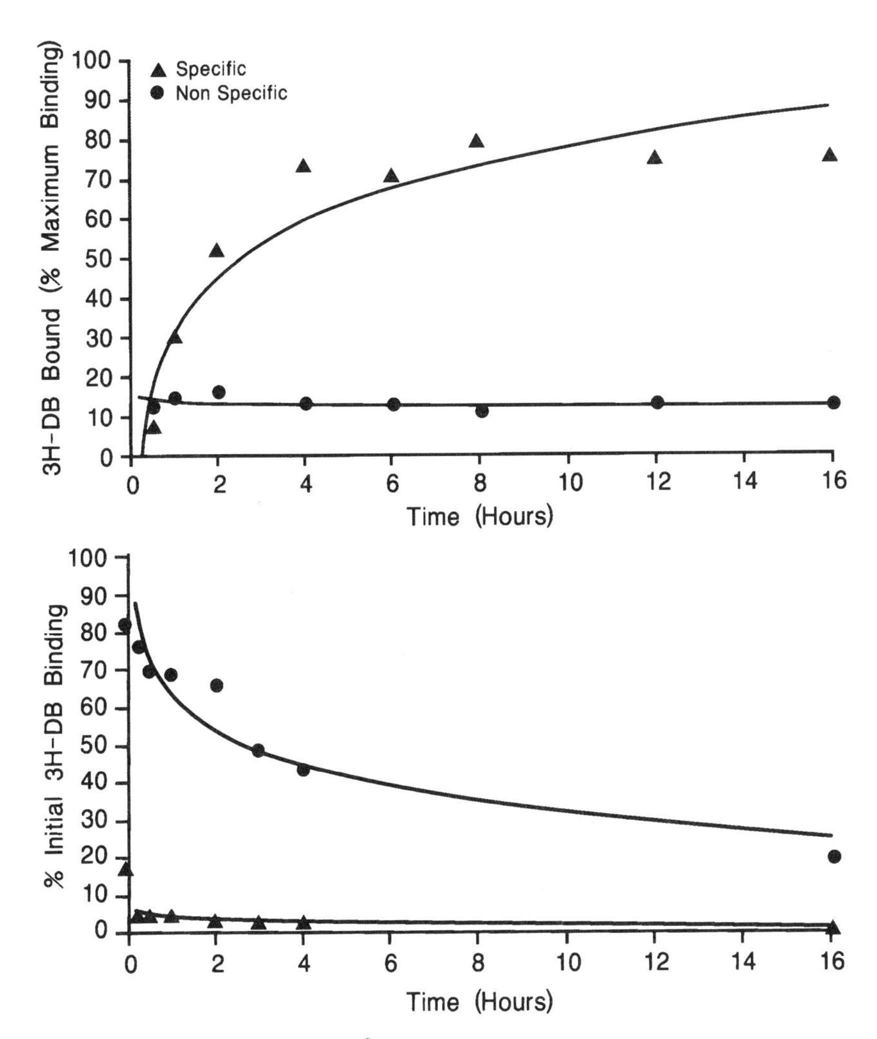

FIGURE 1. Time dependence of [^{3}H]DB binding and dissociation in splenocytes. To determine binding kinetics (*top panel*), freshly harvested cells were incubated at 37°C with 50,000 cpm [^{3}H]DB alone or [^{3}H]DB plus 1 μg/tube unlabeled drug. A set of three tubes of each of these was harvested at the times shown, washed onto glass fiber filters, and cell-associated [^{3}H]DB counted by liquid scintillation spectrometry. Specifically bound [^{3}H]DB was calculated as the difference between total cpm bound ([^{3}H]DB only) and that in the presence of unlabeled competitor. To determine the dissociation rate (*lower panel*), cells were preincubated with 50,000 cpm [^{3}H]DB or [^{3}H]DB plus 1 μg/tube unlabeled DB for 3 hours. After washing to remove unbound drug, cells were incubated at 37°C and harvested at the times shown. Data were analyzed as in the association studies.

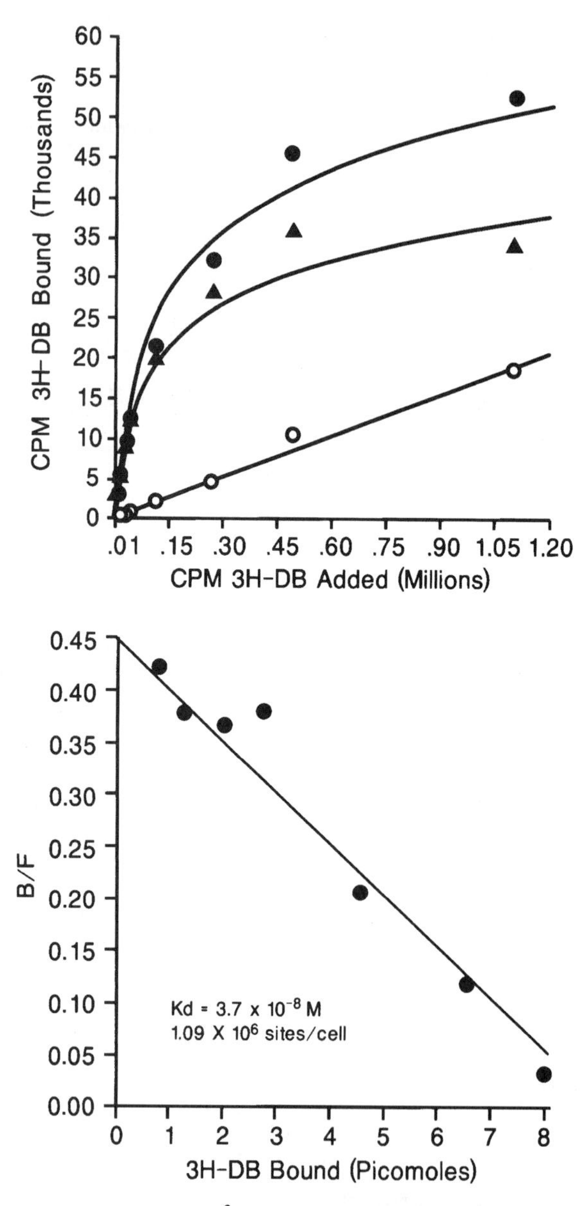

FIGURE 2. Saturation isotherm of [^{3}H]DB binding to mouse splenocytes. Dilutions of [^{3}H]DB were incubated with 2×10^6 cells at 37°C for 4 hours with or without the presence of a 100-fold molar excess of unlabeled DB. Cells were harvested, and specific binding was computed as in FIGURE 1. Scatchard analysis (*lower figure*) was performed on data linearized by plotting as bound/free versus bound. Data from a representative experiment are shown. Symbols: ●, total binding; ▲, specific binding; O, nonspecific binding.

constant, $K_d = 3.7 \times 10^{-8}$ M. At saturation, approximately 1.09×10^6 specific binding sites were computed to be present in each cell. These values are similar to those reported for CSA binding to lymphocytes. Similar data were obtained for [^{3}H]DB binding to human thymocytes (FIG. 3). Like the [^{3}H]DB in murine splenocytes, this was found to be specific and saturable, with a dissociation constant in this experiment of 4.4×10^{-8} M. The results of three such analyses with splenocytes and thymocytes are shown in TABLE 1, in comparison to our previous observations in the Nb 2 node lymphoma cell.[15] The average K_ds of [^{3}H]DB binding to murine splenocytes ($3.56 \times 10^{-8} \pm 1.0$ M) and to human thymocytes ($1.03 \times 10^{-8} \pm 1.04$ M) were virtually identical. Similarly, the number of binding sites per cell, 1.08×10^6 and 1.0×10^6 for splenocytes and thymocytes is identical. In both cell types the percent [^{3}H]DB specifically bound was quite high in comparison to total binding, 85.6 ± 1.6 and 77.9 ± 5.6 for splenocytes and thymocytes, respectively. In comparison to results with normal lymphocytes, the Nb 2 node lymphoma cell has a lower affinity binding site ($K_d = 2.1 \times 10^{-7}$ M), almost 10-fold the number of binding sites per cell (1.0×10^7), and a slightly lower proportion of specific binding (70.0 ± 10%).

Subcellular Location of [^{3}H]DB Binding Site

To determine the subcellular compartment(s) containing the [^{3}H]DB binding site, murine splenocytes pretreated with radioligand were homogenized and the amount of [^{3}H]DB bound measured in the homogenate, the 20,000 × *g* pellet (mitochondria), the 100,000 × *g* pellet (membranes), and the 100,000 × *g* supernatant (cytosol). The data from three such experiments are shown in TABLE 2. Of 142,281 ± 8,965 cpm [^{3}H]DB specifically bound in the homogenate, 2,113 ± 567 cpm were present in the mitochondrial fraction (1.5%), 183 ± 61 cpm in the membrane fraction (0.13%), and 108,710 ± 4,591 cpm in the cytosol fraction (77%). Of the total cpm [^{3}H]DB present in the cytosol fraction, 88.5 ± 1.1% was specifically bound. Therefore, the DB binding site in normal lymphocytes is cytosolic, similar to those reported for CSA, FK-506, and glucocorticosteroids.

Competitive Binding Studies

Although DB is the only didemnin available as a radioligand, immunological testing has been performed on two other didemnins, didemin A (DA) and the acyclo form of DB (acyclo-DB), in which the ring structure of DB is absent. DA possesses a spectrum of immunologic activity similar to DB, but is one-tenth as potent. The acyclo dB, in contrast, has no immunologic activity. In order to determine whether the immunologic activities of the drugs correlate with their ability to interact with the specific [^{3}H]DB binding site, we performed competitive binding studies in which the ability of DB, DA, and acyclo-DB to inhibit [^{3}H]DB binding in splenocytes was assessed (FIG. 4). In comparison to DB, much higher concentrations of DA were required to inhibit binding of [^{3}H]DB. At 1000 ng/tube, however, DA blocked binding of the radioligand by more than 70%, evidence that it interacts directly at the

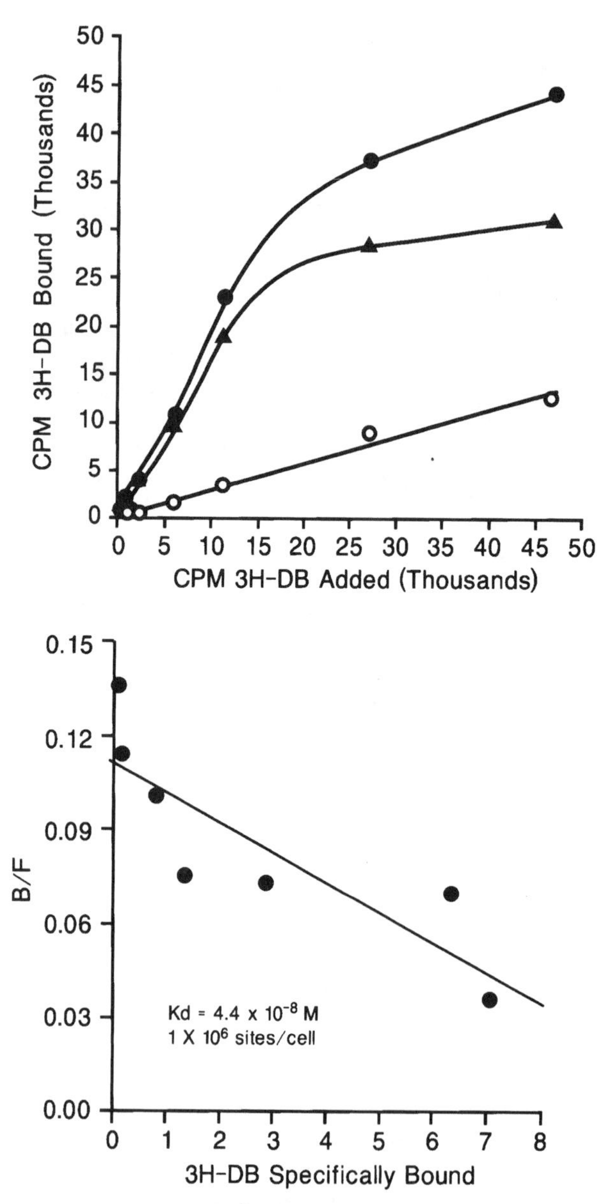

FIGURE 3. Saturation isotherm of [^{3}H]DB binding to human thymocytes. Freshly harvested thymocytes were treated exactly as in FIGURE 2. Data from a representative experiment are shown. Symbols: ●, total binding; ▲, specific binding; O, nonspecific binding. *Lower figure* shows Scatchard analysis.

TABLE 1. Characteristics of the [^{3}H]DB Binding Site in Lymphocytes

Cell Type[a]	Parameter		Percent Specific[c,d]
	K_d (M)[b]	B_{max} (sites/cell)[b]	
Mouse splenocytes	$3.56 \times 10^{-8} \pm 1.0$	$1.08 \times 10^{6} \pm 0.05$	85.6 ± 1.6
Human thymocytes	$5.08 \times 10^{-8} \pm 1.04$	$1.03 \times 10^{6} \pm 0.03$	77.9 ± 5.6
Nb2 lymphoma	$2.10 \times 10^{-7} \pm 0.4$	$1.00 \times 10^{7} \pm 0.2$	70 ± 10

[a]Freshly harvested balb/c splenocytes, human thymocytes, or Nb2 cells cultured 3 hr at 37°C with [^{3}H]DB with or without 2 μg/ml unlabeled DB.
[b]From Scatchard plot of saturation isotherm experiments.
[c]Specific CPM bound/total CPM bound at [^{3}H]DB ≤ K_d.
[d]Mean ± SE, three experiments.

DB binding site. In contrast, the immunologically inactive acyclo-DB exerted minimal effects (10–14% inhibition) at this concentration.

Because DB possesses a cytosolic binding site, as do the immunosuppressive drugs CSA, FK-506, and dexamethasone, we examined the possibility that these agents might compete for the [^{3}H]DB-binding site (FIG. 5). Using drug concentrations of up to 1000 ng/tube, a 200-fold molar excess failed to exert appreciable affects on [^{3}H]DB binding. In contrast, DB effectively competed with radioligand for binding sites. Therefore, the cytosolic binding site for DB appears to be distinct from those previously described for CSA,[19] FK-506, and dexamethasone.

DISCUSSION

The data presented here demonstrate that normal murine splenocytes and human thymocytes possess a specific, saturable, and reversible cytosolic binding site for [^{3}H]DB, similar to that previously reported in the Nb 2 node lymphoma tumor cell

TABLE 2. Subcellular Location of Specific[a] [^{3}H]DB Binding[b]

Fraction	Experiment			Mean ± SEM
	1	2	3	
Homogenate	120,650	149,825	156,370	142,281 ± 8,965
20,000 × *g* pellet	2,623	2,977	740	2,113 ± 567
100,000 × *g* pellet	170	60	320	183 ± 61
Cytosol	100,077	106,785	119,68	108,710 ± 4,591
% total[c]	82.9	71.3	76.3	76.8 ± 2.7
% specific[d]	90.1	89.6	85.7	88.5 ± 1.14

[a]Specific binding = CPM [^{3}H]DB bound ([^{3}H]DB only) − CPM [^{3}H]DB bound ([^{3}H]DB plus unlabeled DB); 3 hr at 37°C.
[b]Cpm [^{3}H]DB binding per 3×10^6 balb/c mouse splenocytes.
[c]Cpm specifically bound [^{3}H]DB in cytosol/cpm [^{3}H]DB in homogenate × 100.
[d]Cpm [^{3}H]DB specifically bound/to total cpm [^{3}H]DB bound × 100.

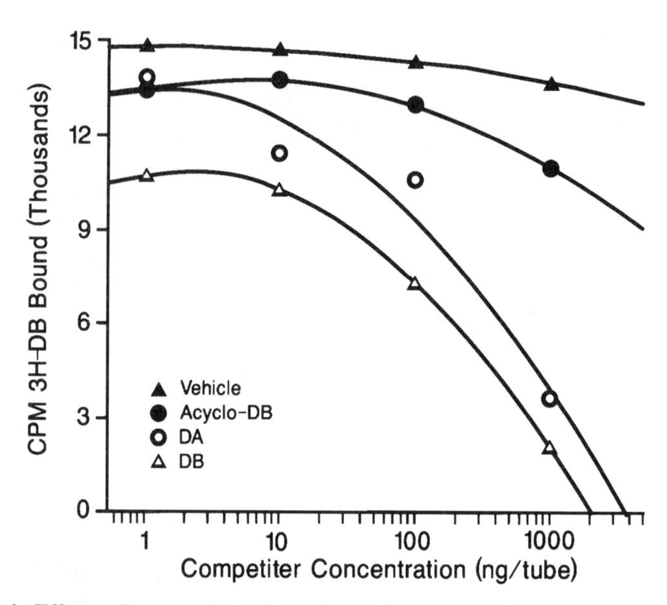

FIGURE 4. Effects of immunologically active and immunologically inactive didemnins on binding of [^{3}H]DB. Splenocytes were incubated for 4 hours at 37°C with either 20,000 cpm (5 ng) [^{3}H]DB, or [^{3}H]DB plus 1, 10, 100, and 1000 ng/tube unlabeled DB, DA, or acyclo-DB. Cells were harvested and data calculated as in FIGURE 1. Data are from a representative experiment. Error bars are omitted for clarity; error did not exceed 10% of the mean values.

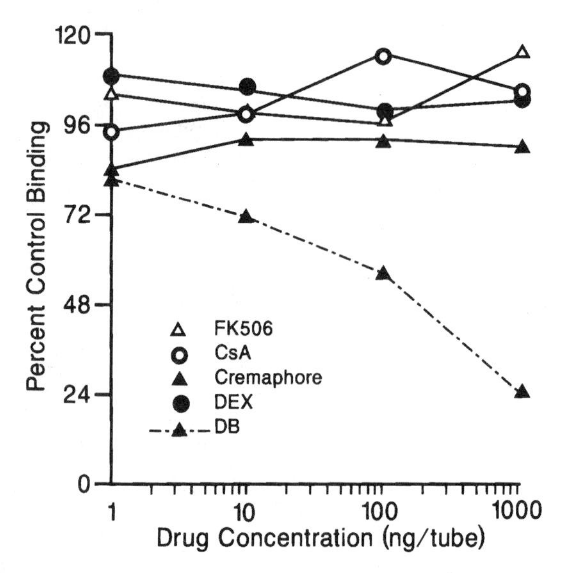

FIGURE 5. Effects of CSA, FK506, and dexamethasone on [^{3}H]DB binding to murine splenocytes. These experiments were run exactly as described in FIGURE 4, except that the CSA vehicle cremaphore was run in parallel to control for nonspecific effects. Data are from a representative experiment, and error did not exceed 10% of the mean for any point.

line.[15] This is consistent with the earlier observations by Phillips *et al.*[20] that DB preferentially associated with the cellular fraction of blood. It is interesting to compare the characteristics of this binding site in normal lymphocytes and the transformed Nb 2 node lymphoma cell. Nb 2 cells express 10-fold more binding sites than do normal cells. Conversely, the affinity of the binding site in normal cells is approximately four- to five-fold higher than in Nb 2 cells. Thus, a reciprocal relationship appears to exist between binding-site affinity and number of sites per cell. Because in both cases there is a single class of site present, this may be due to structural differences in the binding site itself, rather than a different proportion of various binding-site species. The answer to this must await purification of these cytosolic proteins. The relative affinities of the DB-binding site in normal lymphocytes and Nb 2 cells also correspond to the antiproliferative potency of DB in each cell type, with the drug being 10-fold more potent in normal lymphocytes (Shen and Montgomery, unpublished observations). In both cases, maximal inhibitory effects are produced at less than 10% binding-site occupancy. This does not exclude a role of this putative receptor in the antiproliferative actions of the didemnins. Shen *et al.*[15] have found that the rank order of didemnin potency to inhibit Nb 2 cell proliferation is identical to that of their ability to compete for the DB-binding site. Likewise, in normal cells, DA was a less potent competitor than was DB, while acyclo-DB was inactive. This again corresponds to their effects on lymphocyte proliferation. Similar observations have been made for CSA and FK-506 binding proteins. Only a low level of receptor occupancy is required for the expression of their immunologic activity. Moreover, recent studies clearly demonstrate that drug–protein complexes are directly involved in mediating the immunosuppressive effects of these agents.[19,21]

The following observations support the concept that the DB-binding site is a novel species of immunophilin, similar but distinct from those that bind and mediate the actions of CSA, FK-506, and rapamycin. The DB-binding site is exclusively cytosolic, as are immunophilins, and DB possesses a leucyl–prolyl amide bond that is important in immunophilin–drug interactions.[21] Nevertheless, the DB binding site does not interact with CSA and FK-506, since neither of these agents compete with [^{3}H]DB binding. In addition, Geschwendt *et al.*[14] have reported that DB fails to compete with binding of radiolabeled CSA. This may be because the three-dimensional structures of DB and CSA are quite different, with CSA being planar while DB is highly globular. On this basis, Hossain *et al.*[22] pointed out that it would be unlikely for DB and CSA to interact with a common binding site. The ability of didemnin analogues to interact with the DB-binding site directly correlates with their antiproliferative effects on lymphocytes, supporting the view that this binding site is immunophilin-like. Although DB binds to a cytosolic site that mediates its actions, however, the nature of its immunosuppressive effects are quite different from those of CSA and FK-506. This will be discussed further below.

A large amount of recent data is available that has begun to clarify the mechanisms of action of CSA and FK-506 at the cellular, biochemical, and molecular levels,[16,19,21,23,24] but relatively little is known about the intracellular actions of didemnins. In spite of this, sufficient evidence is available to suggest that DB has a different site of action than CSA and FK-506. For example, although CSA, FK-506, and DB appear to block early events in lymphocyte activation, only DB inhibits ongoing proliferation of normal lymphocytes and lymphocyte cell lines such as Jurkat (unpublished results) and L1210 leukemia.[2] Nevertheless, DB is much more potent when added at the time of normal lymphocyte activation, as opposed to several hours post activation, suggesting that different processes are effected to inhibit activation and ongoing proliferation. CSA and FK-506 inhibit transcription of the IL-2 gene and synthesis of IL-2 protein, but have no effect on responsiveness to this lymphokine.[24] Furthermore, it is well documented that the inhibition of lymphocyte mitogenesis by these agents is at least partially reversed by addition of exogenous IL-2, substantiating IL-2 as a principal target for the immunosuppressive actions of these drugs.[24] In contrast, IL-2 synthesis is unaffected by DB[7] (and our unpublished observations), DB inhibition is not reversed by large amounts of exogenous IL-2, and the proliferation of IL-2-driven cell lines such as CTLL are potently inhibited.[7] Therefore, it is unlikely that DB-binding protein complex acts upon the same calcium-dependent signaling pathways implicated in CSA and FK-506 inhibition of calcineurin phosphatase with subsequent blockade of nuclear transcription factor activation.[23] Rather, the actions of DB appear to be more similar to those of rapamycin, which also fails to alter IL-2 synthesis but does inhibit proliferation stimulated by this lymphokine.[24] Intriguingly, this drug competes with FK-406 for binding to its immunophilin, but antagonizes the immunosuppressive effect of FK-506.[24] Rapamycin has no effects on CSA actions.[24] Although DB binding is unaffected by FK-506, we have not tested to obtain results from either radiolabeled FK-506 or rapamycin for these types of interactions with DB. When taken together with the arguments posed above, these observations strongly suggest that the mechanisms of DB action, and the nature of its putative immunophilin-binding site, are quite different from those of CSA and FK-506.

An additional aspect of DB action at the cellular level has recently been reported by Teunissen *et al.*,[6] which may be important in understanding its immunosuppressive and antitumor effects. At subtoxic concentrations, DB was found by flow cytometry to inhibit the cell surface expression by human lymphocytes of several adhesion molecules. ICAM-1 and LFA-3 were totally inhibited by DB, and LFA-1 was significantly reduced. Similar results were obtained with macrophages. This was corroborated by the observation that cluster formation between T cells and antigen presenting cells/accessory cells was strongly inhibited by DB. In contrast, CSA exerted no effects. Further, DB also reduced the expression of HLA-D antigens and slightly reduced levels of the IL-2 receptor on lymphocytes. Therefore, it appears that the intracellular signaling pathway(s) affected by DB is directly involved in the cell surface expression of important immunoregulatory proteins.

In summary, the didemnins have potent immune effects and act by different mechanisms than other known immunosuppressive agents, yet appear to do this by interaction with a novel form of an immunophilin-like molecule. Thus, these marine natural products may, at the very least, be of great value as probes to investigate the mechanisms of lymphocyte activation and immune function. This potential is greatly enhanced by the availability of more than 30 structural analogues of the didemnins, and a large body of data on the structure–activity relationships of these compounds.

REFERENCES

1. RINEHART, K. L., JR., J. B. GLOER, J. C. COOK, S. A. MIZAK & T. A. SCAHILL. 1981. Structures of the didemnins, antiviral and cytotoxic depsipeptides from a Caribbean tunicate. J. Am. Chem. Soc. **103:** 1857–1859.
2. RINEHART, K. L., JR., J. B. GLOER, R. G. HUGHES, JR., H. E. RENIS, J. P. MCGOVERN, E. B. SWYNENBERG, D. A. STRINGFELLOW, S. L. KUENTZEL, & L. H. LI. 1981. Didemnins: Antiviral and antitumor depsipeptides from a Caribbean tunicate. Science **12:** 933–935.
3. RINEHART, K. L., JR., J. B. GLOER, G. R. WILSON, R. G. HUGHES, JR., L. H. LI, H. E. RENIS & J. P. MCGOVERN. 1983. Antiviral and antitumor compounds from tunicates. Fed. Proc. **42:** 87–90.
4. JIANG, T. L., R. H. LIU & S. E. SALMOM. 1983. Antitumor activity of didemnin B in the human tumor stem cell assay. Cancer Chemother. Pharmacol. **11:** 1–4.
5. MONTGOMERY, D. W. & C. F. ZUKOSKI. 1985. Didemnin B: A new immunosuppressive cyclic peptide with potent activity *in vitro* and *in vivo.* Transplantation **40:** 49–56.
6. TEUNISSEN, M. B. M., F. H. M. PISTOOR, H. A. H. RONGEN, M. L. KAPSENBERG & J. D. BOS. 1992. A comparison of the inhibitory effects of immunosuppressive agents cyclosporine, tetranactin, and didemnin B on human T-cell responses *in vitro.* Transplantation **53:** 875–881.
7. LEGRUE, S. J., T. C. SHEU, D. D. CARSON, J. L. LAIDLAE & S. K. SANDUJA. 1988. Inhibition of T-lymphocyteproliferation by the cyclic polypeptide didemnin B: No inhibition of lymphokine stimulation. Lymphokine Res. **7:** 21–29.
8. RUSSELL, D. H., A. R. BUCKLEY, D. W. MONTGOMERY, N. A. LARSON, P. W. GOUT, C. T. BEER, C. W. PUTNAM, C. F. ZUKOSKI & R. A. KIBLER. 1987. Prolactin-dependent mitogenesis in Nb2 node lymphoma cells: Effects of immunosuppressive cyclopeptides. J. Immunol. **138:** 276–284.
9. ALFREY, E. J., C. F. ZUKOSKI & D. W. MONTGOMERY. 1992. Didemnin B prolongs survival of mouse skin allografts. Transplantation **54:** 188–189.
10. YUH, D. D., R. P. ZURCHER & R. E. MORRIS. 1989. Efficacy of didemnin B in suppressing allograft rejection in mice and rats. Transplant. Proc. **21:** 1141–1143.
11. STEVENS, D. W., R. M. JENSEN & L. E. STEVENS. 1989. Didemnin B prolongs rat heart allograft survival. Transplant. Proc. **21:** 1139–1140.
12. MONTGOMERY, D. W., A. CWELNIKER & C. F. ZUKOSKI. 1987. Didemnin B—an immunosuppressive cyclic peptide that stimulates murine hemagglutinating antibody responses and induces leukocytosis *in vivo.* Transplantation **43:** 133–139.
13. CRAMPTON, S. L., E. G. ADAMS, S. L. KUENTZEL, L. H. LI, G. BADINER & B. K. BHUYAN. 1984. Biochemical and cellular effects of didemnins A and B. Cancer Res. **44:** 1796–1801.
14. GSCHWENDT, M., W. KITTSTEIN & F. MARKS. 1987. Didemnin B inhibits biological effects of tumor-promoting phorbol esters on mouse skin, as well as phosphorylation of a 100-kD protein in mouse epidermis cytosol. Cancer Lett. **34:** 187–191.
15. SHEN, G. K., C. F. ZUKOSKI & D. W. MONTGOMERY. 1992. A specific binding site in Nb2 node lymphoma cells mediates the effects of didemnin B, an immunosuppressive cyclic peptide. Int. J. Immunopharmacol. **14:** 63–73.
16. SCHREIBER, S. L. 1991. Chemistry and biology of the immunophilins and their immunosuppressive ligands. Science **251:** 283–287.
17. MONTGOMERY, D. W., G. K. SHEN, E. D. ULRICH, L. L. STEINER, P. R. PARRISH & C. F. ZUKOSKI. 1992. Human thymocytes express a prolactin-like messenger ribonucleic acid and synthesize bioactive prolactin-like proteins. Endocrinology **131:** 3019–3026.
18. SCATCHARD, G. 1949. The attraction of proteins for small molecules and ions. Ann. N.Y. Acad. Sci. **51:** 660–672.
19. HANDSCHUMACHER, R. E., M. W. HARDING, J. RICE & R. J. DRUGGE. 1984. Cyclophilin: A specific cytosolic binding protein for cyclosporine A. Science **226:** 544–547.
20. PHILLIPS, J. L., R. SCHWARTZ & D. D. VON HOFF. 1989. In vitro distribution of diacetyl didemnin B in human blood cells and plasma. Cancer Invest. **7:** 123–128.
21. ROSEN, M. K., R. F. STANDQAERT, A. GALAT, M. NAKATSUKA & S. L. SCHREIBER. 1990.

Inhibition of FKBP rotamase activity by immunosuppressant FK506: Twisted amide surrogate. Science **248:** 863–866.

22. Hossain, M. B., D. Helm, J. van der Antel, G. M. Sheldrick, S. K. Sanduja & A. J. Weinheimer. 1988. Crystal and molecular structure of didemnin B, an antiviral and cytotoxic depsipeptide. Proc. Natl. Acad. Sci. USA **85:** 4118–4122.
23. Schreiber, S. L. & G. R. Crabtree. 1992. The mechanisms of action of cyclosporine A and FK506. Immunol. Today **13:** 136–141.
24. Sigal, N. H. & F. J. Dumont. 1992. Cyclosporine A, FK-506, and rapamycin: Pharmacologic probes of lymphocyte signal transduction. Annu. Rev. Immunol. **10:** 519–560.

A Novel Solid-Phase Assay for Lectin Binding

Comparative Studies on β-Galactoside-Binding S-Type Lectins from Fish, Amphibian, and Mammalian Tissues[a]

H. AHMED, N. E. FINK,[b] AND G. R. VASTA[c]

Center of Marine Biotechnology
Maryland Biotechnology Institute
University of Maryland System
Baltimore, Maryland 21202

A family of β-galactoside-binding S-type lectins (galaptins or S-lac lectins) with a subunit mass of about 14 kDa occurs in a variety of animal tissues. Studies on carbohydrate recognition by rat lung galaptin,[1] human lung galaptin,[2] calf heart galaptin,[3] and human spleen galaptin[4,5] have shown that type 1 (Gal β1,3GlcNAc) and type 2 (Gal β1,4GlcNAc) oligosaccharide backbone sequences are potent inhibitors of their binding. Although the primary structures of galaptins from the aforementioned species exhibit considerable similarity, these structures show extensive differences when compared with galaptins from phylogenetically more distant vertebrate species.[6,7] Because differences in primary structure may be reflected in the carbohydrate binding properties, we characterized the fine carbohydrate specificity of galaptins from striped bass and toad ovaries and compared them with that from the bovine spleen galaptin, through a sensitive, novel solid-phase inhibition assay we developed for that purpose: The purified lectins were alkylated and conjugated with horseradish peroxidase (HRP), and the labeled lectins purified by size-exclusion HPLC. The galaptin–peroxidase conjugates were preincubated with a variety of saccharides and tested for binding to desialylated fetuin-coated plates by ELISA. This assay is very sensitive, highly reproducible, and requires very low concentrations of HRP-galaptin conjugate and sugar inhibitors.

MATERIALS AND METHODS

Galaptins from bovine spleen, striped bass (*Morone saxatilis* Walbaum) ovaries and toad (*Bufo arenarum* Hensel) ovaries were purified as described elsewhere (Fink

[a]Supported by National Science Foundation Award MCB-91-05875-02 (GRV) and the Consejo Nacional de Investigaciones Clientíficas y Técnicas de la República Argentina (NEF).
[b]On temporary leave from Departamento de Ciencias Biológicas, Cátedra de Hematología, Universidad Nacional de La Plata, Argentina.
[c]Address for correspondence: G. R. Vasta, Center of Marine Biotechnology, Maryland Biotechnology Institute, University of Maryland System, 600 E. Lombard St., Baltimore, MD 21202.

et al.[8] and Ahmed *et al.*, this volume). For the preparation of galaptin–peroxidase conjugates, each purified galaptin (1 mg) was adsorbed on a DEAE-Sepharose column (0.5 ml) and alkylated with 1 ml of iodoacetamide (0.1 M) in the presence of lactose (0.1 M) at 4°C for one hour. The column was washed with PBS (diluted 1 : 10), and the bound alkylated protein was eluted with 1 ml of PBS/0.5 M NaCl. Lactose (34 mg) and horseradish peroxidase (2 mg) were added to the alkylated galaptin, and the mixture was cross-linked with 120 μl of 1% glutaraldehyde at 4°C for 19 hr. The mixture was diluted 40-fold and loaded on a DEAE-Sepharose column, and the washed and the bound conjugate was eluted with PBS/1 M NaCl. The active HRP-galaptin conjugate was isolated on a lactosyl-Sepharose column and was finally separated by FPLC on a Superose 6 column. The protein content of the conjugate was determined with Coomassie blue dye; the agglutination titers were determined by agglutination with pronase-treated, fixed rabbit erythrocytes; and the specific activity was calculated. The conjugates were stored with 1% BSA/50% glycerol at −20°C until use. For the solid-phase binding-inhibition assay, the wells were coated with asialofetuin (0.5 μg/100μl/well) at 37°C for 3 hr, fixed with 2% formaldehyde (37°C, 30 min) and washed with PBS/0.05% Tween 20 (pH 7.5). Galaptin conjugates (10 ng) were incubated with various concentrations of each sugar inhibitor at 4°C for 1 hr, and the mixtures were added to the asialofetuin-coated plates. After an incubation at 4°C for 1 hr, the washed and bound conjugate was detected with ABTS. The concentrations of sugars producing 50% inhibition of galaptin binding relative to the control (inhibitor substituted by PBS) were graphically determined and normalized relative to lactose (TABLE 1).

TABLE 1. Carbohydrate Inhibition of Galaptin Binding to Asialofetuin Expressed as I_{50} Relative to Lactose[a]

Compound	Relative Activity		
	Striped Bass	Toad	Bovine
Gal β1,4Glc	1	1	1
Gal β1,4GlcNAc	3.1	3.8	5.5
Gal β1,3GlcNAc	2.8	4.4	ND
Gal β1S1 βGal	10.0	3.9	5.9
Glc β1,4Glc	<0.004	<0.02	<0.003
NeuAc α2,6Galβ1,4Glc	<0.3	<0.2	<0.1
GlcNAc β1,6Galβ1,4Glc	<0.05	<0.09	<0.04
Gal β1,3GalNAc	<0.03	<0.06	<0.02
Fuc α1,3[Galβ1,4]Glc	<0.08	<0.08	<0.04
MeO-2Gal β1,4Glc	3.3	2.5	4.1
Fuc α1,2Galβ1,4Glc	0.3	0.2	0.5
NeuAc α2,3Gal β1,4Glc	0.2	0.2	0.4
Gal	0.008	0.005	0.007
Gal α-OMe	0.027	0.03	0.008
Gal β-OMe	0.008	0.006	0.004

[a]Lac I_{50} for striped bass varied from 0.03 to 0.05 mM; for toad varied from 0.05 to 0.08 mM; for bovine varied from 0.06 to 0.09 mM. All sugars except Fuc are in D form.

RESULTS AND DISCUSSION

The solid-phase inhibition data revealed that the galaptins tested have binding sites at least as large as a disaccharide. For all three galaptins, lactose and its glycosides, *N*-acetyllactosamine, Gal β1,3GlcNAc, and thiodigalactoside were better inhibitors than galactose and its galactosides; disaccharides that have nonreducing terminal β-galactosyl residues linked (1,3), (1,4), and (1,6) to Glc, GlcNAc, Ara, Man, Fru_f, glucitol, or gluconic acid were better inhibitors than free Gal or its glycosides. The disaccharide Glc β1,4Glc, as compared to Gal β1,4 Glc, had no inhibitory activity, indicating the importance of 4′-OH groups of the nonreducing sugar. Substitution of the 6′-OH groups on the Gal residue of Gal β1,4Glc interferes with the binding because NeuAc α2,6Gal β1,4Glc and GlcNAc β1,6Gal β1,4Glc were poor inhibitors. The four-OH groups from GlcNAc (in Gal β1,3GlcNAc) and 3-OH groups from Glc (in Gal β1,4Glc) are also very important for lectin–sugar interaction since Gal β1,3GalNAc and Fuc α1,3[Gal β1,4] Glc (3-fucosyllactose) do not inhibit the binding. The hydroxyls at 2′ and 3′ of Gal β1,4Glc are not critical, inasmuch as substitutions of these hydroxyls (MeO-2Gal β1,4Glc, Fuc α1,2Gal β1,4Glc, NeuAc α2,3Gal β1,4Glc) do not affect binding significantly. For the monosaccharides, the α-anomer was more effective than its β counterpart. Hydrophobic interaction might occur in the striped bass and bovine galaptin, inasmuch as the nitrophenyl-galactosides were better inhibitors than methyl-galactosides. That effect was not observed with toad galaptin. Our data indicate that the overall carbohydrate binding properties of these S-type lectins are very similar, suggesting that despite substantial differences in overall primary structure, the architecture of the binding site may have been highly conserved during vertebrate evolution. In contrast, vertebrate and invertebrate C-type lectins exhibit a wide variety of carbohydrate specifities.[9,10]

REFERENCES

1. Leffler, H. & S. H. Barondes. 1986. J. Biol. Chem. **261:** 10119–10126.
2. Sparrow, C. P., H. Leffler & S. H. Barondes. 1987. J. Biol. Chem. **262:** 7388–7390.
3. Abbot, W. M., E. F. Hounsell & T. Feizi. 1988. Biochem. J. **252:** 283–287.
4. Lee, R. T., Y. Ichikawa, H. J. Allen & Y. C. Lee. 1990. J. Biol. Chem. **265:** 7864–7871.
5. Ahmed, H., H. J. Allen, A. Sharma & K. L. Matta. 1990. Biochemistry **29:** 5315–5319.
6. Paroutaud, P., G. Levi, V. I. Teichberg & A. D. Strosberg. 1987. Proc. Natl. Acad. Sci. USA **84:** 6345–6348.
7. Hirabayashi, J., M. Satoh & K. Kasai. 1992. J. Biol. Chem. **267:** 15485–15490.
8. Fink, N. E., H. Ahmed & G. R. Vasta. 1993. Submitted for publication.
9. Drickamer, K. 1988. J. Biol. Chem. **263:** 9557–9560.
10. Vasta, G. R. 1992. *In* Glycoconjugates: Composition, Structure and Function. H. J. Allen & E. C. Kisailus, Eds.: 593–634. Marcel Dekker. New York.

Elasmobranch and Teleost Fish Contain Thiol-Dependent β-Galactoside-Binding Lectins That Are Cross-Reactive with Those Identified and Characterized in Bovine Spleen[a]

H. AHMED, N. E. FINK,[b] AND G. R. VASTA[c]

Center of Marine Biotechnology
Maryland Biotechnology Institute
University of Maryland System
Baltimore, Maryland 21202

The S-type animal lectins (galaptins or S-lac lectins) require the presence of thiols for binding activity but not Ca^{2+} or other divalent cations.[1] Although galaptins have been characterized in several vertebrate models (human, mouse, chicken, and frog),[2] there is little information about the presence of related molecules in vertebrate species below the level of the amphibians[3,4] or their biological role and evolution as recognition molecules. Here we describe the subunit structure, molecular properties, and the identification of amino acid residues involved in sugar recognition of a galaptin from bovine spleen, and the presence of similar molecules in tissues of various fish species.

MATERIALS AND METHODS

The bovine spleen galaptin was purified as described elsewhere.[5] Briefly, bovine spleen tissue was extracted with 0.1 M lactose/0.01 M β-mercaptoethanol (ME)/PBS (diluted 1 : 10), centrifuged, and the supernatant absorbed on DEAE-Sepharose pre-equilibrated with ME/PBS (diluted 1 : 10). After washing the ion-exchange bed, the bound protein was eluted with ME/PBS/0.5 M NaCl and loaded on a lactosyl-Sepharose or an asialofetuin-Sepharose column. The column was washed with ME/PBS/NaCl followed by ME/PBS (diluted 1 : 10), and bound protein was eluted with 0.1 M lactose/ME/PBS (diluted 1 : 10). The lactose eluate was passed through a DEAE-Sepharose column and stored with 50% glycerol at −20°C. Isoelectric

[a]Supported by National Science Foundation Award MCB-91-05875-02 (GRV) and the Consejo Nacional de Investigaciones Científicas y Técnicas de la República Argentina (NEF).

[b]On temporary leave from Departamento de Ciencias Biológicas, Cátedra de Hematología, Universidad Nacional de La Plata, Argentina.

[c]Address for correspondence: G. R. Vasta, Center of Marine Biotechnology, Maryland Biotechnology Institute, University of Maryland System, 600 E. Lombard St., Baltimore, MD 21202.

focusing was performed as described by Vasta *et al.*[6] Fluorescence quenching was carried out following Mitra and Sakar.[7] Chemical modification of amino acid residues was performed as described in Ahmed and Gabius.[8]

RESULTS AND DISCUSSION

The purified galaptin gave a single sharp peak on size-exclusion chromatography (Superose 6 FPLC) corresponding to 28.5 kDa and a single band on SDS-PAGE (both under reducing and nonreducing conditions) corresponding to 14.5 kDa. On isoelectric focusing gels it showed five bands corresponding to pls 5.0 (3.4%), 4.88 (16.6%), 4.73 (41.2%), 4.63 (33.0%), and 4.56 (4.5%). The galaptin is relatively thermostable, and after incubation at 100°C for 30 minutes, it still retains 10% of the original activity as determined by agglutination of pronase-treated fixed rabbit erythrocytes. Quenching of fluorescence of 4-methylumbelliferyl-α-D-galactopyranoside by galaptin shows that the homodimeric protein has two sugar-binding sites with an association constant of 0.3×10^5 M^{-1} at 30°C. The association constant increases at lower temperatures (1×10^5 at 15°C and 3.4×10^5 at 5°C), suggesting that the binding is exothermic. Chemical modification of the galaptin by group-specific reagents in the presence or absence of the specific ligand allowed us to emphasize the role of histidine, tryptophan, arginine, and carboxylic acid for sugar binding. Application of reagents specific for lysine and tyrosine residues failed to affect lectin activity. We now have detected similar cross-reactive lectins in a variety of tissues from shark (*Squalus acanthius*), monk fish (*Squatina* sp.), striped bass (*Morone saxatilis*), and rainbow trout (*Onchorhynchus mykiss*), as determined by Western blot using anti-

TABLE 1. Presence of Galaptin-Like Molecules in Tissues from Teleost and Elasmobranch Fish Species[a]

Species/Tissue	Anti-Bovine Galaptin Antiserum	Anti-Toad Galaptin Antiserum
Striped bass		
Muscle	++	++
Spleen	–	–
Ovary	+	+
Rainbow trout		
Muscle	+	+
Spleen	–	+
Pyloric ceacae	ND	+
Shark		
Spleen	–	+
Pancreas	ND	+
Monk fish		
Spleen	+	–
Muscle	ND	++
Liver	ND	+

[a]Symbols: ++, strong; +, weak; –, no reaction; ND, not determined. Data from ELISA and Western blot experiments.

bovine galaptin and anti-toad galaptin–specific antisera (TABLE 1). Subsequently, we purified the striped bass muscle and ovary galaptins by procedures similar to those described above, including affinity chromatography. This suggests that not only subunit structure and molecular properties may be similar but also the sugar recognition properties of fish and mammalian galaptins are shared.

Within the mammals, the S-type lectins are developmentally regulated and constitute a family of structurally related molecules.[9] Their primary structure differs significantly from those of lower vertebrates, although these still preferentially bind to lactose or *N*-acetyl-lactosamine.[4] Their location is mostly intracellular, in the cytoplasmic compartment, but can be secreted in the extracellular space, although they may remain membrane associated through their carbohydrate-binding sites.[10] Potential ligands, such as polylactosaminoglycans from the extracellular matrix, are ubiquitous in vertebrates, suggesting that protein–carbohydrate interactions may constitute basic phenomena in tissue differentiation and organization common to all taxa, most likely within an extensive functional diversity.

REFERENCES

1. DRICKAMER, K. 1988. J. Biol. Chem. **263:** 9557–9560.
2. HARRISON, F. L. 1991. J. Cell Sci. **100:** 9–14.
3. TEICHBERG, V. I., I. SILMAN, D. D. BEITSCH & G. RESHEFF. 1975. Proc. Natl. Acad. Sci. USA **72:** 1383–1387.
4. HIRABAYASHI, J., M. SATOH & K. KASAI. 1992. J. Biol. Chem. **267:** 15485–15490.
5. FINK, N. E., H. AHMED & G. R. VASTA. 1993. Submitted for publication.
6. VASTA, G. R., J. C. HUNT, J. J. MARCHALONIS & W. W. FISH. 1986. J. Biol. Chem. **261:** 9174–9181.
7. MITRA, D. & M. SARKAR. 1989. Biochem. J. **262:** 357–360.
8. AHMED, H. & H.-J. GABIUS. 1989. J. Biol. Chem. **264:** 18673–18678.
9. BARONDES, S. H. 1984. Science **223:** 1259–1264.
10. COOPER, D. N. & S. H. BARONDES. 1990. J. Cell Biol. **110:** 1681–1691.

Lectins from the Colonial Tunicate *Clavelina picta* Are Structurally Related to Acute-Phase Reactants from Vertebrates[a]

M. T. ELOLA AND G. R. VASTA[b]

Center of Marine Biotechnology
Maryland Biotechnology Institute
University of Maryland System
Baltimore, Maryland 21202

The presence of carbohydrate-binding molecules in invertebrates is well documented.[1] Nevertheless, few agglutinins have been purified and fully characterized in their biochemical properties and molecular structure. From those taxa for which structural information is available, namely molluscs, arthropods, and sponges, most lectins are multivalent, oligomeric proteins or glycoproteins with binding sites for relatively small carbohydrate moieties.[2] Considerably less is known about the protochordates and echinoderms, despite their close affinity to the vertebrates. We have concentrated our efforts on examining the structural and functional features and evolutionary relationships between lectins from selected protochordate and echinoderm species and "nonspecific" recognition molecules from vertebrates such as acute-phase reactants, in particular C-type lectins; members of the pentraxin family such as C-reactive protein (CRP); serum amyloid P (SAP) and serum amyloid A; and coagulation factor VIIIc. In this paper we describe the purification and biochemical characterization of four distinct fucosyl-binding lectins from the colonial tunicate *Clavelina picta,* and their structural and serological relationships with acute-phase proteins.

MATERIALS AND METHODS

The methods for the purification of *Clavelina* lectins, their biochemical characterization, peptide preparation, amino acid composition and sequence, and preparation of rabbit antisera were carried out as described earlier.[3,4] Briefly, the purification procedure consisted of a series of steps in the following sequence: ammonium sulfate fractionation, affinity chromatography (L-fucose agarose), size-exclusion chromatography (Sephacryl S-300 and Superose 6), and ion-exchange chromatography (DEAE Sepharose CL-6B). Purity was assessed by immunoelectrophoresis, PAGE, and size exclusion.

[a]Supported by National Science Foundation Award MCB-91-05875-02 (GRV).
[b]Address for correspondence: G. R. Vasta, Center of Marine Biotechnology, Maryland Biotechnology Institute, University of Maryland System, 600 E. Lombard St., Baltimore, MD 21202.

RESULTS AND DISCUSSION

Three of the four components isolated (CPL-II, -III, and -IV) yielded single peaks by analytical size-exclusion HPLC. CPL-I consisted, as assessed by size-exclusion HPLC and PAGE under nondenaturing conditions (in a Tris-borate-EDTA buffer), of a series of molecular species with values ranging from 540 kDa to 950 kDa. CPL-II and CPL-III resulted in single components of 72 kDa and 64 kDa, respectively. CPL-IV exhibits a major component of 260 kDa. Under denaturing conditions, changes in mobility of all four lectins suggested that the native molecules result from noncovalent associations of smaller species: CPL-I resolved into components ranging from 157 to 300 kDa; CPL-II, CPL-III, and CPL-IV presented single components of 57, 32, and 175 kDa, respectively. Under reducing conditions the electrophoretic mobility of CPL-I, CPL-II, and CPL-III was decreased, indicating the presence of intrachain disulfide bonds; CPL-IV exhibited a single subunit of 34 kDa. Under reducing conditions all native species of CPL-I yielded similar subunit profile proportions of the four major components, suggesting that the native molecular species are not random combinations but rather organized structures of the non-covalently bound subunits. All four lectins are acidic, in agreement with the amino acid composition analysis, and heterodisperse. The most striking result was the finding of one peptide from CPL-I that showed more than 90% homology to human factor VIIIc. CPL-III exhibits a certain degree of heterogeneity and several regions of internal homology. Interesting stretches of identities with invertebrate and vertebrate acute-phase, Ca^{2+} binding and recognition proteins have been identified. CPL-I showed the strongest cross-reactivity with factor VIIIc but also with CRP and SAP; CPL-III cross-reacted with CRP and SAP only (TABLE 1). The analysis of the serological cross-reactivity suggests that *Clavelina* lectins are related to acute-phase reactants from vertebrates, in particular to the pentraxins CRP and SAP and to the pro-coagulant factor VIIIc. The cross-reactivity of CPL-I with human pro-coagulation factor VIIIc is further substantiated by the finding of the aforementioned CPL-I peptide homologous to F VIIIc. The similarities with CRP may reside in the binding sites for phosphorylcholine since a monoclonal antibody that recognizes those sites

TABLE 1. Cross-Reactivities[a] of Conventional Antisera Developed against Human CRP, Human Factor VIIIc, *Clavelina picta* Lectins, and *Didemnum candidum* Lectin DCL-I

	Clavelina Lectins					
Antiserum Anti-	I	II	III	IV	CRP	SAP
C. p. plasma	+	+	+	+	+	ND
CPL-I	+	−	+	−	+	ND
CPL-II	+	+	−	−	+	ND
CPL-III	+	−	+	+	+	+
DCL-I	+	−	+	−	+	+
CRP	+	+	+	ND	+	ND
SAP	−	+	+	ND	ND	+
F VIIIc	+	−	−	−	ND	ND

[a]Results from Western blot and ELISA.

in CRP (kindly provided by Dr. J. Volanakis, Birmingham, Alabama) cross-reacts with *Clavelina* lectins in a phosphocholine-inhibitable fashion. Our earlier results obtained on the galactosyl binding lectins from the tunicate *Didemnum candidum*[5,6] suggest that some tunicate lectins may be related to the vertebrate acute-phase reactants.

The similarities in primary structure observed suggest that CPL-I and CPL-III are mosaic molecules with fucose- and Ca^{2+}-binding sites that are related to vertebrate Ca^{2+}-modulated self/non-self recognition molecules, including surprising similarities with the mammalian pro-coagulant factor VIIIc. A recent report on the primary structure of a lectin from the tunicate *Polyandrocarpa misakiensis* suggests homology with recognition molecules from vertebrates.[7] The mosaic structure of *Clavelina* lectins contributes to support the hypothesis that multiple biological roles in self/non-self recognition, such as embryonic development, wound repair, and defense against pathogenic bacteria, are accomplished by a single molecular species.

REFERENCES

1. Vasta, G. R. 1991. *In* Phylogenesis of Immune Functions. N. Cohen & G. W. Warr, Eds.: 73–101. CRC Press. Boca Raton, FL.
2. Vasta, G. R. 1992. *In* Glycoconjugates: Composition, Structure and Function. H. J. Allen & E. C. Kisailus, Eds.: 593–634. Marcel Dekker. New York.
3. Vasta, G. R. 1993. Submitted.
4. Elola, M. T., J. Pohl & G. R. Vasta. 1993. Submitted.
5. Vasta, G. R., J. Hunt, J. J. Marchalonis & W. W. Fish. 1986. J. Biol. Chem. **261:** 9174.
6. Vasta, G. R. & J. J. Marchalonis. 1986. J. Biol. Chem. **261:** 9182.
7. Suzuki, A., T. Takagi, T. Furukohri, K. Kawamura & M. Nakauchi. 1990. J. Biol. Chem. **265:** 1274.

Hemolymph Lectins of the Blue Crab, *Callinectes sapidus*, Recognize Selected Serotypes of Its Pathogen *Vibrio parahaemolyticus*[a]

F. J. CASSELS,[b] E. W. ODOM,[c] AND G. R. VASTA[c,d]

[b]*Department of Gastroenterology*
Walter Reed A.I.R.
Washington, D.C.

[c]*Center of Marine Biotechnology*
Maryland Biotechnology Institute
University of Maryland System
Baltimore, Maryland 21202

Because lectins appear to be ubiquitous in fluids and tissues of both invertebrates and vertebrates, they have been proposed as likely candidates for a carbohydrate-based self/non-self recognition system.[1–3] For many years the blue crab *Callinectes sapidus* has been employed as a model for the study of many aspects of invertebrate biology and pathobiology,[4,5] not only due to intrinsic interest in this species, but also to provide understanding of certain basic biological mechanisms that may operate in other taxa, for example, the molecular basis of non-self recognition in species that lack the adaptive immunity generated through the antibody-based immune system. Initial experiments[6] showed that *C. sapidus* serum possessed lectin activity toward a variety of vertebrate RBC and lymphoid cell lines, both untreated and enzyme treated, and that this activity could be fractionated by crossed absorption experiments in at least three distinct lectin reactivities. This study was designed to investigate whether carbohydrate binding proteins present in *C. sapidus* serum and hemocytes may have a role in non-self recognition. In this report we briefly describe the biochemical properties of two serum lectins from *C. sapidus* and provide preliminary evidence that these lectins may be involved in defense mechanisms against *Vibrio parahaemolyticus,* a known pathogen for *C. sapidus.*

MATERIALS AND METHODS

The purification of *C. sapidus* serum lectins was carried out as described elsewhere[7] by ammonium sulfate precipitation (35% saturation) of *C. sapidus* serum

[a]Supported by NOAA Grant NA90AA-D-SG063 through the University of Maryland Sea Grant College to GRV and National Institutes of Health MARC Award 5 F31 GM14903-02 to EWO/GRV.

[d]Address for correspondence: G. R. Vasta, Center of Marine Biotechnology, Maryland Biotechnology Institute, University of Maryland System, 600 E. Lombard St., Baltimore, MD 21202.

followed by affinity chromatography on colominic acid conjugated to epoxy-activated Sepharose 6B. Elution was carried out with GlcNAc (200 mM) and EDTA (30 mM) at pH 7.6 and purity assessed by SDS-polyacrylamide gel electrophoresis. Eleven different *V. parahaemolyticus* serotypes were streaked onto nutrient agar plates (2.3% nutrient agar, 2% NaCl) and grown at 37°C for 24 hours. Bacteria were suspended in Tris-buffered saline (TBS)(2% NaCl), washed three times (3 min, 3500×*g*), and resuspended to approximately 6×10^9 cells/ml in TBS (2% NaCl, 0.5% BSA). Bacterial agglutination and agglutination–inhibition assays with *C. sapidus* serum and purified lectin were performed by adapting the method of Dunsford and Bowley,[8] with one-hour incubations, and scored macroscopically or with the aid of a Jena inverted microscope. The substitution of serum or purified lectin by TBS served as controls for all titrations.

RESULTS AND DISCUSSION

Two serum lectins from the blue crab (*C. sapidus*) were purified using a colominic acid–conjugated Sepharose 6B matrix. Each lectin was eluted specifically using a monosaccharide (GlcNAc) or EDTA. This procedure resulted in high purification factors and specific activity. Under reducing SDS-PAGE both lectins, CSL-I and CSL-II, displayed two distinct subunits (CSL-I: 37 kDa and 38 kDa; CSL-II: 31 kDa and 33 kDa). Crude *C. sapidus* serum and CSL-I, a lectin specific for *N*-acylamino sugars, were tested for their ability to agglutinate 11 serotypes of *V. parahaemolyticus,* a known pathogen for *C. sapidus* and a causative agent of human gastroenteritis. Ten of the 11 serotypes tested were agglutinated by serum, although at relatively low titers (1:2–1:64). From those, only the ones yielding the highest titers (1:32–1:64) (O4:K11, O4:k13, and O4:K63) were agglutinated by CSL-I (Titers 1:400–1:800). All serotypes agglutinated by the lectin exhibit the O4 antigen. Of particular interest was the fact that the serotypic O1:K38 (ATCC 27969), which was not agglutinated by either *C. sapidus* serum or CSL-I, had been originally isolated from *C. sapidus* hemolymph. *N*-acetylneuraminic acid, *N*-acetylglucosamine and *N*-acetylmuramic acid behaved as the most effective inhibitors for the binding of CSL-I to the serotypes O4:K63 and O4:K11 but glucose and *N*-acetylglutamic acid did not inhibit at all. It has been shown that none of the K (capsular) antigens contain *N*-acetylated sugars but O4 (LPS) contains *N*-acetyl glucosamine and glucose. CSL-I would bind to *N*-acyl residues exposed on the O4 antigens of *V. parahaemolyticus.* Our results suggest that *C. sapidus* lectins are able to specifically recognize and agglutinate most (but not all) *V. parahaemolyticus* serotypes and would function as opsonins by facilitating phagocytosis or encapsulation and melanization of the agglutinated bacteria. *V. parahemolyticus* is a pathogen of *C. sapidus*[9,10] and is also a causative agent of human gastroenteritis, that primarily results from ingesting improperly prepared seafood.[11] Although the investigation of the interactions of *C. sapidus* hemolymph and hemocyte lectins with *V. parahaemolyticus* are of considerable biological interest, human health concerns may bring added relevance to these studies.

REFERENCES

1. LIS, H. & N. SHARON. 1986. *In* The Lectins: Properties, Functions and Applications in Biology and Medicine. I. E. Liener, N. Sharon, I. J. Goldstein, Eds.: 266–293. Academic Press. Orlando, FL.
2. COOMBE, D. R., P. L. EY & C. R. JENKIN. 1984. Q. Rev. Biol. **59:** 231–255.
3. VASTA, G. R. 1991. *In* Phylogenesis of Immune Functions. N. Cohen & G. W. Warr, Eds.: 73–101. CRC Press. Boca Raton, FL.
4. JOHNSON, P. T. 1980. Histology of the Blue Crab, *Callinectes sapidus*: A Model for the Decapoda. Praeger Publishers. New York.
5. PROVENZANO, A. J. 1983. The Biology of Crustacea, Vol. 6: Pathobiology. Academic Press. New York.
6. CASSELS, F. J., J. J. MARCHALONIS & G. R. VASTA. 1986. Comp. Biochem. Physiol. **85B:** 23–30.
7. CASSELS, F. J., E. ODOM CRESPO & G. R. VASTA. 1993. Submitted for publication.
8. DUNSFORD, I. & C. C. BOWLEY. 1955. Techniques in Blood Grouping. Thomas. Springfield, IL.
9. KRANTZ, G. E., R. R. COLWELL & E. LOVELACE. 1969. Science **164:** 1286–1287.
10. SINDERMANN, C. J. 1977. Disease diagnosis and control in North American marine aquaculture. Elsevier. New York.
11. FEKETY, R. 1983. Rev. Infect. Dis. **5:** 246–257.

A Dimeric Lectin from Coelomic Fluid of the Starfish *Oreaster reticulatus* Cross-Reacts with the Sea Urchin Embryonic Substrate Adhesion Protein, Echinonectin[a]

A. M. SNOWDEN[b] AND G. R. VASTA[c]

Center of Marine Biotechnology
Maryland Biotechnology Institute
University of Maryland System
Baltimore, Maryland 21202

Species of Deuterostome invertebrates, echinoderms and tunicates among others, are considered to be the closest extant representatives of lineages that gave origin to the chordata. Our interest in these taxa resides in the fact that they may constitute adequate models for the study of evolutionary aspects and structure/function relationships of protein–carbohydrate interactions that mediate internal defense and cell–cell/substrate adhesion mechanisms.[1] Only a few carbohydrate-binding proteins from echinoderms have been characterized in their biochemical properties so far,[2–6] and little is known regarding their phylogenetic relationships and biological role. One of the earliest reported examples of echinoderm carbohydrate-binding molecules, "bindin," from the sea urchin *Strongylocentrotus purpuratus,* has been postulated to mediate sperm adhesion to eggs.[3] Another protein isolated from eggs and embryos from the sea urchin *Lytechinus variegatus* is "echinonectin." Isolated as a fibronectin-like molecule, echinonectin appears to mediate functions of cell adhesion to the extracellular matrix[5] and behave like a galactose-binding lectin. A lectin from coelomic fluid of the sea urchin *Anthocidaris crassispina*[4] shares a number of amino acid residues in the carbohydrate-binding domain with mammalian C-type lectins such as the serum mannose receptor, a human acute-phase opsonin.[7] Due to its large size, abundance, and easy maintenance in the laboratory, the Caribbean "cushion starfish" *Oreaster reticulatus* constitutes a suitable model for these studies. We have purified a galactosyl-binding protein from the coelomic fluid of *Oreaster reticulatus* and determined its subunit structure and biochemical properties.[8] We have now characterized its serological cross-reactivity to "echinonectin," a sea urchin cell–extracellular matrix adhesion molecule; several lectins from tunicates; and a mammalian pentraxin.

[a]Supported by National Science Foundation Award MCB-91-05875-02 (GRV).
[b]This work is part of Ms. Snowden's Biology Graduate Dissertation, Johns Hopkins University.
[c]Address for correspondence: G. R. Vasta, Center of Marine Biotechnology, Maryland Biotechnology Institute, University of Maryland System, 600 E. Lombard St., Baltimore, MD 21202.

MATERIALS AND METHODS

The galactosyl-binding lectin from coelomic fluid of the starfish (*Oreaster reticulatus*), collected at two different locations of Andros and Lee Stocking Islands, Bahamas, was purified to homogeneity through a combination of methods that includes ammonium sulfate precipitation and affinity and ion exchange and size-exclusion chromatography and was biochemically characterized as reported elsewhere.[8] Western blots with conventional antisera and monoclonal antibodies were carried out following our standard protocols. Purified "echinonectin" and a specific rabbit antiserum were a generous gift of Dr. M. C. Alliegro (Baltimore Biotech, Inc.).

RESULTS AND DISCUSSION

The *Oreaster reticulatus* lectin was present at comparable titers in all immature and mature specimens examined. The molecular weight of the native protein was approximately 70,000, as determined by analytical size-exclusion FPLC (TABLE 1). Two distinct subunits of molecular weights 17,000 (**a**) and 18,400 (**b**) in equimolar amounts were observed under reducing conditions, suggesting that the native protein is a disulfide bond linked tetramer $\alpha_2\beta_2$ (TABLE 1). The lectin from coelomic fluid of the sea urchin *Anthocidaris crassispina,* one of the few lectins that have been characterized in echinoderms, exhibits a very different subunit structure: The multimeric lectin is constituted by 22 to 24 disulfide-bonded subunits of 13 kDa and exhibits a complex carbohydrate specificity, possibly recognizing internal sugar residues from O-linked oligosaccharide structures. The *Oreaster reticulatus* lectin is acidic, with an isoelectric point of 5.3. The lectin did not discriminate blood groups from the human ABH system and strongly agglutinated protease- and *Vibrio cholerae* neuraminidase (VCN)-treated human and rabbit erythrocytes. The most effective carbohydrate inhibitors for *Oreaster* lectin were disaccharides with nonreducing terminal β-linked galactosyl residues, such as Galβ1-4Glc and Galβ1-3Ara. However, the β-glycosides with noncarbohydrate aglycon such as Galβ-OMe were also effective inhibitors, indicating that the subterminal residue does not contribute to the binding. An acetamido group in C2 greatly reduced the affinity. Galactans, porcine mucin, and desialylated glycoproteins inhibited at concentrations as low as 500 ng/ml. The binding activity was lost by treatment with EDTA (50 mM) and only partially restored after subsequent incubation with Ca^{2+} (64% at 40 mM), Co^{2+} (34% at 4 mM) and Cd^{2+} (16% at 0.5 mM). Moreover, several cations such as Sr^{2+}, Cd^{2+},

TABLE 1. Molecular Properties of *Oreaster reticulatus* Lectin

Native MW (FPLC)	70,000
Subunit MW (SDS-PAGE; reducing conditions)	17,000 (α) 18,400 (β)
Subunit structure	$\alpha_2\beta_2$
Covalently bound carbohydrate	None
Minimum concentration for agglutination	0.48 μg/ml

and Mg^{2+} inhibited the active lectin. Sr^{2+} competitively inhibited the lectin even in the presence of exogenous Ca^{2+}.

The cross-reactivity between *Oreaster reticulatus* lectin and human C-reactive protein (CRP), lectins from the colonial tunicates *Clavelina picta* (lectins CPL-I, CPL-II, CPL-III and CPL-IV) and *Didemnum candidum* (lectins DCL-I and DCL-II) and the fibronectin-like cell adhesion protein "echinonectin" from sea urchin embryo *Lytechinus variegatus* was tested by Western blot using conventional antisera and monoclonal antibodies against all tested proteins. *Oreaster reticulatus* lectin did not cross-react with human CRP or any of the tunicate lectins (CPL-I, CPL-II, CPL-III, CPL-IV, DCL-I, or DCL-II) but reciprocally cross-reacted with "echinonectin." Our preliminary results indicate that *Oreaster reticulatus* lectin recognizes numerous marine bacterial strains isolated from the starfish's environment (Snowden and Vasta, unpublished). A related species, the starfish *Asterina pectinifera,* has been recently found to contain distinct lectins that recognize marine bacteria and mammalian carbohydrate moieties.[6] As proposed elsewhere,[7] C-type lectin domains contain highly conserved residues that are shared by a number of mammalian and invertebrate lectins that function as opsonins, as well as molecules that mediate cellular adhesion to the extracellular matrix.

REFERENCES

1. Vasta, G. R. 1992. *In* Glycoconjugates: Composition, Structure and Function. H. J. Allen & E. C. Kisailus, Eds.: 593–634. Marcel Dekker. New York.
2. Finstad, C. L., G. W. Litman, J. Finstad & R. A. Good. 1972. J. Immunol. **108:** 1704–1711.
3. Vacquier, V. D. & G. W. Moy. 1977. Proc. Natl. Acad. Sci. USA **74:** 2456–2460.
4. Giga, Y., K. Sutoh & A. Ikai. Biochemistry **24:** 4461–4467.
5. Alliegro, M. C., C. A. Ettensohn, C. A. Burdsal, H. P. Erickson & D. R. McClay. 1988. J. Cell Biol. **107:** 2319–2327.
6. Kamiya, H., K. Muramoto, R. Goto & M. Sakai. 1992. Dev. Comp. Immunol. **16:** 243–250.
7. Drickamer, K. 1988. J. Biol. Chem. **263:** 9557–9560.
8. Snowden, A. M. & G. R. Vasta. 1993. Submitted.

A Comparative Study of the Respiratory Burst Produced by the Phagocytic Cells of Marine Invertebrates

KAREN L. BELL AND VALERIE J. SMITH

School of Biological and Medical Science
Gatty Marine Laboratory
University of St. Andrews
St. Andrews, Fife, Scotland KY16 8LB

Studies of phagocytosis in vertebrates have shown that the phagocytes generate intracellular superoxide (O_2^-) and hydrogen peroxide (H_2O_2) following cell membrane stimulation by non-self materials. This process has been termed the respiratory burst.[1] In contrast, little work has been done to investigate the metabolic events associated with phagocytosis in invertebrates. We have previously shown that the phagocytic hyaline cells of the shore crab, *Carcinus maenas,* produce O_2^- following stimulation.[2] In this study, we have extended the work to examine non-self recognition by hemocytes of the solitary tunicate, *Ciona intestinalis,* using the ferricytochrome *c* reduction assay.

The detailed experimental method has been reported previously.[2] Briefly, for both *C. maenas* and *C. intestinalis* the phagocytic cells were separated using density gradient centrifugation. The assay was carried out in microtiter plates, and the phagocytes were incubated with an elicitor, catalase, and ferricytochrome *c.* Phorbol 12-myristate 13-acetate (PMA) (Sigma, Poole, Dorset) or concanavalin A (Con A) (Sigma) were used as elicitors, and for controls buffer was substituted for the elicitor. The plates were incubated at 20°C and read at 550 nm at 5-minute intervals on a microplate reader (Dynatech MR5000). To confirm that reduction of ferricytochrome *c* was due at least in part to O_2^-, superoxide dismutase (SOD) (Sigma EC1.15.1.1 from bovine erythrocytes) was included in the reaction mixture.

From TABLE 1, it is apparent that stimulation of the phagocytes from *C. maenas* and *C. intestinalis* with PMA results in a significant reduction of ferricytochrome *c*

TABLE 1. Absorbance Values at 550 nm after 30-Minute Treatment of the Cells with MS, PMA (5 mg·ml^{-1}), or con A (5 mg·ml^{-1})

	Absorbance at 550 nm		
Species	PMA	Con A	MS
C. maenas	0.048 ± 0.003	0.053 ± 0.003	0.034 ± 0.002
C. intestinalis	0.023 ± 0.003	0.023 ± 0.004	0.013 ± 0.004

NOTE: Values are means ± SE; *n* = 4 and 3 for *Carcinus maenas* and *Ciona intestinalis,* respectively.

TABLE 2. Effect of SOD on Ferricytochrome *c* Reduction by Separated Phagocytes

	Absorbance at 550 nm		
Species	PMA	PMA + SOD	MS
C. maenas	0.047 ± 0.004	0.038 ± 0.004	0.034 ± 0.002
C. intestinalis	0.026 ± 0.008	0.020 ± 0.008	0.013 ± 0.006

NOTE: Values are means in absorbance ± SE; n = 3 and 4 for *Carcinus maenas* and *Ciona intestinalis,* respectively.

compared to the buffer-treated controls ($p = 0.002$, $n = 4$ and $p = 0.028$, $n = 3$, respectively). After incubation of the phagocytes harvested from *C. maenas* or *C. intestinalis* with Con A, there was also a significant reduction in ferricytochrome *c* ($p = 0.004$, $n = 4$ and $p = 0.045$, $n = 3$, respectively). Incubation of the cells with SOD decreased the amount of ferricytochrome *c* that was reduced, thus indicating that the reduction was due in part to O_2^- (TABLE 2).

These results indicate that the phagocytic cells of a crustacean and a tunicate have the metabolic capability to generate a respiratory burst following non-self stimulation. From this work we can speculate that the ability to generate reactive oxygen moities has been conserved during evolution. Further study of this phenomenon in invertebrate species may provide a simple model for understanding the evolution of this immune response in vertebrates.

REFERENCES

1. BAEHNER, R. L., S. K. MURRMANN, J. DAVIS & R. B. JOHNSTON. 1975. The role of superoxide anion and hydrogen peroxide in phagocytosis-associated oxidative metabolic reactions. J. Clin. Invest. **56:** 571–576.
2. BELL, K. L. & V. J. SMITH. 1993. *In vitro* superoxide production by hyaline cells of the shore crab *Carcinus maenas* (L.). Dev. Comp. Immunol. **17:** 211–219.

Blood Cell–Mediated Cytotoxic Activity in the Solitary Ascidian *Ciona intestinalis*

CLARE M. PEDDIE AND VALERIE J. SMITH

School of Biological and Medical Sciences
Gatty Marine Laboratory
University of St. Andrews
St. Andrews, Fife, KY16 8LB, Scotland

Nonfusion reactions between allogeneic colonial ascidians and rejection responses to tissue grafts by solitary ascidians are associated with infiltration of the area by blood cells and subsequent *in vivo* cytotoxic activity.[1,2] *In vitro* cytotoxic activity by circulating blood cells has been investigated in mammals, lower vertebrates, and some invertebrates. However, there have been few detailed *in vitro* analyses of cytotoxic responses by ascidian hemocytes. This study examines *in vitro* blood cell–mediated cytotoxic activity in the solitary ascidian, *Ciona intestinalis,* towards mammalian tumor cells using a modification of the fluorochromasia cytotoxicity assay.[3]

MATERIALS AND METHOD

The cytotoxic blood cell populations were enriched by density gradient centrifugation of blood harvested from *C. intestinalis.*[4] Target cells were allowed to stabilize in 10 ml of low-salt marine saline (MS I) (740 mOsm) (12 mM $CaCl_2{\cdot}6H_2O$; 11 mM KCl; 26 mM $MgCl_2{\cdot}6H_2O$; 45 mM Tris; 38 mM HCl; 0.3 M NaCl; pH 7.4) for 30 min at 20°C before labeling with 5-carboxyfluorescein diacetate (CFDA) (Sigma).[3] The labeled targets were then washed and resuspended in MS (940 mOsm) (12 mM $CaCl_2{\cdot}6H_2O$; 11 mM KCl; 26 mM $MgCl_2{\cdot}6H_2O$; 45 mM Tris; 38 mM HCl; 0.4 M NaCl; pH 7.4). Cytotoxic activity was detected by incubating labeled target cells (2×10^6 ml^{-1} in MS) with effector suspensions (2×10^7 ml^{-1} in MS depending on the effector to target cell (E:T) ratio) in a 96-well microtiter plate for fluorometric use (Dynatech, Billinghurst, Sussex, England). For controls, heat-inactivated effector cells (15 min, 46°C in water bath) were substituted for untreated effector cells.[3] At the end of the incubation period, the microtiter plate was centrifuged (250 g, 5 min), and the pellets were resuspended in MS. The remaining fluorescence was measured using a microplate attachment to a luminescence spectrometer (Perkin Elmer LS50) in fluorescence mode (excitation wavelength, 490 nm; emission wavelength, 518 nm). The percentage specific release of CFDA (%SR) for each assay was calculated, as shown below, using the fluorescence data from the control (F_c) and experimental wells (F_e) with each of four well replicates.[3]

$$\%SR = (1 - F_e/F_c) \times 100$$

The assay was performed at increasing effector to target cell ratios and using a range of mammalian tumor cells previously used in assays for vertebrate natural killer cell activity.

RESULTS

This study demonstrates a population of nonspecific cytotoxic effector cells in the blood of *C. intestinalis* that kill mammalian target cells *in vitro.* Preliminary experiments showed that the assay provided optimal conditions for the functioning of the effector cells while maintaining low background leakage from the target cells. Hemocyte populations, enriched by density gradient centrifugation, exhibited cytotoxic activity that increased with the effector to target cell ratio (FIG. 1). Both human (K562) and mouse (P815, WEHI [3B] and L929) target cell lines were killed by the tunicate effector cells, indicating nonspecific activity (FIG. 2).

DISCUSSION

Preliminary results are presented that show that hemocytes, enriched by continuous density gradient centrifugation of the blood from the solitary tunicate, *Ciona intestinalis,* effect spontaneous, nonspecific cytotoxic activity against mammalian

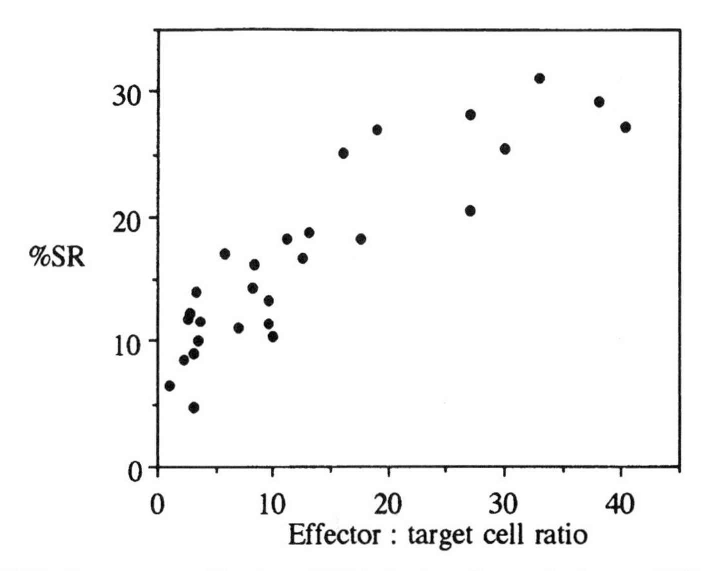

FIGURE 1. Percentage specific release (%SR) of carboxyfluorescein diacetate (CFDA) from the target cells (WEHI 3B) when incubated with enriched effector hemocytes from *Ciona intestinalis* at different effector to target cell ratios. Target cells were incubated with enriched effector cells from *Ciona intestinalis* for 40 min at 20°C. Each experiment was repeated four times.

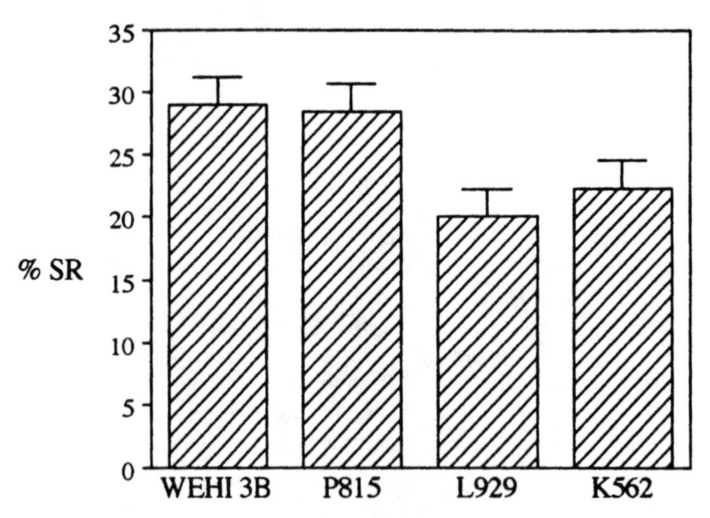

FIGURE 2. Percentage specific release (%SR) of CFDA from different mammalian target cell lines when incubated with enriched effector hemocytes from *Ciona intestinalis.* Target cells were incubated with enriched effector cells from *Ciona intestinalis* for 60 min at 20°C at an E:T ratio of 10:1. Each experiment was repeated twice; results are means ± SEM. WEHI, a mouse myelomonocytic leukemic cell (strain 3B); K562, a human erthyromyeloid leukemia cell; L929, an adherent murine fibroblast; P815, a methylcholantrene-induced mastocytoma cell from DBA/2 mouse.

tumor cells *in vitro.* This activity may represent a primordial form of the nonspecific spontaneous cytotoxic activity described in lower vertebrates.[5] The method developed here could be applied to investigate blood cell–mediated cytotoxicity in a range of marine invertebrates.

REFERENCES

1. Raftos, D. A., N. N. Tait & D. A. Briscoe. 1987. Allograft rejection and alloimmune memory in the solitary urochordate, *Styela plicata.* Dev. Comp. Immunol. **11:** 343–351.
2. Rinkevich, B. 1992. Aspects of the incompatibility nature in botryllid ascidians. Anim. Biol. **1:** 17–28.
3. Bruning, J. W., M. J. Kardol & R. Arentzen. 1980. Carboxyfluorescein fluorochromasia assays I. Non-radioactively labeled cell-mediated lympholysis. J. Immunol. Methods **33:** 33–44.
4. Smith, V. J. & C. M. Peddie. 1992. Cell cooperation during host defense in the solitary ascidian *Ciona intestinalis* (L). Biol. Bull. **183:** 211–219.
5. McKinney, E. C., L. Haynes & A. L. Droese. 1986. Macrophage-like effector of spontaneous cytotoxicity from the shark. Dev. Comp. Immunol. **10:** 497–508.

Induction of a Cellular Defense Reaction Is Accompanied by an Increase in Sensory Neuron Excitability in *Aplysia*

ANDREA L. CLATWORTHY,[a] GILBERT A. CASTRO, AND EDGAR T. WALTERS

Department of Physiology and Cell Biology
University of Texas Medical School at Houston
Houston, Texas 77225

In recent years, bidirectional interactions between the immune and nervous systems have received much attention.[1,2] The complexity of vertebrate systems, however, has made it difficult to study these interactions at a mechanistic level. To address this issue, we have developed a simple invertebrate model that is well suited to cellular analysis. Invertebrates and vertebrates share similar basic cellular defense responses to wounded-self or non-self, that is, the directed migration towards and aggregation of defense cells at the target site and phagocytosis of foreign materials small enough to be engulfed by phagocytic cells.[3,4] Therefore, fundamental principles underlying basic immune function may be common to invertebrate and vertebrate systems.

We chose the marine mollusc *Aplysia californica* as our model because it has been used extensively in studies of neural plasticity. Many of its large, accessible neurons have been characterized both electrophysiologically and functionally, which facilitates analysis at the cellular level. To induce an immune reaction, *Aplysia* were anesthetized with ice-cold magnesium chloride (Sigma), and a narrow strip of cotton gauze tied loosely around the peripheral nerves innervating one side of the animal's tail. One to 30 days later, the excitability of sensory neurons having axons in loosely ligated nerves was tested and compared to contralateral control sensory neurons innervating the other half of the tail. Sensory neurons were paired by location in the ganglia and examined sequentially, alternating between left and right sides.

The cotton gauze tied around the peripheral nerves induced a cellular defense response evidenced by encapsulation of the gauze. Sensory neurons having axons in ligated nerves were hyperexcitable compared to control cells: Sensory spike threshold to an intracellular depolarizing pulse was reduced, spike afterhyperpolarization was reduced, and spike amplitude and duration were increased. Furthermore, spike accommodation was decreased such that sensory cells with axons in ligated nerves fired more spikes to a 1-sec intracellular depolarizing pulse than cells on the control nonligated side (TABLE 1).

Sensory hyperexcitability was expressed with a latency of about five days. When tested one to four days after ligation, the excitability of sensory neurons having axons in ligated nerves was not significantly different from that recorded in control sensory

[a]Address for correspondence: Andrea L. Clatworthy, Department of Physiology and Cell Biology, University of Texas Medical School, P.O. Box 20708, Houston, TX 77225.

cells. However, 5 to 30 days after ligation, sensory cells with axons in ligated nerves were hyperexcitable compared to control cells. The effect was highly robust, being expressed in 37 of 41 animals.

When tying the cotton gauze around the peripheral nerves, it was impossible to avoid mildly stretching the nerves. However, this is unlikely to account for the hyperexcitability because sham ligation (the cotton thread was tied loosely around the nerves and then immediately removed) failed to induce hyperexcitability.

Sensory hyperexcitability is also unlikely to be a consequence of direct axonal injury because histological examination revealed no axonal damage and electrical stimulation of nerves distal to the ligation evoked action potentials that were conducted through the ligated region and recorded in the soma. This is an important finding because Walters *et al.*[5] have reported that crushing the axons of the same population of sensory neurons induces qualitatively similar hyperexcitability (TABLE 1). Conceivably, injury to a peripheral nerve could trigger the animal's cellular defense system, thereby unleashing a variety of potential signals such as cytokines released from the activated immunocytes. Stefano *et al.*[6] found preliminary evidence for this hypothesis by showing that injuring a peripheral nerve in the mollusc *Mytilus* stimulates the directed migration and adherence of immunocytes to the cut end of the nerve in a naloxone-reversible manner. It will be interesting to test the hypothesis

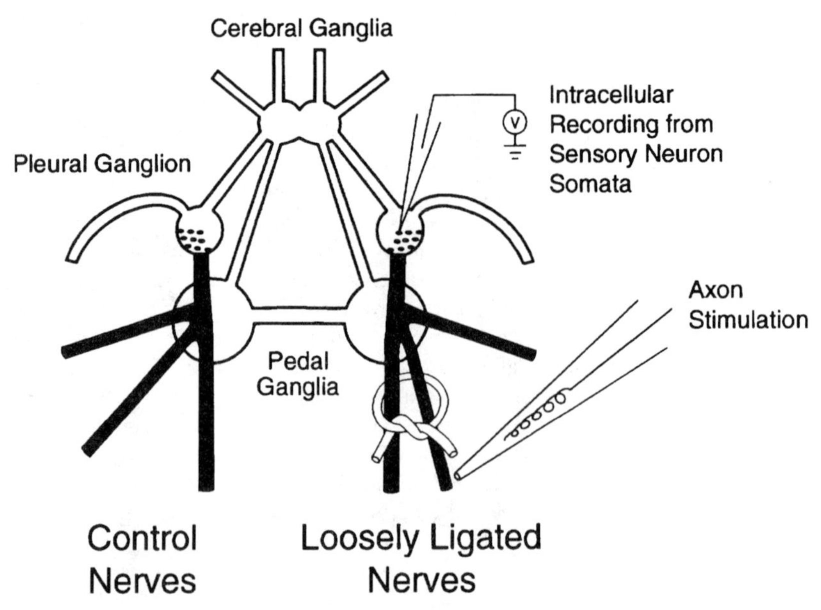

FIGURE 1. Schematic of experimental design. Peripheral nerves innervating one side of the body were loosely ligated with a narrow strip of cotton gauze. One to 30 days later, sensory neurons located in the pleural ganglia were exposed and impaled with microelectrodes, and their excitability tested and compared to paired, control sensory cells with axons in nonligated nerves. The viability of sensory axons in ligated nerves was tested by electrically stimulating the nerves distal to the ligation site and recording whether evoked action potentials were conducted through the ligated region to the soma.

TABLE 1. Summary of the Effects of Loose Ligation and Axonal Crush on Sensory Spike Characteristics and Sensory Excitability[a]

Action Potential	Comparison to Contralateral Control	
	Periaxonal Loose Ligation	Axonal Crush
Threshold	↓	↓
AHP	↓	↓
Amplitude	↑	↔
Duration	↑	↑
Accommodation	↓	↓

[a]See Clatworthy *et al.*[7]

that the immune system plays an important role in injury-induced sensory plasticity and other issues related to neural–immune communication using simple invertebrate models that are well suited to detailed analysis at the cellular level.

REFERENCES

1. TANCREDI, V., C. ZONA, F. VELOTTI, F. EUSEBI & A. SANTONI. 1990. Brain Res. **52:** 149–151.
2. WEINREICH, D. & B. J. UNDEM. 1987. J. Clin. Invest. **79:** 1529–1523.
3. TRIPP, M. R. 1961. J. Parisitol. **47:** 745–751.
4. PROWSE, R. H. & N. N. TAIT. 1969. Immunology **17:** 437–443.
5. WALTERS, E. T., H. ALIZADEH & G. A. CASTRO. 1991. Science **253:** 797–799.
6. STEFANO, G. B., M. K. LEUNG, X. ZHAO & B. SCHARRER. 1989. Proc. Natl. Acad. Sci. USA **86:** 626–630.
7. CLATWORTHY, A. L., G. A. CASTRO, B. U. BUDELMANN & E. T. WALTERS. 1994. J. Neurosci. In press.

Quantitation of Lectin Binding by Cells Harvested from the Spleen and Anterior Kidney of the Japanese Medaka (*Oryzias latipes*)

VICTORIA HENSON-APOLLONIO[a]
AND VALERIE JOHNSON

Purdue University North Central
Westville, Indiana 46391

We are interested in using the Japanese medaka (*Oryzias latipes*) as an animal model to study the immune system of lower vertebrates. This fish has been used as a developmental model for many years. The availability of many wild-derived medaka strains makes this animal an attractive immunogenetic model. This set of experiments was performed in order to determine if lectin binding could be used to discriminate various subpopulations of cells in the anterior kidney and spleen of the Japanese medaka. Spleen cells and cells harvested from the anterior kidney of the fish were obtained from a highly inbred population. Cells were stained with a variety of fluorescein isothiocyanate (FITC)-conjugated lectins and then analyzed for intensity of staining by flow microfluorimetry. Results of these experiments show that cells from both organs exhibit differential, but intense, binding by wheat germ agglutinin (WGA), concanavalin A (Con A), and ricin agglutinin (RCA_{120}). Staining of kidney and spleen cells by soybean agglutinin (SBA) and kidney cells by peanut agglutinin (PNA) was less intense. Few cells were stained by Dolichos agglutinin (DBA) or Ulex agglutinin (UEA II). Both Con A and RCA_{120} staining patterns allowed the identification of two cellular subpopulations in the anterior kidney. Scatter plots derived by use of electronic gates and set to identify homogeneously staining cells revealed that cells seen in each of the Con A peaks could also be distinguished on the basis of homogeneity of size and granularity (FIG. 1). However, the same type of analysis of data obtained by staining kidney cells with RCA_{120} did not allow discrimination of cells that were homogeneous for size and granularity, within subpopulations or peaks of fluorescence intensity (FIG. 2).

Results of flow cytometric analysis of spleen cells stained with FITC-conjugated lectins are shown in TABLE 1. In nine experiments, WGA stained 60 to 90% of the cells, with an intensity of approximately 100 times that of the phosphate-buffered saline (PBS) control or background. Both Con A and RCA_{120} also stained a large proportion of spleenocytes with a medium intensity. In addition to this medium-intensity peak, another less numerous population of spleenocytes stained very brightly with both Con A and RCA_{120}. The fluorescence intensity was in the order of 1000 times brighter than the control. Both PNA and SBA bound to a smaller proportion

[a]Address for correspondence: Purdue University North Central, 1401 South U.S. Highway 421, Westville, IN 46391.

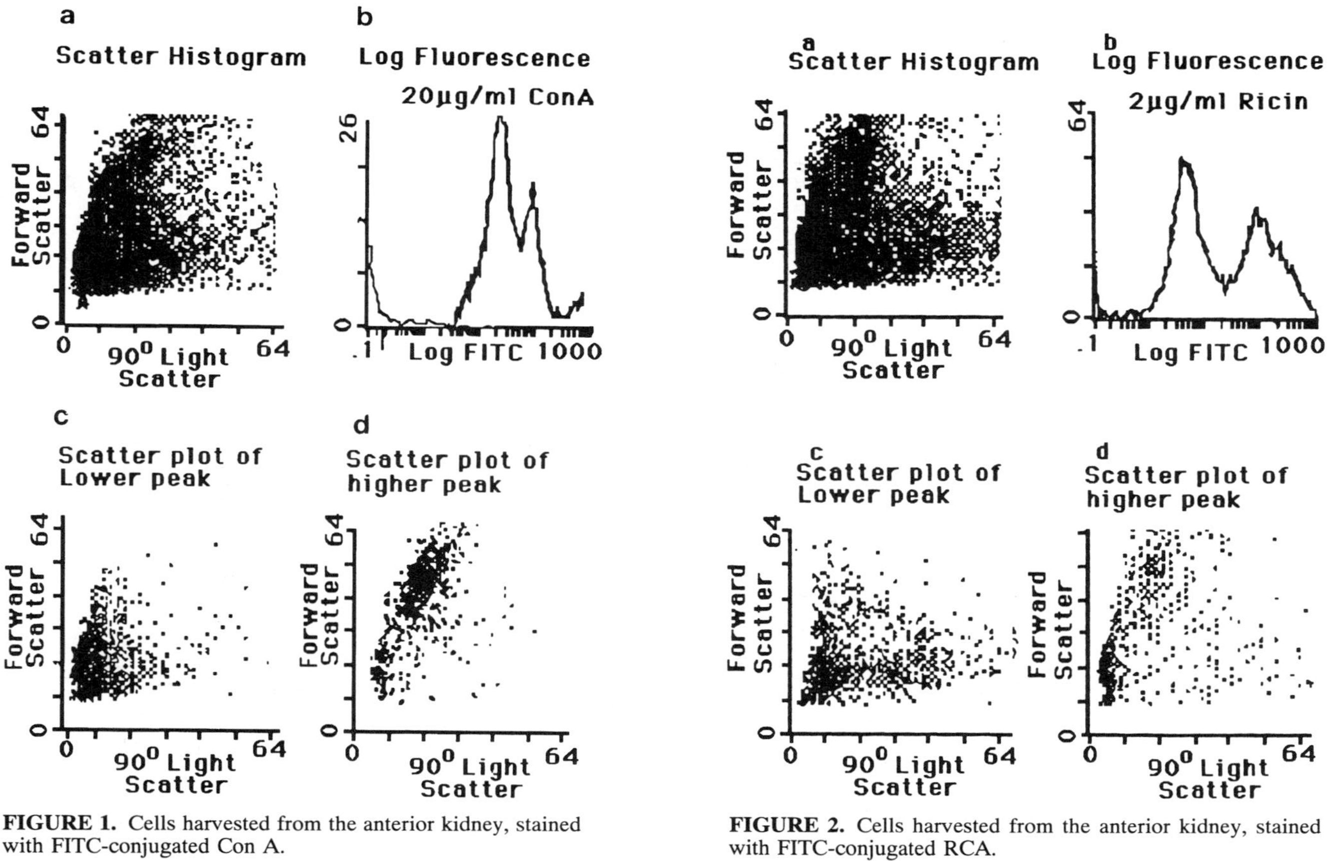

FIGURE 1. Cells harvested from the anterior kidney, stained with FITC-conjugated Con A.

FIGURE 2. Cells harvested from the anterior kidney, stained with FITC-conjugated RCA.

TABLE 1. Lectin Staining of Cells from the Spleen of the Japanese Medaka (*Oryzias latipes*)

Lectin (mg/ml)	Percent of Cells Staining with a High Intensity (1000 ×)	Percent of Cells Staining with a Medium Intensity (100 ×)
PNA (20, 2.0, 0.2)		6.8
Con A (20)		95.4
Con A (2.0)		54.1
DBA (20, 2.0, 0.2)		<1.0
SBA (20, 2.0, 0.2)		7.1
RCA (2.0)	14.6	82.9
WGA (2.0)		64–92
UEA II (20, 2.0, 0.2)		<1.0
(20, 2.0, 0.2)		

of spleenocytes, and the staining was of medium intensity. Neither DBA nor ULEX II seemed to bind to medaka spleenocytes.

Results of lectin staining of anterior kidney cells are shown in TABLE 2. The lectin, WGA, consistently stained 50 to 60% of the cells from the anterior kidney at an average level of fluorescence that was approximately 100 times that of the PBS background staining. Staining of anterior kidney cells with either Con A or RCA_{120} resulted in a bimodal distribution of staining intensity. With both of these lectins, approximately 50% of the anterior kidney cells had a mean fluorescence intensity of 100 times background. However, Con A bound 24% of the anterior kidney cells with an intensity 1000 times that of the PBS control, while the lectin RCA_{120} stained 43% of the anterior kidney cells with an intensity of 1000 times that of background. Initial data analysis has shown that subpopulations identified by Con A and RCA_{120} binding are not the same with regard to size and granularity of the cell types stained by these lectins. The lectins, PNA and SBA, stained a small (6.0 to 8.0%) proportion of the anterior kidney cells with a medium (100 times background) intensity. The lectins, DBA and ULEX II, bound only to a small (2.0 to 2.9%) percentage of the anterior kidney cells.

TABLE 2. Lectin Staining of Cells from the Anterior Kidney of the Japanese Medaka (*Oryzias latipes*)

Lectin (mg/ml)	Percent of Cells Staining with a High Intensity (1000 ×)	Percent of Cells Staining with a Medium Intensity (100 ×)
PNA (20, 2.0, 0.2)		6.4–8.1
Con A (20, 2.0)	24.2	51.2–61.7
DBA (20, 2.0, 0.2)		2.0–2.6
SBA (20, 2.0, 0.2)		7.0–8.7
RCA (0.2)	43.4	51.6
WGA (2.0)		51.5–63.5
UEA II (20, 2.0, 0.2)		2.1–2.9

Use of multivariate analysis (FIG. 1) demonstrates the utility of flow microfluorimetry to assess the homogeneity of cells located within a fluorescence peak. A large proportion of anterior kidney cells stained with Con A bind this lectin either at a medium or at a bright intensity. As can be seen from FIGURE 1c, cells that stain with a moderate intensity with Con A are small and less granular than the population of cells that stain with a greater intensity (FIG. 1d). This correlation is not seen when cells from the anterior kidney are stained with RCA_{120}. As can be seen in FIGURE 2, cells that stain with a medium amount of fluorescence are of medium size and, for the most part, are not very granular, while cells within the higher staining peak are agranular and are of small size.

These experiments have identified differential subpopulations of cells from these medaka organs. These results will allow the separation of various types of cells for *in vitro* studies. Histochemistry experiments with intact and dispersed tissue, from the spleen and anterior kidney, are in progress.

ACKNOWLEDGMENTS

We would like to acknowledge assistance from Dr. J. Paul Robinson of the Purdue University Flow Cytometry Laboratory and Dr. Christopher A. Bidwell, Purdue University, West Lafayette, Indiana.

Rapid Allogeneic Recognition in the Marine Sponge *Microciona prolifera*

Implications for Evolution of Immune Recognition

TOM HUMPHREYS[a]

Cancer Research Center and
Kewalo Marine Laboratory
University of Hawaii
Honolulu, Hawaii 96813

My laboratory has studied allorecognition in *M. prolifera* by apposing pieces of sponge tissue or small aggregates of dissociated cells.[1] Severed branches from the same or different individuals were tied together laterally on a glass slide and maintained in running sea water. Branches from the same sponge always join functionally by 24 to 48 hours. Tissue from different sponges (n = 81) carry out an immediate rejection reaction first evident at two to three hours as a visible yellow line along the boundary of contact in the otherwise orange sponge tissue. Microscopic examination shows that the yellow line represents an accumulation of gray cells, which are wandering, multigranular cells found in many sponge species.[2] In *M. prolifera,* gray cells are autofluorescent.[3] Epidermis dissected from zones of contact and examined under UV fluorescent microscopy illustrates graphically the extensive accumulation of these cells at the boundary of contact (FIG. 1). The gray cells line up, possibly with other cells, and in three to four days lay down 5- to 10-μm extracellular lamellae, which separate the incompatible tissue. Cytotoxic processes are not evident. This formation of a collagenous barrier represents a general effector mechanism for histoincompatibility of many sponge[4] species to both allogeneic and xenogeneic contact.

The immediate reaction evidenced by accumulation of gray cells suggests that recognition operates at a cellular level. This idea was examined by apposing individual-specific aggregates of dissociated cells from two different sponges. The cells of two sponges were dissociated and reaggregated separately for 6 hours in shaker flasks[5] to produce round aggregates of about 0.4 mm diameter containing 5×10^5 total cells, including a few hundred gray cells. An aggregate from each sponge was placed in individual conical-bottomed wells of microtiter plates and allowed to settle together. Upon contact, the two aggregates adhere and by 4 to 6 hours round up into a single ball of cells. Meanwhile, in most combinations, a reaction becomes evident by 2 to 4 hours as a yellow line marking the accumulation of gray cells in the orange cell mass at the boundary where the aggregates initially adhered. As the formation of the barrier proceeds, the aggregates from the two individuals round up separately and pull apart, exposing the forming barrier. These rapid recognition reactions by small aggregates indicate that many *M. prolifera* cells are capable of immediately recognizing foreign cells upon contact. The intimate involvement of

[a]Address for correspondence: Cancer Research Center and Kewalo Marine Laboratory, University of Hawaii, 1236 Lauhala St., Honolulu, HI 96813.

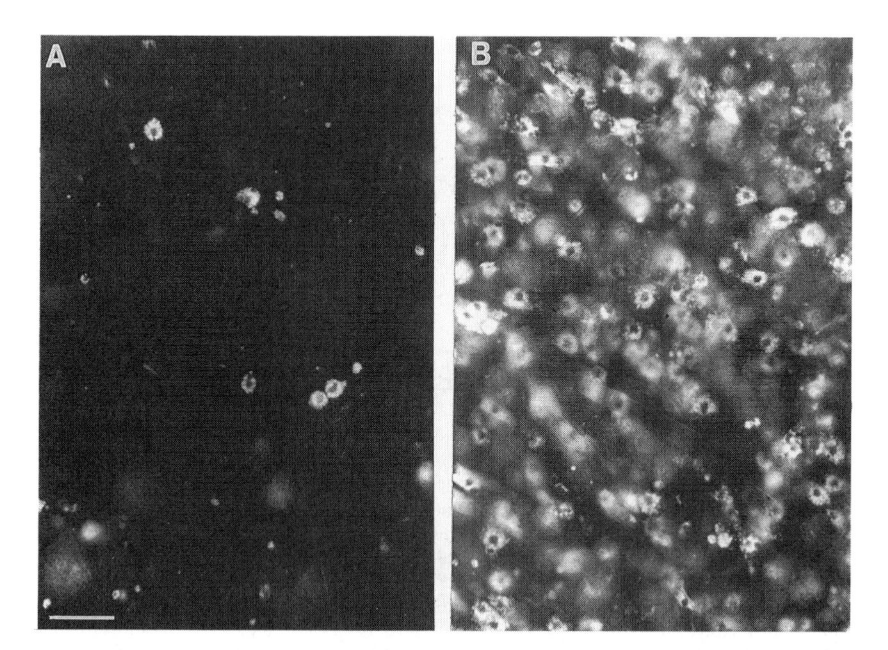

FIGURE 1. Fluorescent microscopic examination of sponge epidermis dissected from areas between apposed branches of sponges. After 9 hours of autogeneic (**A**) and allogeneic (**B**) contact, the adhering epidermides were peeled away from underlying tissue, mounted under a cover slip in sea water, and observed directly. The fluorescent images represent gray cells whose autofluorescent granules surround the nucleus. (Most cells in the multilayer epidermides are out of focus.) Photographed with a Zeiss Axiophot microscope with UV epi-illumination and filter set #48 77 02 using Ektachrome film. Line equals 40 μm.

gray cells in this process suggests that these cells of otherwise unknown function may be the functional immunocytes of sponges. The course of events indicates that recognition activates the release of cytokine-like activities that attract more gray cells and organize the rejection reaction at the boundary of contact.

The rapid cellular recognition without prior sensitization exhibited by *M. prolifera* provides an exciting opportunity to examine the fundamental cell biology of a primordial immune recognition reaction. Burnet[6] argued that invertebrate immune recognition could best be understood as a reaction guided by the failure of immunocytes to detect a self histocompatibility marker on any foreign cell. He noted the conceptual difficulty of devising a scheme whereby an individual can randomly inherit histocompatibility markers and simultaneously have receptors that can specifically recognize only the histocompatibility markers the individual has inherited. This problem of selection of receptors that recognize self histocompatibility allotypes is approachable given more recent precedents from T-cell biology of generation of diversity and positive selection for T-cell receptor specificity.[7] Invertebrate immunocytes that recognize self could be generated by positive selection from a population of preimmunocytes, expressing a diversity of receptor molecules

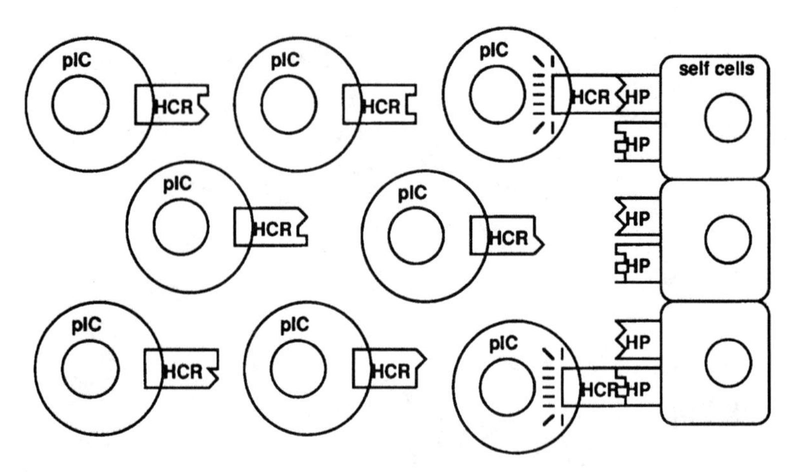

FIGURE 2. Positive selection model for differentiation of sponge immunocytes that recognize self. Self cells *(right)* express two histocompatibility proteins (HP) encoded by two histocompatibility gene alleles inherited by the individual. Preimmunocytes (pIC) express a diversity of self histocompatibility protein receptors (HCR). Most diverse HCR on pIC do not recognize self HP's *(left)* and are deleted. Preimmunocytes expressing histocompatibility receptors that specifically recognize one or the other histocompatibility protein carried by the individual *(right)* undergo positive selection, clonal expansion, and differentiation to functional immunocytes. Such immunocytes can specifically recognize self cells and immediately reject cells not displaying the self marker.

capable of recognizing the various histocompatibility allotypes carried in the genome of the species. The histocompatibility receptor diversity, like T-cell receptor diversity, could be encoded by multiple germline sequences or produced by somatic genetic processes.[7] A model of this type is outlined in FIGURE 2.

Although the relationships of invertebrate and vertebrate immunity are obscure, the idea of self receptor molecules that undergo generation of diversity and positive selection has the potential of placing invertebrate immune recognition in the evolutionary scheme leading to the more complex vertebrate foreign recognition established for T cells and B cells. In invertebrates the generation of diversity and positive selection could be directed inward to recognize self, while in vertebrates it has been redirected outward to recognize foreignness. Our results with sponges allow the examination of a rapid recognition reaction in the simplest metazoans at the cellular level to explore these and other fundamental ideas about the evolution of immunity.

REFERENCES

1. HUMPHREYS, T. & S. ZEA. Submitted.
2. SIMPSON, T. L. 1984. The Cell Biology of Sponges. Springer-Verlag. New York.
3. KUHNS, W. J., S. BRAMSON, T. L. SIMPSON, W. BURKART, J. JUMBLATT & M. M. BURGER. 1980. Eur. J. Cell. Biol. **23:** 73–79.

4. Van de Vyver, G. 1980. *In* Phylogeny of Immunological Memory. M. J. Manning, Ed. Elsevier. Amsterdam.
5. Rice, D. & T. Humphreys. 1983. J. Biol Chem. **258:** 6394–6399.
6. Burnet, F. M. 1971. Nature **232:** 230–235.
7. Kronenberg, M., G. Siu, L. E. Hood & N. Shastri. 1986. Ann. Rev. Immunol. **4:** 529–591.

The Isolation of Putative Major Histocompatibility Complex Gene Fragments from Dogfish and Nurse Shark

SIMONA BARTL[a] AND IRVING L. WEISSMAN

Hopkins Marine Station
Stanford University
Pacific Grove, California 93950
and
Department of Pathology
Stanford University School of Medicine
Stanford, California 94305

All available information suggests that cartilaginous fish may possess an immune system quite different from that of higher vertebrates. Thus far, no T-cell-dependent immune responses have been identified in cartilaginous fish (reviewed in Kaufman *et al.*[1]). They also display slower and more varied allograft responses in comparison to bony fish and mammals.[2] As a first step to understanding the unique nature of shark immunity, we have been characterizing the molecules encoded in the major histocompatibility complex (MHC) of shark.

In order to isolate shark MHC homologues, we designed three highly degenerate oligonucleotides to use as primers in the polymerase chain reaction (PCR). Primers 1 and 2 correspond to regions around two conserved cysteine codons found in the membrane proximal domain-encoding portion of class I and II genes. Primer 3 was internal to primers 1 and 2. At the position corresponding to the codon downstream of the first cysteine codon, we choose to use three inosines (I) in primer 1 to allow for complete variation in the template sequence. Primers 1, 2, and 3 can anneal to 196,608; 2,097,152; and 8192 potential target sequences, respectively. Primer 1 is CCAAGCTTTG(TC)IIIG TI(AT)(CAG)IGGITT(TC)TA(TC)CC. Primer 2 is CCG-GATCCA(GA)I(GC)(AT)I(GC)(AT)(AG)TGI(TA)(CT)IACI(TC)(TG)(AG)CA. Primer 3 is AGGATCCAIG TIC(CA)(GA)TCI(CG)(CG)(GA)TTIGG. Nucleotides in parentheses are degenerate. Restriction sites were added at the 5′ ends of each primer.

DNA from dogfish, *Squalus acanthias,* and nurse shark, *Ginglymostoma cirratum,* was used as template. The first five PCR cycles were 94°C for 1 min, 40°C for 2 min, and a slow rise over 2 min to 72°C for 3 min. The next 35 cycles were 94°C for 1 min, 45°C for 2 min, and 72°C for 3 min with a final extension step of 72°C for 15 min. PCR products were digested with the appropriate restriction enzymes, gel purified, cloned, and sequenced as single-stranded DNA. With this technique, we isolated 10 PCR products containing open-reading frames through both primers and

[a]Present address: Department of Biological Sciences, University of North Carolina, 601 South College Road, Wilmington, NC 28493-3297.

```
            ********                                                 *********
carp I       CLVTGFYPRDIEMNIRLNRINIESQ--ISSGIRPNDDESFQMRSSVKIDRNHRGSYDCHVIHSSL
carp IIβ     CSAYDFYPKPIKLTWMRDDKE-VTTDVTSTEELADGDWYYQIHSHLEYFPKPGEKISCVVEHASS

dogfish I    CHVHGFYPSGINATWLHNGGT-VQQEVLSSRILPNTDGTFQTTLQISVTPQSRDTYT
n.shark IIα  CFADGFYPPHITMKWRRNNEPMTDCD-NITEFYIKDDFTYRRFSYLSIVPSPGDMYSCHVEHSSL


dogfish 9A   CGVTGFYPRFIEVVWQRN-EDFVS-DVESPGLLPNHDGTYQIKKTLILAAEDEAEYLCHVDHSCL
dogfish 9C   CGVSGFYPRFIEVVWQRN-EDFVS-DVESPGLLPNHDGTYQIKITLTLAAEDEAEYWCHVNHTCL
n.shark 41D  CGVRGFYPPEIDVMWLKNGAPIPD-GVINTVLLSDGDWTYQVEDLLQYQPVSGDPYTCRVNHSTL
n.shark 41A  CGVRGFYPAKISVTWLKNGQKVSDADV-TVELLSNGDWTYQVRQYLQYEPVYGDKYTCRVDHSSL
             ********                       ********
n.shark 43B  CGVWGFYPQSISVTWLKD-GEHAPET-KSTGLLPNPDGTW
n.shark 43E  CGVRGFYPQSISVTWLKN-GEHAPET-KSTGLLPNRDCTW
n.shark 44B  CGVWGFYPHAVEVTLWRE-GKALDYT-QSSGVLPNRDCTW
n.shark 44E  CGVRGFYPQAIEVTLWRE-GKALDYT-QSSGILPNPDGTW
n.shark 44C  CGVTGFYPRDIEVTLLRN-GQPITDT-ESTGILPNGDCTW
n.shark 44D  CGVRGFYPRFIEVVWQRN-EDFVSDV-ESPGLLPNGDGTW
```

FIGURE 1. The deduced amino acid sequences of PCR products isolated from dogfish and nurse shark DNA. Longer products were generated with primers 1 and 2 while shorter products were amplified with primers 1 and 3. Regions encoded by the primers are indicated with stars. For comparison, amino acid sequences for carp class I and class IIβ,[3] dogfish class I,[4] and nurse shark class IIα[5] are shown at the top. Dashes indicate gaps in the alignments.

```
dogfish 9A    TCGGTTTATTGAGGTGGTTTGGCAGAGAAATGAAGATTTCGTCTCGGATGTGGAATCGCCTGGCCTATTGCCGAAC
              CATGACGGAACCTATCAGATAAAGAAAACTCTAATACTCGCAGCTGAGGATGAAGCTGAGTATTTG            142
dogfish 9C    TCGGTTTATTGAGGTGGTTTGGCAGAGAAATGAAGATTTCGTTTCGGATGTGGAATCGCCTGGCCTATTGCCGAAC
              CATGACGGAACCTATCAGATAAAGATAACTCTAACGCTCGCAGCTGAGGATGAAGCTGAGTATTGG            142
n.shark 41A   AGCAAAGATCAGTGTCACGTGGCTGAAAAATGGACAGAAGGTTTCTGATGCTGATGTCACAGTTGAGTTGTTATCA
              AATGGTGACTGGACATATCAGGTTCGCCAGTATCTCCAGTATGAGCCAGTGTATGGTGATAAATACACC         145
n.shark 41D   TCCTGAGATTGATGTCATGTGGCTGAAAAATGGAGCACCAATTCCTGATGGTGTCATCAACACTGTCCTGTTATCA
              GATGGTGACTGGACCTATCAGGTGGAGGATCTCCTCCAGTATCAGCCAGTGTCTGGGGATCCGTATACC         145
n.shark 43E   ACAGTCCATCAGCGTGACCTGGTTGAAGAATGGAGAACATGCCCCTGAAACCAAATCTACTGGGCTATTA         70
n.shark 43B   ACAGTCCATCAGCGTGACCTGGTTGAAGGATGGAGAACATGCCCCTGAGACCAAATCTACTGGGCTATTA         70
n.shark 44B   TCATGCCGTCGAGGTGACTCTGTGGAGAGAGGGGAAAGCCCTCGATTACACCCAGTCCAGTGGAGTCCTG         70
n.shark 44C   TCGCGATATCGAAGTCACCCTCCTGAGAAATGGCCAGCCGATCACGGACACTTAATCCACTGGCATCCTG         70
n.shark 44D   TCGGTTTATTGAGGTGGTTTGGCAGAGAAATGAAGATTTTGTCTCGGATGTGGAATCGCCTGGCCTATTG         70
n.shark 44E   TCAAGCCATCGAGGTGACTCTGTGGAGAGAGGGGAAAGCCCTCGATTACACCCAGTCCAGTGGGATCCTG         70
```

FIGURE 2. The nucleotide sequences of two PCR products isolated from dogfish DNA and eight PCR products cloned from nurse shark DNA. The top four sequences were generated using primers 1 and 2. The bottom six sequences were amplified with primers 1 and 3. Primer sequences are not included. The total length of each clone is shown on the right.

exhibiting similarity to MHC genes (FIG. 1). Surprisingly, no two identical clones were found. The nucleotide sequences of the longer clones (142–145 bp without the primer sequences) were used to search GenBank 75 (FIG. 2). In each case, the sequences found with the greatest similarity were MHC genes. The shorter clones did not contain enough sequence information to successfully pull out MHC genes. Because we found eight nurse shark clones that potentially contain MHC gene fragments, it is possible that the nurse shark may have several MHC loci and/or MHC pseudogenes.

To determine which of the nurse shark clones were expressed, we amplified cDNA from nurse shark peripheral blood lymphocyte RNA (kindly provided by Dr. Martin F. Flajnik, University of Miami School of Medicine). We isolated multiple copies of two clones, one of which was identical to clone 41A. This clone was used to successfully isolate two nurse shark MHC class IIB genes.[6]

Because MHC class I and class IIA homologues have been isolated from shark, and clone 41 is a class II A/B fragment, it remains to be determined what MHC molecules clones 41D, 9A, and 9C represent. In addition, the shorter clones isolated from the nurse shark may represent other MHC loci, MHC pseudogenes, or possibly non-MHC immunoglobulin-gene superfamily members.

REFERENCES

1. KAUFMAN, J., M. FLAJNIK & L. DU PASQUIER. 1990. *In* Phylogenesis of Immune Functions. G. W. Warr & N. Cohen, Eds.: 125–149. CRC. Boca Raton, FL.
2. BORYSENKO, M. & W. H. HILDEMAN. 1970. Transplant. Proc. **10:** 545–551.
3. HASHIMOTO, K., T. NAKANISHI & Y. KUROSAWA. 1990. Proc. Natl. Acad. Sci. USA **87:** 6863–6867.
4. HASHIMOTO, K., T. NAKANISHI & Y. KUROSAWA. 1992. Proc. Natl. Acad. Sci. USA **89:** 2209–2212.
5. KASAHARA, M., M. VAZQUEZ, K. SATO, E. C. MCKINNEY & M. F. FLAJNIK. 1992. Proc. Natl. Acad. Sci. USA **89:** 6688–6692.
6. BARTL, S. & I. L. WEISSMAN. 1994. Proc. Natl. Acad. Sci. USA, in press.

Neuroendocrine Markers Expressed on Sea Urchin Coelomocytes and Other Immunoregulatory Parameters May Be Used to Monitor Environmental Changes[a]

AURELIA M. C. KOROS[b] AND ANN PULSFORD[c]

[b]*University of Pittsburgh Graduate School of Public Health*
Pittsburgh, Pennsylvania 15261

[c]*Plymouth Marine Laboratory*
Citadel Hill
Plymouth, United Kingdom PL1 2PB

Human "natural killer" (NK) cells, small-cell lung cancer cells, and sea urchin coelomocytes share evolutionarily conserved antigens NKH1 (N901) (CD56) and HNK1 (Leu7) (CD57). The shared reactivity, defined by murine monoclonal antibodies anti-NKH1 (Coulter Immunology, Hialeah FL), anti-Leu7, and anti-Leu19 (Becton Dickinson, Sunnyvale CA), was reported previously.[1] Using indirect immunofluorescence and flow cytometry, the reactivity of about 5 to 10% of coelomocytes from two genera of sea urchins, *Strongylocentrotus purpuratus* and *Eucidaris,* was described. In the current study, similar reactivity of coelomocytes from *Echinus esculentus* is reported.

Echinus were harvested from 50 meters depth off the coast of Plymouth, England and maintained in tanks of running sea water at the Plymouth Marine Laboratory in April 1990. Coelomocytes were obtained by aspirating 3 ml of coelomic fluid into 2 ml of 0.1M EGTA in sterile sea water in a 5-ml syringe. About 1 million coelomocytes were obtained per milliliter of coelomic fluid. Aliquots of 10^5 coelomocytes in 100 μl were stained by adding 50 μl of monoclonal antibody as either hybridoma cell culture supernatant or mouse ascites fluid diluted in Hanks' balanced salt solution containing 0.1% sodium azide and 2% fetal bovine serum (monoclonal wash) at 4°C for 1 hr. Isotype controls were included. Cells were washed twice by centrifugation at 400 × *g* for 5 min. $(Fab')_2$ goat anti-mouse IgG and IgM conjugated with fluorescein isothiocyanate (Tago, Burlingame CA) diluted 1:100 in monoclonal wash was added at 4°C for 1 hr. Cells were washed twice more and resuspended in 1% paraformaldehyde. Cell samples were sent with frozen cool packs to Pittsburgh, Pennsylvania, and analyzed by using a single argon laser flow cytometer (Epics C, Coulter Electronics, Hialeah, FL). Bit maps were drawn around populations of cells that appeared unclumped and free of debris; 5000 cells were examined for each sample. Data were plotted as cell frequency versus log fluorescence intensity.

[a]This work was funded by the Marine Biological Association of the United Kingdom, the Cancer Federation Inc. of the United States, and the University of Pittsburgh Central Research Development Fund.

TABLE 1. Reactivity of *Echinus esculentus* Coelomocytes with Monoclonal Antibodies

Urchin Number	Not Stained	$(FAB')_2$	Isotype Controls IgG	Isotype Controls IgM	αNKH1	αCD3	αHNK1	αNKH1A
1	1.68	0.44	2.44	0.88	0.52	0.96	0.40	6.82[a]
2	0.96	0.96	4.12	2.12	2.56	7.48[a]	2.07	12.70[a]
3	0.90	3.82	2.40	6.86	4.81[a]	15.98[a]	4.22	8.72
4	0.93	3.84	15.84	15.82	7.52	14.66	7.72	10.86
5	0.80	8.66	14.20	15.90	8.91	16.06	8.30	10.64
6	0.77	3.96	6.56	16.00	13.94[a]	65.29[a]	14.77	18.48[a]
7	0.91	2.66	10.02	12.64	2.76	18.80[a]	2.94	11.08
8	0.94	1.84	2.54	9.01	1.86	10.74[a]	1.92	2.20
9	0.99	0.20	1.14	1.28	1.16	1.54	5.16[a]	5.16
10	0.98	4.28	3.00	1.40	3.52[a]	2.14	1.22	1.94

NOTE: Experiment of April 1990. The monoclonals have the following isotypes: αHNK1, αNKH1A are IgM; αCD3, αNHK1 are IgG; αHNK1 identifies CD57; αNKH1 and αNKH1A identify CD56.

[a]Positive reactivity above isotype controls.

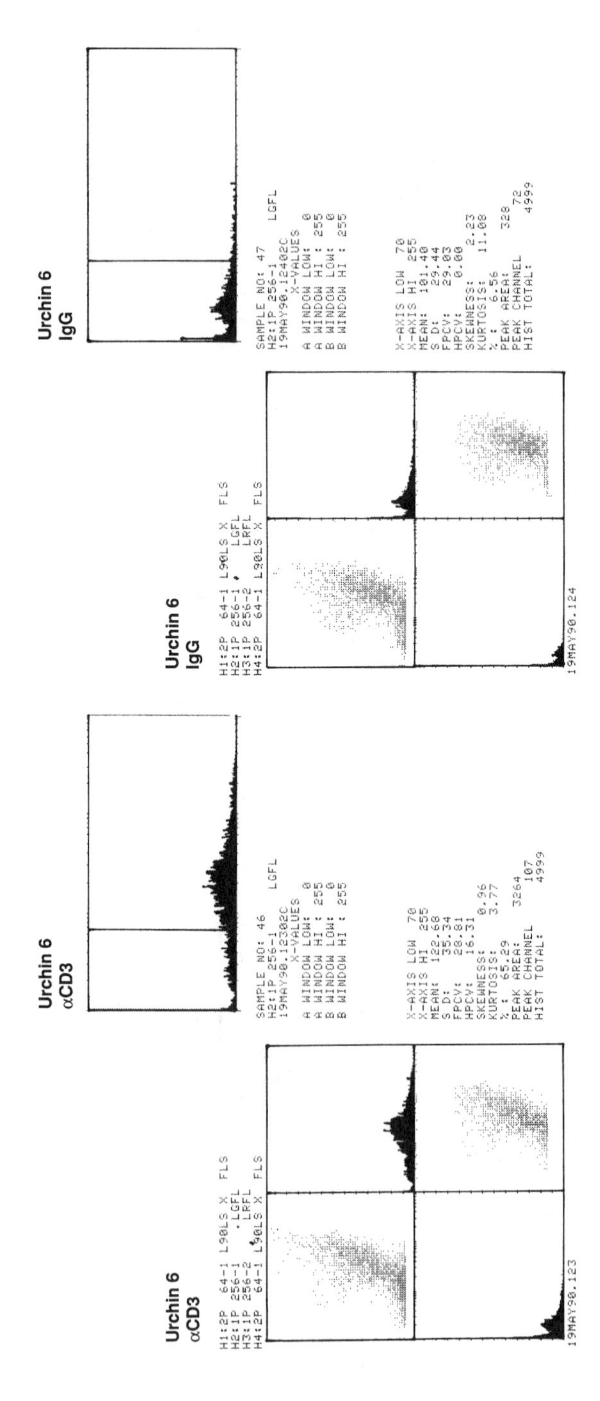

FIGURE 1. Flow cytometry data from a single *Echinus esculentus* urchin. Cell frequency plotted versus log fluorescence intensity. (Taken from Koros and Pulsford[2]; reprinted with permission from *Marine Environmental Research*.)

Results obtained from 10 *Echinus esculentus* are shown in TABLE 1 and are similar to results obtained from the two other genera of urchins. Of the 10 *Echinus,* 5/10 reacted with anti-CD3, 3/10 reacted with anti-CD56 above isotype controls, 4/10 reacted with anti-CD57 but also reacted with the isotype controls. It is not known if this is reactivity with Fc receptors. The reactivity of urchin 6 with anti-CD3 is shown in FIGURE 1. Although the functional significance of the small percentage of coelomocytes reacting with monoclonal antibodies directed as neuroendocrine markers is not known, the fact that human NK cells that have similar phenotypes have functional activity that decreases when humans are "stressed" suggests that periodic sampling of coelomic fluids may provide a measure of environmental "stress" to sea urchins in a variety of habitats.

REFERENCES

1. KOROS, A. M. C., M. J. TAYLOR, J. C. HISERODT, B. STEPHENS, R. J. LAKOMY, M. DALBOW, R. RAIKOW, J. LANGE & M. WAGNER. 1987. Sea urchin coelomocytes, human peripheral blood NK cells and human small lung cancer lines may share evolutionarily conserved antigens HNK1 (Leu 7) and N901 (NKH1). *In* Leukocyte Typing III. White Cell Differentiation Antigens. A. J. McMichael *et al.,* Eds.: 866–869. Oxford University Press. New York.
2. KOROS, A. M. C. & A. PULSFORD. 1993. Marine Environ. Res. **35:** 137–140.

Phylogeny of the Third Complement Component, C3, and Conservation of C3-Ligand Interactions[a]

JOHN D. LAMBRIS, MANOLIS MAVROIDIS, JOHN PAPPAS, ZHEGE LAO, AND YANG WANG

Department of Pathology and Laboratory Medicine
University of Pennsylvania
Philadelphia, Pennsylvania 19104

Of the 30 distinct complement proteins recognized to date, C3 is probably the most versatile and multifunctional molecule known, interacting with at least 20 different proteins.[1,2] It plays a critical role in both pathways of complement activation and participates in phagocytic and immunoregulatory processes.[1,2] The study of C3 molecules from different species gives insight into the structural elements involved in its different functions, the structural features constrained by selection, and the evolutionary history of C3 and complement in general. C3-like activity has been reported in a variety of species, including invertebrates, yet thus far, C3 has been purified only from chordates and has been found to be present in representatives of each of the seven living classes of vertebrates.[1,2] The complete primary structures have been deduced for C3 of human, guinea pig, mouse, rat, hagfish, lamprey, and cobra; and partial primary structures have been determined for C3 of rabbit and *Xenopus* (for review see Lambris[2]).

In this study we purified C3 from different species, analyzed the conservation of its structural and functional features, and obtained the cDNA sequence of frog *(Xenopus gilli),* trout *(Salmo gairdneri),* and chicken C3. C3 was purified from swine (Po), rabbit (Rb), mouse (Mo), cobra (Co), *Xenopus (Xe),* axolotl (Ax), chicken (Ch), and trout (Tr) serum.[3] All C3s tested were composed of two chains (α/β-chain) and contain a thioester bond within the α-chain. The two N-linked high-mannose carbohydrates found in human C3 were only conserved in Rb C3. In contrast, Xe, Ax, and Tr C3 have this moiety only in the β-chain and Po and Mo C3 only the α-chain. Co C3, in contrast to cobra venom factor (CVF), lacks Con A binding carbohydrates in either chain. The NH_2-termini of the *Xe* and Ax C3 β-chains were found to be blocked. TABLE 1 summarizes the features of C3 from different species characterized in this and other studies (for review see Lambris[2]).

The cDNA sequences of Tr, *Xe,* and Ch C3s were obtained and the amino acid sequence similarity between the different C3s is shown in TABLE 2. Several C3 regions have been found to be highly conserved including that of the thioester bond, which is crucial to the function of C3. The thioester site, GCGEE, is 100% conserved in all species, and the amino acid sequences of the surrounding regions are highly similar. This conservation of the thioester bond and its surrounding hydrophobic amino acids emphasizes the functional importance of this region in maintaining,

[a]This work was supported by National Science Foundation Grant DCB90-18751 and National Institutes of Health Grant 30040.

TABLE 1. Amino Acid Sequence Conservation between Trout C3 and Other Related Proteins

	Hu C3	Rb C3[a]	Ra C3	Mo C3	GP C3	Ch C3[a]	*Xe* C3[a]	Co C3	Tr C3	La C3	Ha C3	Hu C4	Mo C5	Hu α_2M
Human C3	100/100	79.6/87.0	78.2/85.5	77.1/85.3	78.1/86.3	56.3/70	52.2/66.3	51.3/66.1	44.1/58.9	32.3/48.1	30.9/46.8	28.4/43.3	28.0/45.0	21.3/35.4
Rabbit C3[a]		100/100	79.8/87.5	78.5/86.1	78.4/87.5	53.1/67.1	50.7/65.3	49.7/66.0*	43.1/58.3	33.3/48.2	28.4/44.8	26.7/41.7	26.3/45.4	18.0/35.5
Rat C3			100/100	90.1/94.1	81.0/87.6	57.3/70.6	53.1/67.6	52.2/66.7	44.2/58.5	32.7/48.3	28.7/45.1*	27.4/42.4	28.6/45.4	21.4/35.7
Mouse C3				100/100	79.8/87.1	57.2/70.5	53.5/67.6	51.5/66.3	43.3/59.6	32.2/48.0	29.1/45.2*	27.4/42.6	28.3/44.9	21.0/35.2
Guinea Pig C3					100/100	55.6/70.2	53/66.9	52.5/67.4	44.9/59.7	33.1/49.0*	30.2/46.4	30.4/46.0	29.0/45.7	24.8/39.8
Chicken C3[a]						100/100	55.9/68.5	58.4/70.3	46.3/60.7	34.5/52.0	32.5/48.4	31/46	28.1/44.9	22.7/37.7
Xenopus C3[a]							100/100	54.8/68.2	46.7/60.0	35.2/51.6	32.3/49.5	30.2/45.8	28.5/45.3	23.9/39.5
Cobra C3								100/100	43.8/60.2	32.3/47.0	30.6/45.6	27.3/42.1	27.5/43.4	21.2/35.4
Trout C3									100/100	33.1/50.2	29.9/46.1	27.5/42.0*	28.2/44.7	22.9/37.2
Lamprey C3										100/100	32.8/48.4	29.0/44.6	29.1/44.8	21.6/35.2
Hagfish C3											100/100	27.9/42.2	28.3/43.3	20.6/35.3
Human C4												100/100	27.3/42.2	21.7/34.9
Human C5													100/100	11.4/26.8
Human α_2M														100/100

ABBREVIATIONS: Hu, human; Mo, mouse; Ra, rat; Rb, rabbit; *Xe, Xenopus;* Ax, axolotl; Co, cobra; GP, guinea pig; Tr, trout; Po, pig; Ch, chicken; La, lamprey; Ha, hagfish.

NOTE: Percent identity/identity plus similarity. All alignments were performed using open gap cost 7 and unit gap cost 2 except the ones indicated with an asterisk for which open gap cost 15 was used.

[a]Partial amino acid sequence.

TABLE 2. Summary of Structural and Functional Properties of C3 from Different Species

Species	Hu	Po	Rb	Ra	GP	Mo	Ch	*Xe*	Co	Ax	Tr	Ha	La
Number of residues	1663	nd	nd	1663	1666	1661	nd	nd	1651	nd	1640[c]	1620	1660
% Identity/similarity to human C3	100/100	nd	80/87[a]	78/86	78/86	77/85	56/70[a]	52/66[a]	51/66	nd	44/59	31/47	32/48
Chain structure	α/β	α/β	α/β	α/β	α/β	α/β	α/β	α/β	α/β	α/β	α/β	α/β	α/β/γ
M_r α/β × 10^3	115/75	112/68	119/72	115/65	115/65	113/66	118/68	112/83	107/64	110/73	112/70	115/77	84/74/32
Thioester in	α	α	α	α	α	α	α	α	α	α	α	α	α
Con A binding to	α/β	α	α/β	nd	α	α	α	β	–	β	β	nd	nd
Human CR1 binding	+	+	+	nd	+	+	–	+	–	–	–	nd	nd
Human CR2 binding	+	+	+	+	nd	+	–	+	–	–	–	nd	nd
Human C5 binding	+	nd	+	+	nd	+	+	+	nd	+	+	nd	nd
Human H binding	+	+	+	+	+	+	+	–	–	+	+	nd	nd
Human B binding	+	+	nd	nd	+/–	nd	nd	–	nd	nd	nd	nd	nd
Human P binding	+	nd	+	+	nd	+	nd	+	nd	nd	–	nd	nd
Human MCP binding	+	nd	+	nd	–	nd	nd	nd	nd	nd	nd	nd	nd
Classical pathway	+	+	+	+	+	+	+	+	+	nd	+	–	–
Alternative pathway	+	+	+	+	+	+	+	+	+	nd	+	+	+[b]

[a]Based on partial sequence.
[b]Functioning only in opsonization.
[c]Without the signal peptide.

throughout evolution, the capacity of C3 to attach to surfaces.

Binding studies using the different C3s and Hu C3-ligands showed that H binds to Hu, Rb, Po, Ch, and Tr iC3; CR1 to Hu, Rb, Po, and *Xe* iC3; CR2 to Hu, Rb, Mo, and *Xe* iC3; and B to Hu and Po C3b (see TABLE 1). These data suggest that the conservation of the sequences comprising the binding sites for H, P, B, CR1, and CR2 is important in maintaining binding to these molecules.[1,2] Hu C3 was recently found by our laboratory to have multiple interaction sites for human C3b/C4b binding proteins CR2, H, and B.[1,2] The differential binding of the human factor H, factor B, CR1, and CR2 to different C3s made possible the conclusion that although these four molecules bind to the same domain in human C3, the exact binding sites are different.[1,2]

The interactions of C3 from different species with autologous C3-binding proteins and the C3 convertase and factor I cleavage specificities were investigated by analyzing electrophoretically the C3 fragments fixed on zymosan after activation of Tr, Ax, and *Xe* complement. We found that the degradation pattern is similar to that observed for human C3. NH_2-terminal amino acid sequence of the 68-kDa and 43-kDa fragments showed that these fragments are analogues of human C3 fragments generated by C3 convertase and factors I and H. Although similar fragments were found in the presence or absence of EGTA (which inhibits classical complement pathway), no fragments were detected when the activation of the different sera was performed in the presence of EDTA (which inhibits both complement pathways). These data suggest that the species so far tested possess proteins with functions similar to those of human factors B, D, I, and H. The C3 convertase cleavage site is conserved within all the C3s tested, and their C3 convertases have the same specificity (Arg-Ser) as those of human complement. In relation to the factor I cleavage sites, not all are conserved in all species. The factor I cleavage specificity is Arg-X.

REFERENCES

1. LAMBRIS, J. D. 1990. The Third Component of Complement: Chemistry and Biology. Springer-Verlang. Berlin.
2. LAMBRIS, J. D. 1993. Chemistry, biology and phylogeny of C3. *Complement Profiles* **1:** 16–45.
3. ALSENZ, J., D. AVILA, H. HUEMER, I. ESPARZA, D. BECHERER, T. KINOSHITA, Y. WANG, S. OPPERMANN & J. D. LAMBRIS. 1992. Phylogeny of the third component of complement, C3: Analysis of the conservation of human CR1, CR2, H and B binding sites, Con A binding sites and thiolester bond in the C3 from different species. Dev. Comp. Immunol. **16:** 63–76.

Effects of Repeated Immunization and Wound Trauma on Changes in Hemolymph Agglutinin Levels in Brown Shrimp *(Penaeus aztecus)*

B. L. MIDDLEBROOKS, YEUK-MUI LEE, MIN LI,
AND R. D. ELLENDER

Department of Biological Sciences
The University of Southern Mississippi
Hattiesburg, Mississippi 39406

Induction of circulatory lectins by administration of antigen or by injury has been reported for a number of invertebrates,[1–3] including crustaceans.[1] Summarized here are results of a study of changes in lectin (agglutinin) levels in *Penaeus aztecus* (brown shrimp) over a period from 0 to 144 hours after primary and secondary injection of an erythrocyte or bacterial antigen, and over a similar period of time following injury of (nonimmunized) animals.

After primary injection of *P. aztecus* with sheep erythrocytes, agglutinin levels decreased slightly in the first several hours, reaching the lowest titers by 16 hours post-injection, followed by an increase in titer to normal levels from 24 hours to 72 hours post injection (FIG. 1A). An identical secondary injection made 30 days after the primary injection produced a more rapid and more pronounced drop in agglutinin titers, with lowest titers reached within 4 to 8 hours post injection. This drop was followed by a rapid increase in titer to levels well above the baseline level, with the highest titers reached by 48 hours. Agglutinin titers then decreased slowly to return to levels near the original levels over the next 100 hours. Control animals held for 30 days in aquaria before administration of the primary injection showed a pattern of titers similar to animals given the primary injection immediately after being placed in aquaria. Shrimp receiving the secondary injection thus showed a modified response compared to animals receiving only a single response. A similar experiment using *Vibrio alginolyticus* as antigen produced comparable patterns of changes in agglutinin titers (FIG. 1B), except that the baseline level of titers in all animals held for 30 days was elevated above those in animals at the beginning of the experiment.

Wounding shrimp (puncturing hemocele, removing one eyestalk, removing two pereiopods) caused an increased in agglutinin levels in the first 16 to 24 hours after injury, with the amount of increase directly proportional to the degree of injury (FIG. 2). Titers then dropped to normal levels in all cases by 120 hours.

The results reported suggest that a form of adaptive response exists in *P. aztecus.* The results also demonstrate the need to distinguish between such a response and changes caused by stress, injury, or even the physiological state of the animals.

A

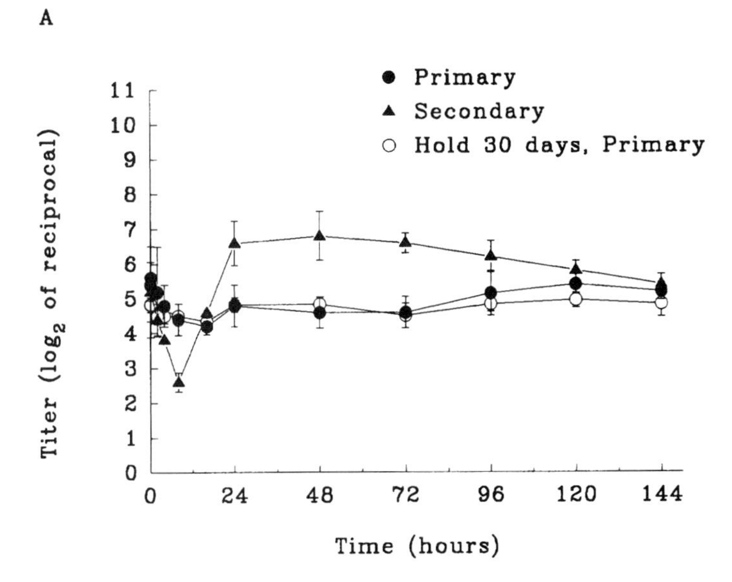

B

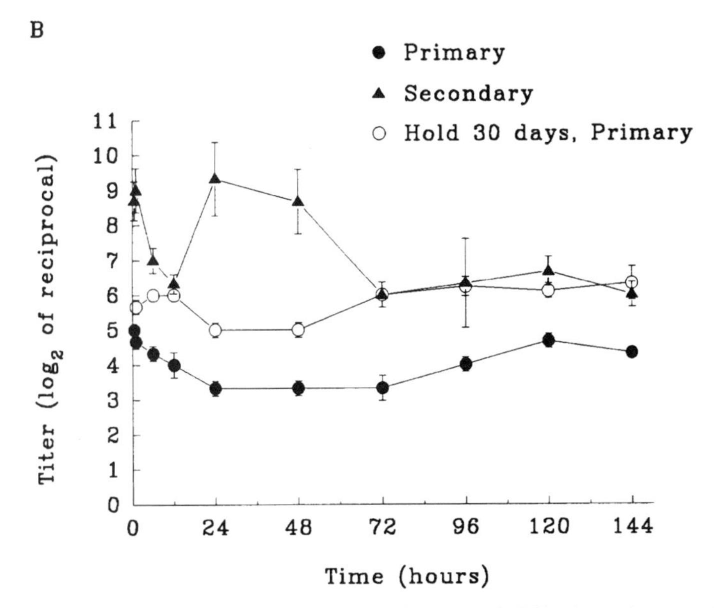

FIGURE 1. Agglutinin levels in *Penaeus aztecus* hemolymph following primary and secondary injection with **(A)** sheep erythrocytes and **(B)** *Vibrio alginolyticus* cells. Agglutination titers were determined using the homologous antigen. Control animals were held 30 days in aquaria before administration of primary injection.

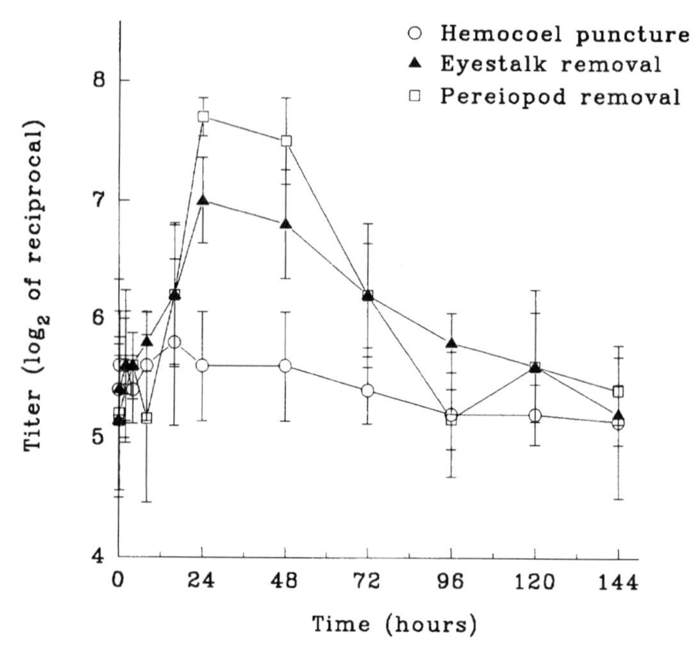

FIGURE 2. Effect of injury on hemagglutinin levels in *Penaeus aztecus* hemolymph. (Agglutination performed using sheep erythrocytes.)

REFERENCES

1. SARBADHIKARY, S. B. & R. BHADRA. 1990. Immunomodulatory stimulation of an invertebrate circulatory lectin by its haptenic molecules of pathogenic origin. Dev. Comp. Immunol. **14:** 31–38.
2. OLAFSEN, J. A., T. C. FLETCHER & P. T. GRANT. 1992. Agglutinin activity in Pacific oyster *(Crassostrea gigas)* hemolymph following in vivo *Vibrio anguillarum* challenge. Dev. Comp. Immunol. **16:** 123–138.
3. MCKENZIE, A. N. J. & T. M. PRESTON. 1992. Biological characteristics of *Calliphora vomitoria* agglutinin. Dev. Comp. Immunol. **16:** 85–93..

Hemocyte Involvement in Muscle Cell Death in Flies[a]

JALEEL A. MIYAN,[b] SHAHID N. CHOHAN, PAUL EVANS, AND N. MARK TYRER

Department of Biochemistry and Applied Molecular Biology
University of Manchester Institute of Science and Technology
Manchester M60 1QD, United Kingdom

To escape from the puparium (eclosion), flies use special muscles that degenerate over 48 hours in the adult once their task has been accomplished.[1] We have investigated the thoracic ventral longitudinal eclosion muscle (vlem), the dorsal anterior eclosion muscle (daem), and the ptilinum eclosion muscles in *Glossina morsitans, Sarcophaga bullata,* and *Calliphora vomitoria* using light and electron microscopy, gel electrophoresis of muscle protein and DNA, and electrophysiology in order to characterize the mode of cell death. Unlike other muscles that degenerate during metamorphosis,[2] they do not appear to respond to hormonal signals. One striking characteristic of eclosion muscle death in flies is the lack of involvement of vast numbers of hemocytes in phagocytosis of the dying muscle, as seen for example in eclosion muscle death in moths.[3] EM evidence shows that a single hemocyte is associated, in a non-phagocytic role, with the vlem and extends processes to each muscle fiber coincident with its death. Each process penetrates deep into fibers through invaginations of the membrane. A neural input is also required for death to occur, inasmuch as cutting the nerve to the muscle results in inhibition of process growth from the hemocyte and survival of that muscle.[4]

Phase-contrast microscopy of living muscles shows that fiber striations are lost by 7 hours after eclosion, and no change in appearance is detectable after this loss. EM evidence shows that sequential dismantling of intracellular components occurs beginning with the myofilaments. Up to 46 hours into degeneration, the remaining muscle fibers do not have the ultrastructure of dead or dying cells. Individual fibers are lost sequentially within the muscle, becoming vacuolated and finally losing membrane integrity. A healthy membrane potential is maintained well past the loss of contractility and myofilament breakdown (FIG. 1).[5] Throughout the process of fiber dismantling and death, the ultrastructure of nuclei and mitochondria does not change. Furthermore, electrophoretic gels of genomic DNA show no signs of degradation. There is a considerable increase in rough endoplasmic reticulum throughout the process, suggesting synthetic activity is associated with muscle dismantling (FIG. 2). Administration of actinomysin D or cycloheximide results in failure of degeneration, indicating a direct role for transcription and translation in the process.

To date, three mechanisms of cell death have been identified: (A) necrosis, a

[a]This work is funded by The Royal Society (London), The Wellcome Trust, and The Science and Engineering Council (U.K.). JAM is a Royal Society University Research Fellow.

[b]Address for correspondence: Department of Biochemistry and Applied Molecular Biology, University of Manchester Institute of Science and Technology, P.O. Box 88, Manchester M60 1QD, United Kingdom.

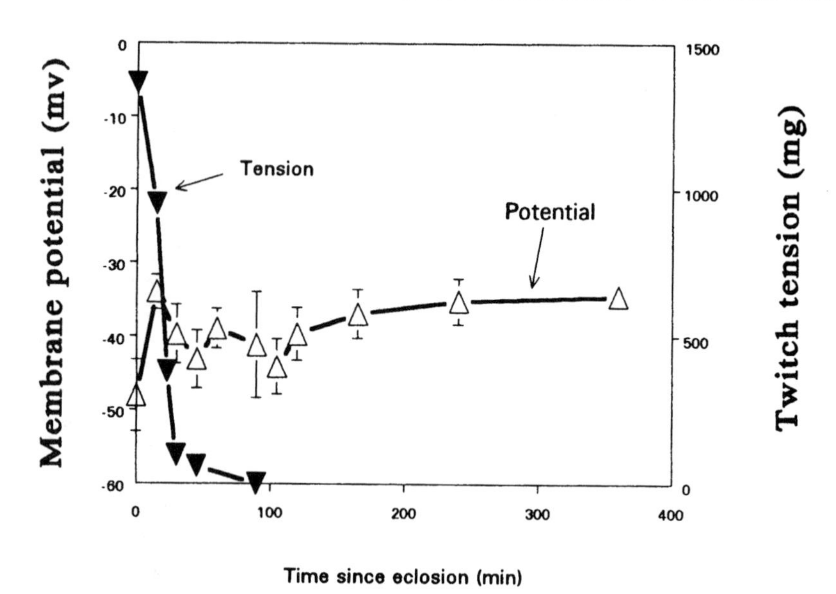

FIGURE 1. Plots of resting membrane potential and nerve-evoked twitch tension in the thoracic dorsal anterior eclosion muscle of *Sarcophaga bullata.* Flies were sampled at different time points after eclosion. The anterior tendon of the muscle was attached to a miniature force transducer using microforceps. The motoneurons to this muscle were stimulated by suction electrode attached to the neck connective. It is clear from these results, and from ultrastructural evidence (see FIG. 2), that dismantling of the contractile apparatus occurs within an intact cell membrane.

passive, damage-induced autolysis; (B) apoptosis, the classical (vertebrate) "programmed" cell death mechanism requiring the active participation of the cell in its own death; and (C) "programmed" cell deletion, an (initially) active mechanism of death requiring synthetic activity in the cell for its own death. Clearly, the death of fly eclosion muscles is not necrotic but programmed. However, it shows none of the characteristics of apoptosis (chromatin condensation, DNA laddering on gels, formation of apoptotic bodies), and indeed apoptosis has not been demonstrated in any invertebrate system apart from a single report of chromatin condensation in snail hemocytes.[6] Nor is degeneration by programmed cell deletion demonstrated, since the nuclei do not show ultrastructural breakdown, and protein synthesis is indicated throughout the process rather than in a burst.[2] This mechanism must represent another category of programmed cell death that may be related to the (perhaps cytotoxic) role of the hemocyte.

The death of eclosion muscles in flies is thus unusual in three respects: It involves single hemocytes in a non-phagocytic role, requires intact innervation, and exhibits continuous synthetic activity. Although the evidence to date is circumstantial, we believe that hemocyte involvement is central to muscle death. Any interference in the system results in muscle survival. Tissue/organ culture experiments will unravel the components of this fascinating system.

A

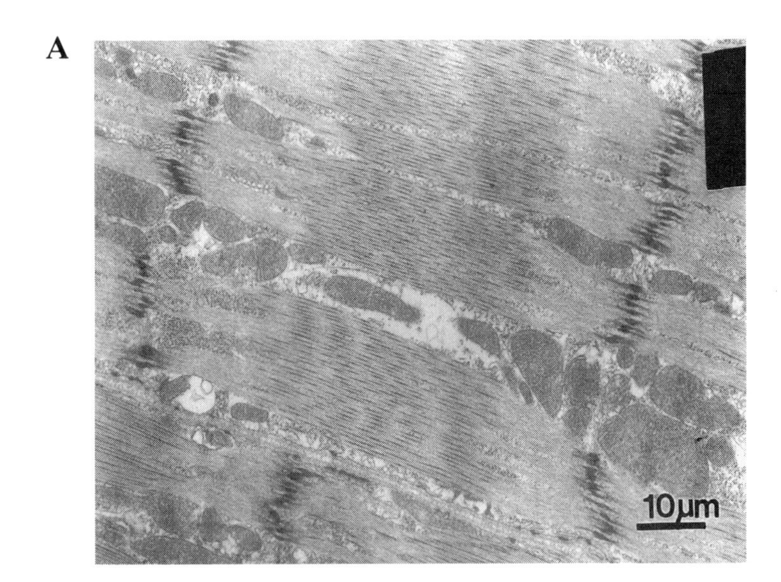

B

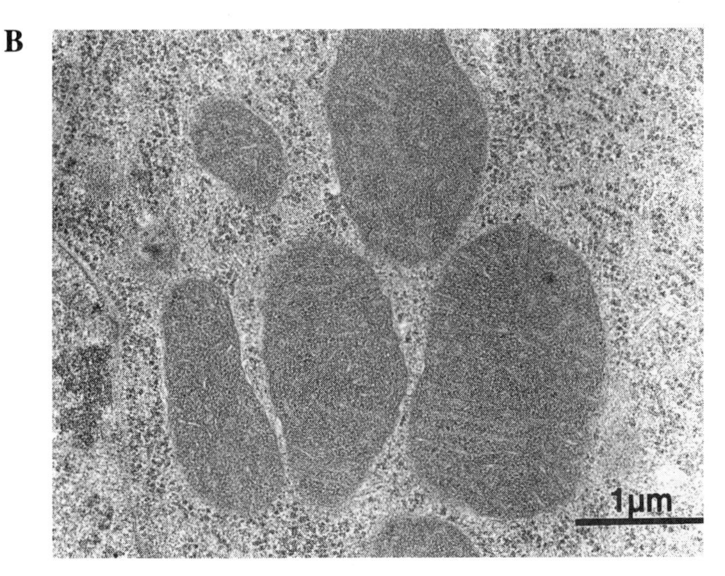

FIGURE 2. Transmission electron micrographs of the thoracic ventral longitudinal eclosion muscle of *Glossina morsitans.* **(A)** Section of a muscle taken immediately after eclosion to show the active intracellular structure of the fibers. The perforated Z-disks are characteristic of supercontracting muscles. This section should be compared with **(B),** which is taken from a muscle 46 hours after eclosion. There is a complete absence of myofilaments and a striking proliferation in rough endoplasmic reticulum. The nucleus and mitochondria show no signs of degeneration even at this late stage in the process. No damage can be detected in either the plasma membrane or basement membrane, and there is no evidence for phagocytic attack.

REFERENCES

1. MIYAN, J. A. 1989. Proc. R. Soc. London B **236:** 91–100.
2. SCHWARTZ, L. M., S. W. SMITH, M. E. E. JONES & B. A. OSBORNE. 1993. P.N.A.S. **90:** 980–984.
3. RHEUBEN, M. B. 1992. J. Exp. Biol. **167:** 91–117.
4. MIYAN, J. A. 1990. Tissue Cell **22(5):** 673–680.
5. BOTHE, G. & J. A. MIYAN. 1992. J. Physiol. **452:** 146*P*.
6. OTTAVIANI, E. & A. FRANCHINI. 1988. Acta Zool. **69(3):** 157–162.

A Possible Immunomodulatory Role of Endozepine-like Peptides in a Tunicate[a]

M. PESTARINO[b]

Istituto di Anatomia Comparata
Università di Genova
16132 Genova, Italy

The endogenous benzodiazepine-related substances, or endozepines, are regulatory peptides able to displace benzodiazepines and the β-carboline-3-carboxylate derivatives from their specific synaptic binding sites. Consequently, the precursor polypeptide has been called diazepam-binding inhibitor (DBI).[1] Trypsin digestion of rat DBI generates different neuroregulatory peptides such as the octadecaneuropeptide DBI 33-50 (ODN) and the triakontatetraneuropeptide DBI 17-50 (TTN). Endozepines not only modulate $GABA_A$ receptor activity, but also stimulate mitochondrial steroid biosynthesis[2] and inhibit insulin and glucagon secretion.[3] Moreover, endozepines have been found not only in the central nervous system,[1,4] but also in peripheral tissues.[5]

MATERIALS AND METHODS

Specimens of the solitary ascidian *Styela plicata* (Protochordata, Tunicata) were collected inside the port of Genoa and quickly transferred to the laboratory. After narcotization with tricaine (MS222) (Serva, Germany), the neural complex (cerebral ganglion and neural gland), branchial sac, and gut were dissected out and fixed with Bouin's fluid for 5 hours. Then, the samples were dehydrated in ethanols and embedded in Paraplast Plus (Monoject Scientific Inc., Ireland). Consecutive sections (5 μm thick) were mounted on glass slides and processed for immunohistochemistry. The immunohistochemical technique used rabbit antiserum to synthetic ODN (diluted 1:200 to 1:800) (UCB, Belgium), FITC-labeled swine anti-rabbit IgG (diluted 1:200) (Dako, DK), and rabbit peroxidase–antiperoxidase (PAP) (diluted 1:200) (Dako, DK). Control experiments were performed on adjacent sections by substituting normal serum or ODN antiserum preabsorbed overnight at 4°C with the homologous antigen (10^{-7} M).

RESULTS

ODN-like immunoreactivity was found in lymph nodules within the connective tissue of the branchial wall and in particular hemoblasts which show a strong immunofluorescence (FIG. 1a). Moreover, immunoreactive circulating lymphocytes

[a]This work was supported by a grant from the Italian M.U.R.S.T. (fund 60%) and C.N.R. (102CT92.02710.04/115.03769).
[b]Address for correspondence: Istituto di Anatomia Comparata, Università di Genova, Viale Benedetto XV 5, 16132 Genova, Italy.

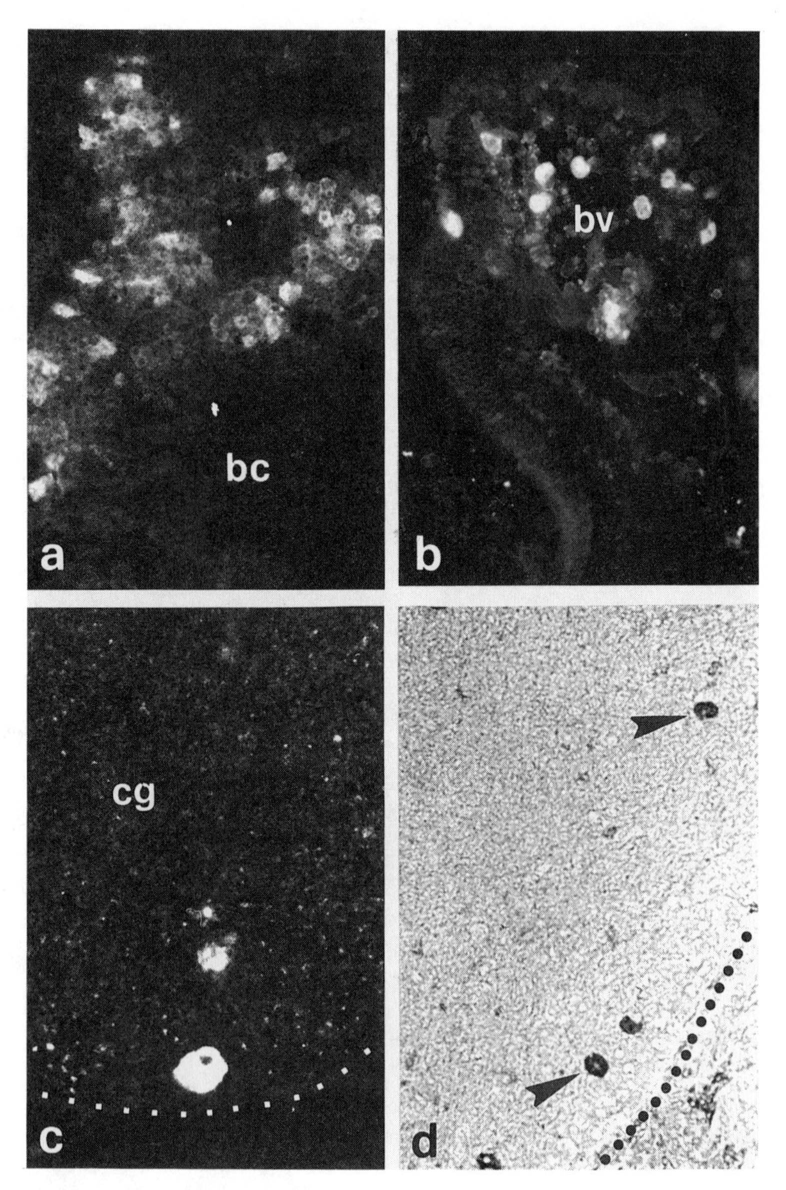

FIGURE 1. Lymph nodes are visible in the connective tissue of the branchial wall (bc, cavity of the branchial sac). Clusters of hemoblasts show an ODN-like immunofluorescence in the cytoplasm around the nucleus. Some large lymphocytes are scattered in the hemopoietic tissue and are strongly immunoreactive (**a:** × 250). Circulating blood cells are visible in a blood vessel (bv) of the branchial wall and ODN-like containing leukocytes are present (**b:** × 250). Immunofluorescent (**c:** × 450) and PAP-immunostained neurons *(arrows)* (**d:** × 250) are located in the cortex of the cerebral ganglion; the dotted lines mark the border of the ganglion.

and leukocytes were visible in the branchial blood vessels (FIG. 1b). Some nerve cell bodies localized mainly in the cortex of the cerebral ganglion contained ODN-like peptides (FIG. 1c,d). Finally, when the primary antiserum was preabsorbed with ODN, no immunostaining was detected either in cerebral ganglion or blood cells.

DISCUSSION

The distribution and characterization of hormone peptides in tunicates has been studied by biochemical, immunocytochemical, and chromatographic techniques (reviewed by Pestarino[6]). Recently, the production of immunoreactive neuropeptide Y (NPY)[7] and enkephalin-like materials[8] has been demonstrated in blood cells of the ascidian *Styela plicata.* At the same time, it is well established that tunicates have humoral and cellular immune responses (reviewed by Cooper[9]). Further, lymphocyte-like cells display a cytotoxic activity responsible for transplantation rejection[10] and proliferate in response to allogeneic stimuli.[11] The expression of ODN-like antigen determinants by tunicate hemoblasts and lymphocytes supports the involvement of endozepine-like peptides in immunomodulatory pathways. At the same time, the production of ODN-like materials by neurons of *S. plicata* further supports the relationships between the immune and neuroendocrine systems in tunicates. These results indicate that tunicates are a useful experimental model for studies on the phylogeny of the neuro-endocrine-immune axis.

REFERENCES

1. GUIDOTTI, A., C. M. FORCHETTI, M. G. CORDA, D. KONKEL, C. D. BENNETT & E. COSTA. 1983. Proc. Natl. Acad. Sci. USA **80:** 3531–3535.
2. PAPADOPOULOS, V., A. BERKOVICH, K. E. KRUEGER, E. COSTA & A. GUIDOTTI. 1991. Endocrinology **129:** 1481–1488.
3. OSTENSON, C. G., B. AHREN, S. KARLSSON, E. SANDBERG & S. EFENDIC. 1990. Regul. Peptides **29:** 143–151.
4. ALHO, E., E. COSTA, P. FERRERO, M. FUJIMOTO, D. COSENZA-MURPHY & A. GUIDOTTI. 1985. Science **229:** 179–181.
5. ROUTLET-SMIH, F., M. C. TONON, G. PELLETIER & H. VAUDRY. 1992. Peptides **13:** 1219–1225.
6. PESTARINO, M. 1991. Adv. Neuroimmunol. **1:** 114–123.
7. PESTARINO, M. 1992. Boll. Zool. **59:** 191–194.
8. PESTARINO, M. 1992. Presence of immunocyte-derived neuroendocrine peptides in a protochordate ascidian. *In* Recent Advances in Cellular and Molecular Biology. R. J. Wegmann & M. A. Wegmann, Eds. Vol. 1: 115–119. Peeters Press. Leuven, Belgium..
9. COOPER, E. L., B. RINKEVICH, G. UHLENBRUCK & P. VALEMBOIS. 1992. Scand. J. Immunol. **35:** 247–266.
10. RAFTOS, D. A., N. N. TAIT & D. A. BRISCOE. 1987. Dev. Comp. Immunol. **11:** 343–351.
11. RAFTOS, D. A. & E. L. COOPER. 1991. J. Exp. Zool. **260:** 391–400.

Transplantation Immunity in the Sea Anemone *Condylactis gigantea*

CHARLES F. SHAFFER[a,b,c] AND DAVID L. MASON[b]

[b]*Department of Biology*
Wittenberg University
Springfield, Ohio 45501
and
[c]*Keys Marine Laboratory*
Layton, Florida 33001

Graft rejection reactions have been observed in a number of invertebrate species[1] including several groups of cnidaria.[2] However, sea anemones have heretofore proven resistant to attempts at transplantation. This study was designed to develop (a) a method for grafting *Condylactis gigantea,* an anemone abundantly available in the Florida Keys and the Caribbean and (b) to observe the macroscopic and microscopic events surrounding graft rejection in this species.

MATERIALS AND METHODS

Anemones of the species *Condylactis gigantea* obtained in the Bay of Florida within a five-mile radius of the Keys Marine Laboratory received half-centimeter square "allografts" of column epithelium tucked into incisions in their collars and held in place by loosely closing the incision with a mattress stitch of 4-0 silk suture. Epithelium taken from pigmented regions was used in order to facilitate visual assessment of rejection through the translucent collar tissue. Grafts were observed macroscopically, and one was removed every 24 hours, fixed in gluteraldehyde, and embedded in Spurr plastic to be sectioned for light and electron microscopy.[3] For each graft made, a comparable piece of tissue from the same donor was placed in a plastic Petri dish where it was maintained by daily changes of sea water as a control against ischemic death of the implanted grafts. A total of 24 grafts on 24 different hosts and a comparable number of control tissues were observed. Donors and hosts were collected from sites separated by several miles of open ocean in order to increase the possibility of genetic diversity.

RESULTS

All control tissues were alive and essentially undamaged after 10 days of observation in sea water. All grafts were dead after 4 days residence in collar tissue.

[a]Address for correspondence: Department of Biology, Room 301 Science Hall, Springfield, Ohio 45501.

Macroscopic signs of rejection included decolorization, erosion, and finally a caseous-like necrosis. Microscopically, granulated cells of unknown origin were found in abundance within the area of and in the grafts as soon as 24 hours after transplantation (FIG. 1). High-resolution electron microscopy of the granules within these cells revealed "lamellar-like" structures (FIG. 2).

DISCUSSION

The results reported here represent a cursory and preliminary examination of transplantation immunity in anemones. Studies of the specificity of the reaction observed by us, using autografts as indicators of possible self-tolerance is essential. Furthermore, histochemical analysis of the granules present in the cells found in the rejection exudate is indicated. Finally, the question of whether the granulated cells are active mediators of the rejection response or merely "bystanders at the scene of an accident" must be resolved before a clearcut picture of transplantation immunity in the sea anemone will emerge.

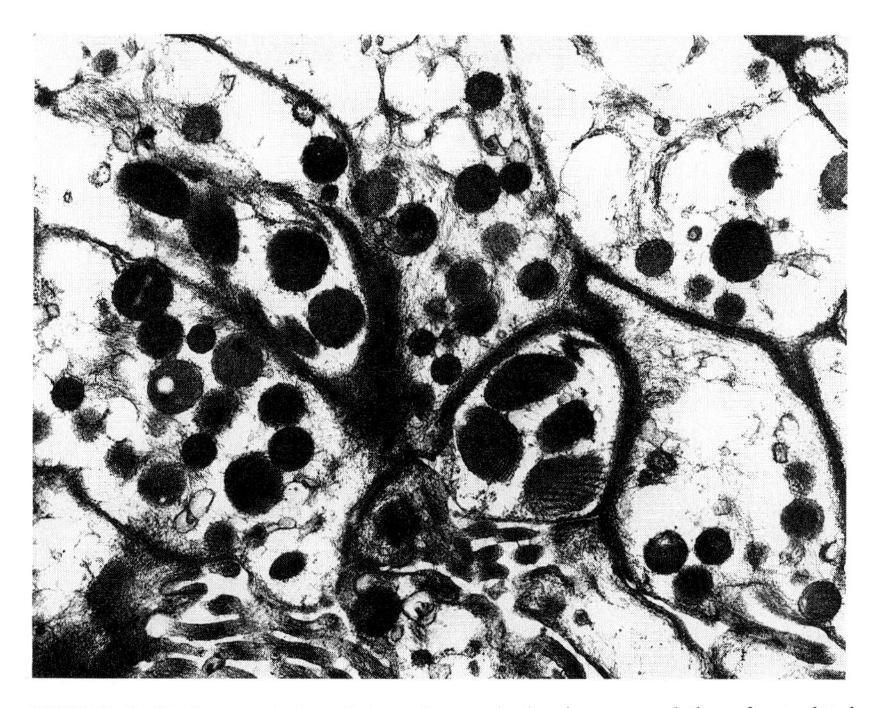

FIGURE 1. Medium-resolution electronmicrograph showing accumulation of granulated cells at the graft–host interface.

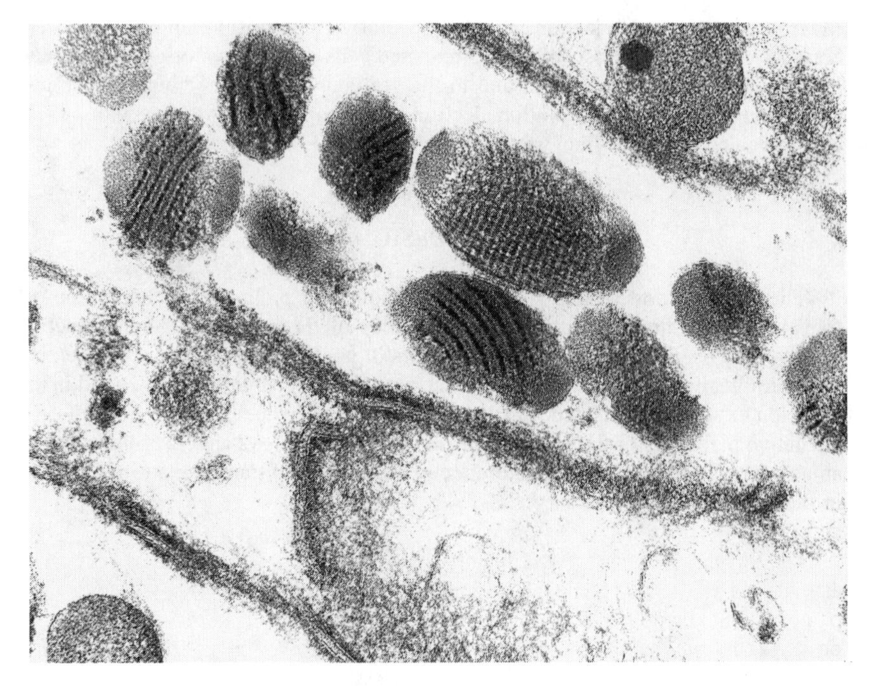

FIGURE 2. High-resolution electronmicrograph showing "lamellar-like" structures found in the granules of cells present in rejection exudate.

REFERENCES

1. BIGGER, C. H. 1988. Historecognition and immunocompetence in selected marine invertebrates. *In* Invertebrate Historecognition. R. Grosberg, D. Hedgecock & K. Nelson, Eds. Plenum Press. New York.
2. BIGGER, C. H. & W. H. HILDEMANN. 1982. Cellular defense systems of the coelenterata. *In* Reticuloendothelial System, Vol. 3. N. Cohen & M. Segal, Eds. Plenum Press. New York.
3. SPURR, A. R. 1969. J. Ultrastruc. Res. **26:** 31–42.

The Complement System of the Teleost Fish *Sparus aurata*

O. SUNER & LL. TORT

Facultad de Ciencies
Departamento de Fisiologia
University Autonoma Barcelona
Barcelona B-08193, Spain

The complement system is one of the most important mechanisms of the immune system in recognizing and eliminating non-self substances. Upon activation, it generates a group of inflammatory molecules and a potent lytic complex, acting in the clearance of potentially harmful agents.

Phylogenetically, the complement system seems to have developed first in fish, although there is circumstantial evidence for some kind of primeval alternative pathway in some invertebrate phyla.[1] Lamprey, a cyclostom that is one of the most primitive of living vertebrates, appears to have alternate complement activity.[2] Later in evolution, the cartilaginous fish appeared, followed by the teleosteans, which seem to have developed both the alternative (ACP) and the classical pathway (CCP).[3]

Very little data is available concerning to the complement system in fish. Thus, this study was undertaken in order to study the complement system of the telostean fish *Sparus aurata.*

Serum incubation with zymosan, LPS, inulin, (activators of the ACP of mammalian complement) effectively depleted its hemolytic activity. Chelation with EDTA completely abrogated its hemolytic activity, while chelation of serum in a Mg^{2+} EGTA buffer allowed the hemolytic reaction to occur. Hydrazyne (destroys C3 and C4 components of mammalian complement) treatment destroyed the hemolytic activity. This activity was abolished after heating the serum at 45°C for 20 min (TABLE 1). These results suggested the presence of the ACP in this fish.

The levels of complement are expressed in ACH_{50} and CH_{50} titers, which are defined as the reciprocal of the serum dilution giving 50% hemolysis of unsensitized and sensitized RaRBC, respectively. The mean ACH_{50} ± SD for n = 30 was 120 ± 30 ACH_{50} units/ml. Serum was absorbed with RaRBC to remove the naturally occurring antibodies. After successive absorptions, the ACH_{50} titers did not show any significant decrease (TABLE 2).

To determine the presence of a classical pathway, fish were immunized with RaRBC, in order to observe whether, upon RaRBC sensitization with fish antibodies, hemolytic titers could be increased or not. Sera from 30 fish was pooled, and the titer found was of 400 CH_{50} units/ml.

A poor correlation (r^2 = 0.126) was observed between ACH_{50} and CH_{50} titers

TABLE 1. Hemolytic Activity of Sea Bream Serum before and after Serum Treatments

Serum Treatment	Hemolytic Activity[a]
Untreated[b]	150 ± 8
Heat 45°C	0
Zymosan	6 ± 1
LPS	10 ± 2
Inulin	15 ± 1
Hydrazyne	0
EGTA	123 ± 5
EDTA	6 ± 2

NOTE: Numbers represent the mean ± SEM (n = 3).

[a]The sera from 30 fish were pooled, and the hemolytic activity was detected using a modification of the technique of Platt-Mills and Ishizaka.[4] Numbers express the reciprocal of the serum dilution causing 50% lysis of RaRBC.

[b]The serum was serially diluted in a gelatin veronal buffer (GVB), pH 7.5, containing 10 mM Mg^{2+}. In the rest of treatments, serum was diluted in a GVB containing 10 mM Mg^{2+} and 10 mM EGTA.

from 35 different fish, indicating that these two types of hemolytic activities mainly reflects different activation patterns.

The ratio CH_{50}/ACH_{50} observed is much lower in this fish than that of mammals, since the ACH_{50} titers of sea bream are much higher than those of mammals. Hence, it is concluded that the ACP functions as an important nonspecific defense mechanism in *S. aurata.* It suggests a greater importance for this pathway in fish than in higher vertebrates since the specific response in fish is still poorly evolved when compared with that of mammals.[5] The antibody response in fish has low affinity, limited heterogeneity, and poor anamnestic qualities; and the response time or lag-phase is longer compared with that of mammals. Thus, powerful nonspecific mechanisms were required to equilibrate the balance between specific and nonspecific response.

TABLE 2. Absorption of Naturally Occurring Antibodies

	Serum Activity	
Serum Absorptions	Hemolytic Activity[a]	Hemagglutinating Activity[b]
Unabsorbed serum	126 ± 6	12
First absorption	122 ± 5	8 ± 1
Second absorption	119 ± 6	1 ± 1
Third absorption	117 ± 4	0
Fifth absorption	112 ± 7	0

[a]Numbers are expressed in ACH_{50} units/ml and represent the mean ± SEM (n = 3).

[b]Values are expressed as the reciprocal of the last dilution giving hemagglutination.

REFERENCES

1. Lambris, John D. 1993. The chemistry, biology and phylogeny of C3. Complement Today. Complement Profiles. Vol. I: 16–45. Karger. Basel.
2. Nonaka, M. *et al.* 1984. Purification of a lamprey complement protein homologous to the third component of the mammalian complement system. J. Immunol. **133:** 3242–3249.
3. Sakai, D. K. 1992. Repertoire of complement in immunological defense mechanism of fish. Annu. Rev. Fish Dis. 223–247.
4. Platt-Mills, T. A. E. & K. Ishizaka. 1974. Activation of the alternative pathway of human complement by rabbit cells. J. Immunol. **113:** 348.
5. Ellis, A. E. 1989. The immunology of teleosts. *In* Fish Pathology. R. J. Roberts, Ed.: 135–152.

Index of Contributors

Middlebury College
0 00 02 0801636 0